T0338266

Seismic Retrofit of Existing Reinforced Concrete Buildings

# Seismic Retrofit of Existing Reinforced Concrete Buildings

*Stelios Antoniou*
*Seismosoft ltd.*
*Pavia*
*Italy*

This edition first published 2023


*Registered Office*
John Wiley & Sons, Inc., 111 River Street, Hoboken, NJ 07030, USA
John Wiley & Sons Ltd, The Atrium, Southern Gate, Chichester, West Sussex, PO198SQ, UK

For details of our global editorial offices, customer services, and more information about Wiley products visit us at www.wiley.com.

Wiley also publishes its books in a variety of electronic formats and by print-on-demand. Some content that appears in standard print versions of this book may not be available in other formats.



*Library of Congress Cataloging-in-Publication Data is Applied for:*

Hardback ISBN: 9781119987321

Cover Design: Wiley
Cover Image: Courtesy of Dr. Stelios Antoniou

Set in 9.5/12.5pt STIXTwoText by Straive, Pondicherry, India

SKY10045377_040323

*To Eleni, Lydia and Miley*

# Contents

# Foreword by Rui Pinho

It is undeniable that seismic assessment and retrofitting of the existing building stock in many parts of the world has become the primary focus of structural engineers. The reasons behind this pressing need are multiple and multifaceted, as clearly explained and discussed in this book. However, the vast majority of currently active structural engineering practitioners did not receive any formal training on this particular topic, which only in recent years has started to find its way into the syllabus of civil engineering degrees. There is therefore a great need for seismic retrofitting guidance within the practicing community – guidance that should be not only of a conceptual and theoretical nature but also, or perhaps mainly, practice oriented.

This textbook addresses perfectly such undeniable need, as a result of the rather unique and special technical background of its author. Indeed, Dr. Stelios Antoniou combines an impeccable academic training in structural earthquake engineering, obtained from the National Technical University of Athens and from Imperial College London, with a two decades of hands-on experience in the seismic assessment and retrofitting of existing structures (he is partner and technical director of a construction company specializing in this field, Alfakat) and with an equally long, accomplished career of developing earthquake engineering software tools that are employed by thousands of users around the globe (he is co-founder and chief developer of Seismosoft).

It is therefore easy and immediate to appreciate how the present textbook will inevitably read very differently from other publications on the same topic, some of which I am co-author of, which tend to feature a more academic and formal writing-up and discussion, as opposed to the very practical and colloquial style adopted by Dr. Antoniou. The latter renders this volume not only extremely rich and informative in terms of contents and actual application, but also rather easy and pleasant to follow.

The book is logically organized in two main parts. The first of these introduces the current status quo on the common weaknesses found in existing reinforced concrete structures (Chapter 2), the methods available to gain knowledge on a building's properties and characteristics (Chapter 3), and the typical techniques at the disposal of practitioners to retrofit this type of structures (Chapter 4). Unlike other books, however, such overview is given and described with the assistance of several tens of photographs from actual buildings and retrofitting applications undertaken by Dr. Antoniou, which renders it truly unique and clear. This first part of the book is then closed by a precious section where the author provides his experienced insight on the type of criteria one should have in mind when selecting one retrofitting strategy over another (Chapter 5).

The second part of the book is instead fully focused on the procedure that practitioners need to follow in order to assess the seismic response of an existing reinforced concrete building and then define an appropriate and code-compliant seismic retrofitting intervention. It thus naturally covers not only the selection of appropriate seismic performance targets (Chapter 6) and structural analysis techniques (Chapter 7), but also discusses advanced structural modeling issues (Chapter 8) and the necessary structural performance code-compliance checks (Chapter 9). The manner in which these four steps can and should be brought together in the process of seismic assessment and retrofitting is illustrated by a start-to-finish application to an actual case-study (Chapter 10), which perfectly leverages the very practical software tools developed by Dr. Antoniou.

In short, this is a volume that I believe all structural engineering practitioners, as well as students and academics, should have on their bookshelves, given not only the invaluable and unique insight that it provides on the type of challenges one is faced with when dealing with the seismic performance of existing reinforced concrete buildings, but also the very clear and practical guidance it conveys on how to potentially intervene in such structures.

*Rui Pinho*
Professor of Structural Engineering
University of Pavia, Italy

## Acknowledgments

Writing this book was harder than I initially thought, but more rewarding than I could have ever imagined. None of this would have been possible without so many supportive people in my life. I wish to express my sincere gratitude to all those who have directly or indirectly contributed to this endeavor. In particular, I would like to express my special thanks to:

- Dr Rui Pinho for being a great friend and partner (in Seismosoft), for providing continuous support and unfailing assistance, and for inspiring me in many different ways. His contribution to my work has been invaluable, not least because he had the initial idea for this book back in the summer 2019.
- Kostas Antoniou, who has been extremely helpful, supportive, and resilient throughout all this time. Apart from being my brother and a lifetime friend, he is a fantastic partner (in Alfakat), who keeps tidying up all the messes that I leave behind, when I move forward to the next "big" thing.
- Zoe Gronti from Seismosoft and Giouli Liaskou from Alfakat, who made the initial reviews of the chapters and gave me extremely valuable first feedback on several aspects of the book.
- My friends and colleagues in Seismosoft and Alfakat, who have contributed significantly to the book in many different ways. Special thanks to Dr Thanasis Farantos, Panagiotis Doulos, Thodoris Rakintzis, Evi Visviki, Thanasis Karatzas, Yiannis Spilios, Vaggelis Trikkas, Dr Fanis Moschas, Artan Xhemalallari, Apostolis Economou, George Kalfas, Nancy Gouma, Vassilis Samaras, Nikos Modes and Marios Basoukos.
- Christos Giannelos and Christos Giarlelis for providing useful material and photographs, but more importantly for their constructive comments, which helped me in crystallizing several points I make throughout the book.
- Odysseas Verroios, Nikos Zarkadoulas, George Kyriakou, Jose Poveda, and Christos Varelas for providing useful material and photographs.
- Sara Kaufman for the corrections in the initial English text, but also for her insightful suggestions and positive comments.
- My parents for their continued love, support, and patience, but also for their guidance and encouragement at the different stages of my academic and professional life.
- Everyone on the Wiley team who helped in getting this book out in the market.
- Last, but certainly not least, my family – and in particular, my wife, Eleni, and my daughter, Lydia, for their patience and understanding all this time, and more importantly, for making this life journey as good as it gets (and even better).

# 1 Introduction

## 1.1 General

The vast majority of existing buildings, even in the most developed countries, have been built with older provisions, with low or no seismic specifications. As a result, their ability to withstand earthquake loads is considerably lower with respect to modern standards, and they suffer from significant irregularities in plan and/or elevation, low ductility, and low lateral strength and stiffness. They exhibit increased vulnerability to seismic loading and often have a critical need for strengthening.

Seismic assessment and strengthening is a promising field of civil engineering. It requires special knowledge and often poses great challenges to the engineer, both in the design and the construction phase of the strengthening interventions. Because many older structures are vulnerable to seismic activity, this constitutes an exciting new field of the construction industry that is far from saturated and is expected to gain importance and exhibit significant development in the years to come. This significance is highlighted by the publication of several documents and standards worldwide that are dedicated to this subject from organizations such as the American Society of Civil Engineers (ASCE), the Federal Emergency Management Agency (FEMA) in the United States, New Zealand Society for Earthquake Engineering (NZSEE) in New Zealand, the Earthquake Planning and Protection Organization (EPPO) in Greece, The European Committee for Standardization in the EU, and other regulatory agencies in Europe.

The documents and standards provided by these agencies include ASCE 41 (ASCE 2017) and its predecessors FEMA 273 (FEMA 1997) and FEMA 356 (FEMA 2000) in the United States, Eurocode 8, Part 3, in Europe (CEN 2005), NZSEE in New Zealand (2017), KANEPE (2022) in Greece, and large dedicated chapters in NTC-18 (NTC 2018) in Italy and the Turkey Building Earthquake Regulation (TBDY 2018) in Turkey.

The main incentive for writing this book has been the realization that, despite the importance of the subject and the publication of thousands of papers on the strengthening of existing structures, there are very few complete books or reports with specific guidelines on the strategy for structural retrofit. This book attempts to provide structural engineers a thorough insight on seismic assessment and strengthening, specifically for existing

*Seismic Retrofit of Existing Reinforced Concrete Buildings*, First Edition. Stelios Antoniou.

reinforced concrete buildings, providing information on available strengthening techniques and on the methodologies and procedures that should be followed to assess an existing or strengthened structure. More importantly, it gives detailed directives on the strategy for the strengthening interventions – that is, which method(s) to use, when, how, and why.

This chapter provides a brief introduction on the main international and national standards employed for the seismic evaluation and strengthening of existing structures. The remaining book is divided in two parts.

In the first part (Chapters 2–5), issues of a more practical nature will be discussed, as well as all the field works related to structural assessment and strengthening. In Chapter 2, the main problems affecting existing reinforced concrete (RC) buildings are described, such as bad detailing, poor workmanship, corrosion, and bad construction practices. In Chapter 3, the challenges of monitoring RC buildings are explained, and the methods for the measurement of the required structural quantities, and the nondestructive and destructive testing involved are presented. In Chapter 4, the available strengthening techniques for the retrofit of a reinforced concrete building are described in detail, and the most important design and construction issues related to them are discussed, together with their main advantages and disadvantages. A simple example with the seismic retrofit of an existing building with the most important techniques is also presented. Finally, in Chapter 5 the criteria for selecting the most appropriate strengthening method and the basic principles of conceptual design are discussed, and more than 15 actual case studies from practical applications of the retrofit of existing buildings are presented, explaining in each case which strengthening method was selected and why.

The second part (Chapters 6–10) constitutes a detailed description of the code-based seismic assessment and retrofit procedures, which should be followed for the structural upgrading of existing RC buildings. In Chapter 6 the performance levels, the limit states, the seismic hazard levels, and the performance objectives are explained, and a detailed discussion is given on how to make an appropriate selection of performance objective. In Chapter 7 the linear and nonlinear methods for structural analysis are presented, and in Chapter 8 general issues on the modeling of reinforced concrete structures are discussed, with a particular focus on the modeling for nonlinear analysis. Chapter 9 describes the main safety verification checks that are performed in a structural evaluation methodology, while in Chapter 10 an application example of the structural evaluation and strengthening of an existing reinforced concrete building is presented, following all the steps of the procedure: selection of performance and seismic hazard levels, modeling, structural analysis and safety verifications.

Appendix A presents a succinct description of the code-based assessment methodologies and procedures of the most common standards worldwide: ASCE 41 (US standard), and Eurocode 8, Part 3 (EC8, European standard). It is noted that, although reference to other codes will also be made in different sections of the book, the main focus will be on ASCE 41 and EC8, Part 3, which are the best known and most used worldwide.

Appendix B provides a large set of pictures from bad construction and design practices in existing RC buildings. These are some characteristic examples I have come across during my professional life, and they give a rough idea of what an engineer should expect when working on existing buildings.

Finally, Appendix C presents a selection of characteristic photographs on the various methods of strengthening, so that the reader can appreciate most of the construction details of these methods, and understand the challenges and difficulties related to their construction.

In the examples of Chapters 4, 5, and 10, all the structural analyses and the verification checks are performed with the use of the SeismoBuild (2023) and SeismoStruct (2023) software packages. These are two finite element programs by Seismosoft Ltd. capable of performing linear and nonlinear structural analysis, and carrying out the entire code-based assessment procedures for different structures. SeismoBuild focuses on reinforced concrete buildings and concentrates on the code-based methodologies, while SeismoStruct allows more freedom, and can also model other types of structures (e.g., steel, composite structures, bridges).

## 1.2 Why Do Old RC Buildings Need Strengthening?

Old reinforced concrete buildings have been typically designed and constructed without considerations for seismic loads and the lateral resisting system. Even in the cases when a seismic code existed (buildings constructed after the 1960s or 1970s), usually the prescribed earthquake load was just a fraction of today's standards with design ground accelerations less than 0.10g, and often close to or equal to zero. This means that a large proportion of the existing building stock have been designed without consideration for any seismic forces or against very low horizontal loads (Figure 1.1).[1]

These buildings suffer from bad construction practices, low material grades, lack of stirrups, short lap splices, bad detailing (e.g., the hoops were never bent to 135° angles inside the concrete core), lack of correct supervision, and poor workmanship. As a result, the ductility in older RC buildings is very low and unreliable. Moreover, the aging framing system, the carbonation of concrete, and the corrosion of reinforcing steel further degrade the buildings' capacity to sustain earthquake loads. As a result, the lateral capacity of older reinforced buildings is significantly lower (often less than 50%) than the capacity of similar buildings designed with today's standards.

It is noted that with other types of construction, namely steel and composite structures, this difference is not as accentuated. Steel is a more ductile material, steel buildings do not suffer so much from brittle types of failure (such as shear in RC structures), and they retain an adequate level of ductility even if they were not designed specifically for it. Furthermore, anti-corrosion measures were taken during the construction of steel buildings even in older times, and this has prevented the rapid degradation of the buildings' strength during their lifetime.

1 It should be noted that with older seismic codes these small seismic coefficients were not further reduced with the application of a behavior q factor, as is done in today's standards. However, in the general case the difference between the horizontal forces imposed now and then is still very high (typically 100% or more). After all, older RC buildings lacked ductility; even if a q-factor was applied to them, this could not have assumed a high value.

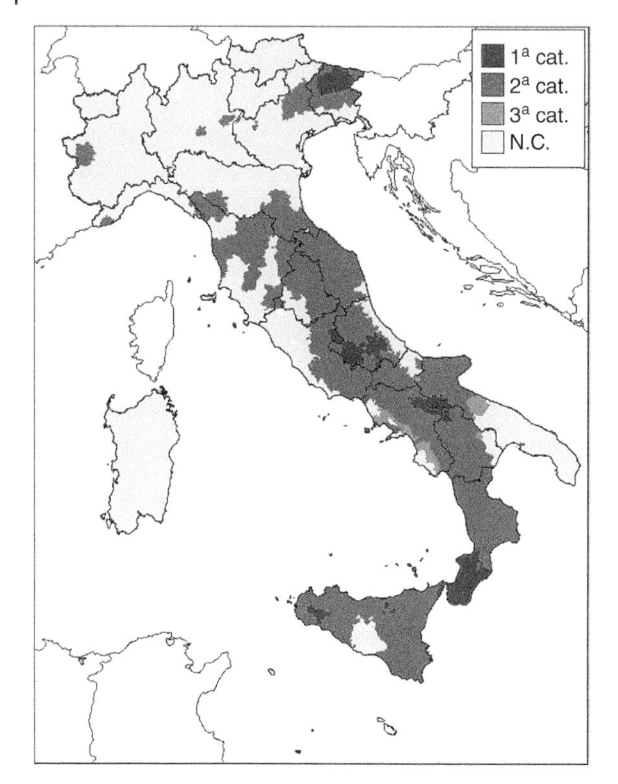


**Figure 1.1** Classification of seismic zones in Italy in 1984. The seismic coefficient was 0.10 in zone 1 (darkest areas), 0.07 in zone 2, 0.04 in zone 3, and 0.00 elsewhere (Pinho et al. 2019; Meletti et al. 2006).

Finally, with respect to existing masonry buildings, which also suffer from significant structural deficiencies and are as vulnerable as RC buildings (if not more so), the importance of seismic evaluation and strengthening in RC buildings is greater not because of the technical challenges involved but, rather, because RC buildings constitute the vast majority of the buildings in most countries, and typically they are larger and higher.

## 1.3 Main Differences Between Assessment and Design Methodologies

Nowadays, civil engineers are well accustomed to the standard design methodologies and procedures for reinforced concrete buildings, which are taught in all universities and civil engineering courses and are often applied in everyday practice. However, structural assessment – and more importantly, structural strengthening, which is usually done following the assessment methodologies[2] – constitute more complicated processes than engineers are generally familiar with, and they pose significant challenges both in the design and the construction of the interventions.

2 The new structural configuration that consists of both the existing and the new parts is conceived by the designer, and is then checked following the code-based assessment procedures.

During the design of the retrofit, the engineer should consider all the architectural and operational restrictions imposed by the existing building and find compromises between structural safety, economy, and the social and operational consequences of the strengthening interventions (e.g., disturbance to the residents, suspension of operations). Moreover, the presence of both existing and new materials in the same structure and the fact that these have different properties (mean strength and standard deviation from tests for the existing materials, nominal strength for the new materials) and cannot be treated in a universal manner often cause confusion. In the construction phase, several locations are difficult to access, are completely inaccessible (e.g., structural components below heavy mechanical installations), or require serious nonstructural damage for their strengthening (e.g., footings).

Furthermore, good, average, or bad knowledge of the existing structure is a new concept, which, although very important in the assessment procedures, is not encountered in the design methodologies; the assumption is made that, with the exception of the material strengths, engineers have full knowledge of the building and it will be constructed as designed. In existing buildings, however, even when the engineer has full access to the as-built drawings, there is always a level of uncertainty about whether these drawings reflect the actual state of the building and whether there are deviations that affect the response of the structural system. Consequently, depending on the accuracy and reliability with which the structure under investigation is known, the standards define different knowledge levels that indirectly affect the safety factors employed in the safety verifications. Hence, the engineer should decide if conducting a thorough investigation of the structural configuration is worthwhile, and if it could allow for lighter and less intrusive interventions with economic and operational benefits, and thus adapt the retrofit design strategy accordingly.

Recognizing all these difficulties and restrictions, the different standards worldwide have introduced additional options for the performance objectives, with more seismic hazard levels and more structural performance levels (referred to as limit states in the European codes), from which the engineer may choose. The objective of this is to allow for a more flexible framework in which engineers can provide viable technical solutions more easily. On the downside, it significantly increases the options that are available to the designers, further complicating their role by adding parameters such as the selection of different levels of risk or the remaining life of the structure. Obviously, specialized knowledge, experience in both design and construction, and correct engineering judgment are needed at the different stages of the process in order to impartially consider all the advantages and disadvantages of the different options and make optimal decisions.

It should be recalled that the objective of the assessment methodologies is somehow different from the design methodologies. In the latter, the engineer designs structural members in a new configuration so that the building is safe under the prescribed loading situations, whereas with the former the engineer determines whether an existing structural configuration is safe or not. This is a fundamental difference. In the design procedures the main output of the process is the CAD drawings to be passed to the construction site of the new building (Figure 1.2a). By contrast, in the assessment procedures, the main output is a set of checks that verify whether the existing structure is safe under the prescribed loading, and if it is unsafe, how and why it is unsafe – that is, which members are inadequate and need upgrading, and under which specific actions (Figure 1.2b).

(a)

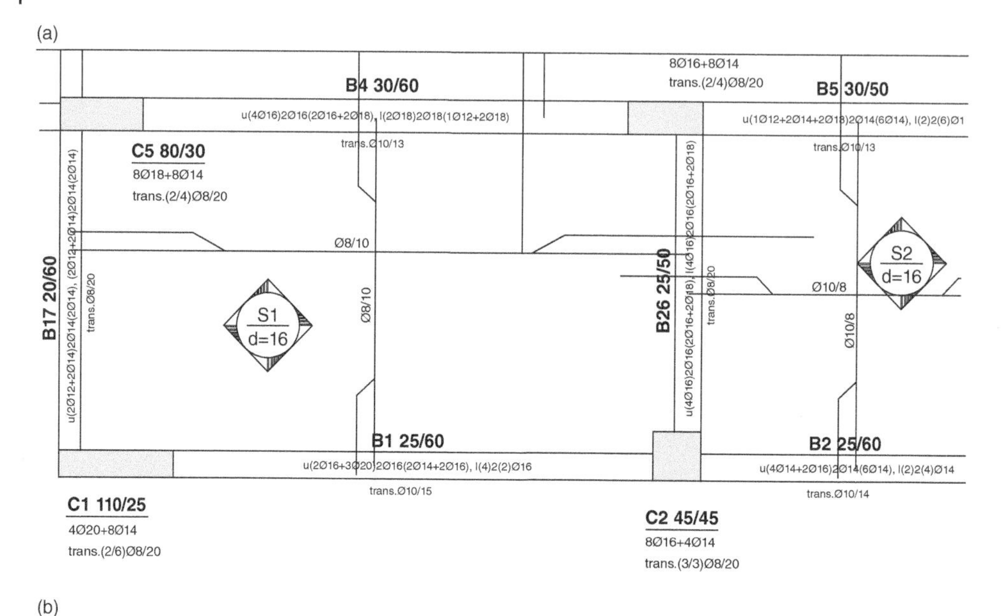


(b)

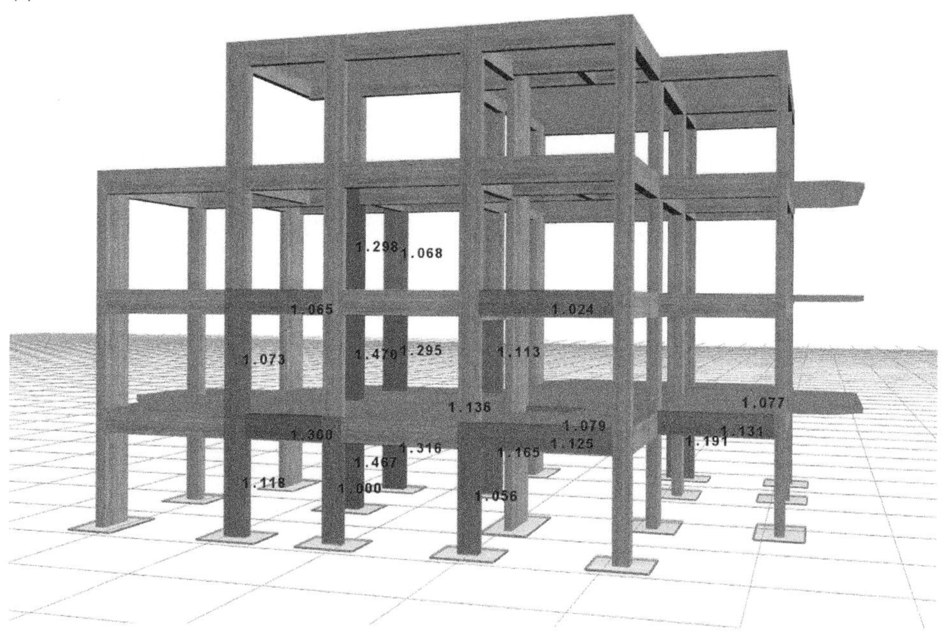


**Figure 1.2** Basic output of (a) the design procedures and (b) the assessment procedures. *Source:* Stelios Antoniou.

In the assessment methodologies, both the geometry and the reinforcement of the building should be defined and modeled in the structural model. This constitutes an important difference with respect to the typical design procedures, in which the user specifies the geometry (location and dimensions of the members) and the software package does the necessary calculations and checks to estimate the reinforcement needed for each member. In contrast, in

**Design Procedures**

1. geometry
2. member dimensions

member reinforcement so that structure is safe

**Assessment Procedures**

**Figure 1.3** Design vs. assessment procedures. *Source:* Stelios Antoniou.

assessment procedures the complete building configuration (geometry, components' dimensions, and reinforcement – longitudinal and transverse) is introduced, and what is checked is whether the existing building can sustain the specific loading level (Figure 1.3).

These differences between the two approaches are reflected in the structure of the seismic design and seismic evaluation codes worldwide. In most codes, different documents are introduced for the two cases, e.g. ASCE 41 vs. ACI 318 (US standards; ASCE 2017; ACI 2019), Eurocode 8, Part 3, vs. Eurocode 8, Part 1 (European standards; CEN 2005; CEN 2004), NZSEE:2017 vs. NZS 3101.1&2:2006 (New Zealand standards; NZSEE 2017; New Zealand standards 2006a, b), KANEPE vs. EAK 2000 (Greek standards; KANEPE 2022; EPPO 2001). Even in the cases when the assessment and the design procedures are combined in a single document, there are different chapters for the former, with significant differences even in the definition of limit states and the seismic hazard levels (e.g., Italian standard NTC-18 and Turkish standard TBDY; NTC 2018; TBDY 2018).

## 1.4 Whom Is this Book For?

In my humble opinion, there is a profound lack of documents in the literature that fully describe the concepts and procedures for a successful structural evaluation and more importantly for effective structural strengthening. There are many books, reports, and papers that provide details on certain aspects of the entire process, such as fiber-reinforced polymers (FRPs) strengthening of beams, comparison of RC jackets with FRPs, and analysis of framed structures with or without dampers, and with or without steel braces. Very few documents, however, have attempted to describe the general framework for the entire process for the strengthening of an existing vulnerable building (e.g., FIB 2022, FEMA 2006; FEMA 2018; Pinho et al. 2019; Tsionis et al. 2014), and even fewer provide specific guidelines on the strategy for structural upgrading.

This book includes a dedicated chapter (Chapter 5) with the main concepts of structural upgrading, together with more than 15 cases studies of the retrofitting of real buildings. This is the main objective of this manuscript: to describe the most common strengthening techniques that an engineer can employ, to make suggestions on the selection of the best retrofit strategies for each building type, and to present examples from practice that explain these suggestions.

The book is mainly targeted to engineers who have a background in the design of new reinforced concrete buildings and who want to expand their qualifications to structural assessment and strengthening and gain an in-depth understanding of the strengthening techniques and how these can be employed in the case of a retrofit project. It is also useful to scholars who want to acquire a comprehensive understanding of the retrofit of reinforced concrete buildings. Finally, field engineers can also find useful information on the application of the different strengthening techniques in the first part of the book.

## 1.5 Main Standards for the Seismic Evaluation of Existing Structures

Although there are numerous seismic standards worldwide regarding the seismic design of new structures, the number of regulations and guidelines (as separate norms or as parts of larger seismic codes) on the assessment and strengthening of existing reinforced concrete buildings is relatively small; such documents are only available in more developed countries with high seismicity. The two assessment regulations most used worldwide are ASCE 41-17 (ASCE 2017) in the United States (in particular, Tier 3 of ASCE 41[3]) and Eurocode 8, Part 3, in Europe (CEN 2005). Special reference should also be made to the New Zealand Technical Guidelines for Engineering Assessment (NZSEE 2017), which, although not so extensively used, has exerted considerable influence, similar to other New Zealand guidelines in the past.

These three documents have influenced the other guidelines for the seismic assessment and strengthening of buildings in different countries. In Italy, the NTC-18 (Norme Tecniche per le Costruzioni) Technical Standards for Construction (NTC 2018) and the accompanying Commentary (Circolare n. 7 2019), which is the continuation of the previous standard NTC-08 (NTC 2008; Circolare n. 617 2009), has a dedicated chapter on structural assessment and retrofit. In Greece, KANEPE (2022), the national Code of Structural Interventions, is a large, detailed document fully dedicated to the assessment and strengthening of reinforced concrete buildings. Both the Italian and the Greek standards have strong influences from the Eurocode 8, since in both countries there is a gradual attempt to make all construction guidelines fully compliant to the Eurocodes framework. In the new Turkish Building Seismic Code (TBDY 2018), there is also a large chapter on seismic assessment and strengthening, but in this case there are numerous references to different US standards, and in particular to ASCE-41.

The assessment methodologies prescribed in these standards have striking similarities and analogies. In all cases, the evaluation of an existing structure is carried out in relation to certain performance levels, which describe structural performance and the maximum allowable damage to be sustained, and specific seismic hazard levels and demands that are represented by target response acceleration spectra that correspond to specific return periods (or equivalently probabilities of exceedance in 50 years).

---

3 The seismic evaluation and retrofit process per ASCE 41-17 consists of three tiers, including Tier 1 screening procedure, Tier 2 deficiency-based evaluation procedure, and Tier 3 systematic evaluation procedure.

All normative documents raise awareness of the uncertainty with which existing structures are known, and in this regard they introduce additional safety factors[4] in the evaluation methodologies that reflect our uncertainty regarding the structural configuration.

The same four methods are employed for structural analysis. These are: the static linear, the dynamic linear, the static nonlinear (pushover), and the dynamic nonlinear procedure. There are small variations, such as the lateral load profile and the target displacement calculation in pushover analysis, the number of selected records in nonlinear dynamic analysis or the naming conventions. However, the basic concepts and the general philosophy are the same, and the framework for the systematic evaluation and retrofit approach of the standards is very common.

Within this detailed and systematic framework, more simplified analytical procedures are permitted by some codes, which allow for a rapid seismic evaluation for reinforced concrete buildings with hand calculations or with the aid of simple spreadsheets. For instance, ASCE-41 proposes a Tier 1 process for screening and a Tier 2 process for a simpler deficiency-based evaluation and retrofit, NZSEE proposes the SLaMA (Simple Lateral Mechanism Analysis) methodology, and NTC-18 proposes a simplified evaluation method but only for masonry buildings.

Regarding the safety verifications and the calculation of the capacity of the components, although there are several variations, the general framework is again very similar in all standards. There is a distinction between ductile and brittle components and failure mechanisms and between primary and secondary members, the former being more important in resisting lateral seismic forces. The checks are performed in terms of forces or moments, with the exception of ductile failure modes, for which the checks may be done in terms of deformations. Capacity design considerations are usually employed to calculate the demand in brittle failure modes in the linear methods of analysis, contrary to the nonlinear methods, where the demand is directly extracted from the analytical calculations. There are, of course, differentiations between the different approaches. Examples include the expressions in the calculation of the members' shear capacity or the beam-column joint capacity, and the deformation quantity that is checked in bending in the nonlinear procedures (chord rotation in all European codes, plastic hinge rotation in ASCE 41, curvature in the NZS, and strains in TBDY). However, conceptually these variations do not affect the general philosophy of the verifications, which is the same in all documents.

Finally, two important features introduced in the NZ guidelines are worth mentioning. The assessment procedure is based on one evaluation parameter, defined as the percentage of new building standard (% NBS), which is used for quantifying the safety level against seismic actions and is expressed as the ratio between the seismic capacity of the existing building (or of an individual member of it) under the specified seismic action, and the ultimate limit state shaking demand for a similar new building on the same site:

$$\%NBS = \frac{\text{Ultimate capacity}\left(\text{seismic}\right)}{\text{ULS seismic demand}} \times 100\%$$

4 They are referred to as knowledge or confidence factors, but in practice they are additional safety factors that are employed when the knowledge of the structural configuration is limited.

**Table 1.1** Assessment outcomes and potential building status (NZSEE 2017 – Part A, Table A3.1).

| Percentage of new building standard (%NBS) | Alpha rating | Approx. risk relative to a new building | Life-safety risk description |
|---|---|---|---|
| >100 | A+ | Less than or comparable to | Low risk |
| 80–100 | A | 1–2 times greater | Low risk |
| 67–79 | B | 2–5 times greater | Low to medium risk |
| 34–66 | C | 5–10 times greater | Medium risk |
| 20 to <34 | D | 10–25 times greater | High risk |
| <20 | E | 25 times greater | Very high risk |

Furthermore, the guidelines present a table (shown as Table 1.1), in which the relative risk level of an existing building is quantified on the basis of the value assumed by the %NBS coefficient, and a seismic capacity rating is assigned (A+ to E).

Similar considerations regarding the classification of the seismic risk of buildings are made in the Italian guidelines of NTC-18, which also require that the assessment of the safety level of existing buildings be carried out in relation to the safety required for new buildings (Pinho et al. 2019). The New Zealand %NBS coefficient corresponds to the $\zeta_E$ parameter in NTC-18, which is the ratio between the maximum seismic action bearable by the structure and the maximum seismic action that would have been used in the design of a new construction on the same ground and with the same characteristics (Figure 1.4).

$$\zeta_E = \frac{PGA_C}{PGA_D}$$

Similarly, a seismic risk classification system for existing buildings has been established by combining considerations on both the value of human life (through the appropriate safety levels) and the possible economic and social losses. There are eight seismic risk classes, which indicate an increasing risk level from class A+ to class G (Figure 1.5).

The method is applicable to any type of building and makes use of two parameters, PAM (Perdita Annua Media Attesa), the expected average annual loss, and IS-V (Indice di Sicurezza), the safety index. The PAM index represents the relationship between the cost of the repairs caused by the design earthquake and the cost of reconstruction, expressed as a percentage that is distributed annually during the lifetime of the building. The safety index IS-V expresses the relationship between the peak ground acceleration (PGA) for which the limit state of life safety is reached for the existing building and that which would be used for a new building with the same characteristics on that specific site – in other words, it is the $\zeta_E$ parameter for the life safety limit state.

A seismic risk class from A+ to G is assigned for each of the two parameters, and the final seismic risk class of the building is the lower of the two.

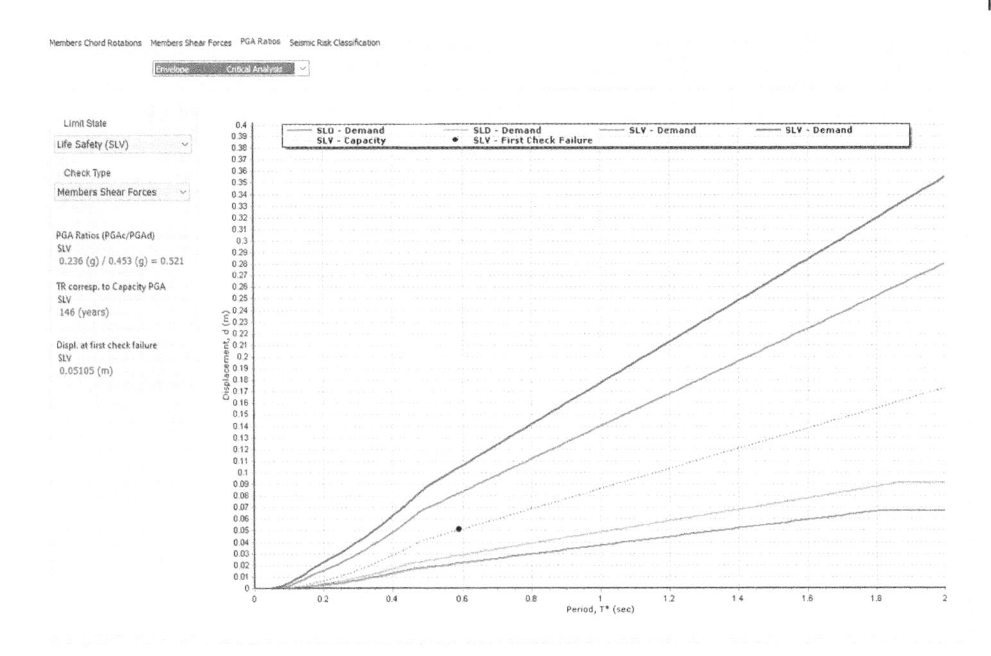


Figure 1.4 Calculation of the $\zeta_E$ parameter in SeismoBuild. *Source:* Stelios Antoniou.

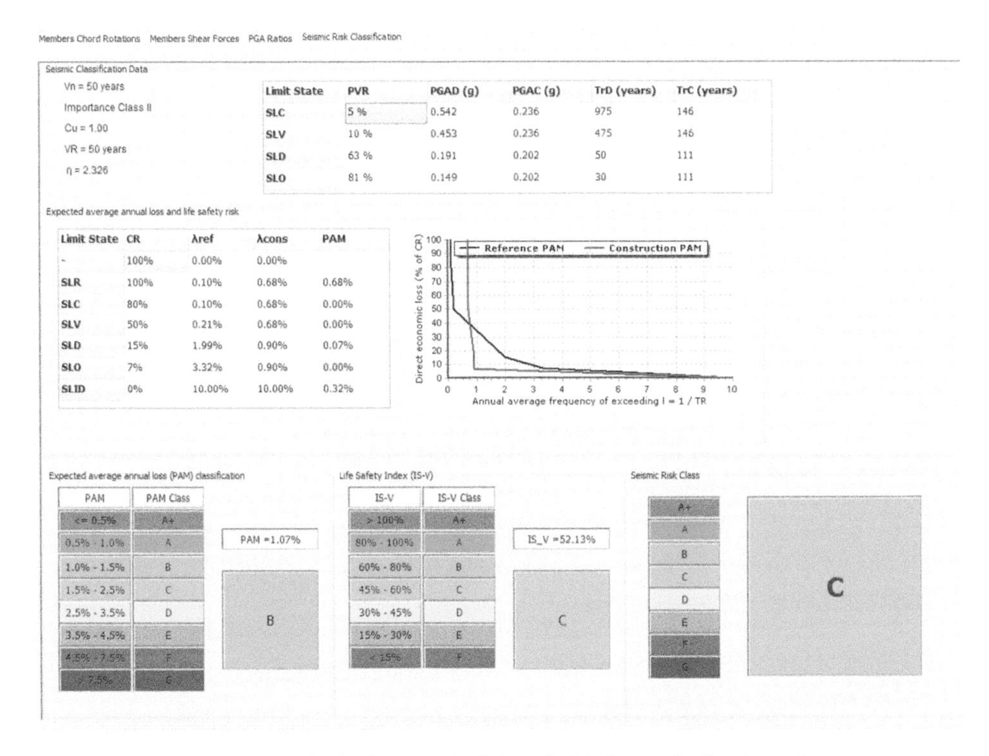

Members Chord Rotations  Members Shear Forces  PGA Ratios  Seismic Risk Classification

Seismic Classification Data

Vn = 50 years
Importance Class II
Cu = 1.00
VR = 50 years
η = 2.326

| Limit State | PVR | PGAD (g) | PGAC (g) | TrD (years) | TrC (years) |
|---|---|---|---|---|---|
| SLC | 5 % | 0.542 | 0.236 | 975 | 146 |
| SLV | 10 % | 0.453 | 0.236 | 475 | 146 |
| SLD | 63 % | 0.191 | 0.202 | 50 | 111 |
| SLO | 81 % | 0.149 | 0.202 | 30 | 111 |

Expected average annual loss and life safety risk

| Limit State | CR | λref | λcons | PAM |
|---|---|---|---|---|
| - | 100% | 0.00% | 0.00% | |
| SLR | 100% | 0.10% | 0.68% | 0.68% |
| SLC | 80% | 0.10% | 0.68% | 0.00% |
| SLV | 50% | 0.21% | 0.68% | 0.00% |
| SLD | 15% | 1.99% | 0.90% | 0.07% |
| SLO | 7% | 3.32% | 0.90% | 0.00% |
| SLID | 0% | 10.00% | 10.00% | 0.32% |



Expected average annual loss (PAM) classification

| PAM | PAM Class |
|---|---|
| <= 0.5% | A+ |
| 0.5% - 1.0% | A |
| 1.0% - 1.5% | B |
| 1.5% - 2.5% | C |
| 2.5% - 3.5% | D |
| 3.5% - 4.5% | E |
| 4.5% - 7.5% | F |
| > 7.5% | G |

PAM =1.07%

B

Life Safety Index (IS-V)

| IS-V | IS-V Class |
|---|---|
| > 100% | A+ |
| 80% - 100% | A |
| 60% - 80% | B |
| 45% - 60% | C |
| 30% - 45% | D |
| 15% - 30% | E |
| < 15% | F |

IS_V =52.13%

C

Seismic Risk Class

A+
A
B
C
D
E
F
G

C

Figure 1.5 Calculation of seismic risk class in SeismoBuild. *Source:* Stelios Antoniou.

## References

[ACI] American Concrete Institute ACI 318 (2019). *ACI CODE-318-19: Building Code Requirements for Structural Concrete and Commentary. ACI Committee 318*. American Concrete Institute.

[ASCE] American Society of Civil Engineers (2017). *Seismic Evaluation and Retrofit of Existing Buildings (ASCE/SEI 41–17)*. Reston, Virginia: ASCE.

CEN (2004). *European Standard EN 1998-1: 2004. Eurocode 8: Design of Structures for Earthquake Resistance, Part 1: General Rules, Seismic Actions and Rules for Buildings*. Brussels: Comité Européen de Normalisation.

CEN (2005). *European Standard EN 1998-3: 2005. Eurocode 8: Design of Structures for Earthquake Resistance, Part 3: Assessment and Retrofitting of Buildings*. Brussels: Comité Européen de Normalisation.

Circolare n. 617 (2009). Circolare 02/02/2009 n. 617, Istruzioni per l'Applicazione Nuove Norme Tecniche Costruzioni di cui al Decreto Ministeriale 14 gennaio 2008, G.U. del 26/02/2009, n. 47 (in Italian).

Circolare n. 7 (2019). Circolare 21/01/2019 n. 7 C.S.LL.PP. Istruzioni per l'applicazione dell'«Aggiornamento delle 'Norme tecniche per le costruzioni'» di cui al decreto ministeriale 17 gennaio 2018. (19A00855), G.U. Serie Generale del 11-02-2019, n.35 - Suppl. Ordinario n. 5 (in Italian).

EPPO (Earthquake Planning and Protection Organisation) (2001). Greek Code for Seismic Resistant Structures, EAK 2000 (in Greek).

[FEMA] Federal Emergency Management Agency (1997). *NEHRP Guidelines for the Seismic Rehabilitation of Buildings, FEMA 273*. Report, prepared by the Applied Technology Council and the Building Seismic Safety Council for the Federal Emergency Management Agency, Washington, D.C.

[FEMA] Federal Emergency Management Agency (2000). *Pre-standard and Commentary for the Seismic Rehabilitation of Buildings. FEMA 356*. Report, prepared by the American Society of Civil Engineers for the Federal Emergency Management Agency, Washington, D.C.

[FEMA] Federal Emergency Management Agency (2018). *Seismic Evaluation of Older Concrete Buildings for Collapse Potential, FEMA P-2018*. Report, prepared by the. Washington, D.C.: Applied Technology Council for the Federal Emergency Management Agency.

[FEMA] FEMA NEHRP (2006). *Techniques for the Seismic Rehabilitation of Existing Buildings, FEMA 547*. Building Seismic Safety Council for the Federal Emergency Management Agency.

[FIB] Fédération Internacionale du béton (2022). *Guide for Strengthening of Concrete Structures*. Fib bulletin 103. Technical report. 316 pages, ISBN 978-2-88394-157-1, May 2022.

KANEPE (2022). *Greek Code of Structural Interventions, Revision 3*. Earthquake Planning and Protection Organization, EPPO (in Greek).

Meletti, C., Stucchi, M., and Boschi, E. (2006). Dalla classificazione sismica del territorio nazionale alle zone sismiche secondo la nuova normativa sismica. In: *Norme Tecniche per le Costruzioni* (ed. D. Guzzoni). Il Sole 24 Ore In Italian.

New Zealand Standards (2006a). NZS 3101 – Part 1: Concrete Structures Standard – The Design of Concrete Structures.

New Zealand Standards (2006b). NZS 3101 – Part 2: Concrete Structures Standard – Commentary.

NTC 2008. D.M. Infrastrutture e Trasporti 14/01/2008, Norme tecniche per le costruzioni, G.U. 04/02/2008, n.29, NTC 2008 (in Italian).

[NTC 2018]. D.M. Infrastrutture e Trasporti 17/01/2018, Aggiornamento Norme tecniche per le Costruzioni, G.U. 20/02/2018, Suppl. Ord n. 8, NTC 2018 (in Italian).

NZSEE (2017). *The Seismic Assessment of Existing Buildings*. Technical Guidelines for Engineering Assessment. New Zealand Society of Earthquake Engineering, Wellington.

Pinho, R., Bianchi F. and Nascimbene R. (2019). Valutazione sismica e tecniche di intervento per edifici esistenti in c.a. Maggioli Editore (in Italian).

SeismoBuild (2023). SeismoBuild – A computer program for the linear and nonlinear analysis of Reinforced Concrete Buildings. www.seismosoft.com.

SeismoStruct (2023). SeismoStruct – A computer program for static and dynamic nonlinear analysis of framed structures. www.seismosoft.com.

TBDY (2018). *Türkiye Bina Deprem Yönetmeliği, Turkish Seismic Building Code*. Ankara: Disaster and Emergency Management Presidency in Turkish.

Tsionis, G., Apostolska, R., and Taucer, F. (2014). *Seismic Strengthening of RC Buildings*. JRC Science and Policy Reports. Joint Research Center.

# 2

# Know Your Building: The Importance of Accurate Knowledge of the Structural Configuration

## 2.1 Introduction

Back in 2017, I was in Jordan to give a course on the strengthening and retrofit of RC buildings. During one of my presentations I was going on and on, talking about the significance of accurate knowledge of the structural configuration, and how important it is to know the building that is going to be analyzed, to know its geometry, to know the exact member dimensions, and to dig and uncover the existing footings. Above all, I was iterating the need to know the exact reinforcement of the members, or at least of the vertical members, columns, and walls.

When the audience got tired of me focusing on "trivial" things (compared to "important" subjects like the advanced nonlinear analysis, which was the main reason they had attended the course in the first place), one of the attendees, who was one of the most capable and knowledgeable engineers in the group, interrupted me and complained: What is the need for all this? Why on earth can't we use the minimum reinforcement ratio provided by the code that was in force at the period when the building was designed?

This is a fair question. It is a known fact that the majority of existing buildings have been designed without provisions for seismic resistance. However, even older codes had provisions on the minimum design requirements, most importantly the "minimum reinforcement ratio." In theory, assuming and employing a minimum reinforcement, instead of a detailed survey on site, would be enough. Furthermore (again in theory), this is also on the safe side.

This chapter describes all the reasons why one should leave the comfort of the office and spend a significant amount of time getting the best possible knowledge of the building under consideration. Several factors, related to incomplete information, bad detailing, poor workmanship, corrosion, and bad practices and deficiencies during construction, are analyzed and explained. At the end of this chapter, readers should understand that getting an adequate knowledge of the structural configuration of existing buildings is not as simple and straightforward as it might seem at first sight, but it is a crucial stage in the process nonetheless.

*Seismic Retrofit of Existing Reinforced Concrete Buildings*, First Edition. Stelios Antoniou.

## 2.2 What Old RC Buildings Are Like

The obvious thing that one could say about older buildings is that they are different from modern ones, and have not been built with the stringent, high standards that we follow nowadays, both in design and construction. This is clear to everyone, there is a general consensus about it, and there is not much to comment on this. After all, this is why there is a need to upgrade and strengthen older buildings. Yet, the most important question is not whether they are different from modern construction, but rather *how* they are different. This is where things start to get interesting.

When studying older buildings for an adequate amount of time, one realizes that the design and construction practices in a specific region and at a specific period have a lot in common. The material grades, the member sizes, both columns and beams, the existence or not of shear walls, the type and dimensions of the foundations, the reinforcement (the rebar and stirrup diameters, the detailing and the steel classes) are all very similar in buildings of the same region constructed during the same period. The time periods with common characteristics are typically determined by changes in the structural code imposed by the authorities. For instance, since the significant changes in the seismic design in Greece occurred with the 1959 and 1984 seismic codes, a building designed in 1979 is expected to be much more similar to a building of the 1960s, than to one constructed in 1986.

In contrast to the relative ease with which the period of construction can be identified, it is much more difficult to identify the regions with common characteristics, and there is by no means a direct correlation between a specific region and the country it is in. For instance, the construction practice in western Greece has always been different (more conservative) than the practice in Athens and the central and eastern part of the country – due in part to the higher seismicity of the region and also to the significant psychological effect of the 1953 earthquake in Kefalonia that left the island in ruins. Similarly, the construction practices in southern Italy are different (more conservative) compared to northern Italy (again, due to the higher seismicity, but also because of the 1908 Messina earthquake that devastated Sicily).

Another interesting observation is that the structural drawings have generally been followed in the superstructure of existing buildings; however, many variations are observed in how the design was actually implemented at the foundation level. Drawings are predominately followed fairly well in the cases of conspicuous works, whereas in works that are hidden and covered throughout the lifetime of the building, poor workmanship is the rule, rather than the exception. Not surprisingly, the most striking bad construction examples in Appendix A come from footings and the foundation level.

It is noteworthy that, as explained below, the large variations that exist in the workmanship of the underground works do not always reveal poor quality. For instance, I have seen several cases where the cone of a footing has been cast with inclined formwork, contrary to today's practice of placing formwork only at the sides of the footing, pouring the concrete, and leaving its slump to form the cone that is described in the drawings. However, in the majority of the cases, the lack of thorough control and deviation from the specifications resulted in building foundations with significant (and very dangerous) deficiencies.

Another equally interesting and important thing about older construction is that older structural drawings lack many details that are extremely important, if seen from the current perspective with the knowledge about earthquakes that we have nowadays. For instance, in older drawings there is no or limited mention about lap splices, reinforcement detailing, and, most importantly, stirrups.

The factors discussed above, together with the surprising (or more precisely shocking) negligence and poor workmanship and supervision in several extreme cases, are the main reasons why older buildings deviate significantly from today's construction practices, and why so often older structures need upgrading and retrofit. Generally speaking, in the majority of cases there are common features, problems, and deficiencies associated with existing buildings. Below is a list with the most important issues.

### 2.2.1 Lack of Stirrups

The lack of transverse reinforcement is the most important problem in the majority of the cases of existing buildings, and the most common reason why a building needs strengthening.

In new construction, the codes enforce strict regulations on the hoops' layout, imposing limits for the minimum hoop diameter, the maximum hoop spacing, and their correct anchorage with a 135° bend inside the concrete core. With the capacity design approach that was introduced in the 1980s, we can now be sure that a ductile failure mechanism in bending will precede any catastrophic brittle failure in shear.

Things were completely different before the 1980s, even in countries with more developed seismic design like the US or New Zealand (Niroomandi et al. 2015). Engineers (not only practicing engineers but also the academic community) were not aware of confinement, ductility, or the capacity design philosophy, and usually designed the members with the minimum transverse reinforcement to withstand gravity loading. Seismic loading was not considered a major load case for design, and engineers paid much more attention to the shear reinforcement of beams (where the shear demand from the dead loads is larger), rather than columns. The placement of shear reinforcement in the beams was relatively better and tidier with constant spacing between the hoops at least in the critical regions, whereas in columns the transverse reinforcement was often smaller with lower steel class, and spacing between the hoops that was larger and often variable even along the same member (e.g., from ∅6/30 to ∅6/50 cm). Cases of very large hoop spacing (up to 1.00 m or more) in the vertical members were not so rare, either. Last but not least, the hoops were not anchored properly, without the 135° bending inside the concrete core (Figure 2.1).

It is noteworthy that very often in buildings in the 1970s and earlier, the shear reinforcement is *not specified at all* for the columns in the drawings, and only the longitudinal reinforcement is described (Figure 2.2), which is a good indication of the low importance that the transverse reinforcement had in the design. It is stressed again that this problem is observed in developing and developed countries alike (Figure 2.3), which highlights the lack of knowledge and awareness with respect to the importance of the transverse reinforcement in reinforced concrete buildings.

It is noted that even in the relatively decent cases with good placing and constant spacing of stirrups without corrosion, the total shear reinforcement in existing columns (e.g., ∅8/30 cm with S220, C12/15 and 4∅20 mm longitudinal reinforcement) is approximately 10–20% of what would be placed nowadays (typically ∅8/10 or ∅10/10 cm in seismic prone regions with S500, C25/30 and 8∅20 mm longitudinal reinforcement), resulting in a very reduced shear capacity, even for the same member sizes (Figure 2.4).

Figure 2.1 Shear reinforcement (or the lack of it) in existing RC columns. *Source:* Stelios Antoniou.

## 2.2.2 Unconventional Reinforcement in the Members

Because the consensus on how to reinforce a concrete column or beam has only been reached three to four decades ago, it is not uncommon to find members being reinforced in different and unconventional ways. Some not-so-old theories on the transfer of shear forces proposed the non-uniform placement of stirrups in a member. From my early days in university, I still

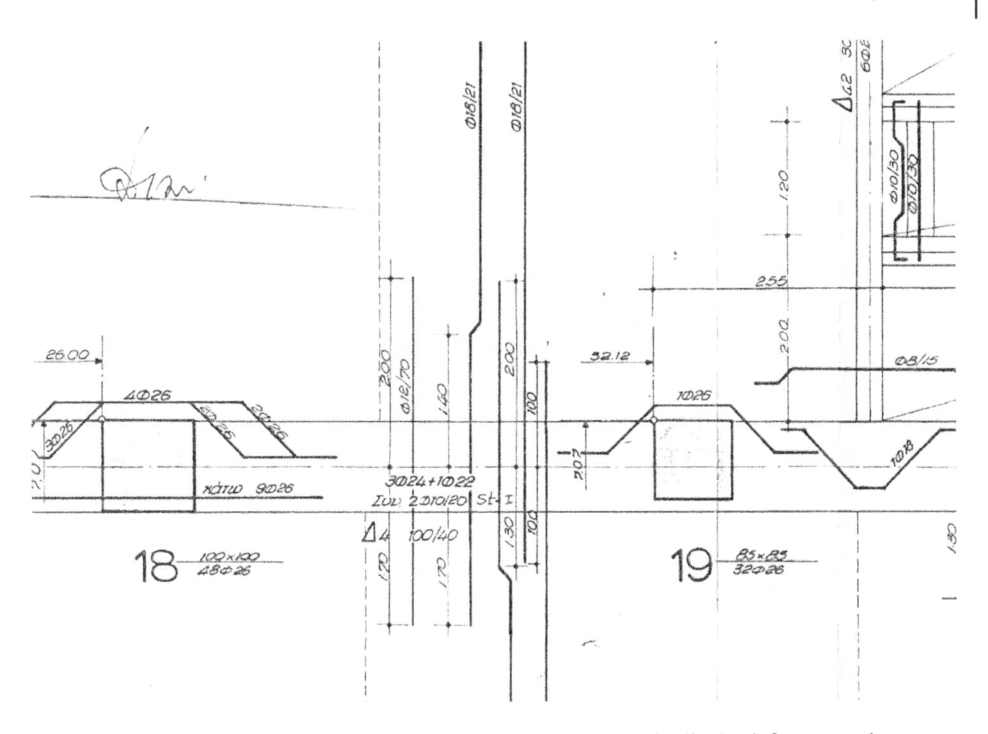


Figure 2.2 Plan view of an existing building, where only the longitudinal reinforcement is indicated in the columns. *Source:* Stelios Antoniou.

remember the story about a well-known professor who had proposed a theory about placing hoops only at two thirds or half of the column and got really upset when the experiment did not confirm his theory – obviously, the specimen failed at the exact point where there were no stirrups. And that was in the early 1990s! Of course, such unconventional practices may also be encountered in the existing building stock. In one case that I came across, the entire basement of a large building (80 columns in total) was reinforced in a very consistent way as in Figure 2.5: Ø8/25 cm stirrups for the first 1.50 m and then 130 cm without any stirrup. That was not bad practice in construction; it was bad design from a designer who simply did not know how buildings fail under seismic loading, when the building was constructed in the early 1970s.

### 2.2.3 Large, Lightly Reinforced Shear Walls or Lack of Shear Walls

Similar to transverse reinforcement, the engineering community in the 1950s, 1960s, and 1970s lacked the scientific knowledge about the importance of large and heavily reinforced shear walls with pseudo-columns that are very common nowadays. Consequently, large, stiff shear walls are rare in existing buildings, resulting in flexible buildings that sustain large deformations during significant seismic events. Furthermore, because of the lack of shear walls, the effects of stiffness irregularities in plan or in elevation (e.g. pilotis at the ground level) are much more pronounced, as well as the effect of short columns.

It is noteworthy that in the cases when shear walls are present in the construction (e.g. shear walls around the perimeter of the elevator shaft), they are lightly reinforced and thin (typically 20 cm thick), increasing rather than decreasing the overall vulnerability. Indeed, such lightly

Figure 2.3 Damage to columns and beam-column joints at West Anchorage High School, Anchorage, Alaska during the 1964 Alaska earthquake (NISEE e-Library n.d., Karl V. Steinbrugge Collection: images S2279 and S2263).

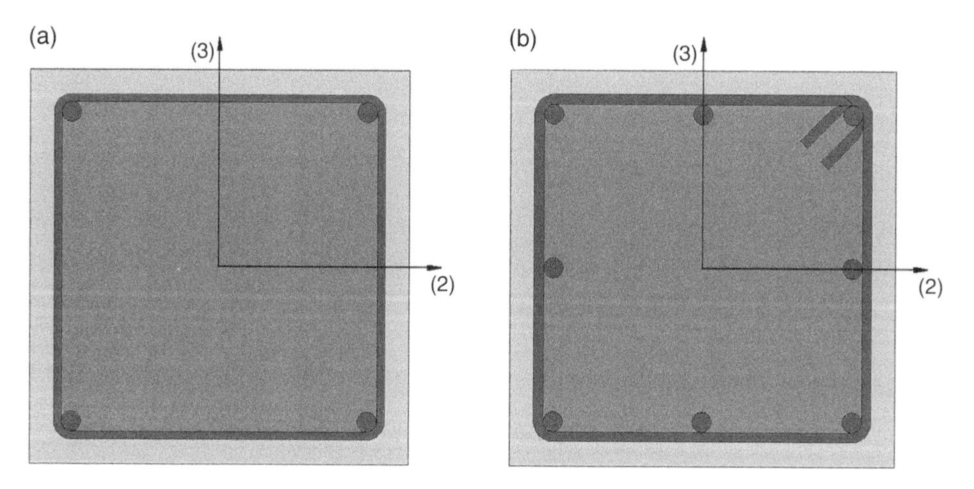


Figure 2.4 Comparison of the shear capacity for typical existing and new RC column: (a) existing column C40/40: 4⌀20 mm, ⌀8/30 cm, materials S220 & C12/15: Shear capacity ≈ 35 kN (b) new column C40/40: 8⌀20 mm, ⌀10/10 cm, materials S500 & C25/30: Shear capacity ≈ 275 kN. *Source:* Stelios Antoniou.

Figure 2.5 A highly unconventional (and vulnerable) way to apply transverse reinforcement. *Source:* Stelios Antoniou.

reinforced members are typically reinforced with one or two reinforcement grids (e.g. #∅8/25 to #∅10/20 cm), and are usually the first to sustain damage even in moderate seismic events, typically failing in shear with diagonal cracks (Figure 2.6). Moreover, they are often located asymmetrically, at the corners or the sides of the building (usually where the elevator shaft and the stairs are located) without any consideration to the stiffness distribution and the center of stiffness, causing increased deformations to some of the vertical members, due to torsion.

Figure 2.6 A large, lightly reinforced shear wall that failed prematurely during a moderate seismic event.

### 2.2.4 Lap Splices

Unlike the case of shear reinforcement, engineers in previous decades recognized the need for lap splices. However, there were no strict rules in the codes, and typically the splices were considerably smaller than the current standards. Usually, one could expect in existing members lap splices of not more than 50 or 60 cm for rebar ∅14 mm to ∅20 mm, but very often lengths as low as 20 cm may be found (Figure 2.7a). Rebars bent at 180° at their ends, in order to achieve better bonding and transformation of the forces, are commonly found. However, in such cases the length of the splice is even smaller, leading to worse rather than better transformation of the rebar forces (Figure 2.7b). Furthermore, there are several cases of bad workmanship – e.g., the lower rebars are not anchored inside the concrete core, but rather in the cover outside the stirrups (Figure 2.7c), or rebars that are bent with angles as high as 90° inside the slab level to decrease the size of the column in upper floors (Figure 2.7d).

### 2.2.5 Corrosion

Another problem commonly found in existing RC buildings is the corrosion of the reinforcement, typically caused by carbonation and/or bad casting of concrete. Corrosion affects both the longitudinal rebars and the stirrups, with the effects on the latter being more serious, due to their smaller size and the fact that their degradation affects the shear capacity, which is a brittle type of failure. It is not uncommon to find hoops that are completely degraded by corrosion, or even structural members that have failed completely because of it (Figure 2.8).

If left unchecked, corrosion can cause significant decrease in both the bending and the shear capacity of members. What is more, because corroded rebars "swell," increase in volume, and cause cracking to the cover concrete, corrosion is a phenomenon with gradually increasing intensity that can quickly lead to serious degradation of the members' strength. On the positive side, corrosion can be easily identified by cracks on the surface of

**Figure 2.7** Reinforced concrete members with problematic lap splices: (a) very short lap lengths, (b) rebars bent at 180o angles but with very short lap lengths, (c) rebars anchored outside of the concrete core and (d) rebars bent at 90o angles inside the slab. *Source:* Stelios Antoniou.

the concrete or the plaster coating; structural members without significant cracking are guaranteed to have no problems with corrosion. In other words, if a member has corroded reinforcement, cracks will be spotted on its surface; if there are no cracks on the member, one can be sure that its reinforcement is not corroded.

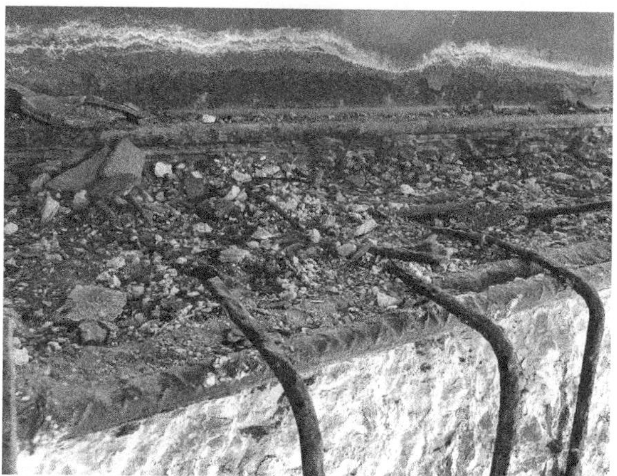

**Figure 2.8** Reinforced concrete members with corrosion problems. *Source:* Stelios Antoniou.

### 2.2.6 Geometry: Location of Structural Members

Before 1980, engineers mainly designed buildings for gravity forces, without much consideration about the lateral loading (seismic or wind loads). For this reason, issues such as the lateral stiffness distribution or the torsional stiffness were largely irrelevant. As a consequence, it is very common that existing buildings were constructed to be highly asymmetric in plan view, with large shear walls concentrated at one side or one corner of them, causing significant torsional effects. In such cases, the location of the large, stiff walls are rarely determined by structural and safety considerations, but rather by architectural and operational issues; for instance, in the majority of cases, the RC walls are located around the stairs and the elevator shafts, without similar stiff members at the other parts of the building.

### 2.2.7 Geometry: Bad Alignment of the Columns

I have encountered several cases of badly constructed RC buildings, where the columns are not correctly aligned vertically, and the column of the upper floor is displaced by 10 or even 20 cm, with respect to the underlying column. These cases, which constitute examples of extremely bad workmanship, are not as rare as one would expect. Obviously, the lap splices of the rebars are nonexistent, and considering also the small interaction surface between the two members, this practically means that the connection has a very limited capacity to transfer the lateral loading to the ground (Figure 2.9).

**Figure 2.9** Bad alignment of the columns: in both cases the upper column is displaced by 10 cm in one direction and almost 20 cm in the other with respect to the lower column. *Source:* Stelios Antoniou.

### 2.2.8 Geometry: Arbitrary Alterations During Construction or During the Building's Lifetime

When engineers hear about alterations in a building, they almost immediately think of alterations in use. Indeed, there are cases when an alteration in use can end up with much larger forces applied to the structure, both dead loads, but more importantly inertia forces that develop during a seismic event, for instance changing a regular residential floor to a warehouse full of paper in the entire story height. These cases should be thoroughly checked and usually measures are required for supporting the new loads.

Nevertheless, in the most common cases, these alterations of use are related to small changes in the applied load, such as change from a residential area ($q = 2.0–3.0\,\text{kN/m}^2$) to a commercial area ($q = 4.0–5.0\,\text{kN/m}^2$) and the required interventions are mostly related to the deficiencies in the current structural configuration (e.g., lack of reinforcement, corrosion, low material properties), rather than the increase in the loads themselves.

There is, however, one case of alterations that is very important, highly dangerous, and that considerably increases structural vulnerability: arbitrary, unauthorized alterations that have been carried out during the lifetime of the building without proper design and checks. The list of such bad (very bad!) practices is endless:

i) Arbitrary change in the reinforcement, the dimensions, or the location of structural members during construction.
ii) Avoidance of the construction of some members, because of last-minute changes in the architectural layout.
iii) Demolition of structural members during the lifetime of the building, because of architectural changes. These changes are sometimes accompanied by some countermeasures that very often consider only the gravity loading, without any consideration regarding the seismic response of the structure. Among these changes, the removal of columns or load-bearing masonry walls are the most common, e.g. Wikipedia (2022). All the same, the removal of beams or the creation of large opening in the slabs (which often cancels the diaphragmatic action of the remaining slab) may also be found.
iv) Demolition of load-carrying masonry walls, in order to create openings.
v) Construction of new lofts inside the building without proper design and consideration about the increased loads applied to the existing structural members and the change in the dynamic characteristics of the building. Sometimes the lofts are constructed with steel so as to *not increase the loads on the existing building too much*.
vi) As with the previous case, sometimes additional floors are constructed without consideration to the load-bearing capacity of the existing structure. Such interventions have led in the recent past to several collapses of buildings under the increased gravity loading, especially in developing countries.

### 2.2.9 Bad Practices with Respect to the Mechanical and Electrical Installations

Since the technicians for the mechanical and electrical installations work on a building after the completion of the load-bearing system, it is not uncommon that they demolish parts of the RC members in order to pass their cables and pipes through. Most of the time,

**Figure 2.10** Arbitrary concrete demolition for passing the mechanical and electrical installations. *Source:* Stelios Antoniou.

they understand and respect the significance of the framing system and they concentrate on opening small holes in the slabs without causing damage to columns or walls. However, cases of significant damage to the beams or the columns are not uncommon (Figure 2.10).

In this category lies the terrible practice of placing sewage pipes at the center of RC columns, mainly for architectural reasons – that is, to hide them from view (Figure 2.11). This practice was very common in southern Europe in the 1980s. Apart from reducing the volume of concrete of the member, there are two more serious negative side effects of the presence of the pipe inside the column. First, possible damage to the pipe can result in the leakage of the waste and the fast corrosion of the column reinforcement. Second, and more importantly, when technicians need to fix some damage to the pipe, typically when the pipe is clogged, they have to demolish part of the concrete of the column, significantly

Figure 2.11 Pipes inside reinforced concrete column.

reducing its capacity. After the repair, the damage to the column could be easily fixed with repair mortars; however, this is rarely the case. Usually, the technicians (and the owners!) leave the column unrepaired and weakened.

### 2.2.10 Soft Ground Stories

In certain regions the construction system of pilotis is widely used for architectural reasons, in order to increase the architectural volume and to create free space for circulation or car parking under the construction. Although practical and (under certain conditions) highly aesthetic, pilotis create a very weak, soft story at the ground level and significantly increase the vulnerability of the building under seismic loading. In modern design, this is taken into consideration by introducing large, stiff and heavily reinforced shear walls. However, this is not the case in older construction, where the vertical members are typically small and lightly reinforced, especially in the transverse reinforcement.

Pilotis, combined with all the other bad practices presented in the current section, constitute one of the main reasons for the increased vulnerability of older construction that often justify the retrofit of a structure. It is noteworthy that the strengthening of pilotis, apart from being technically very important, can also be done relatively easily with interventions only at the ground and foundation levels.

### 2.2.11 Short Columns

This is another important factor in the increased vulnerability of existing buildings. Short columns are very often found in industrial buildings, where the infill walls are stopped at a certain height in order to create an opening and provide light inside the building. In modern design, short columns do not constitute a very important issue, due to the presence of large, stiff shear walls. On the contrary, older buildings lack shear walls and are more flexible, making the short columns an extremely weak element of the construction; these short columns typically fail early in shear and lead to the collapse of the entire building (see Figure 2.13, but also Figure 2.12).

Figure 2.12 Soft ground story in a RC building. *Source:* Stelios Antoniou.

Figure 2.13 Short reinforced concrete columns. *Source:* Stelios Antoniou.

### 2.2.12 Different Construction Methods

In new buildings typically a single type of construction method is employed (RC, masonry, composite, or steel). Even in the cases where different methods are combined (e.g., RC columns, walls and beam with a steel roof), special care is taken to correctly connect the different structural parts, and the design and construction of the different parts is carried out simultaneously, considering the entire structure as a single entity.

Unfortunately, this is not always the case for older buildings, and there are many cases with mixed systems, such as unreinforced masonry and reinforced concrete, RC and steel, or different types of stone and brick masonry, without special provisions for the appropriate connection of the different parts and materials. In such cases, one has to be extremely careful with two things:

i) The connections of newer parts of the building to older ones, such as the connections of RC columns to the supporting masonry walls. Most often these connections are very poor without any consideration of the transfer of horizontal forces between the two parts, since only the gravity loading has been considered.
ii) The irregularities in elevation and/or in plan, due to the fact that different construction methods have very different stiffness distribution and dynamic characteristics. The effect of these stiffness irregularities can be highly significant (e.g., a steel framing system on top of a reinforced concrete frame) and needs to be thoroughly analyzed and checked.

The most striking case that I have ever encountered is the case of a six-story building with basement that was built gradually since the 1890s. The basement and the ground floor were constructed with stone masonry walls and wooden beams and slabs before 1900, and in the 1920s two more similar masonry floors were added. In the 1950s, two additional stories were constructed with RC members (columns, beams, and slabs) without any special care for the connection of the columns to the supporting masonry walls. Finally, a sixth story was constructed in the 1960s with brick masonry walls and reinforced concrete slabs. The design of the construction of the bearing system seems to have been done empirically in all periods, considering only the gravitational loads, and without any consideration regarding the seismic performance of the building as a whole. For instance, the lateral stiffness of the fourth and fifth floors is significantly smaller with respect to the lower floors, but also with respect to the sixth floor, a factor that would probably give rise to significant second-order effects in the case of a large earthquake event. Furthermore, the diaphragmatic action of the slabs is significant only in the upper floors, since in the lower floors the contribution of the wooden slabs is negligible.

### 2.2.13 Foundation Conditions

As explained at the beginning of this chapter, one of the most striking observations that one makes when dealing with existing buildings is that very often the construction drawings were not followed at all at the foundation level, with many and significant variations from the original design. The bad practices range from lightly reinforced footings (e.g., often with corroded rebar) or footings with small dimensions, to the total lack of reinforcement, or even the absence of the footing altogether (Figure 2.14). It is noteworthy that what is

**Figure 2.14** Problematic footings in RC buildings. *Source:* Stelios Antoniou.

shown in Figure 2.14b is not actually a footing. The rock below the building had been dug in the shape of a cone and just 5 cm of concrete has been cast, in order to make it look like a footing! Surprisingly, the building stood without problems for almost four decades before it was strengthened (the foundation and the superstructure alike).

Even when the design has been correctly followed at the foundation level, the requirements of older standards were significantly lower with respect to today's practices, leading to footings significantly smaller to what one would expect to find in a modern building[1] (strip footings or raft foundations were extremely rarely back then).

### 2.2.14 Discussion

One striking absence in the list above is a particular reference to the longitudinal reinforcement, even for the vertical structural members. In general, bending does not constitute a significant challenge in the majority of existing buildings, at least if the corrosion of the rebar is not significant. This is because it is a ductile failure mechanism, which is not so dangerous for the overall structural integrity, hence less strict checks are permitted, especially for the nonlinear analysis types (employing deformation-based, rather than force-based, acceptance criteria).

The most common problem is the lack of adequate shear reinforcement, which is practically the case in every building that has not been designed with modern code standards. This lack affects the majority of the vertical and the horizontal members and poses the most important challenge to the designer of the strengthening interventions.

It is noteworthy that, among the factors mentioned in the previous sections, the most striking deficiencies that are attributed to poor workmanship and supervision (lack of footings, lack of alignment of columns, lack of reinforcement) occur perhaps in up to 5% of all the buildings; what is more, such problems usually occur simultaneously, i.e. at the same structure. This is not surprising; the building that lacks footings is the most probable candidate for the bad alignment of the columns and the poor anchorage of the rebars. For instance, Figure 2.14c, which shows the lack of a footing, and Figure 2.9b, which shows the lack of alignment in the vertical members, were taken from the same building.

Not all buildings have all the deficiencies described above. Some have simpler and more specific problems, some have more general and complicated issues. Furthermore, and more importantly, these are not the only problems that one can encounter in existing buildings. Bad construction plus irresponsible engineers and owners can be a dangerous – even lethal – combination. Be distrustful and suspicious! If something (anything!) is described in the drawings, it does not mean that it has been applied in the construction. It is your job to check and confirm it, and to come up with high-quality record drawings (which will then be passed to the analyst in the office to create an accurate structural model to be analyzed in the computer).

1 All the same, whether all these problematic foundations require strengthening is questionable; this issue will be discussed later in Chapter 4.

The immediate question that pops up after the terrifying observations described in this chapter is the following: *Does this mean that we have to check everything in an existing building? And how can we do this, especially in buildings that are occupied and in operation?* The answer to the first question is that in theory, yes we need to check everything. In a seismic event, the concept of the "weakest link" is prevalent. Among the 20 columns at the ground level, the one that will sustain the most damage and cause the collapse of the building is obviously the weakest one. Even if there are 19 strong and robust columns, this is not enough. The damage (and possibly the collapse) will be inflicted by the twentieth weak and poorly reinforced column. Therefore, although it can be a daunting task, indeed everything (dimensions, geometry, reinforcement, lap splices, corrosion) must be checked and confirmed.

In practice, however, things can be a bit easier. Good or bad workmanship is something that applies more on the structure level rather than on the member level. Or to be more precise, it applies on the level of each structural part (columns, footings, beams, longitudinal reinforcement, transverse reinforcement, or lap splices) in a quite consistent manner. Try to imagine a crew of laborers working on a building in the mid-1970s, e.g. on the concrete formworks or on the placement of reinforcement. These are 5, 10, 20, or 50 people "doing their jobs" in a way they have learnt from experience, with the same guidance and the same supervision throughout the duration of the construction. Hence, the chances are that the good or bad practices encountered in one or two members are common in all similar members. If you uncover two or three footings and they have been constructed according to the as-built drawings, there is no reason to keep on digging to confirm things for all the other footings. Similarly, if the footings are small, thin, or shallow, and/or lack adequate reinforcement, it is reasonable to assume that all the footings are inadequate.

Note also that construction practices can be expected to be the same for buildings of the same type, in the same region (not necessarily country), and for the same period of construction (e.g. the 1960s, 1970s, or the 1980s). After all, the same engineers and craftsmen have worked and built a series of projects employing the same (good or bad) practices and habits. From experience I have concluded that any important improvement in the employed construction practices usually occurs after the introduction of a new stricter design code. For instance, in Greece (a country that I know well), construction practices, in terms of column sizes, stirrup diameters and spacing, concrete strength, and steel grade (Stahl I, III, or IV) are very common in the periods between the years of introduction of the new seismic codes or their updates in 1959, 1984, 1995, and 2003.

Finally, one should be very suspicious regarding building constructed during periods of real-estate booms and rapid development. These are the periods when standards fall, contractors carry out the projects hastily (in order to move on to the next project), and many less competent engineers are called to do the job, resulting in serious construction problems due to the lack of good workmanship and supervision.

In any case, when checking an existing building, apart from adequate knowledge, some level of experience is usually required; it is very important to be prepared and to know what to expect. After completing four to five cases, an engineer will be able to understand the design, the practices, the habits and the problems of existing buildings.

**The Role of the As-Built Drawings**

When analyzing existing buildings, having the construction or the as-built drawings is always useful and helps engineers understand and measure the building. Yet it is never wise to fully trust such drawings! In the majority of cases, they have been followed during construction by responsible engineers and owners. However, there is also a surprisingly large proportion of buildings with significant deviations from the authorized drawings, and this proportion is by no means negligible! Furthermore, in several buildings there have been arbitrary alterations during their lifetime. Consequently, one should always be cautious and insist on checking things before accepting the as-built drawings as correct.

### 2.2.15 One Final Example

To conclude this section and to become definitely and ultimately convincing about the unpleasant surprises that one can find in existing RC buildings, I will depict the most extreme case that I have ever encountered. It is a building that had been analyzed and strengthened by a competent and experienced colleague using SeismoBuild (2023), which is how we learned of its existence. It is 10-story building built in the 1970s in the center of a large European city. Its market value is currently estimated at several million euro. When it was constructed, the owner of the ground floor and the mezzanine wanted to open a bakery shop, and the central column of the front view posed a big hindrance in the operation of the shop (this column was supposed to support the gravity loads from the eight stories above). A very "efficient" engineer (a safety maniac could also say that he was irresponsible) found the solution! He removed the column at the ground floor and the mezzanine, and constructed a large inverted beam of 1.50 m height at the second story, just above the mezzanine (beam in yellow in Figure 2.15). The large beam was designed to carry the load from the eight columns on top of it. To this large beam he connected a series of crossbeams that transferred the entire gravity loading of the third floor. Finally, and as if all this was not enough, the large beam had a direct support by a column only at one of its edges. At the other edge the beam was suspended (!) from a column connected to beams of the third floor (in orange in Figure 2.15). The building sustained a couple of relatively small earthquakes during its lifetime without damage; however, I seriously doubt if it would be able to remain intact (or sustain only light damage) during a really large earthquake.

## 2.3 How Come Our Predecessors Were So Irresponsible?

How come our predecessors were so irresponsible? This is the question that crosses one's mind after reading the long list of bad practices and problems of the previous section. Of course, some of the problems are related to the age of the building (e.g., corrosion) – after all, we are discussing structures that are 30, 40, or 50 years old. However, most of the problems in the list are related to bad workmanship (e.g., lap splices, arbitrary changes in the structural system), or – even worse – to bad design (reduced reinforcement, lack of stirrups, large lightly reinforced shear walls, in plan or in elevation irregularities). Why were older RC buildings so badly constructed?

Figure 2.15 The achievements of a very irresponsible engineer, as modeled in SeismoBuild; the large inverted beam is in yellow and the column from which it is suspended, is in orange. *Source:* Stelios Antoniou.

The main reason is the lack of adequate knowledge regarding the dynamic behavior under earthquake loading. Earthquake engineering, contrary to other fields of civil engineering, is a relatively new field of science. Back in the 1960s or 1970s, terms like *capacity design* of structures or *confinement* were unknown. The significance of stirrups was confined to withstanding a limited shear force demand on the columns and the beams imposed by the gravity loading, as well as to hold together the longitudinal rebar when casting the concrete (although it is hard to believe this now, it is something that I heard from several older engineers and craftsmen, who were active in the 1970s and explained that back then, this was considered to be the main contribution of stirrups!).

What is more, our knowledge about the earthquake loading itself was also limited. For decades, the research in the field was guided by the El Centro earthquake, which was not a weak record, however much stronger seismic events have been recorded since then.

Because of this limited knowledge, the seismic loads that were considered in all older standards were a fraction of today's requirements. For instance, nowadays in Greece three different values for the design ground acceleration are imposed, 0.16, 0.24, and 0.36 g, with the larger part of the country employing the 0.24 coefficient. On the contrary, before 1984 the values of the corresponding seismic coefficients (also considering the differences in the safety factors employed by the different codes) were lower than 0.10, throughout the country. Similarly, in Ecuador the current seismic coefficient in Quito is currently 0.40, considerably larger than the 0.10 value, which was employed until 1990. Note that in the regions of higher seismicity in the western part of the country seismic coefficients as large as 0.50 are employed nowadays, values that were simply inconceivable 30–40 years ago.

One final factor is more social, rather than technical, and plays an important role in the decisions that our societies make regarding safety standards. The significance of human lives, and in general the issue of safety, both public and private, was regarded differently a few decades ago as compared to today's requirements. If one looks closer at societies of the recent past, it can be easily understood that people were more accustomed to turbulence, instability, damage, and loss. Let us sit back and think for a moment about our world – say, at the end of one of the most interesting years of the previous century, 1968. WWII had finished just 23 years prior and people's memories of the destruction and the atrocities were as vivid as our present memories of the end of the 1990s. In 1968, the USA was stuck in the Vietnam War, and the casualties in the 365 days of the previous year had amounted to a total of 16 899 soldiers killed, which comes out to almost 50 soldiers dead every day. One of the most prominent candidates for the presidency of the US had been assassinated in June, as was his brother five years before, who was then the US president. Six years before 1968, during the Cuban Missile Crisis, the world was on the brink of a nuclear war. On an individual level, wearing a seat belt in cars was not compulsory; it had been offered as an accessory just 10–12 years prior to that. And more than 40% of the adults in the US were regular smokers – in fact, smoking was still considered cool. Last but not least, the outbreak of the Hong Kong flu pandemic in 1968 and 1969 killed between 1 million and 4 million people globally, but nobody really noticed.

With respect to public or private safety, it was a completely different world than today. Societies and people were better prepared to accept material or financial losses, and human casualties. Consequently, safety standards were not (and were not expected to be) as strict as nowadays, not only in building construction but in every aspect of human activity; societies simply had other priorities. From that perspective, it can be understood that the psychological impact of the San Fernando earthquake that struck three years later, in 1971, was significant, causing severe damage and killing around 60 citizens. However, it was by no means as big as the impact that a similar event would have on the Californian population today. This difference was reflected in the building standards enforced back then, as is reflected in the standards nowadays.

## 2.4 What the Codes Say – Knowledge Level and the Knowledge Factor

The comprehensive or limited knowledge of an existing structure is a very important concept for all the assessment procedures, but it is something that is not encountered in design methodologies. When designing a new structure, engineers have other types of uncertainties (e.g. the material strengths and the corresponding bell-shaped curves of their Gaussian distributions) but a full knowledge of the building that is going to be built in the future is implicitly assumed.

On the contrary, in existing buildings, even when we have full access to the construction or the as-built drawings, there is always a level of uncertainty about whether these drawings reflect the actual state of the structure, and whether there are deviations that considerably affect the response of the structural system. The factors determining the appropriate knowledge level are: (i) *the geometry*, the geometrical properties of the structural system and the nonstructural components, (ii) *the details*, the amount and the detailing of the reinforcement in reinforced concrete buildings, the connections between members, or the connection of floor diaphragms to the lateral resisting structure, and (iii) *the materials*, the mechanical properties of the constituent materials.

Depending on the accuracy and reliability with which we know the structure under investigation (i.e., whether we have good, average, or poor knowledge of it), most standards define different knowledge levels for the purpose of choosing the admissible type of analysis and the appropriate knowledge or confidence factor values. It is very important to note that the selected level of knowledge should always be supported by technical data relevant to the structure; the higher levels of knowledge are restricted to those conditions where it can be technically supported. Based on the selected knowledge level, different knowledge factors (or confidence factors for the European codes) are employed. The knowledge factors are used as additional safety factors to account for our lack of knowledge of the structural configuration. They operate either on the demand (in which case they are equal or larger to unity, e.g. in the Eurocodes) or on the capacity (in which case they are equal or smaller to 1.0, as in ASCE 41).

The different standards provide similar categorization processes with small variations and slightly different nomenclature. For example, Eurocode 8 provides a unified knowledge level for the entire structure, whereas ASCE-41 accepts different knowledge levels for the different components, e.g. we could choose the minimum knowledge level at the foundation, which we do not know well, but a usual knowledge level for the columns and beams, which we are able to measure with larger confidence. Similarly, there can be two or (in most standards) three levels of knowledge. Table 2.1 summarizes the main features of the knowledge levels for each code.

The main variations between the standards are the following:

i) In general, there are three levels of knowledge with the exception of the Turkish code, whereby only two levels can be defined. All the same, it is noted that effectively there are also only two levels in ASCE-41, since the value of the knowledge factor is the same (1.00) for both the usual and the comprehensive levels. In the Greek code, the three levels of knowledge are called data reliability levels, DRL.
ii) Depending on the knowledge level, different values for the knowledge factors (for the US and Turkish codes) or confidence factors (for Eurocode 8 and the Italian code) are assumed. In the Greek code, a slightly different approach is proposed, whereby different partial material factors are provided for existing and new materials, depending on the selected DRL.
iii) ASCE 41 and the Turkish code propose knowledge factors that have values equal to or smaller than 1.00, and operate on the capacity side of the inequality (effectively reducing the capacity values for the minimum level of knowledge by 25%). In contrast, in

**Table 2.1** Knowledge levels and knowledge (or confidence) factors for the different standards.

| Standard | Eurocode 8, Part 3 | ASCE-41 | Italian code NTC-18 | Greek code KANEPE | Turkish code TBDY |
|---|---|---|---|---|---|
| Levels of knowledge | KL1<br>KL2<br>KL3 | Minimum<br>Usual<br>Comprehensive | KL1<br>KL2<br>KL3 | Tolerable DRL<br>Sufficient DRL<br>High DRL | Minimum,<br>—<br>Comprehensive |
| Knowledge or confidence factors | 1.35<br>1.20<br>1.00 | 0.75–0.90<br>1.00<br>1.00 | 1.35<br>1.20<br>1.00 | (different partial material factors for existing and new materials, depending on the DRL) | 0.75<br>—<br>1.00 |
| Operate on. . . | Structural level | Component level | Structural level | Different parts of the structure | Component level |

*Source:* Stelios Antoniou.

Eurocode 8 and the Italian code the confidence factors assume values equal to or larger than 1.00 and operate on the demand (effectively increasing the demand by 20% or 35% for the usual and the minimum levels, respectively). For the Greek code, the partial material factors have lower values for better DRLs, for instance the partial factor $\gamma_c$ for concrete is 1.15 for the high DRL, 1.30 for the sufficient DRL and 1.45 for the tolerable DRL. Note also that knowledge factors different than the default values can be accepted, depending on the confidence with which the engineer feels that he/she knows the structural configuration.

iv) In the Greek code and for only the new materials (e.g., shotcrete and reinforcing steel in a jacket), some custom multipliers, which are larger than 1.00 (i.e., 1.05–1.20), are provided. The multipliers operate on the partial factors for concrete and steel and account for the difficulty of accessibility and inspection, when applying the retrofit techniques, and the subsequent deviations in uniformity and quality.

v) Whereas for the European codes (EC8, Italian, Greek) the knowledge level is defined on the structural level, in the American and the Turkish standards the knowledge level is defined on the component level, an approach that seems to be more correct. For instance, the level of knowledge of the foundation components is in the majority of cases smaller than the knowledge level in the upper structure.

One important note regarding the knowledge levels is that it is forbidden by all codes to use the more advanced nonlinear methods of analysis, if the level of knowledge is minimum (KL1, minimum, or tolerable DRL). Obviously, there is no reason to use sophisticated methods, when there is no good knowledge of the building. The more approximate, but computationally less demanding, linear methods should be enough, and there is no need to waste time, effort, and money on the more advanced nonlinear methods. In such cases, increased safety factors are employed, in order to account for the loss of accuracy due to both the lack of good knowledge of the building and the approximations of the linear analytical methods.

Note, however, that there is one other, more subtle, reason for this limitation, which is probably more important. Advanced nonlinear methods can give a false impression of accuracy in the calculations – an accuracy that does not exist if the knowledge of the building is limited. One can start with a vague knowledge of the structural configuration, the geometry, the foundation, the soil conditions, the reinforcement, and the materials. Then one passes approximate (and very often assumed) information to the structural analysis software in a very deterministic fashion, as if they were 100% correct, runs refined nonlinear analysis and comes up with a set of analytical results that seem and feel very accurate (at least more accurate than those from linear analysis). One gets compensated for the extra computational effort with checks that are easier to pass (the checks in bending are carried out using deformations in nonlinear methods but moments in linear methods, whereas for the brittle types of failure, such as shear, a capacity design philosophy is employed in the linear methods for the calculation of the demand, which is always more conservative). However, the entire approach is wrong. The overall accuracy of the process is controlled by the weakest link, i.e. by the part of the process that is more approximate, which in this case is the limited knowledge of the structure (see also Kam and Jury 2015). From this perspective, it is clear that the accuracy of the system both with linear and nonlinear methods is exactly the same. Hence, using the nonlinear methods is not only pointless, but it can also compromise safety.

## 2.5 Final Remarks

This chapter makes clear why the minimum reinforcement or some other default reinforcement pattern cannot be used in order to avoid all the hassle of a detailed survey. Very often the buildings under investigation are constructed differently from the proposed design drawings, and in most of the cases these deviations are against safety (absence of members, lack of reinforcement, unauthorized changes, bad detailing, etc.).

Considering all the difficulties posed by existing buildings, especially those that are occupied and operational, the measurement of a building, its geometry, the dimensions, and the reinforcement is very often the most difficult and daunting task throughout the entire assessment process. All the same, it is one of the most important parts, especially in those (not so rare) cases where there are signs of poor workmanship and supervision.

## References

Kam, W. and Jury, R. (2015). *Performance-based Seismic Assessment: Myths and Fallacies. Paper Number S-03, 2015*. Rotorua, New Zealand: New Zealand Society for Earthquake Engineering Conference.

Niroomandi, A., Pampanin, S., and Dhakal, R.P. (2015). *The History of Design Guidelines and Details of Reinforced Concrete Column in New Zealand*. Rotorua, New Zealand: New Zealand Society for Earthquake Engineering Conference.

NISEE e-Library (n.d.). Earthquake Engineering Online Archive. Karl V. Steinbrugge Collection, NISEE University of California, Berkeley. https://nisee.berkeley.edu/elibrary.

SeismoBuild (2023). SeismoBuild – A computer program for the linear and nonlinear analysis of Reinforced Concrete Buildings. Available at URL: www.seismosoft.com.

Wikipedia (2022). Versailles wedding hall disaster. https://en.wikipedia.org/wiki/Versailles_wedding_hall_disaster (accessed: May 15, 2022).

# 3

# Measurement of Existing Buildings, Destructive and Nondestructive Testing

## 3.1 Introduction

What should be investigated during a building survey? The simple answer to this question is, *everything that is required in order to have correct and reliable measured drawings that depict all the important structural components*. Sure. . . but what does this mean? Does it imply that the engineer has to check every member, every wall, column and beam, every rebar and every stirrup, all the footings and all the partition walls? Should he/she get measurements of the material strength from all the columns, beams, and slabs of the buildings? Obviously, this is not possible within the limitations of all practical applications. To make things worse, in most cases this survey is required to be done in buildings that are occupied – where people operate, live, and work. One needs to find a compromise between an acceptable level of accuracy on the one hand and effectiveness and economy (in terms of time and money) on the other.

This chapter explores the difficulties of the monitoring process, the most common methods to measure the quantities with reasonable reliability, and some rules of thumb so that the engineer understands the limits and limitations for an acceptably accurate survey.

## 3.2 Information Needed for the Measured Drawings

In the process of measuring an existing building, the general concept is that it needs to be investigated and documented to a sufficient extent and depth so as to obtain acceptable data reliability on which to base the assessment or redesign. The information for structural evaluation should cover the following points:

- Identification of the structural system and of its compliance with the regularity criteria.
- Information on any structural changes that have occurred since construction, which may alter the behavior and seismic response of the building.
- Identification of the type of building foundations.
- Identification of the ground conditions (soil classification).
- Information about the overall dimensions and cross-sectional properties of the building elements and the mechanical properties and condition of the constituent materials.

*Seismic Retrofit of Existing Reinforced Concrete Buildings*, First Edition. Stelios Antoniou.

- Information about identifiable material defects and inadequate detailing, lap splices, and anchorages.
- Information on the seismic design criteria used for the initial design, including the value of the force reduction factor (q-factor), if applicable.
- Description of the present and/or the planned use of the building (with identification of its importance class).
- Reassessment of imposed actions, taking into account the use of the building.
- Information about the type and extent of previous and present structural damage, if any, including earlier repair measures.
- The presence of infill walls, which have adequate strength and stiffness to affect the structural response.
- Determination of the potentially harmful environmental exposure class for the structure.
- Information on any identified significant errors in the initial design – information on material defects and their description.
- The distribution of the total building mass on plan view and in elevation.
- Information on the adjacent buildings and their distance from the building under investigation.
- Geometric measurements of:
  - Cross section dimensions, the length of the structural elements and thickness of finishes, as constructed.
  - Leveling, eccentricity measurements, deviation measurements, etc.
  - Crack widths or detachments in concrete or masonry elements.
  - Deformation and discontinuities in joints, displacements, etc.
  - Permanent deformations.
  - Time development of the aforementioned phenomena.

The data on the as-built condition of the structure, components, site, and adjacent buildings must be collected in sufficient detail to perform the selected analysis procedure. As discussed in the previous chapter, a better knowledge of the structural configuration is needed for the more accurate nonlinear methods, whereas for the linear methods we can accept a more approximate description.

The input data can be collected from a variety of sources:

- General information on the history of the building.
- Design or as-built drawings for both the original construction and any subsequent modifications.
- The relevant documentation and technical reports of the building.
- Field investigations.
- In-situ and/or laboratory measurements and tests.
- Relevant data sources (e.g., contemporary codes and standards).

In order to minimize uncertainties, cross-checks should be made between the data collected from the different sources.

The required data to be gathered are the following:

- *Geometry,* the location of the structural members and the distances between them, but more importantly their exact dimensions.

- *Details,* which in most cases are the members' reinforcement, both longitudinal and transverse. Information about the amount and the detailing (e.g., lap splices, development length) should be provided.
- *Materials,* i.e. detailed information on the mechanical properties of the construction materials.

Very often engineers and owners alike focus more on the mechanical properties of the employed materials, and most importantly on the concrete compressive strength. This can also be observed in the different assessment standards, all of which have lengthy parts on the destructive and nondestructive tests (NDT) required for its estimation. Although the concrete strength is by no means of small importance, it seems that the monitoring process should focus more on other parameters, such as the shear reinforcement, the stirrup diameter, the lap splices and the corrosion of steel.

For instance, whether the compressive concrete strength $f_c$ is 12 or 16 MPa is far less important than the hoop diameter (6, 8, or 10 mm), the hoops anchorage, the (lack of) lap splices, or the extent of the corrosion to the building members.

The availability of the construction – or even better, the as-built drawings – contributes significantly to the reliability of the data, but also to the ease and the speed of the operations. Nevertheless, thorough site investigations are always needed, especially for the most critical members, even when these drawings are available. This is because, as was explained in the previous chapter, there are often significant deviations from the original drawings, and important information can be missing (e.g., transverse reinforcement of the columns). Hence, engineers should confirm that the drawings have been implemented adequately, record all important deviations from the original design, and consider them in the assessment. Obviously, in the case where there are significant discrepancies from the original plans, a more thorough survey is required, with respect to the cases of good fit between reality and the drawings.

The choice of measurements and tests is largely based on engineering judgment. Obviously, for higher knowledge levels (KL), the inspection and testing requirements are larger. For example, Eurocode 8 recommends the inspection for details of 20% in KL1, 50% in KL2, and 80% for KL3. In my opinion, all (without exception) the vertical members should be checked in detail, whereas for the beams one can be less stringent, since in general the failure of a beam does not jeopardize the overall structural stability.

One important aspect of the process that always poses a very big challenge to the engineer is the survey of the "hidden" structural elements with limited or no accessibility. These include, for instance, walls, columns, or beams behind gypsum partitions, beams and slabs behind gypsum ceilings, covered footings, and members behind mechanical or electrical installations that cannot be removed. In such cases, especially when reliable construction drawings are not available, identifying the dimensions and reinforcement requires investigative sections that lead to damage to the nonstructural components and hindrance in the building's operation. Although this is not always easy, the extent of the investigation must be sufficient to provide reliable information for the assessment and redesign.

In any case, differences in the member dimensions and reinforcement, variations of the material properties, and modifications of parts of the building during its lifetime are to be expected, even for high-quality constructions. In such cases, the selected knowledge level may be moderated, when there are observations of large deviations from the original design, due to poor workmanship.

## 3.3 Geometry

In order to create a correct and accurate model for structural analysis, the location and the dimensions of all members need to be known and defined. The least that is expected from a complete survey is to accurate measure the geometric quantities of the structure.

Although the process seems straightforward initially, this is rarely the case, especially when the construction drawings are unavailable. For most buildings, the task of identifying the exact structural system and the location and the dimensions of the columns and beams is far from simple. Engineers encounter four main problems in the survey:

1) Many times, walls, columns and beams are hidden behind sheathings made of wood or gypsum boards (Figure 3.1a). This causes problems in the majority of existing buildings that are occupied, because the damage and subsequent repair on these sheathings, necessary for revealing the members, is costly and disrupts the building's operation.
2) In a similar fashion, dropped ceilings made of wood, metal, or gypsum boards prevent access to the beams and slabs, and again damage and repair of parts of the ceilings are required for a correct survey (Figure 3.1b).
3) In the vast majority of buildings, the structural members are covered with plaster, the width of which can vary from 1 cm to 4 or even 5 cm (Figure 3.2). Considering that structural members are covered with plaster on all sides, this can cause an uncertainty of up to 10 cm in the size of the columns or the width of the beams. Although this difference is not very impressive by itself, it does play a rather important role in small columns of 20/20 or 25/25 that are very typical in small or medium-sized buildings of the 1960s or 1970s. For instance, considering that the elastic lateral column stiffness is proportional

(a)

(b)

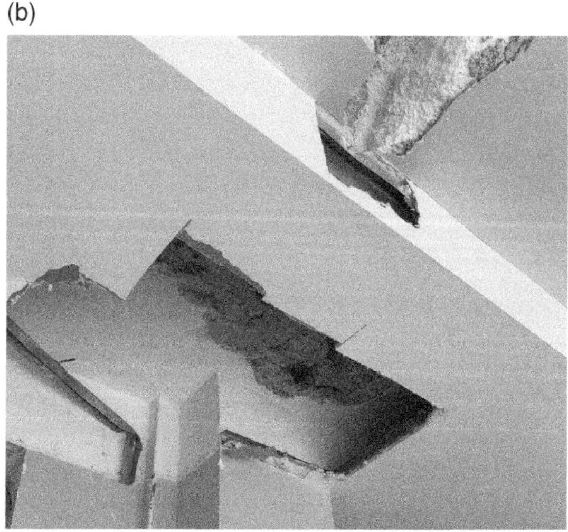

Figure 3.1 Survey in a hotel where the majority of the columns, beams, and slabs were hidden behind gypsum boards: (a) a column behind a gypsum sheathing and (b) a beam behind a gypsum ceiling. *Source:* Stelios Antoniou.

to the section's moment of inertia $I = bh^3/12$ ($b$ = width; $h$ = height), a change in the dimensions of 25% (from 20/20 to 25/25) changes the stiffness of the column in the structural model by almost 150%!

4) As explained in the previous chapter, contrary to the superstructure, the level of workmanship at the foundation level is often extremely poor, with large deviations from the original design, lack of reinforcement, or absence of the footing altogether. Even if the construction drawings are available and complete, one has to always check the footings and whether their size fits the original design. Hence, it is necessary to uncover at least two to three footings in every building so as to be sure that no major errors have been made during construction. The digging necessarily needs to be done until the bottom level of the footing, which typically is at a depth of more than 1.00 m. In order to achieve this, the demolition of the floor and the concrete slab at the ground level is required at an area not less than 4–5 $m^2$ at the perimeter of the column (Figure 3.3). Then, excavation of the backfilling is required, either with a small excavator, or more often by

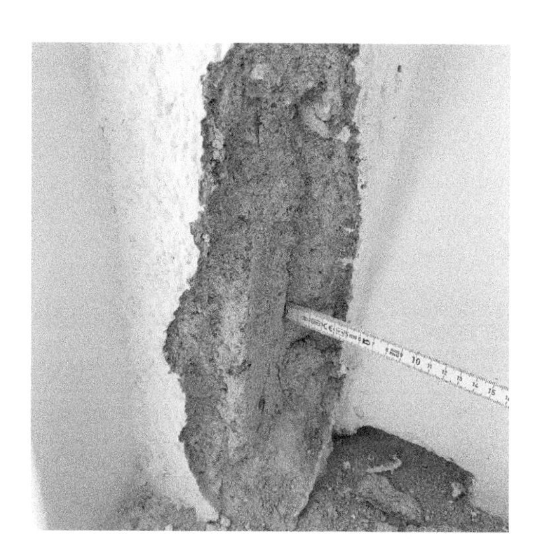

Figure 3.2 A concrete column surrounded by plaster 5 cm thick. *Source:* Stelios Antoniou.

Figure 3.3 The uncovering of the footings can be a very disruptive task. *Source:* Stelios Antoniou.

Figure 3.4 Where are the structural components? *Source:* Stelios Antoniou.

hand. Obviously, this process is expensive, and most importantly very disruptive to the operation of the building.

Because of such problems, it is usually very difficult and time-consuming to identify even the location of the structural members, also considering that in general it is necessary to measure the dimensions of all the components of the building under investigation to have an accurate structural model. To this end, the ferrous scanners described in the next section can be of great assistance in the identification of the location of the vertical components and their reinforcement, when these are embedded in the infills and the walls are covered with plaster, marble or other material (Figure 3.4). It is relatively easy to understand with a scanner whether a metal object lies behind the plaster, which could be either a reinforcing rebar or some other building component, usually a cable or a pipe. The identification of a RC vertical member is done by scanning a complete reinforcing pattern with several longitudinal rebar and several horizontal stirrups, whereas for instance a cable would appear as a single rod. Note also that with some experience it can be understood by simply knocking on a wall whether there is concrete or infills below the plaster, since the sound of the latter is in general more "hollow."

If construction drawings are available, the task gets relatively simpler. Considering that the dimensions and the location of all the members are known from the drawings, the location of the structural members can be easily identified, and a check is required on the overall dimensions of the building, as well as the dimensions of some of the members, to confirm the drawings. However, again things become a bit more intricate in the substructure and the dimensions of the footings, since, as explained already, the existence of the drawings in the general case does not guarantee the correctness of their application and the reliable knowledge of the foundation conditions. Consequently, even in this case the uncovering of some of the existing footings, and the disruption that this brings is necessary.

## 3.4 Details – Reinforcement

Inspecting members' reinforcement is mainly done with nondestructive methods with the use of ferrous or metal detectors, designed to find and estimate the rebar size in concrete structures, and to accurately provide the depth of concrete cover and the plaster above it.

**Figure 3.5** A common ferrous scanner model available in the market. *Source:* Stelios Antoniou.

There is a large variety of scanners, with a wide range of prices, different technologies, and different functionality. Most scanners are hand-guided, cordless, and operate with batteries, so as to enable their use at any building or construction site. The scanners have either integrated LCD screens or can be connected to external screens, tablets, or laptops (Figure 3.5).

Areas of the concrete surface are scanned in lines, and the resulting scans are shown on the screen of the scanner, or the portable monitor unit, as 2D images of the reinforcement for on-the-spot structural analysis and depth cover assessment. Scan data are recordable and can then be transferred to a PC for further analysis, for the creation of assessment reports, and for archiving purposes (Figure 3.6). The engineer can identify from the images the position, number, and size of rebar. The approximate depth at which the rebar are embedded can also be confirmed with most of the scanners on the LCD display. Usually, the scanners, depending on the type and manufacturer, can detect rebar at depths up to 100–150 mm, and can provide accurate measurements at depths up to 50–60 mm.

Metal scanners are extremely useful, and the reinforcement scanning constitutes an integral part of the monitoring of a RC building. However, the entire process is by no means bulletproof and includes serious uncertainties and inaccuracies. Despite how accurate (i.e., expensive) a scanner is, its application has certain limits that hinder the full and accurate representation of the reinforcing patterns of all the structural members. The main difficulties are the following:

- Very frequently, the rebar sizes cannot be reliably estimated, and can only be determined within the order of one rebar size (Figure 3.7). Whereas for large longitudinal rebar this is not a major problem (the bending capacity of a beam does not change significantly if the corner rebar are Ø20 or Ø18 mm), the existence of Ø6 or Ø8 mm stirrups considerably affects the confinement and more importantly the shear strength of the member (note the area of a Ø8 stirrup is approximately double that of a Ø6 stirrup).
- In locations with dense reinforcement, there is an overlap of different rebar and it is not possible to determine either the size or the number of the bars (Figure 3.8). In existing buildings such locations can be the beams at mid-span (lower side) or at the edges (upper side), especially where there is a second layer of longitudinal reinforcement. In newer buildings, columns with large reinforcement percentages can present similar problems.

- Scanning of the upper reinforcement of beams can only be done with significant uncertainties, or cannot be done at all. This is because the upper sides of beams lie below the floor (e.g., wooden floor, floor with ceramic or marble tiles and a layer of cement mortar), the depth of which is generally more than 5 cm. Together with the beam concrete cover, the total depth is beyond the limits of ordinary scanners and prevents the accurate estimation of the reinforcement parameters (Figure 3.9).
- Even when the reinforcement can be detected with scanners, it should be confirmed that there is no significant corrosion or voids in the concrete caused by poor casting, two factors that can significantly affect the member capacity.

Because of such problems, the use of metal scanners is usually supplemented by cross-sections with a small demolition hammer in selected RC members. The members where the

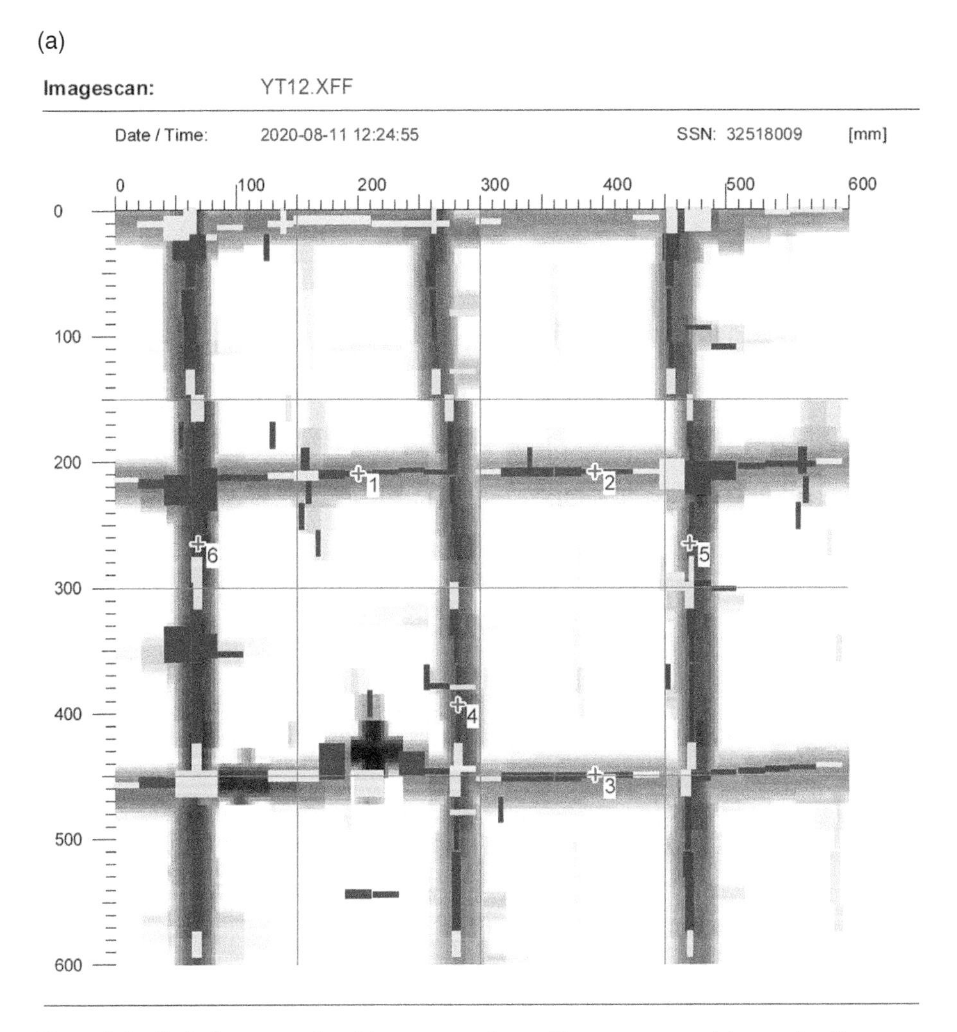


Figure 3.6 Typical scans made with a common type of ferrous scanner (a) scanning the reinforcement grid, (b) scanning for the identification of the stirrup spacing. *Source:* Stelios Antoniou.

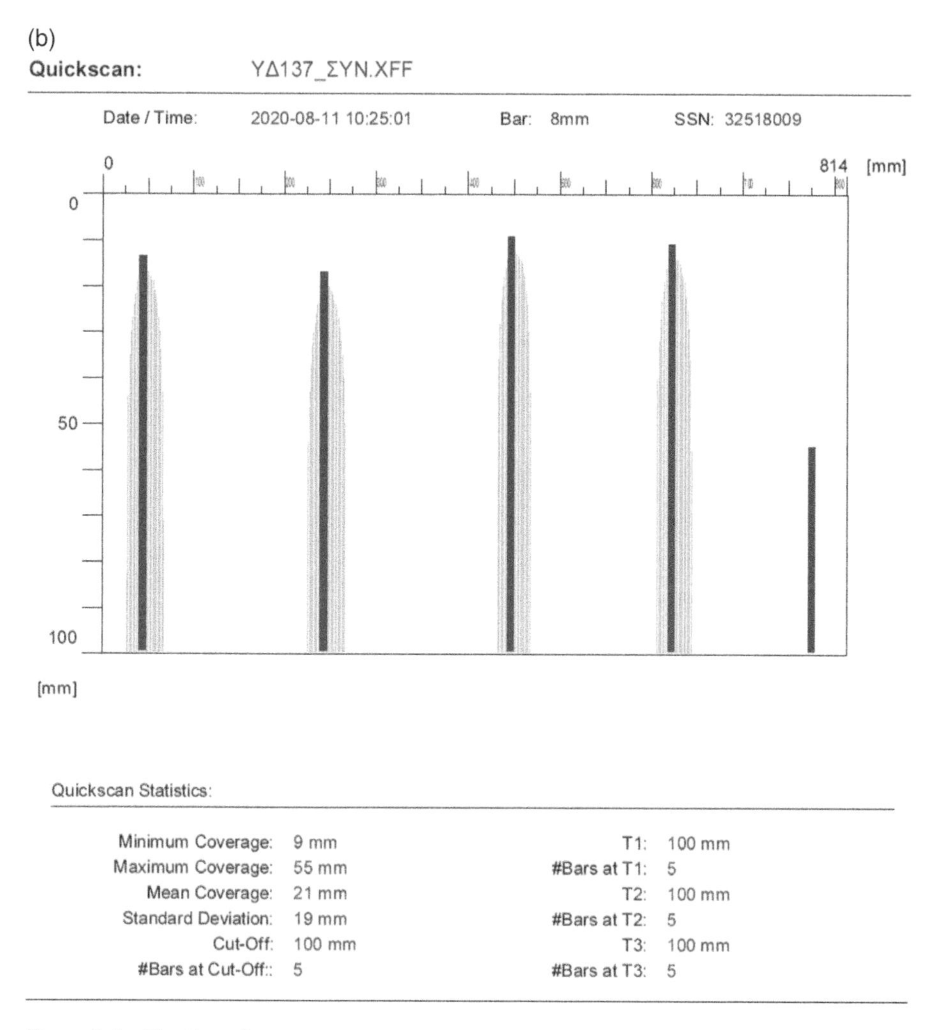


**Figure 3.6** (Continued)

cuts are to be made should be chosen so as to balance the additional information about the structural configuration on the one hand, and the cost of the repair of the damage to the building's components and the disruption to the operation of the building, on the other. For example, one would prefer to demolish a small area of plaster than to destroy a gypsum board. Similarly, creating damage to the exterior of the building is generally preferable to causing damage to the interior.

> One of the parameters that scanners can detect correctly and with great reliability is the spacing between the rebar and most importantly, the stirrups. This is of great importance in the determination of the shear member capacity, considering that the stirrup diameter is usually the same throughout all the members of a building, and it can be easily determined reliably with a cross section in a few members.

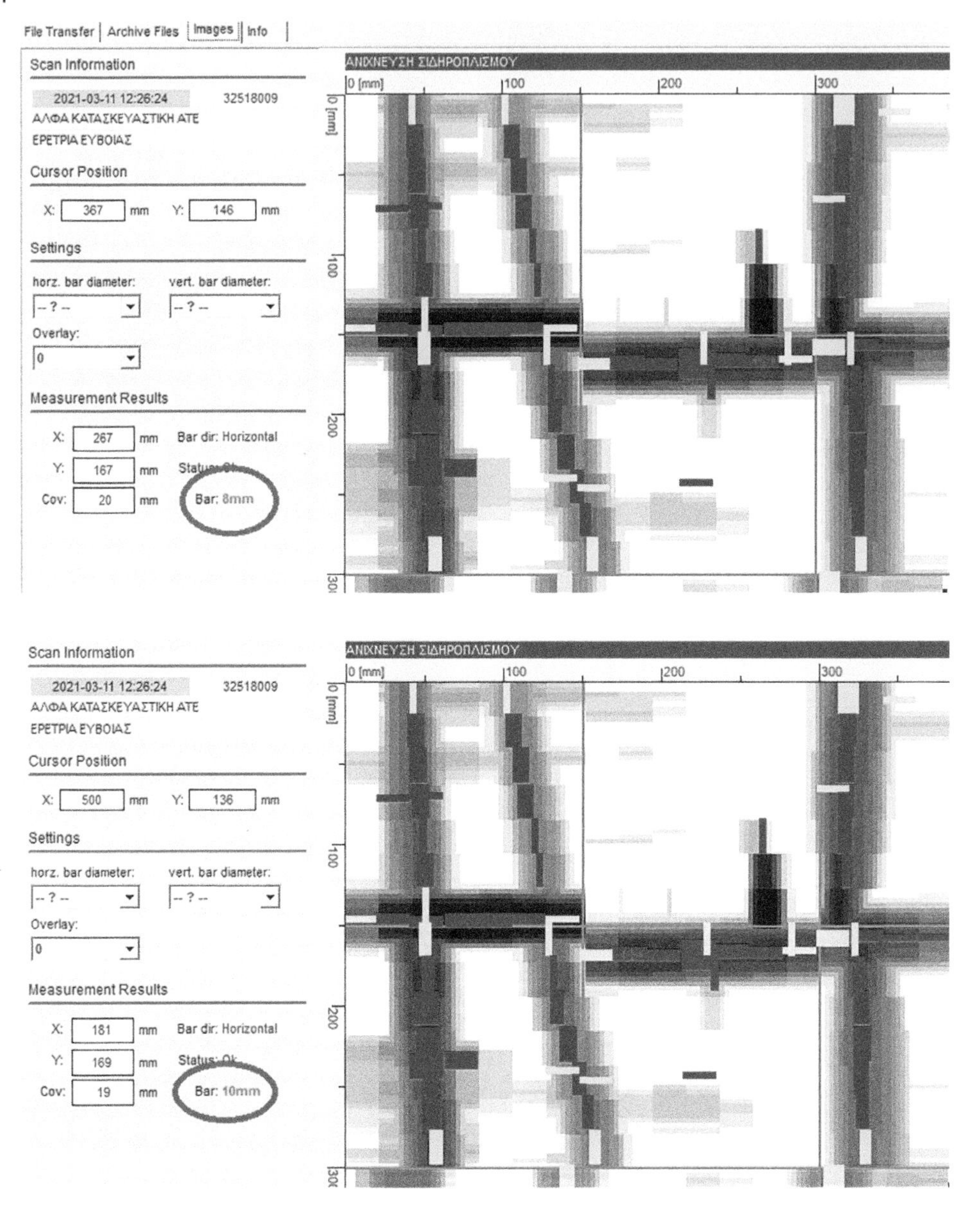


**Figure 3.7** The same stirrup is identified as ∅8 and ∅10. *Source:* Stelios Antoniou.

In several cases, it is costly and disruptive to uncover the reinforcement of certain members. Apart from the case of existing footings that has been discussed in the previous section, the case of the upper reinforcement of the beam edges also presents significant operational and architectural challenges, since the demolition and reconstruction of small areas of a wooden or a tile floor is often problematic. Thirty or forty years after the construction of the building, it is impossible to find tiles and skirting boards of the same size and texture as the existing floor, thus one is left with two unappealing options: (i) lay new tiles that are similar (but not exactly the same) as the existing ones, which has a very negative effect aesthetically, especially if this is done at several locations on the same floor; or (ii) replace the tiles of a larger area (e.g., an entire room), which is unreasonably expensive with respect to the information that this would provide. Similarly, the destruction and reconstruction of a part of a wooden floor usually requires polishing in a larger area (an entire room or an entire floor), which is both expensive and disruptive.

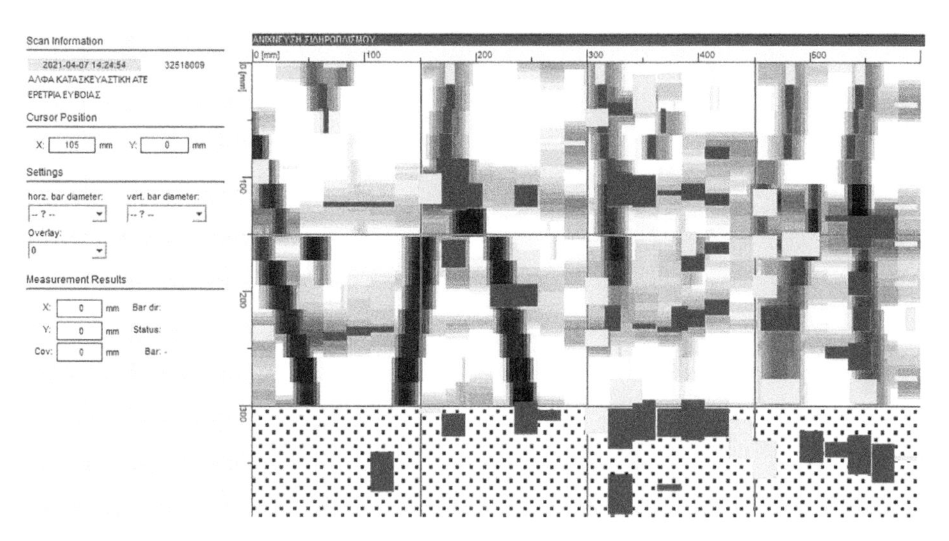


Figure 3.8 Scanner image from a location with dense placement of reinforcement. *Source:* Stelios Antoniou.

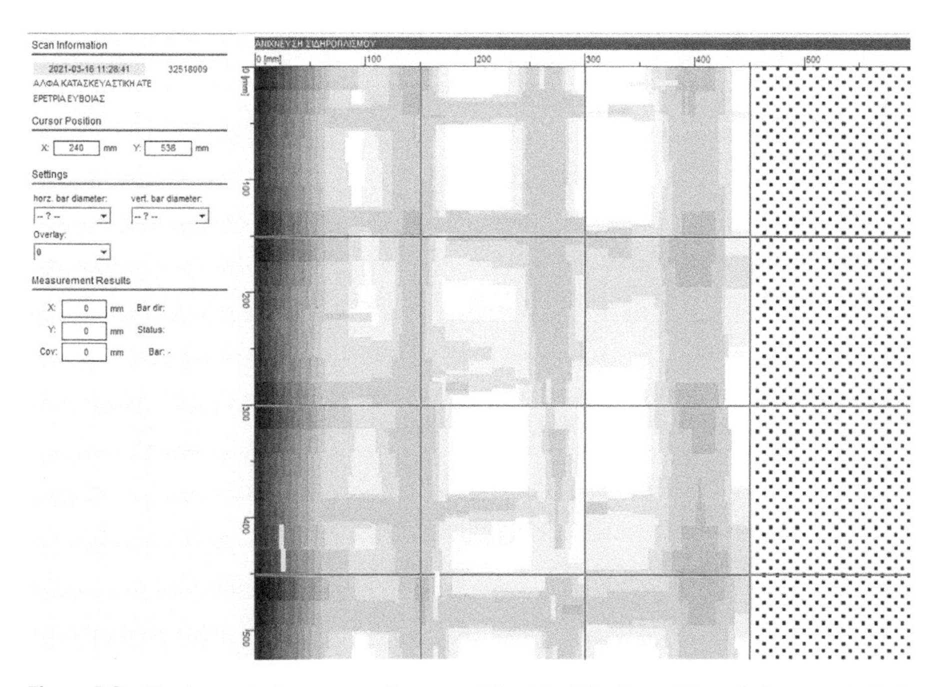


Figure 3.9 The inconclusive scanner image and the identification of the reinforcement of a beam after the demolition and removal of the floor and the beam's cover. *Source:* Stelios Antoniou.

Figure 3.9 (Continued)

Obviously, the monitoring of the reinforcement of RC members in buildings in operation can constitute a significant challenge. In most of the cases, it is economically unacceptable or practically impossible to fully monitor the reinforcement of an entire structure in every position. It is therefore necessary to select the locations, where the scans, the cuts, and the damage are to be done. A compromise between the acquired information on the one hand and the cost and the disruption to the building's operation on the other is always sought, a process in which some experience and sound engineering judgment is required.

## 3.5 Material Strengths

For an accurate structural modeling, the mechanical properties of the construction materials are required. These are usually the concrete, the reinforcing steel, and in some cases infilled walls (bricks and mortar) or other materials that might exist in the structure (e.g., epoxy resins or fiber reinforced polymers from previous intervention works). The required mechanical parameters are mainly (i) the compressive strength for concrete – the elastic modulus and the tensile strength can be estimated indirectly from the compressive strength, whereas the tensile strength is often neglected in order to be on the safe side, (ii) the yield strength for steel – the elastic modulus is equal to 210 GPa for all cases of structural steel, and (iii) the compressive and shear strength of the infills, as well as their corresponding moduli. In all cases, the mean values of the mechanical properties are estimated from the tests, and the lower-bound values (the values that correspond to the characteristic or nominal values of new materials) can be indirectly calculated by multiplying with appropriate modification factors (Table 3.1).

For the estimation of the concrete strength, a combination of destructive and nondestructive tests (NDT) is typically required, and the results of both are used in conjunction.

Table 3.1 Factors to translate lower-bound material properties to expected strength material properties (ASCE 41–17, Section 10.2.2.3).

| Material property | Factor |
|---|---|
| Concrete compressive strength | 1.50 |
| Reinforcing steel tensile and yield strength | 1.25 |
| Connector steel yield strength | 1.50 |

Typically, the smaller set of results from the more accurate destructive methods is used for the calibration of the nondestructive method results, which are more numerous. This approach is followed because the direct estimation of the in-situ resistance of each structural element exclusively through cores would require a large number of tests for the statistical analysis of the results, which is unreasonably disruptive to the building's operation, as well as expensive and time-consuming.

For selecting the number and the position of the materials tests, engineers should apply their judgment to ensure the representativeness of their samples. A minimum number of tests should be determined to allow for the statistical analysis or calibration of the test results, but if there is great variability in the measured quantities, a larger sample should be sought.

Obviously, the amount of required testing is different, depending on the target knowledge level. With the exception of the Greek (KANEPE 2022) and the Turkish (TBDY 2018) codes, the standards do not provide quantitative guidelines for the number of tests required for each level. For instance, for the comprehensive level of knowledge ASCE-41 (ASCE 2017) requires the existence of comprehensive testing and field survey drawings, the verification of the construction drawings by a visual condition assessment, and that the mechanical properties have a coefficient of variation smaller than 20%, without further details on the required number of samples (ASCE-41-17, Table 6.1).

Some standards (ASCE-41 and the Greek code) provide conservative default values, in accordance with the standards at the time of construction of the building. Engineers can use these values without need for further investigations although in certain cases they should be accompanied by limited in-situ testing, at least for the most critical elements. These default values are considered to represent acceptable material qualities and placement practices at the time of construction. If, however, there are clearly poor materials or craft, then lower values should definitely be considered (ASCE-41-17, Section C6.2). In all cases, the engineer is urged to use judgment and sound reasoning.

The hierarchy of the available information is considered to be, from lowest reliability to highest reliability: (i) default values typical of the specific construction in the era of construction for the region and type of construction; (ii) values specified in the available design documentation, which may include drawings, technical reports, and specifications; (iii) values provided in as-built documents and contemporary testing reports for the materials used; and (iv) values determined by destructive and nondestructive testing that reflect the actual values and condition of the materials during the period of the assessment.

## 3.6 Concrete Tests – Destructive Methods

The most common destructive method for the evaluation of the concrete mechanical characteristics of existing buildings is the crushing of cores extracted from representative structural members. Such tests provide the compressive and tensile strengths, as well as the modulus of elasticity.

Concrete cores are usually cylindrical and they are cut by means of a rotary cutting tool with diamond bits (Figure 3.10). In this manner, a cylindrical specimen with a diameter of 10–15 cm is obtained with its ends being uneven. Special care should be taken so that reinforcing rebar are not included in the core, as the presence of reinforcement in the specimen generally decreases its strength. Therefore, scanning the concrete surface with a typical ferrous scanner is always required before the extraction of the core. After the extraction of the core, the hole should immediately be closed with repair concrete or grout of comparable strength and with nonshrinkage properties. The cores can be tested for carbonation with the straightforward use of a chemical indicator. The most commonly used indicator is a solution of phenolphthalein in alcohol and/or water, which turns a strongly alkaline concrete to pink, but if the alkalinity has been lost the concrete will not change color. Other indicators also exist (Figure 3.11).

The core is capped with epoxy resin to make its ends plane and parallel, and then it is tested under compression. The strength of a concrete core test specimen depends on its shape, proportions, and size (e.g., the height/diameter ratio), hence appropriate correction factors are applied, so as to transform the specimen strength to the standard cube strength of concrete.

> Particular attention should be paid not to confuse the cylindrical strength, which is typically used as input in the structural analysis programs, with the cube strength, which is mostly used to describe the concrete strength in the industry and is the typical quantity reported from the tests.

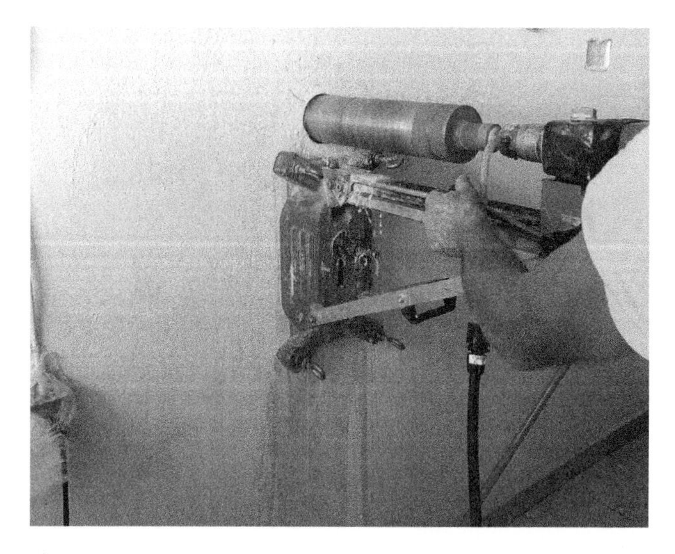

Figure 3.10 Cutting of a cylindrical concrete core. *Source:* Stelios Antoniou.

**Figure 3.11** Two carbonation indicators applied to the same concrete specimen. *Source:* Stelios Antoniou.

The minimum number of cores for a reliable sample is usually three, but different standards provide different specifications for the number of specimens. For instance, ATC 40 (ATC 1996) specifies two cores per floor and eight cores per building (ATC-40, Section 5.4.4.1), and the Greek code (KANEPE 2022) specifies three cores every two floors, but at least three cores at the most critical floor (KANEPE, Section 3.7.1.3). If there is a large dispersion of the obtained concrete strengths, additional samples should be taken.

The method of obtaining and testing concrete cores is the most reliable method to determine the mechanical properties of concrete in existing structures. Its main disadvantages are the relatively high cost and the difficulty in taking samples from vertical members (columns or walls), which are the most critical under seismic loading, in order not to damage them and compromise the overall structural capacity.

## 3.7 Concrete Tests – Nondestructive Methods, NDT

Because destructive methods for concrete testing are expensive, time-consuming, and cause disruption, these tests are conducted to a limited extent in existing buildings, and their results are supplemented by NDT. Several types of NDT have been developed. All these methods are based on the fact that certain physical properties of concrete that can be measured by nondestructive methods can be related to strength. Such properties include the hardness, the resistance to penetration by projectiles, the rebound capacity and the ability to transmit ultrasonic pulses, and X- and Y-rays.

The main NDTs are the following:

i) Rebound hammer test
ii) Penetration resistance test
iii) Pull-off test
iv) Ultrasonic pulse velocity method

### 3.7.1 Rebound Hammer Test

This is probably the most common nondestructive test to concrete. The rebound hammer is an instrument that is used to assess the relative compressive strength of concrete, based on the hardness of its exposed surface. For this reason, an empirical correlation is established between strength and the rebound. The rebound hammer is also called the *Schmidt hammer*, and it is a spring-controlled hammer mass that slides on a plunger within a tubular housing.

During the test, the plunger of the rebound hammer is pressed against the surface of concrete, so that the spring-controlled mass hits the concrete surface with a constant energy, and the distance of its rebound is measured on a scale. This value is called the rebound number or a rebound index and represents the hardness of the concrete surface. The test surface can be horizontal or vertical but the instrument must be calibrated for each specific location.

The Schmidt hammer provides a simple, quick, and inexpensive means of obtaining an indication of the concrete strength in existing members. It is a nondestructive test, since it only requires the removal of the member plaster in a small area of approximately 5 × 5 cm (Figure 3.12). Its results are generally affected by factors that are not directly relevant to the concrete strength, such as the smoothness of surface, the moisture condition of concrete, the type of cement, the aggregates size, and the extent of carbonation on the surface. Hence, calibration of the NDT results with destructive tests that directly estimate the compressive strength of concrete is required.

### 3.7.2 Penetration Resistance Test

The penetration resistance tests estimate the strength of concrete from the depth of penetration of a steel rod. The penetration is inversely proportional to the compressive strength of concrete in the standard test condition. This relationship depends on the type of concrete and type, size, and strength of the aggregates, which are generally unknown and difficult to be determined in existing structures. Furthermore, the derived results are variable

**Figure 3.12** Execution of a rebound hammer test in concrete. *Source:* Stelios Antoniou.

and not very accurate. As a result, similarly to the rebound hammer test, the results should be calibrated with results from the destructive methods. The typical equipment consists of a powder-actuated gun or driver, hardened alloy probes, loaded cartridges, a depth gauge for measuring the penetration of probes and other related equipment.

The probe test provides a quick means of checking the concrete strength in existing members. It is essentially nondestructive, since it requires the removal of plaster in a small area, and it creates a minor hole in the exposed concrete surface.

### 3.7.3 Pull-Off Test

Pull-off tests measure the force required to pull from the surface a standard steel rod inserted in the concrete, and can relate this force to the compressive strength of concrete. Instead of the rod a metallic disc bonded to a surface can also be used.

Similarly to the other methods, its main advantage is that it is inexpensive, fast, and non-destructive, although the plaster must be removed again. All the same, the execution of core crushing is needed to calibrate the results from the pull-off tests.

### 3.7.4 Ultrasonic Pulse Velocity Test, UPV

The ultrasonic pulse velocity (UPV) test, also known as the UPV test, is used to determine the concrete strength by evaluating the homogeneity and integrity of the concrete. It is performed with the help of an ultrasonic pulse velocity tester, which consists of a pulse generator and a pulse receiver, and measures the time of travel of an ultrasonic pulse of 50–54 kHz passing through the concrete (Figure 3.13). The pulse is produced by an electro-acoustical transducer system. The higher pulse velocity indicates higher elastic modulus, strength, density and better integrity of the concrete.

The presence of reinforcing steel in concrete has an appreciable effect on the pulse velocity. It is therefore desirable and often mandatory to choose pulse paths that avoid the

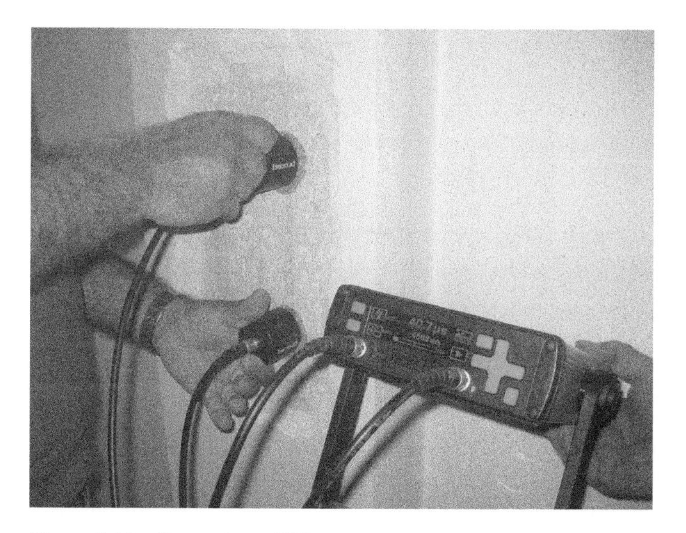

Figure 3.13 Execution of UPV test in concrete.

influence of reinforcing steel. Alternatively, corrections should be made. On the contrary, the UPV tests can achieve better performance, when the concrete member is accessible from more than one side and the transmitter and the receiver can be placed at two opposite sides.

Apart from reporting on the concrete mechanical properties, this method can be used to establish the uniformity of a member and detect discontinuities, cracks, voids, or delamination.

## 3.8 Steel Tests

In most practical applications, the classification of steel can be done by visual inspection. What is checked is whether the surface is smooth or ribbed, as well as any readable markings on the surface of the bars. These, in conjunction with the time of construction of the building and the steel classes that were being used then, are usually all that is required.

If there is doubt about the reliability of the classification through visual identification, the determination of the actual steel mechanical parameters (yield strength, ultimate strength, ductility) can be done through testing. If the rebar are corroded, the expected difference in the characteristics of steel, and the reduced ductility due to corrosion must be taken into account in a conservative fashion. Finally, when welding of new and old reinforcements is to be carried out, an investigation about the steel weldability is also required.

For the measurement of the steel yield and ultimate strength, a specimen of at least 30 cm long should be extracted after the removal of the concrete cover. The continuation of the reinforcement should be immediately restored by welding a new rebar of at least the same diameters to the discontinued parts of the existing steel bar. The specimen is then subjected to uniaxial tensile testing, in order to determine the steel mechanical properties. It is noted that such steel tests are not as common as similar tests for concrete, and more specialized laboratories are required for this task.

## 3.9 Infill Panel Tests

Masonry infill refers to the masonry used to fill the openings in a structural frame. A typical infill is a composite material, which consists of rectangular (solid or hollow) clay or concrete masonry units bonded with mortar. It can be designed to be part of the lateral force-resisting system (participating infill) or to be structurally independent from the building's lateral force-resisting system (nonparticipating infill). In the former case, it can provide an additional strong, ductile system for resisting lateral loads, especially in its plane (ASCE 2011; NCMA n.d.).

In order to determine the behavior of the infill panels, the compressive strength, the shear strength and the corresponding moduli are of interest. These properties can be determined by semi-empirical relations found in the literature, based on geometrical parameters (infill dimensions and thickness, brick dimensions, mortar thickness), the mechanical properties of the individual materials, such as brick strength, mortar strength, and the thickness of the joints.

The main parameter needed for the masonry units is their compressive strength that can be evaluated from a direct compressive test. Their tensile strength can also be estimated with various techniques, namely flexural test, splitting (or indirect tension) test, and direct tension test (Crisafulli 1997). The modulus of elasticity of the bricks depends mainly on the type of material and the compressive strength, and it is determined from empirical relationships.

Regarding the mortar, the direct determination of its compressive strength is not possible, due to the small size of the specimens, since the thickness of a typical mortar joint is 1–2 cm. It is noted that for new infills, the compressive strength test is conducted with 50 mm cube specimens, which are not available in the case of existing buildings. Consequently, the compressive strength of existing walls is calculated indirectly, together with the tensile strength, with a direct tension test by multiplying with the appropriate (compressive strength)/(tensile strength) factor, which generally assumes values of approximately 10. Simultaneously with the strength tests, the specific weight, the density, the porosity, and the granulometric distribution of the mortar is also measured. Finally, the modulus of elasticity is determined from empirical relationships from the compressive strength value.

## 3.10 What Is the Typical Procedure for Monitoring an Existing Building?

Before beginning to measure the members' dimensions and reinforcement, a preliminary survey is required for the identification of the location of the columns, the beams, and the slabs of the building. In this initial phase, it is not necessary to cause any damage to the building, and some parameters might be left unknown (e.g., beams and slabs above a gypsum board ceiling). Yet, the use of a small, light metal detector or scanner is of great help at this stage, in order to identify and confirm the position of the concrete members. It is always a good idea to have several A4 or A3 paper sheets available on site with the plan view of each floor, where notes from the field can be transcribed.

During the main survey, since the role of the transverse reinforcement is dominant in the assessment procedures, and the shear member capacity is considerably affected by it, the monitoring of the stirrups of the beams, and more importantly of the columns, is of great importance. Measuring the distance between the hoops can be accurately and reliably established with most of the good ferrous scanners of the market, and generally it should be carried out in the entire member length for all the columns and for a large proportion of the beams. It is noted that stirrups in older construction were placed one-by-one, by hand and without great attention; hence, in the general case, the spacing of the transverse reinforcement should not be expected to be constant in each member; in order to be on the safe side, the larger spacing should be kept and used in the analytical calculations for each element.

Because scanning can provide reliable results for the rebar diameters within an order of size, it is advisable to first remove with a small demolition hammer the plaster and the concrete cover from an adequate sample of beams and columns, in order to check the stirrups' size. In general, the same hoop sizes are used in the same building for all the columns and for all the beams, although the columns and the beams hoop sizes may be different.

If not equal, the beams size is typically larger, since in previous decades engineers focused more on gravity loading, which causes larger shear demand in the horizontal members.

Simultaneously with the transverse reinforcement, accurate monitoring of the longitudinal reinforcement also takes place. After the preliminary cuts with the demolition hammer, a systematic and thorough scanning of all the vertical and horizontal members of the building is carried out.

Generally speaking, it is not possible to scan all four sides of all the columns of a building, however at least one side is usually exposed and can be scanned. It is then rather safe to assume that the reinforcement is symmetrical on all four sides, which enables us to come up with the reinforcement pattern of the column. Furthermore, in columns that cannot be scanned (e.g., when all four sides are covered by the adjacent infilled walls), the reinforcement can be assumed based on the measured reinforcement of similarly sized columns. In general, the reinforcing patterns in the columns of a building are rather standardized and allow for relatively safe assumptions for the inaccessible elements.

On the contrary, this is not always the case for the horizontal members; the beams in the same building can have completely different reinforcing patterns, with a variable number and size of rebar. In general, the beams' lower reinforcement at mid-span can be detected relatively easily with a scanner, provided that the placement of rebar is not very dense. Some small cuts with a demolition hammer are usually required at a small number of the beams, in order to confirm the scans.

Contrary to the reinforcement at mid-span, the upper reinforcement at the beams' edges is generally difficult to identify, since this would require the demolition of a part of the floor on the upper side. In such cases, a relatively good approximation is to assume that the upper reinforcement at a beam's edges is the same with the lower reinforcement at mid-span. Obviously, this is not always the case; however, very often this is the best that one can achieve – such an approximation is probably preferable to causing significant damage in many locations to the floors of the entire building. What is more, usually this approximation is on the safe side: since under gravity loading the negative design moments at the beams' edges are usually slightly larger than the positive moments at mid-span, the corresponding actual reinforcement is also a bit larger. Another alternative would be to carry out a simple linear analysis just with the actual gravity loads and without safety factors, i.e. $1.0{\cdot}(G + G' + Q)$ and calculate the required reinforcement. The comparison between the calculated reinforcement and the actual reinforcement of the building (in the locations that were checked) can be used to draw conclusions about the reinforcing patterns employed during construction.

In order to estimate the grade of concrete of the building, a combination of destructive and nondestructive methods is typically carried out. Cylindrical specimens are cut from a number of representative members, and tested in an authorized laboratory. These tests are then used to calibrate the results from the NDT that are carried out on a much larger number of members. Note that for the application of the NDT the plaster should be removed in a small area with a diameter of 5–10 cm. Although the number of required tests varies in the different standards, a reasonable number of specimens for the destructive tests is at least three cores every two floors, but at least three cores in the most critical floor. Similarly, a representative number for the NDT could be 30% of all the vertical members and 15% for the beams, in order to achieve a usual knowledge level and 45% of the columns and 25% for the horizontal members for the comprehensive knowledge level. The numbers above are roughly equal to those proposed by the Turkish code, and a bit smaller than those proposed by the Greek code. These are the only two codes that propose specific quantitative criteria. ASCE-41, EC-8 and the Italian code are more vague when specifying the minimum requirements for the classification to each knowledge level.

Estimating the steel grade is usually easier; in the majority of cases, it can be done simply by visual inspection, and it does not require the execution of any laboratory testing. Generally, the grade can be determined from the original drawings, and the technical documentation of the building, since typically the steel specifications have been respected during construction, and relative uniformity is expected in the same building. However, even in the cases when the original plans are not available, the period of construction of the building (which is generally known) and the presence of smooth or ribbed rebar, is usually a good indicator of the steel grade. Finally, note that in many existing buildings, the grade of the longitudinal rebar can be different from that of the transverse reinforcement – for instance, ribbed S400 steel for the longitudinal reinforcement, but smooth S220 for the hoops.

## 3.11 Final Remarks

As explained in this chapter, the monitoring of an existing building with adequate accuracy for the creation of a correct model in the analytical package is an extremely important task, and it is by no means easy. It requires a lot of fieldwork, many times in dirty, evacuated buildings. More importantly, significant experience and engineering judgment is needed in order to make the correct selection of the locations of importance, where the tests on the structural and the nonstructural components are to be done, and to make correct and accurate "guesses" of structural parameters that cannot be investigated on-site (e.g., the reinforcement of columns behind mechanical equipment or behind gypsum board partitions).

Nonetheless, many times the significance of this process is underestimated and engineers choose to focus on other parts of the methodology, which they consider more important, and with which they are more familiar – namely, the creation of the structural model in the FE package and the analytical calculations. However, because of the large deviations that many existing buildings have from their original design, and because of the fact that in the vast majority of cases these deviations jeopardize safety, the correct monitoring of the actual structural configuration is of paramount importance, and failure to do so can lead to a great underestimation of the potential vulnerability of the building, and its seismic risk.

Due to the gradual increase in the significance of structural assessment of existing buildings, the difficulties involved in the process, the relatively expensive equipment required, and the fact that monitoring combines field and office work, I personally believe (or hope) that the monitoring of existing buildings will gradually become a separate specialized field in civil engineering. In fact, such specialized companies are gradually being created, even if they still combine this activity with others. These companies are not to be confused with the traditional laboratories that provide test results on concrete, steel or masonry. They consist of teams of experienced and specialized civil engineers that do mainly field work, with the main objective of providing the measured drawings (and in this process they closely cooperate with a lab that measures the material strengths). This trend is in the right direction; it will increase the accuracy and reliability of the measured results, and ultimately lead to a much better estimation of the structural capacity and the seismic risk of the existing building stock.

## References

[ASCE] American Society of Civil Engineers (2011). *Building Code Requirements for Masonry Structures, TMS 402-1 1/ACI 530-11/ASCE 5-11*. Developed by the Masonry Standards Joint Committee (MSJC).

[ASCE] American Society of Civil Engineers (2017). Seismic Evaluation and Retrofit of Existing Buildings (ASCE/SEI 41-17), 2017, Reston, Virginia.

[ATC] Applied Technology Council (1996). Seismic Evaluation and Retrofit of Concrete Buildings, ATC-40 Report, Applied Technology Council, Redwood City, California.

Crisafulli F.J. (1997). Seismic Behaviour of Reinforced Concrete Structures with Masonry Infills, PhD thesis, University of Canterbury, New Zealand.

KANEPE (2022). Greek Code of Structural Interventions, Revision 2. Earthquake Planning and Protection Organization, EPPO (in Greek).

[NCMA] National Concrete Masonry Association. (n.d.) Design of Concrete Masonry Infill. NCMA TEK Note 14-23. https://ncma.org/resource/design-of-concrete-masonry-infill.

TBDY (2018). *Türkiye Bina Deprem Yönetmeliği, Turkish Seismic Building Code*. Ankara: Disaster and Emergency Management Presidency in Turkish.

# 4

# Methods for Strengthening Reinforced Concrete Buildings

## 4.1 Introduction

This chapter describes the most common techniques for the strengthening of a reinforced concrete building and explains their advantages and disadvantages. Before moving on to the detailed presentation of the methods, some general points regarding the different retrofit techniques are presented in the current section:

- There are only six or seven methods for seismic upgrading. In most practical applications, it is usually apparent from the beginning that some methods cannot be applied due to architectural, operational or geometrical restrictions, lack of knowledge, or unavailability of the appropriate equipment (e.g., base isolation cannot be implemented in buildings that do not have open spaces throughout their perimeter or in regions where the expertise is still limited). Hence, the designer in every building may in fact have to choose from a set of no more than two or three candidate methods. It is the designer's job to weigh the advantages and disadvantages of each method using engineering judgment, preliminary studies, and analytical calculations, and then choose the most appropriate one.
- Special attention should be paid to the fact that, when designing strengthening interventions, not all methods and techniques can decrease the vulnerability of the building. Erroneous implementation can strengthen some parts of the building, simultaneously weakening other parts, actually increasing the overall risk. For instance, a concentration of large shear walls on one side of the building might increase rather than decrease the demand on the vertical members of the other side.
- Different methods have different advantages and disadvantages, and their effect on the global response, the strength, the stiffness, the ductility and the seismic demand may vary considerably, depending on the particular building configuration. There are no standard solutions and recipes that can be applied to any structural type.
- Strengthening techniques can be categorized into two large groups. On the one hand, there are methods that are generally employed at the global level (i.e., considering the entire structure as a single entity) and typically serve to decrease the demand in existing members. The most representative examples in this category are the addition of new

*Seismic Retrofit of Existing Reinforced Concrete Buildings*, First Edition. Stelios Antoniou.

shear walls and base isolation. On the other hand, there are methods that are applied on the member level and are mainly used to upgrade the particular characteristics (e.g., strength and/or the ductility) of individual members (e.g., RC jackets or FRP wraps).

- Very often, a combination of two or more from the available methods might be required. Usually, a more "global" method is first applied (e.g., shear walls or braces), and other methods are applied at a second stage to strengthen individual components or parts of the building that still need upgrading.

After presenting individual retrofit techniques, this chapter evaluates various solutions using a simple example based on strengthening of an existing RC building with a soft ground floor. Chapter 5 then introduces the basic concepts for the strategy of strengthening interventions and analyzes the most common retrofit strategies, together with the criteria for selecting one or more of the available techniques, and their merits and disadvantages.

## 4.2 Literature Review

Considering the importance of the problem of the seismic rehabilitation of the existing building stock, it is not surprising that there is a very large number of documents in the literature that deal with this subject. Yet, most of them focus on specific issues (e.g., assessment of one particular method of strengthening, or proposal of a new variant in the method), and only a small portion of these try to provide a more complete picture of the subject.[1]

All the same, there are enough books, reports, book chapters, and papers to provide a detailed presentation of the possible solutions and to compare their relative merits and disadvantages. One could refer to the numerous US published guidelines (e.g., ATC-40 1996; FEMA 273 1997; FEMA 356 2000; FEMA-547 FEMA NEHRP 2006; FEMA P–420 2009; FEMA P-58-1 2018a; FEMA P-58-2 2018b; FEMA P-2018 2018c; FEMA P-2090/NIST SP-1254 2021), the fib 103 report (FIB 2022) or the work of Pinho et al. (2019), Tsionis et al. (2014), Dritsos et al. (2019), Tassios (2016), Costa et al. (2018), Fardis (2009), Caterino et al. (2008), Okakpu and Ozay (2015), Gkimprixis et al. (2020), Aydenlou (2020), Thermou and Pantazopoulou (2014), Ilki et al. (2009), Gkournelos et al. (2021), Kaplan and Yılmaz (2012), Akbar et al. (2020), Aslam et al. (2016), Ilki and Fardis (2014), Baros and Dritsos (2008), Dritsos (2005a,b), Spyrakos (2004) and Phan et al. (1988), to name but a few.

Some of these documents also provide guidance on when and under what circumstances to apply each specific strengthening method. What is missing, though, from most of these publications is a detailed description of the difficulties and challenges posed by actual projects – that is, economic, architectural, and operational constraints that inevitably affect the final decisions. More importantly, very limited information and guidelines exist on how to overcome all these difficulties.

1 These documents should not be confused with the plethora of books that are available for the design of new seismic resistant RC buildings, or the books that provide general principles on the subject, which again focus on new construction.

Besides these more general documents, there are literally thousands of papers and reports that present specific aspects of the problem (e.g., evaluation of an existing or a new retrofit method, or comparison between two or more of the methods). Naming all of them is beyond the scope of this book; however, I will reference hereby some characteristic papers, so that readers can get more information on specific aspects of the strengthening of RC buildings if they wish to.

- RC jacketing: Bousias et al. (2007a), Chalioris et al. (2014), Karayannis et al. (2008), Varum et al. (2013), Marini and Meda (2009), Waghmare (2011), Vandoros and Dritsos (2008), Bousias et al. (2007b), Chang et al. (2014).
- RC shear walls: Kaltakci et al. (2008), Kaplan and Yılmaz (2012), Marriott et al. (2007), Mori et al. (2008).
- RC Infilling: Poljanšek et al. (2014), Chrysostomou et al. (2014), Fardis et al. (2014), Fardis et al. (2013).
- Wing walls: Ou and Truong (2018), Chang et al. (2014), Kabeyasawa et al. (2011), Kabeyasawa et al. (2010), Yamakawa et al. (2006).
  Fiber reinforced polymers (FRPs):
  - General on FRPs: Fédération Internacionale du béton (FIB) (2019), Fédération Internacionale du béton (FIB) (2001), Pantazopoulou et al. (2016), Triantafillou (2004), Triantafillou (2001), Ravikumar and Thandavamoorthy (2014), Setunge et al. (2002), Irwin and Rahman (2002).
  - FRP wrapping: Parghi and Alam (2015), Megalooikonomou et al. (2012), Seyhan et al. (2015), Esmaeeli et al. (2015), Ma and Li (2015), Zhou et al. (2013), Thermou et al. (2011), Murali and Pannirselvam (2011), Thermou and Pantazopoulou (2009), Harajli and Dagher (2008), Ozcan et al. (2008), Colomb et al. (2008), Pantelides et al. (2008), Thermou and Pantazopoulou (2007), Ghosh and Sheikh (2007), Haroun and Elsanadedy (2005a), Harries et al. (2006), Haroun and Elsanadedy (2005b), Bousias et al. (2004), Memon and Sheikh (2005), Ghobarah and Galal (2004), Sause et al. (2004), Antonopoulos and Triantafillou (2003), Sheikh and Yau (2002), Khalifa and Nanni (2000), Ma et al. (2000).
  - FRP laminates: Elwan et al. (2017), Ahmed et al. (2011), El-Hacha et al. (2001), Mofidi et al. (2013), Balsamo et al. (2005a), Balsamo et al. (2005b), Alferjani et al. (2013), Narmashiri et al. (2010), Jumaat et al. (2010).
  - Near surface-mounted (NSM) FRPs: Chandran and Subha (2018), Moham and Matlab (2018), Jiang et al. (2016), Fahmy and Wu (2016), Seyhan et al. (2015), Li et al. (2013), Sarafraz and Danesh (2012), Islam (2008), Szab and Balzs (2007), Parretti and Nanni (2004), El-Hacha and Rizkalla (2004).
  - FRP anchors: Grelle and Sneed (2013), Sneed (2013), Koutas and Triantafillou (2013), Fagone et al. (2014), Breña and McGuirk (2013), Bournas et al. (2015), Vrettos et al. (2013), Niemitz et al. (2010), Ozbakkaloglu and Saatcioglu (2009), Eshwar et al. (2008), Orton et al. (2008).
  - FRP bars: Kobraei et al. (2011), Abdel-Kareem et al. (2019).
  - Debonding: Jumaat et al. (2011b), Jumaat et al. (2011a).
  - Strengthening of infills with FRPs: Pohoryles and Bournas (2019), Yuksel et al. (2010), Binici et al. (2007), Ilki et al. (2007), Almusallam and Al-Salloum (2007), Yuksel et al. (2005).

- Strengthening of infills and textile reinforced mortars (TRMs): Pohoryles and Bournas (2020), Furtado et al. (2020), Gkournelos et al. (2019), Pohoryles and Bournas (2019), Koutas and Bournas (2019), Akhoundi et al. (2018a, 2018b), Triantafillou et al. (2018), Koutas et al. (2015, 2014a,b), Da Porto et al. (2015), Okten et al. (2015), Vasconcelos et al. (2012), Al-Salloum et al. (2011), Sevil et al. (2011).
- Steel braces: Formisano et al. (2020), Di Lorenzo et al. (2020), Reggio et al. (2019), Harshita and Vasudev (2018), Prasanna Kumar and Vishnu (2017), Shashikumar et al. (2018), Kanungo and Bedi (2018), TahamouliRoudsari et al. (2017), Eskandari et al. (2017), Tiwari and Bhadauria (2017), Dhiraj and Prasad (2016), Formisano et al. (2016), Hyderuddin et al. (2016), Durucan and Dicleli (2010), Dubină (2015), Bhojkar and Bagade (2015), Faella et al. (2014), Chavan and Jadhav (2014), Marneris and Kouskouna (2014), Bergami and Nuti (2013), Görgülü et al. (2012), Mazzolani et al. (2009), Maheri et al. (2003), Ghobarah and Elfath (2001), Formisano et al. (2008).
- Buckling-restrained braces: Almeida et al. (2017), Mahrenholtz et al. (2015), Barbagallo et al. (2014), Della Corte et al. (2011), Di Sarno and Manfredi (2009, 2010, 2012), Tsai et al. (2004).
- Steel Plates: Liu et al. (2018), Swetha and James (2018), Ezz-Eldeen (2015), Tarabia and Albakry (2014).
- Steel jackets: Islam and Hoque (2015), Pudjisuryadi et al. (2015), Belal et al. (2014), Sayed-Ahmed (2012), Nagaprasad et al. (2009), Fakharifar et al. (2016), Daudey and Filiatrault (2000), Aboutaha et al. (1999).
- Damping devices: Mazza and Mazza (2019), Oinam and Sahoo (2019), Mazza et al. (2019), Nakai et al. (2019), Mazza et al. (2018b), Lee and Kim (2017), Bianchi et al. (2015), Guo et al. (2015), Teruna et al. (2014), Valente (2013), Briseghella et al. (2013), Kouteva-Guentcheva et al. (2013), Symans et al. (2008), Amadio et al. (2008), Christopoulos and Filiatrault (2006), Molina et al. (2004), Soong and Dargush (1999), Symans and Constantinou (1999), Constantinou et al. (1998).
- Seismic isolation: Lampropoulos et al. (2021), Giarlelis et al. (2018a), Mazza et al. (2018a), Mazza et al. (2017), Mazza and Pucci (2016), Cardone and Gesualdia (2014), Cardone and Flora (2014), Cardone et al. (2012), Skinner et al. (2011), Melkumyan et al. (2011), Clemente and De Stefano (2011), Ferraioli and Avossa (2012), Ferraioli et al. (2011), Matsagar and Jangid (2008), Christopoulos and Filiatrault (2006), Walters (2003).
- Corrosion: Pugliese et al. (2019).

Furthermore, I was able to find a small number of publications that make comparisons between different methods:

- RC jackets, steel jackets, RC walls, and FRP wrapping: Okakpu and Ozay (2014), Lazaris (2019).
- Steel braces and RC walls: Alashkar et al. (2015)
- RC jackets and RC walls: Fauzan et al. (2018)
- RC Jackets and FRP wraps: Di Ludovico et al. (2008)
- RC Jackets and FRP wraps applied to RC joints: Sharma et al. (2010)
- FRP wraps and steel plates: Bsisu et al. (2011).
- FRP wraps and steel jackets: ElSouri and Harajli (2011), Li et al. (2009).
- Steel braces, steel jackets, and steel plates: Braconi et al. (2013).
- Base isolation and dissipative devices: Olariu et al. (2000).

Finally, there are a couple of publications that propose and assess solutions that combine two or more retrofit methods simultaneously (e.g., steel braces with FRP wrapping) (Mazza and Mazza 2019), FRP wrapping and steel plates (Afshin et al. 2019, Realfonzo and Napoli 2009), or publications that propose selective retrofit schemes, e.g. Pinho (2000).

## 4.3 Reinforced Concrete Jackets

### 4.3.1 Application

Concrete jacketing is probably the most widely used technique for strengthening RC members. It is constructed either with cast-in-place concrete or, more often, with shotcrete. The method involves the addition of a layer of reinforced concrete in the form of a jacket using longitudinal steel reinforcement and transverse steel ties outside the perimeter of the existing member (Figures 4.1 and 4.2).

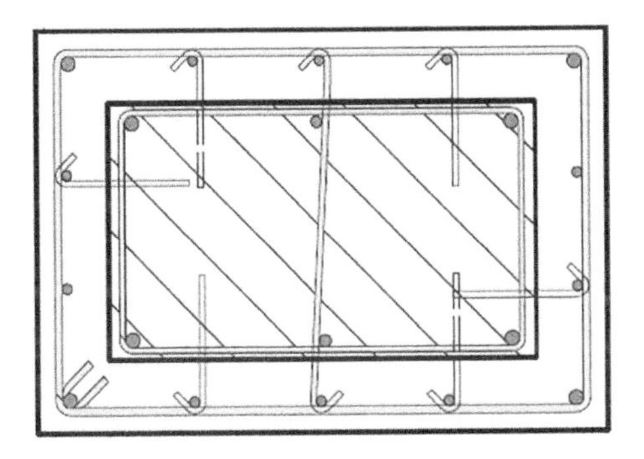
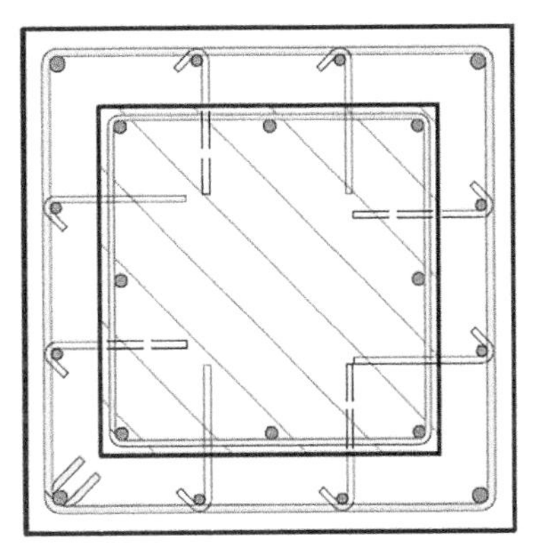
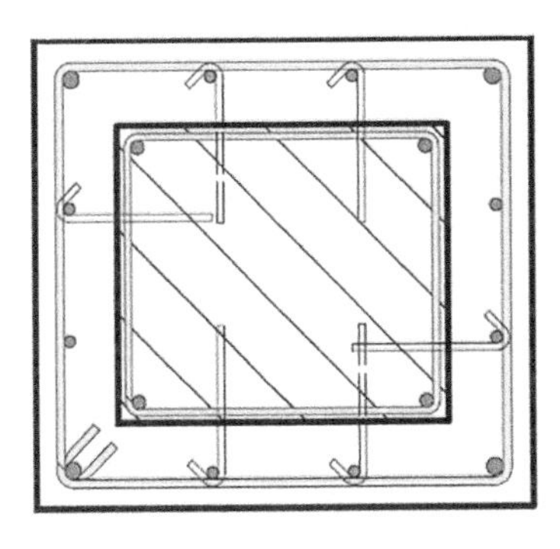
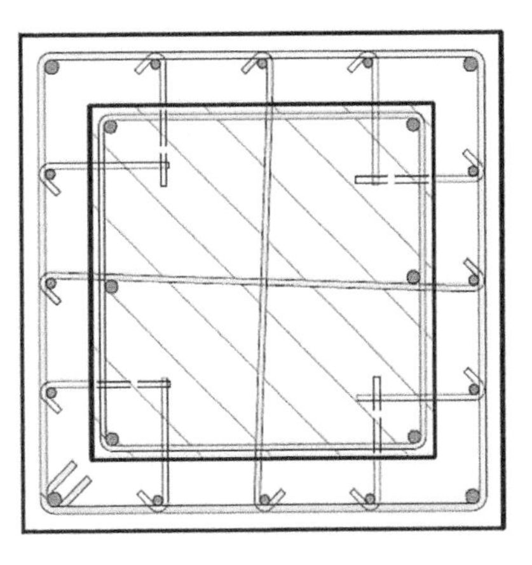

**Figure 4.1** Typical cross sections of reinforced concrete jackets. *Source:* Stelios Antoniou.

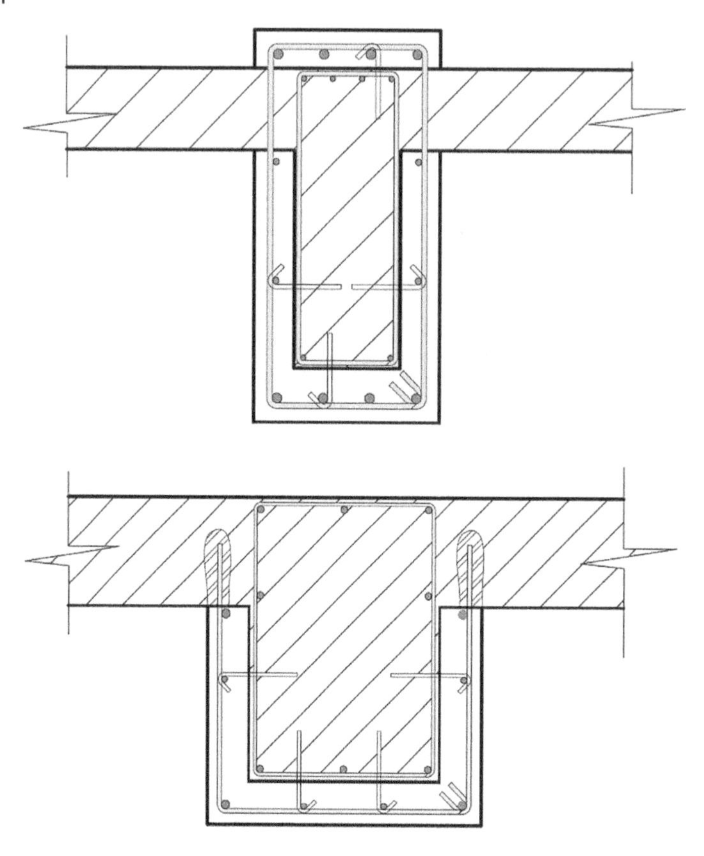

Figure 4.1 (Continued)

Figure 4.2 Reinforced concrete jackets before and after casting of the concrete. *Source:* Stelios Antoniou.

Figure 4.2 (Continued)

The jacketing with cast-in-place concrete demands the installation of formwork around the existing column, on which the formwork is tied in order to withstand the stresses of wet concrete. The thickness of the jacket usually exceeds 10 cm to allow the casting of the concrete without voids and gaps. By contrast, shotcrete allows for jackets of thickness as low as 5 cm. Typically, the jackets are 7.5 cm thick or more, in order to allow for a cover of adequate thickness, the positioning of the longitudinal and transverse reinforcement, and some space between the new rebar and the existing member.

The preparation of the surface of the existing member is critical in jacketing. It is essential that the existing member has a clean, sound concrete base to achieve good bonding conditions with the jacket. The connection of the new and the existing concrete is further enhanced with the roughening of the surface and the introduction of steel dowels (Figure 4.3).

The new vertical steel bars and stirrups of the jacket are then installed according to the designed dimensions and diameters, paying particular attention to the correct closing of the hoops. Since it is often not possible to bend the hoops at 135° angles due to the presence of the existing member and the small thickness of the jacket, welding is often required, as in Figure 4.4.

Special attention should be paid to the beam-column joint regions, where the jackets are generally extended. In order to splice the longitudinal rebar between adjacent floors, typically the creation of small holes in the concrete slab is required, either with a small demolition hammer or a portable electric drill. Vertical holes in the adjacent beams are also needed, and these can only be done with an electric drill to prevent seriously damaging the beam. Special care should be taken in the placement of the four to five stirrups at the level of the beam, since a series of horizontal holes should be made in order to pass through the parts of each individual stirrup. The parts are then welded together (Figure 4.5).

**Figure 4.3** Roughening of the surface of the existing member and introduction of dowels. *Source:* Stelios Antoniou.

Sometimes the new and existing rebar are welded together using U-shaped steel connectors or steel plates (Figure 4.6), and this method can be used as an alternative to the steel dowels. However, nowadays steel dowels are considered to be a better method, since welding the often-corroded, existing reinforcement is unreliable, or even impossible.

Because of the significant increase in the stiffness of the new member with respect to the existing one, and in order to avoid stiffness discontinuities, the jackets need to cover the entire length of the member. This means that the column jackets should not stop at the ground floor level but should extend until the upper surface of the footing, where the longitudinal rebar are anchored inside the existing footing with epoxy resins (Figure 4.7). This often leads to large-scale excavations, especially in buildings with several weak columns that are located close together.

Figure 4.4 Closing of the stirrups through welding. *Source:* Stelios Antoniou.

Figure 4.5 Lap splicing in the jacket between floors and reinforcement arrangement in a beam-column joint. *Source:* Stelios Antoniou.

In a variation of the method, and when only the shear strength and the deformation capacity of the member are of concern, a selective intervention scheme may be applied whereby the jacket (both concrete and reinforcement) could be terminated without being anchored in the beams or the slabs at its ends, leaving a gap of the order of 1–2 cm. This method, however, although effective, is more cumbersome and expensive with respect to similar alternatives (i.e., FRP wrapping) and is not so common in practical applications.

Figure 4.6 Steel connectors between existing and new rebar. *Source:* Stelios Antoniou.

Figure 4.7 Continuation of the jacket until the existing footing. *Source:* Stelios Antoniou.

In cases where the construction of a closed (four-sided) jacket is not feasible (e.g., columns at the perimeter of buildings that are adjacent to other existing properties, or beams in locations without access to the upper floor), three-sided jackets may be applied, provided that the jacket is well connected to the existing member through dowels or welding, and the stirrups are adequately closed or anchored (Figure 4.8). Two-sided and one-sided jackets should be avoided – they are even not permitted by some standards (e.g., Greek

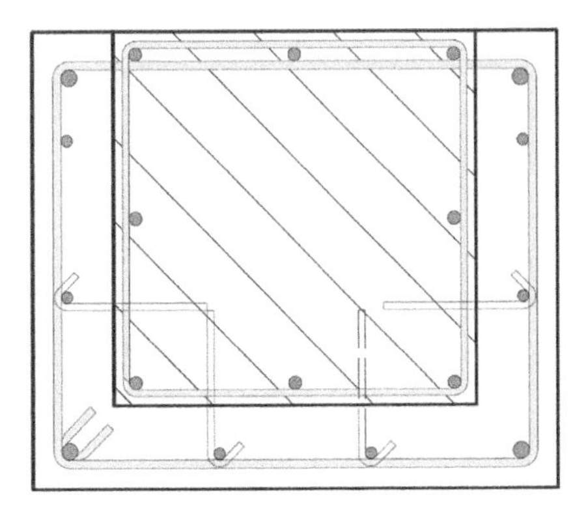

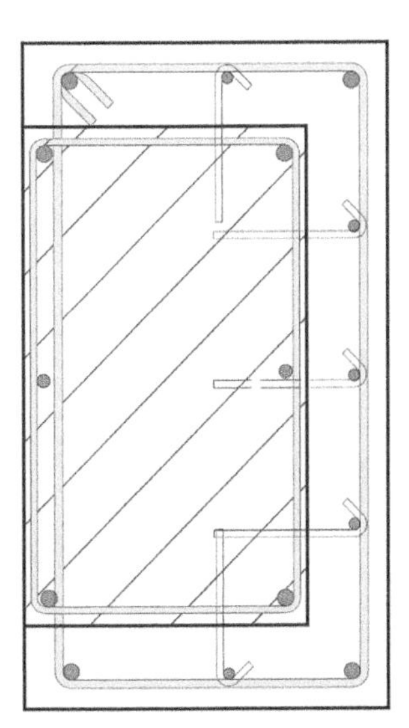

Figure 4.8 Closing the stirrups of the jacket inside the existing member. *Source:* Stelios Antoniou.

Interventions Code). This is because it is not possible to sufficiently anchor the stirrups and effectively connect the jacket with the existing member, and they may not behave monolithically.

It is noted that, when strengthening vertical members with jackets, the sides of the existing member where the jacket is applied must be fully uncovered for the passing of the reinforcement and the construction of the jacket. This requires the demolition and reconstruction of several nonstructural parts, such as infills, floors, tiles, suspended ceiling, doors, or windows. This can cause increased costs, significant disruption, and architectural problems (Figure 4.9). All these factors should be considered when deciding to strengthen a building with jackets, especially when the building is in operation.

Figure 4.9 Demolition of non-structural parts of the building for the construction of a jacket. *Source:* Stelios Antoniou.

### 4.3.2 Advantages and Disadvantages

With reinforced concrete jackets a considerable increase in the strength of the member can be achieved, both in bending and in shear. Furthermore, there is a significant increase in the ductility and the deformation capacity of the member, through the confinement and the anti-buckling action of the new stirrups. The bearing capacity and the flexural strength of the member are enhanced, due to additional longitudinal reinforcement, while the improvement in the shear strength and the ductility are achieved through the additional transverse reinforcement.

The significant increase of the member's strength with the introduction of jackets can be observed by means of a simple example with SeismoStruct (2023) of an existing member and the corresponding jacketed section. Typical member sizes and reinforcement patterns are employed – that is, an existing 25/25 column with 4∅18 rebar and ∅8/30 hoops is strengthened with a 10 cm jacket with 8∅20 and ∅10/10, as in Figure 4.10. The strength under bending increases from 25 to 320 kNm, while the shear strength increases from 20 to 235 kN!

Furthermore, because of the increase in the section's dimensions, an increase of the member stiffness is also achieved. For instance, in the example of Figure 4.10 the increase in the elastic section stiffness is again tenfold. This increase in the stiffness of the strengthened members of an existing building usually causes a decrease in the demand of the members that remain unstrengthened. However, in the typical cases of buildings of the 1960s or

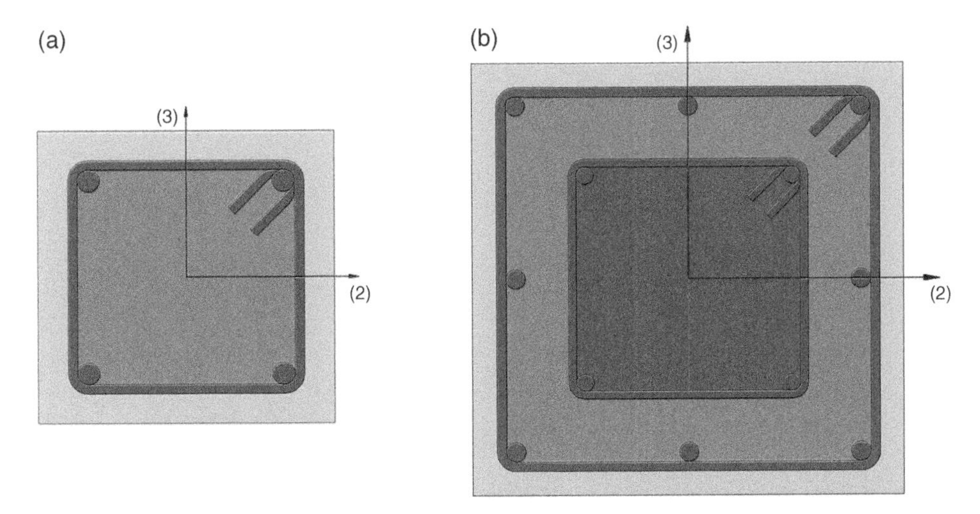


Figure 4.10 Comparison of the shear capacity for a typical C25/25 reinforced concrete column with its jacketed counterpart (a) existing column: 4∅18 mm, ∅8 mm/30, materials S220 and C12/15: Shear Capacity ≈ 20 kN (b) jacket: 8∅20 mm, ∅10 mm/10, materials S500 & C25/30: Shear Capacity ≈ 235 kN. *Source:* Stelios Antoniou.

1970s, this demand decrease is mediocre with respect to the considerable lack of resistance of the existing members, and typically it is not sufficient, so as to leave these members unstrengthened. As a result, when RC jacketing is selected as the main method of retrofit, typically it requires the strengthening of all or almost all the vertical members of a building, at least for the lower floors, where the shear demand on the columns is higher. This means that the use of jackets is advantageous in buildings that undergo radical refurbishment, when the cost for the damage on nonstructural elements from the strengthening interventions is relatively small.

By contrast, in buildings in operation, jacketing all vertical members presents significant (often insurmountable) issues:

- Generally, it is a dirty and disruptive method (Figure 4.11).
- The cost to repair nonstructural damage caused by the jacket construction (e.g., to walls, plaster, ceilings, and tiles) is high.
- Jacketing can impose architectural problems (e.g., jacketing a column adjacent to an opening would require the relocation, replacement, or adaptation of the window or door).

One final disadvantage of jacketing that cannot be ignored is its increased construction cost, as compared to the most widely used alternative, FRP wrapping. All the same, it should be noted that RC jackets feature some significant advantages over FRP wraps, providing enhanced strength and stiffness in both bending and shear. FRP wraps, in their usual application with the fibers perpendicular to the axis of the member, increase only the shear strength, the deformability, and the confinement, but not the bending capacity, as do RC jackets.

Figure 4.11 Jacketing, especially when it is done with shotcrete, is a dirty and disruptive task. *Source:* Stelios Antoniou.

### 4.3.3 Design Issues: Modeling, Analysis, and Checks

The modeling and the checking of the jacketed (nonuniform) sections in a structural model can be easily done, under the following simplifying assumptions:

- The jacketed member behaves monolithically, with full composite action between old and new concrete. Hence, the Euler–Bernoulli hypothesis that plane sections remain plane and normal to the member axis is also valid for the composite (existing concrete + jacket) section.
- The fact that axial load is applied to the old column alone is disregarded, and the full axial load is assumed to act on both the existing and the jacketed part of the section.
- If the two concrete materials cannot be accommodated by the structural analysis package, the concrete properties of the jacket are assumed to apply over the full section of the element.

The member flexural and shear capacities are then multiplied with some modification coefficients that take into account the interaction between the surfaces of the new and the existing concrete, and decrease the total capacity of the (jacketed section) + (existing section) member; for instance, according to EC8 the decrease of the total shear strength is 10% ($V_R^* = 0.90\,V_R$, see EC8-Part 3, A.4.2.2.).

One final key issue in jackets is the ability of the section to transfer forces between the new and the old concrete through the interface between the two concretes, dowels, or welding. Since welding should generally be avoided due to possible corrosion in the existing reinforcement, a good strategy toward this end is to design the dowels between the new and the existing concrete to carry the entire shear force in the interface, ignoring the transfer between the interface, in order to be on the safe side.

## 4.4 Shotcrete

### 4.4.1 Introduction

The term *shotcrete* refers to both the material and the construction method. The material is a concrete or a high-strength mortar, which is literally "shot" into the forms. The method is the application of this material on site (US Army Corps of Engineers 1993, American Concrete Institute ACI 506 (ACI) (2013)).

Strictly speaking, shotcrete (or gunite or sprayed concrete, as it is also called) is not a repair or strengthening method for existing buildings. It is a way of placing and compacting concrete and it has numerous applications, other than retrofit. For example (US Army Corps of Engineers 1993), it can be used (i) to repair spillway surfaces or marine structures that may be damaged by cavitation, abrasion erosion, corrosion of the reinforcement, or deterioration of the concrete; (ii) in underground engineering, as a temporary measure for slope support and stability, for supplementing or replacing conventional support methods such as steel beams, to seal rock surfaces, or to channel water flows; (iii) in tunnel engineering, in mines, subways, and automobile tunnels (Figure 4.12a); and (iv) in new structures, e.g., for the construction of pools, tanks, or domes. However, due to the restrictions imposed in existing buildings by their structural and nonstructural components, cast-in-place concrete in the majority of cases is difficult, expensive, or altogether impossible to apply, which makes shotcrete the usual way of applying concrete in repair and retrofit applications (Figure 4.12b,c). In fact, the use of shotcrete is so common when constructing RC jackets that the two terms are often used interchangeably.

Shotcrete is the official name of the sprayed high-performance concrete, conveyed through a hose and pneumatically projected at high velocity onto a surface. The application of a fine aggregate cement mixture with the use of pneumatic systems was first introduced in the early 1900s, and since then many improvements have been made in the equipment and in the specialized techniques required.

Shotcrete usually contains an increased content of cement and aggregates with a small granulometric gradient. It is placed and compacted at the same time, due to the force, with which it is ejected from the nozzle. It can be sprayed onto any type or shape of surface, including vertical or overhead areas (Figure 4.12). A large proportion of the shotcrete material is expected to bounce off the shooting surface and fall on the ground as rebound waste (Figure 4.13).

**Rebound**

Rebound is aggregate and cement paste that "bounces" off the shooting surface during the application of shotcrete. The rebound waste is larger in dry-mix shotcrete, while the wet-mix shotcrete rebounds somewhat less. Rebound for conventional dry-mix shotcrete, in the best of conditions, can be expected to be 20–30% of the total material that passes through the nozzle for vertical surfaces (walls, columns, sides of beams). For horizontal surfaces that are shotcreted from below (e.g., slabs) the rebound can be up to double. It is noted that the rebound is strongly related to the skill and experience of the operator and the productivity of the shotcrete equipment.

(a)

**Figure 4.12** Application of shotcrete (a) in tunnel engineering , (b) in RC buildings and (c) in masonry buildings. *Source:* Stelios Antoniou.

Figure 4.13 Rebound waste after the application of shotcrete. *Source:* Stelios Antoniou.

### 4.4.2 Dry Mix vs. Wet Mix Shotcrete

Shotcrete can be employed in two variations, wet mix and dry mix; the distinguishing feature is whether the water, which is required for the mixture hydration, is being injected at the nozzle immediately before it is discharged onto the receiving surface (dry mix) or beforehand, during the creation of the mix (wet mix). Shotcrete is usually an all-inclusive term for both versions.

The *dry-mix method* involves mixing the ingredients in dry conditions, placing them into a hopper or a bag, and then conveying them pneumatically with a continuous flow through a hose to the nozzle. The cement and aggregate mixture is prepared on site and the water necessary for the cement hydration is injected at the nozzle by a nozzleman, who controls the addition of water. The water and the dry mixture are often not completely mixed at the nozzle, but the hydration is completed as the mixture hits the receiving surface. This process requires a skilled nozzleman, especially in the case of thick or heavily reinforced sections.

*Wet-mix shotcrete* involves pumping of a previously prepared concrete, typically ready-mixed concrete, to the nozzle. The cementitious material, aggregate, water, and admixtures are thoroughly mixed, as would be done for conventional concrete. Compressed air is introduced at the nozzle to impel the mixture onto the receiving surface. The wet-process procedure generally produces less rebound waste and dust compared to the dry-mix process.

Below is a list of the most common differences between the two types of concrete:

- Dry-mix shotcrete is applied at a much slower rate than wet mix, and the production rates of the latter are considerably higher. Although the production rates depend highly on the in-situ conditions (obstacles, reinforcement, rebound) the maximum productivity of wet mix can be as high as 4–5 $m^3/h$, whereas that of the dry mix is less than 1 $m^3/h$.
- With the dry mix, intermittent use is easily accommodated, as the dry material is easily discharged from the hose; on the contrary, wet mix is better suited to continuous applications.

- The equipment and maintenance cost for the dry mix is generally lower than those for the wet mix.
- Because the dry mix that is conveyed through the hose is lighter than the wet mix, longer hose lengths are possible with the dry mix.
- The rebound percentages are generally higher with the dry mix.
- When properly proportioned and applied, the dry mix has better bond and higher strengths than the wet mix, allowing more effective placement in overhead and vertical applications without the use of accelerators.
- Application of the wet mix is generally easier; the cementitious materials and the aggregates are mixed with the water and additives prior to the shotcrete application, and the nozzleman does not have to be as skilled as in dry-mix case.
- The wet mix can be used with all ordinary admixtures of common concrete, while only accelerators can be added in dry mixes. The use of air-entraining admixtures (AEA) in shotcrete is practical only in wet mix, and the resistance of dry mix to freezing and thawing is poor.

The differences in the equipment cost, maintenance requirements, operational features, placement characteristics, and productivity may make one or the other of the two alternatives more attractive for a particular application. The dry mix process is much more common in repair and retrofit applications, where it is necessary to stop frequently (e.g., to move from one member to the next one), and in general the larger productivity of the wet mix is not required (small to medium projects). On the contrary, the wet-mix process is more common in underground works, where there are larger areas without obstacles, and its continuous application is possible.

### 4.4.3 Advantages and Disadvantages of Shotcrete

Shotcrete can be used in lieu of conventional concrete for reasons of convenience or, less frequently, of cost. Shotcrete is advantageous in the situations when formwork is cost prohibitive, impractical, or altogether impossible, due to limited access to the work area. Very thin layers of shotcrete can be achieved (up to 3–4 cm when reinforcement is included), considerably thinner than normal casting techniques that require at least 10–12 cm of thickness when reinforcement is present. This makes shotcrete ideal for reinforced concrete jackets.

In retrofit, shotcrete has become a material of vital importance, because of its versatility in shape, which enables the application of concrete in areas with difficult access or totally inaccessible to poured concrete (e.g., columns below floors that are in use and cannot sustain damage). It is placed, consolidated, and compacted at the same time, and the small aggregate size helps improve quality and manageability. It adheres to surfaces and it has reduced shrinkage and lower permeability. Shotcrete usually provides significantly higher bond strengths to existing materials than does conventional concrete.

However, shotcrete is generally more expensive than traditional cast-in-place concrete, especially in countries with increased labor costs. Furthermore, in the case of the dry-mix method, the concrete is not created in a controlled industrialized environment, since there is no way of measuring exactly the amount of water that is added at the nozzle, and correlating it in an accurate fashion to the amount of cement. As a result, increased skill and experience is required by the nozzleman and continuous attention should be paid by the supervisor, in order not to have a very dry mix, which leads to large rebound waste, or a high water content, which results in the slumping of the concrete.

Because of these difficulties, even though the physical properties of sound shotcrete are comparable or superior to those of conventional cast-in-place concrete, the improper application of shotcrete may lead to unacceptably low strengths. Furthermore, considerably larger variations with respect to the cast-in-place concrete are commonly found in the quality and strength, even within the same project, and a larger mean strength is required in order to achieve the target concrete class (because of the increased standard deviation).

Finally, shotcrete, especially the dry-mix variation, is a relatively dirty process. It suffers from high dust production, and a large proportion of the materials is rebound waste.

One usual misconception in construction is that shotcrete can be used to replace plaster. This is often proposed for reasons of speed and economy; however, this is generally not possible. Shotcrete is a "hard" material of poor workability, due to the very low water-to-cement ratio. Hence, it is not possible with shotcrete to achieve a smooth, plane surface as can be done with plaster, unless one significantly increases the water-to-cement ratio, which then compromises its strength. Consequently, in typical projects the shotcrete must be covered by plaster coating of at least 1–2 cm.

### 4.4.4 What Is It Actually Called – Shotcrete or Gunite?

Gunite was originally a trademarked name coined by the American taxidermist Carl Akeley in 1909 and patented in North Carolina. Nowadays, in strengthening and retrofit, *shotcrete* and *gunite* mean the same thing, and the two terms are interchangeable. However, this is not the case in other fields of engineering. For example, in pool construction, shotcrete refers to the wet mix and gunite to the dry mix (Jender 2020). In other areas, gunite has been used to denote small-aggregate sprayed concrete, and shotcrete to denote large-aggregate mixtures.

In any case, the preferred and the most commonly used term today for all gunned material is shotcrete, regardless of the aggregate size, in most fields of engineering.

### 4.4.5 Materials, Proportioning, and Properties

The materials, mixture proportions, and properties of shotcrete are similar in many respects to conventional concrete. The main materials that can be found in a shotcrete mix are described in the following subsections (US Army Corps of Engineers 1993, ASTM International 2001).

#### 4.4.5.1 Cement

In general, the cement requirements for shotcrete are similar to those for conventional concrete, and the same standards apply. For instance, portland cement must meet the requirements of ASTM C150 (ASTM International 2020a) and blended cement must meet the requirements of ASTM C595 (ASTM International 2020b).

#### 4.4.5.2 Pozzolans

Pozzolans are a broad class of siliceous or siliceous and aluminous materials, which in themselves possess little or no cementitious value but which, in finely divided form and in the presence of water, react chemically with calcium hydroxide at ordinary temperature to form compounds possessing cementitious properties. The term *pozzolan* embraces a large number of materials with different origin, composition, and properties, and includes both natural and artificial (man-made) materials (Mehta 1987; Wikipedia 2021).

Pozzolans are added to the shotcrete mixes, in order to achieve improved long-term strength performance and lower permeability. When added to a Portland-cement matrix, pozzolan reacts with the calcium hydroxide and water to produce more calcium silicate gel. Pozzolans should conform to the ASTM C618 Standard (ASTM International 2019).

Some pozzolans (e.g., fly ash) are sometimes added to wet-mix shotcrete to enhance workability, facilitate pumping in longer distances, increase resistance to sulfate attack, and reduce expansion caused by the alkali-silica reaction; however, this is not without controversy (for instance the early age strength development is delayed). Natural pozzolans and fly ash are not typically used with dry-mix shotcrete. On the contrary, silica fume is very often used in dry-mix shotcrete and does not delay the strength development.

#### 4.4.5.3 Silica Fume

Silica fume, also known as microsilica, is an amorphous (noncrystalline) polymorph of silicon dioxide, silica. It is an ultrafine powder, collected as a byproduct of the silicon and ferrosilicon alloy production and consists of spherical particles with an average particle diameter of 150 nm (Wikipedia 2022a). The material is over 85% silica dioxide, it is approximately 100 times finer than portland cement, and it has a specific gravity of 2.1–2.6. Because of its chemical and physical properties, it is a very reactive pozzolan, and its main field of application is as an admixture of shotcrete and concrete.

Silica fume is added to a shotcrete mixture as a cement replacement or, more often, as a supplement to cement, in which case its proportion ranges from 7% to 15% by mass of cement. Silica fume significantly increases the concrete strength, decreases permeability, increases the long-term durability in mechanical and chemical attacks, improves the cohesion and bond strength of shotcrete to substrate surfaces, and reduces the shotcrete rebound. Moreover, it enhances the resistance to carbonation and the resistance to highly aggressive chemical environments (e.g., high sulfate concentrations, refineries, or chemical industries) and prevents the "washout" when fresh shotcrete is subject to the action of flowing water. Although its other characteristics are also very important, silica fume is mainly used in shotcrete to achieve high strengths that can reach up to 80 MPa. Because of its extreme fineness, silica fume particles fill the microscopic voids between the cement particles reducing permeability, increasing the density and the strength of the shotcrete.

#### 4.4.5.4 Aggregates

Aggregate should comply with the quality requirements for ordinary concrete, e.g. ASTM C33 (ASTM International 2018), however, due to the nature of shotcrete and to minimize the rebound, aggregates with smaller grain sizes are usually used, and a uniform grading is essential. Similar to ordinary concrete, the aggregate sizes are provided by tables or charts (Table 4.1). Note that finer aggregates generally produce shotcrete with greater drying shrinkage, while coarser sands result in more rebound.

Table 4.1 Acceptable grading limits for the aggregate (US Army Corps of Engineers 1993) Grading Limits for Aggregate.

| | Percent bv Mass Passing Individual Sieves | | |
|---|---|---|---|
| **Sieve size** | **Grading No. 1** | **Grading No. 2** | **Grading No. 3** |
| 3/4-in. | | | 100 |
| 1/2-in. | | 100 | 80–95 |
| 3/8-in. | 100 | 90–100 | 70–90 |
| 0.19 in. (No.4) | 95–100 | 70–85 | 50–70 |
| 0.093 in. (No. 8) | 80–100 | 50–70 | 35–55 |
| 0.046 in. (No. 16) | 50–85 | 35–55 | 20–40 |
| 0.024 in. (No. 30) | 25–60 | 20–35 | 10–30 |
| 0.012 in. (No. 50) | 10–30 | 8–20 | 5–17 |
| 0.006 in. (No. 100) | 2–10 | 2–10 | 2–10 |

#### 4.4.5.5 Water

While there are no special requirements for the curing water, it is highly recommended to use potable water in the shotcrete mix.

#### 4.4.5.6 Fiber Reinforcement

Fibers are used to increase the ultimate strength, particularly the tensile strength of shotcrete, as well as its ductility and energy absorption capacity (ASTM International 2003). Furthermore, fibers allow for better crack control and may decrease the width of shrinkage cracks in the material. Fibers are discontinuous and, unlike conventional reinforcement, they are distributed randomly throughout the concrete matrix. In shotcrete, they are available in the following general forms (US Army Corps of Engineers 1993):

- Steel fibers
- Synthetic macro or micro-fibers (e.g. polypropylene, polyethylene, polyester, or rayon), with polypropylene fibers being the most widely used material. Synthetic fibers are derived from organic polymers.
- Glass fibers that consist of chopped glass fibers with a resin binder.
- Natural fibers; these are not commonly used in shotcrete.

*Steel fibers.* Steel fibers have been used since the late 1950s in shotcrete to increase its mechanical properties. While steel fibers for reinforced concrete are commercially available in various sizes, the typical fiber lengths for shotcrete range from 2 to 4 cm. The typical proportion is between 1% and 2% by volume.

Steel fibers are found in different shapes, round, flat, or irregular, while additional anchorage is provided by deformations along the fiber length or at the ends – e.g., fibers with hooked or flat ends, with crimps, corrugations, or undulated fibers (Figure 4.14). Steel fibers provide a modest increase in the tensile and flexural strength, but limited increase in the compressive strength of shotcrete, however they contribute to improved load carrying capacity after the cracking of the member.

**Figure 4.14** Steel fibers (Walcoom 2022).

**Figure 4.15** Polypropylene fibers *Source:* Stelios Antoniou.

*Synthetic and polypropylene fibers.* The technology of using synthetic fibers in shotcrete is relatively new, compared to steel fibers, but their use is growing rapidly, and nowadays they are preferred to steel fibers (Figure 4.15). A wide range of types is available – aramid, polypropylene, polyethylene, polyester, or rayon – but polypropylene is mostly used in retrofit applications. The most common specified lengths for polypropylene are between 2 and 4 cm; however, longer lengths can also be accommodated. The typical amount to be added to the shotcrete mix is between 1.0 and 3.0 kg/m$^3$, and it does not require any change in the mixing ratios of the other concrete materials.

The most important use of synthetic fibers is to control plastic shrinkage, but synthetic fibers also increase the toughness and the tensile – and to a limited extent the compressive – strength of the shotcrete.

*Glass fibers.* The fibers consist of alkali-resistant glass (designated AR glass), which is protected from the alkalinity of cement. Alkali resistance is achieved by adding zirconia to the glass; the higher the zirconia content, the better the resistance (Wikipedia 2022b). The application of glass-fiber shotcrete cannot be done with the conventional equipment for shotcrete, but requires a special gun and delivery system; hence, glass-reinforced shotcrete is not widely used in strengthening and retrofit operations.

*Natural fibers*. These are natural fibers of different origin, such as bamboo, sisal, and coconut, but they are rarely used in shotcrete today.

#### 4.4.5.7 Chemical Admixtures and Accelerators

As mentioned above, accelerators are the only admixtures that can be added in dry-mix shotcrete. All the other chemical admixtures, including air-entraining (AEA), water-reducing, and retarding admixtures, can only be used with the wet mixes (ASTM International 2013).

Nowadays, there is a large variety of accelerators, both powdered and liquid, that may have different effects depending on their chemistry, their dosage rate, and the chemistry of the cement and the aggregates (US Army Corps of Engineers 1993). Powdered accelerators are mostly used for dry-mix shotcrete, whereas liquid accelerators can be used with both dry and wet mixes, and they are added at the nozzle prior to the application. In the case of the dry-mix process, the admixtures are usually premixed with the water.

Because of the limitations related to the required equipment, the use of admixtures in shotcrete is not the same as in conventional concrete. Furthermore, some admixtures may adversely affect the shotcrete properties; for instance, some accelerators may reduce the compressive shotcrete strength as high as 40%, or reduce its frost resistance. Therefore, the shotcrete mix that contains the admixtures should be tested in the field prior to application, in order to ensure that the desired properties are achieved.

#### 4.4.5.8 Reinforcing Steel

In general, the same specifications as for conventional concrete should be met by the reinforcing rebar in shotcrete. However, because of the sprayed placement method, the use of bars larger than ∅20 mm should be avoided. Similarly, large rebar concentrations interfere with the placement, prevent the correct buildup of good-quality shotcrete, and can leave large voids behind the reinforcement. Rebar spacing of at least 15 cm is recommended in at least one of the directions in the plane of the shotcrete application.

### 4.4.6 Mix Proportions for the Dry-Mix Process

In the wet-mix process, batching and mixing are practically identical to conventional concrete, thus allowing for very good control and more versatile mix designs. On the contrary, with the dry-mix process, the preparation of the cement and aggregates mix and the addition of water is carried out on site with conditions that cannot be strictly controlled. A very good preparation is required so as to minimize the deviations from the correct material proportions.

Initially, the mix proportions are calculated, based on the water-to-cement ratio (usually between 0.30 and 0.50 by weight), the aggregates to cement ratio (usually between 2.50 and 4.50 by weight), and the actual specific weights of the materials used. During the shotcrete application on site, it is preferable to measure the cement and sand by weight rather than by volume.

The most efficient and practical way to do so on site is with the use of buckets. Buckets filled with cement and the aggregates are weighed and the final dry mix is formed by mixing, for instance three buckets of cement, four buckets of sand, and two buckets of coarse aggregates. Because the sand particles should be thoroughly coated with cement, mixing of at least one minute in a drum-type mixer is required. It is noted that the in-place cement proportion will be higher, and the in-place aggregate grading will be finer than the batched grading due to rebound, especially if larger aggregate sizes are used.

An experienced engineer should carefully oversee the application of shotcrete to guard against large rebound (which is an indication of an overly dry mix or aggregate problems) or slumping of the concrete (which is an indication of high water content). Furthermore, field testing of the dry-mix proportions is highly recommended, especially if no field data exist for a given dry mix. Because the shotcrete strength can vary even for the same mix proportions, depending on the aggregates strength, and the abilities and the proficiency of the nozzleman, prior to the final application of the shotcrete it is advisable to test two to three different mixtures, in order to check that an appropriate mix will be applied.

### 4.4.7 Equipment and Crew

#### 4.4.7.1 Dry-Mix Process

The cementitious materials and the damp aggregates are thoroughly mixed and bagged, possibly well in advance of the shotcrete application (provided that they are kept in dry conditions). Prior to the shotcrete gunning, it is often advantageous to premoisturize the mix to 3–6% by dry mass using a premoisturizer, an apparatus that distributes and mixes water to the dry materials.

The cement-aggregate mixture is then fed to the gunning machine and introduced into the delivery hose through a metering device such as a feed wheel to ensure a constant feed is passed. Compressed air generated by an air compressor is added at the gun and the mixture is carried through the delivery hose to the nozzle. A perforated water ring is fitted at the nozzle, through which the water and the admixtures are introduced. The materials are all mixed to concrete, as they go through the nozzle. The concrete is propelled from the nozzle at high velocity onto the receiving surface.

Dry-mix guns are classified in two categories:

- The *double chamber gun* shown in Figure 4.16 was first introduced in the early 1900s, but its use is now limited. The material enters the upper chamber in batches, but the valve arrangement is such that the discharge from the lower chamber is continuous.
- The *continuous feed gun* is more common today and is shown in Figure 4.17 and in Figure 4.18. It was first introduced in the early 1960s. A rotary gun is employed, and it is continuously fed using an open hopper.

Dry-mix nozzles come in a wide variety of nozzle tips, nozzle sizes, and configurations. A typical nozzle consists of a tip, water ring, control valve, and nozzle body arranged, as depicted in Figure 4.19.

An air compressor of ample capacity should be employed. The compressor should maintain a supply of clean, dry, oil-free air, in order to provide the pressure that drives the material from the delivery equipment into and through the hose, and to maintain sufficient nozzle velocity at all parts of the work. The air pressure should be steady (nonpulsating). Typical air compressors are characterized by their capacity in terms of the guaranteed air

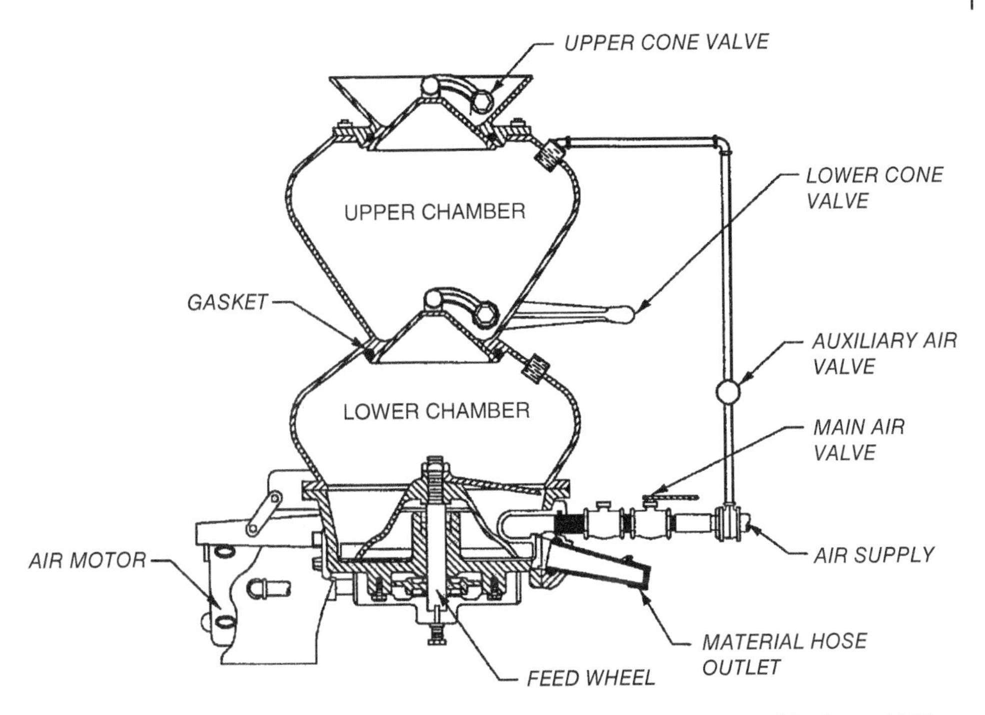


Figure 4.16 Sketch of the double chamber gun for shotcrete (US Army Corps of Engineers 1993; Crom 1966).

delivery (e.g., 275, 400, 600 $ft^3$/min in the Imperial system; $m^3$/min or lt/sec in the metric system) at the norm effective working pressure (typically 100 psi in the Imperial system or 7 bar in SI units or higher). For larger hose diameters and lengths, larger nozzles, and larger productivities, a larger capacity of the compressor is required.

The layout of a typical plant for dry-mix shotcreting is shown in Figure 4.20.

#### 4.4.7.2 Wet-Mix Process

The cement, the aggregates, and the admixtures (except accelerators) are mixed, and the mixture is fed into the wet-mix gun and propelled through the delivery hose to the nozzle by compressed air or pneumatic or mechanical pumping. Air is injected at the nozzle to disperse the stream of concrete and generate the velocity for shotcrete placement.

A typical wet-mix nozzle consists of a rubber nozzle tip, an air injection ring, a control valve, and the nozzle body, as shown in Figure 4.21.

### 4.4.8 Curing and Protection

The curing of shotcrete is extremely important, in order to ensure the proper hydration and bond strength development, and to prevent cracking due to shrinkage. It is noted that the relatively thin sections commonly used in retrofit applications of shotcrete are particularly susceptible to drying shrinkage, and that the development of bond strength is significantly slower than that of compressive or tensile strength.

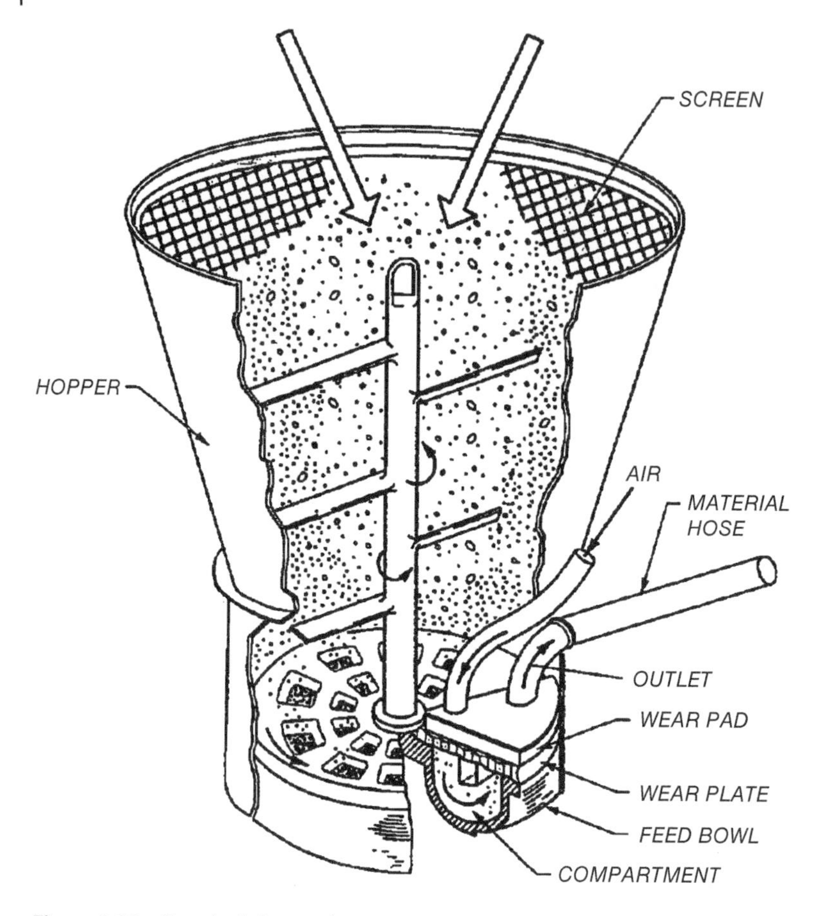


Figure 4.17 Sketch of the continuous feed gun for shotcrete (US Army Corps of Engineers 1993; Mahar et al. 1975).

The shotcrete surfaces should be kept continuously moist for at least seven days; after this time interval the shotcrete has gained sufficient tensile strength to resist shrinkage strains.

### 4.4.9 Testing and Evaluation

The strength of shotcrete should be verified at established intervals (US Army Corps of Engineers 1993, ASTM International 1998). The testing is usually carried out with cylindrical specimens extracted from rectangular or square test panels mounted in a framework (Figures 4.22 and 4.23). The size of the testing panels should be large enough to obtain all the test specimens needed with the same level or uniformity and quality that can be expected in the structure. Square panel of dimensions of at least 70 × 70 cm (preferably even more) should be employed. The thickness of the shotcrete should be no less than 12 cm, in order to allow the extraction of specimens at least 10 cm high. The distance between the cores and the panel edges should be at least 10 cm, and the cores should be extracted after at least seven days of standard curing to attain sufficient strength and allow movement to the testing laboratory.

Figure 4.18 The continuous feed gun for shotcrete on site. *Source:* Stelios Antoniou.

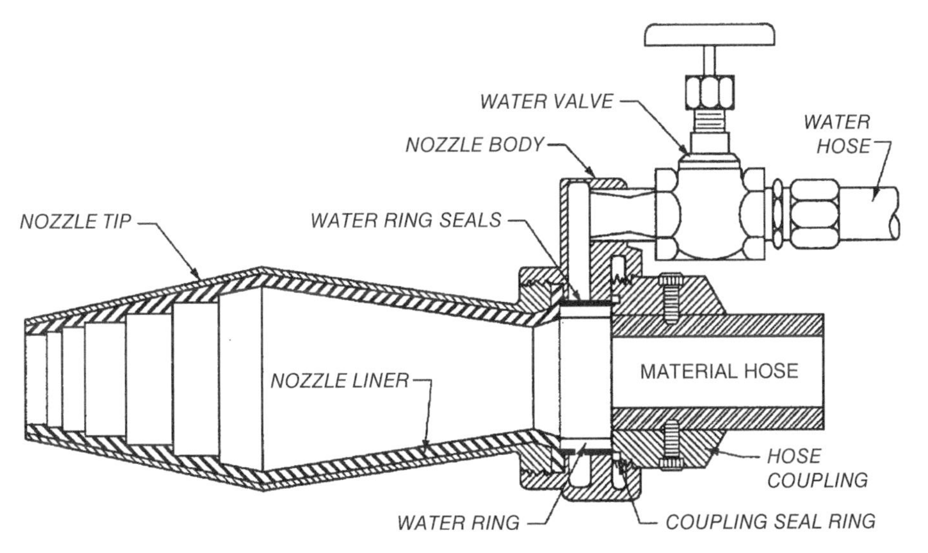


Figure 4.19 A typical dry-mix nozzle (US Army Corps of Engineers 1993).

Since shotcrete quality is highly dependent on the abilities of the nozzleman, a separate panel should be employed for each nozzleman, as well as for each shooting position in the structure.

## 4.5 New Reinforced Concrete Shear Walls

### 4.5.1 Application

This method consists of the construction of new shear walls with large dimensions at selected locations in the building perimeter and/or in the interior of the building. The walls can have a very beneficial effect on the seismic performance of existing buildings,

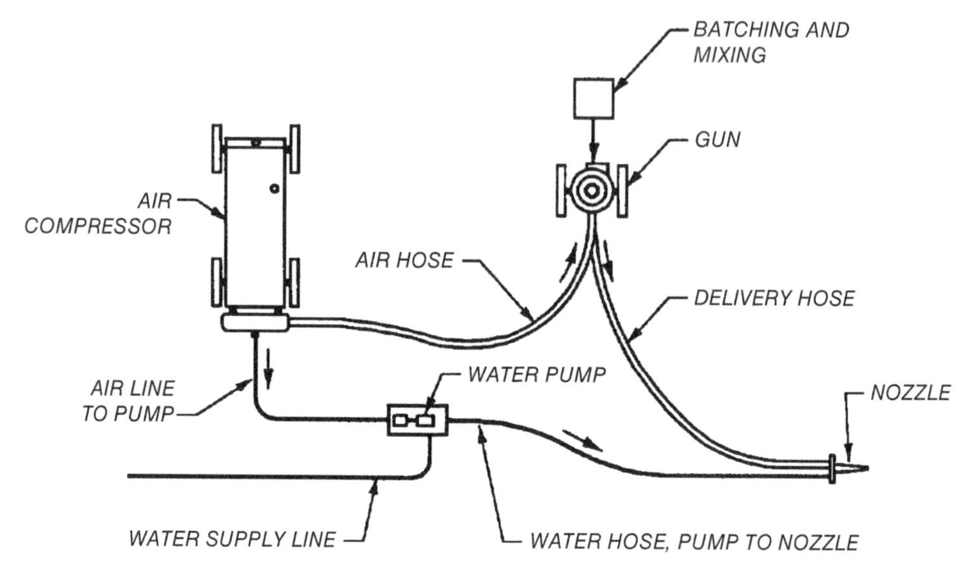


Figure 4.20 A typical plant layout for dry-mix shotcreting (US Army Corps of Engineers 1993; Crom 1966).

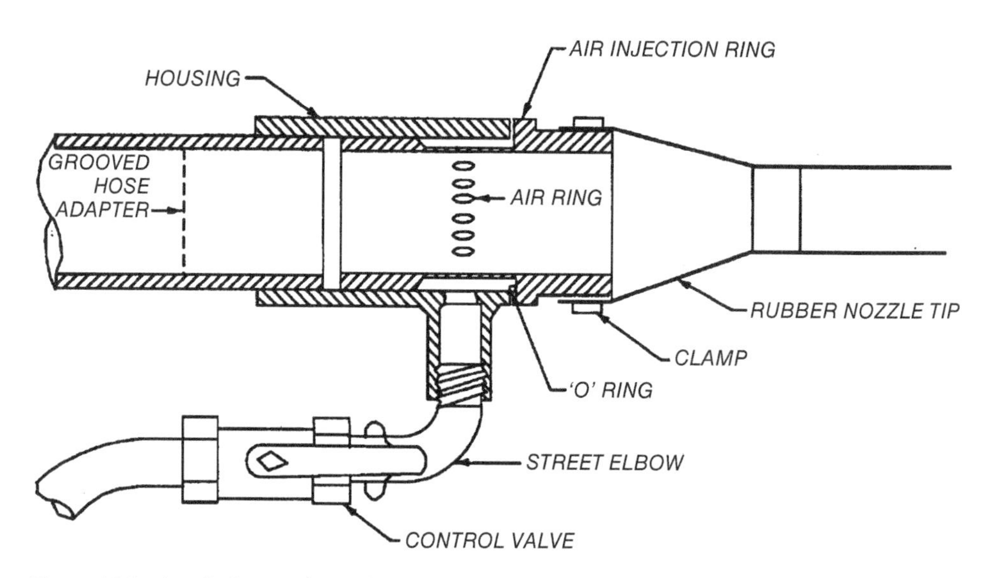


Figure 4.21 A typical wet-mix nozzle (US Army Corps of Engineers 1993).

providing simultaneously a considerable increase in the strength, stiffness, and ductility. One important advantage of the method is the significant decrease in the demand on existing lightly reinforced members of the building, due to the large dimensions and very large stiffness of the new members.

A typical cross-section of a new shear wall added to an existing building is very similar to shear walls of new buildings, with pseudo-columns with closely spaced stirrups at the two edges, and a lightly reinforced web that is expected to sustain damage in a strong seismic

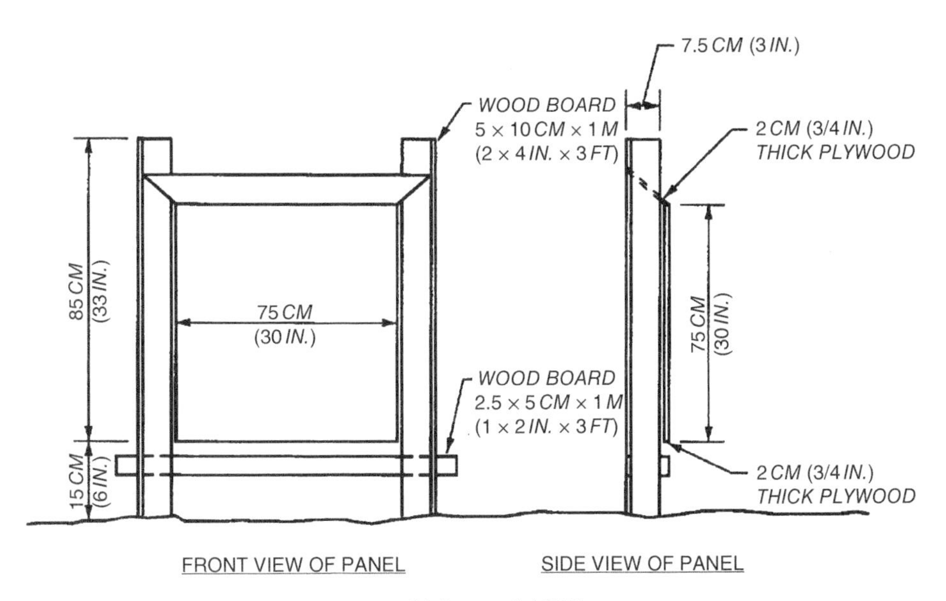


Figure 4.22 Test panel support system (Mahar et al. 1975).

Figure 4.23 Test panel on site. *Source:* Stelios Antoniou.

event. The only significant difference is the large number of dowels that are employed for the connection of the new and the existing members, and the safe transfer of the seismic inertia forces from the existing building to the ground, through the new walls.

It is generally preferable that the wall encapsulates two columns of the existing building, to form strong and ductile jackets, which constitute the pseudo-columns at the edges of the new wall. The jackets can be constructed by cast-in-place concrete together with the wall web (employing formworks around the existing members, columns and beams) or separately using shotcrete. Ideally the new wall is sufficiently thick, so as to also encapsulate the existing beam that connects the two columns. Consequently, wall thicknesses of 35–40 cm are not uncommon (Figure 4.24). In an alternative configuration, a rectangular jacketed column can be extended in one or two sides providing wing walls at the adjacent bays.

The lapping of the longitudinal rebar from one story to the adjacent, and the placement of the stirrups is done in a similar fashion to the jackets' reinforcement; that is by creating holes in the slabs and the beams. Larger holes are also needed for the casting of the concrete. In the typical case, due to the large width of the walls, more than one hole is needed.

Because of the very large stiffness of these new components with respect to the existing vertical members, particular attention should be paid so that the new walls do not unintentionally cause large stiffness irregularities. Hence, it is important that the new walls are added in a symmetric fashion on plan view, in order not to introduce significant torsional

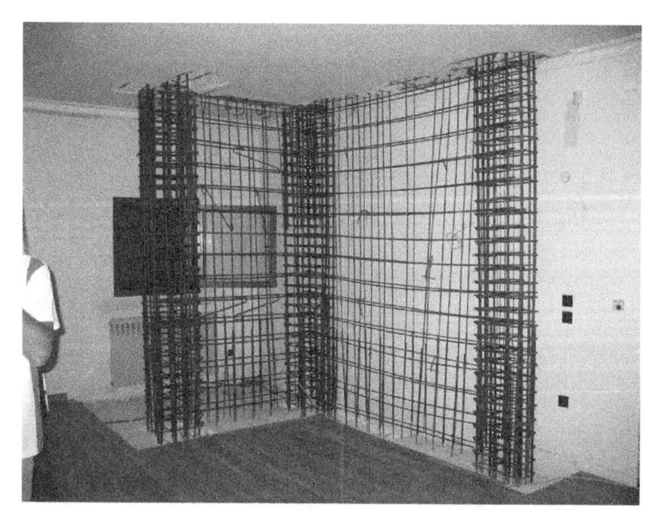

Figure 4.24 New shear walls in existing RC buildings. *Source:* Stelios Antoniou.

Figure 4.24 (Continued)

effects in the seismic behavior. This means that at least two walls should be placed in each horizontal direction symmetrically with respect to the center of mass of the floor (Figure 4.25). Failing to add walls in one of the two directions for architectural or economic reasons means that the building is strengthened only in one direction, and is left unstrengthened in the perpendicular one (Figure 4.26). Note, however, that the construction of walls on the three sides is also acceptable, since it provides adequate translational and rotation stiffness to the building.

Likewise, if the walls stop abruptly at some level along the height of the building (e.g., if they are constructed only at the ground level), this will lead to significant irregularities in elevation and strong, undesirable higher-mode effects, increasing the vulnerability of the floor right above the end of the walls. Therefore, in most of the cases the walls should be extended to the entire height of the building, although a gradual decrease in their width is acceptable (if not desirable).

There are three ways of strengthening a building with RC shear walls:

- The walls are added inside the building bays, usually surrounding the existing columns to form their pseudo-columns (Figure 4.27a).
- The walls are constructed right outside the perimeter of the building. A large number of strong dowels or connectors are required so that the new walls are adequately connected to the existing building. The main advantage of this variation is that the disturbance to the occupants is minimal (Figure 4.27b).
- The walls are added in the form of buttresses at the extremities of the external frames (Figure 4.27c).

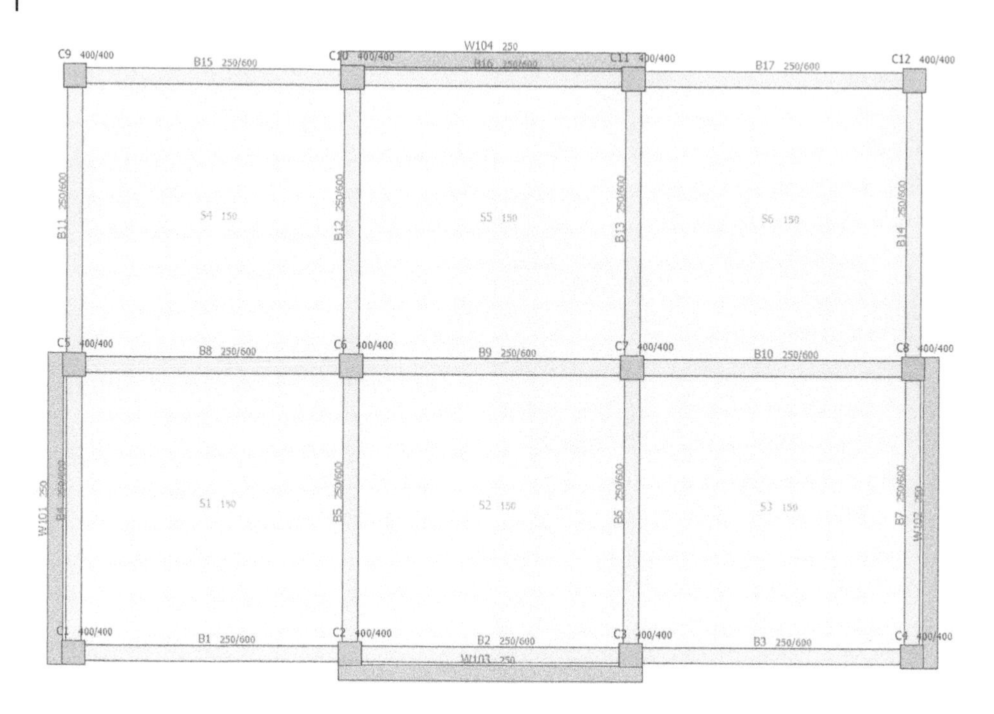


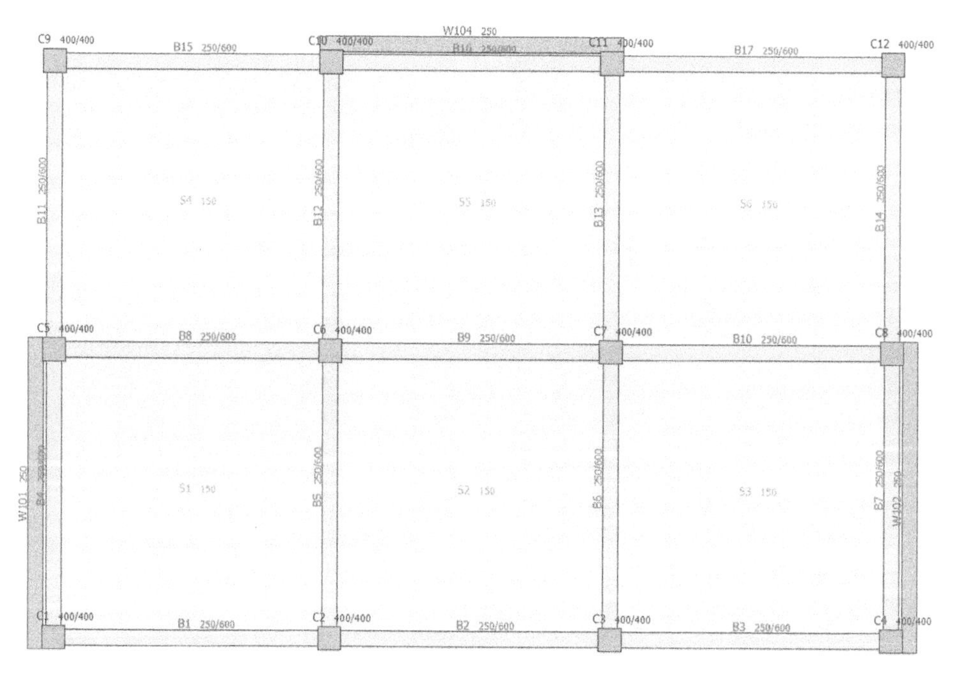


**Figure 4.25** Acceptable placement of new shear walls on plan view. *Source:* Stelios Antoniou.

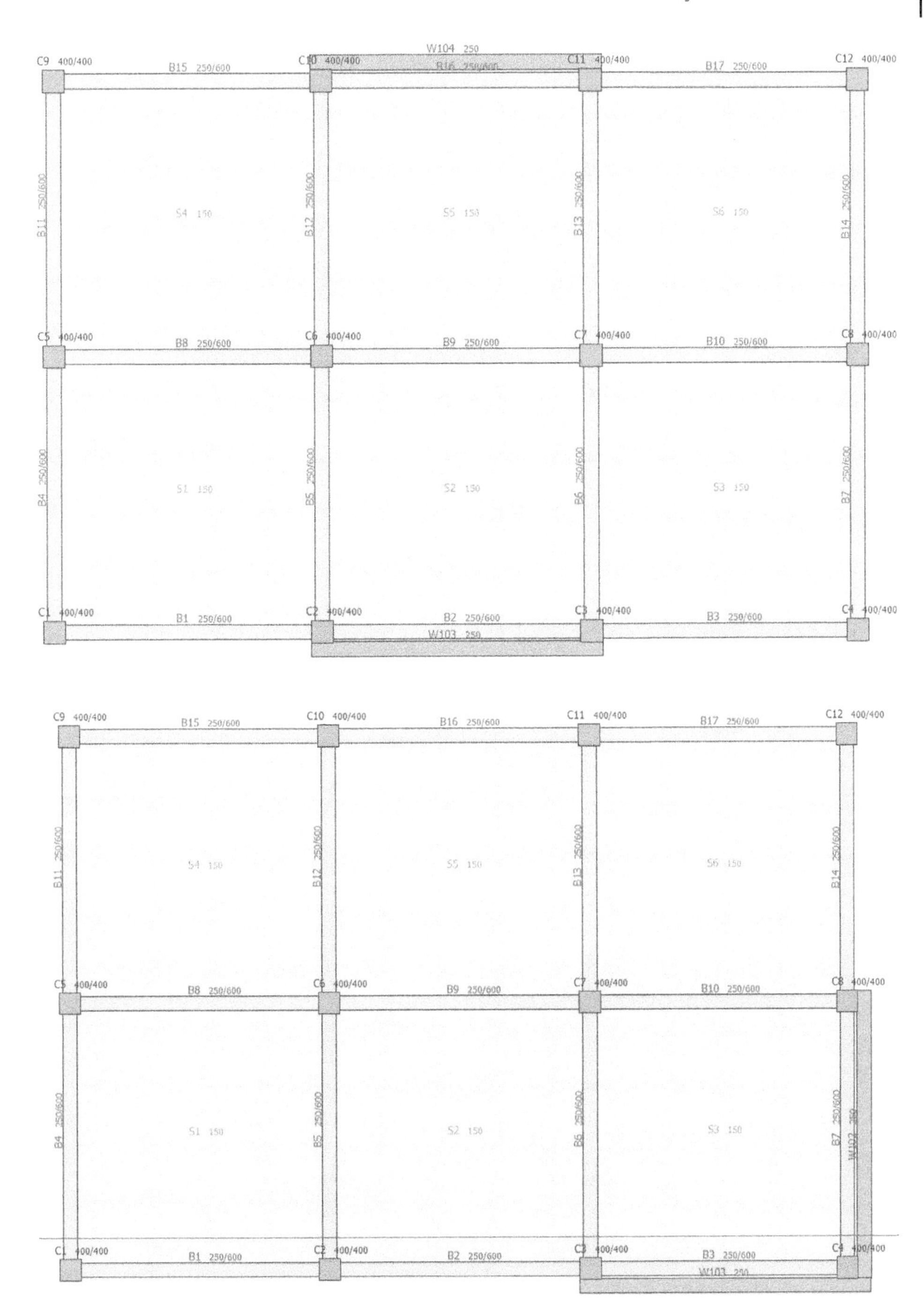


**Figure 4.26** Unacceptable placement of new shear walls on plan view. *Source:* Stelios Antoniou.

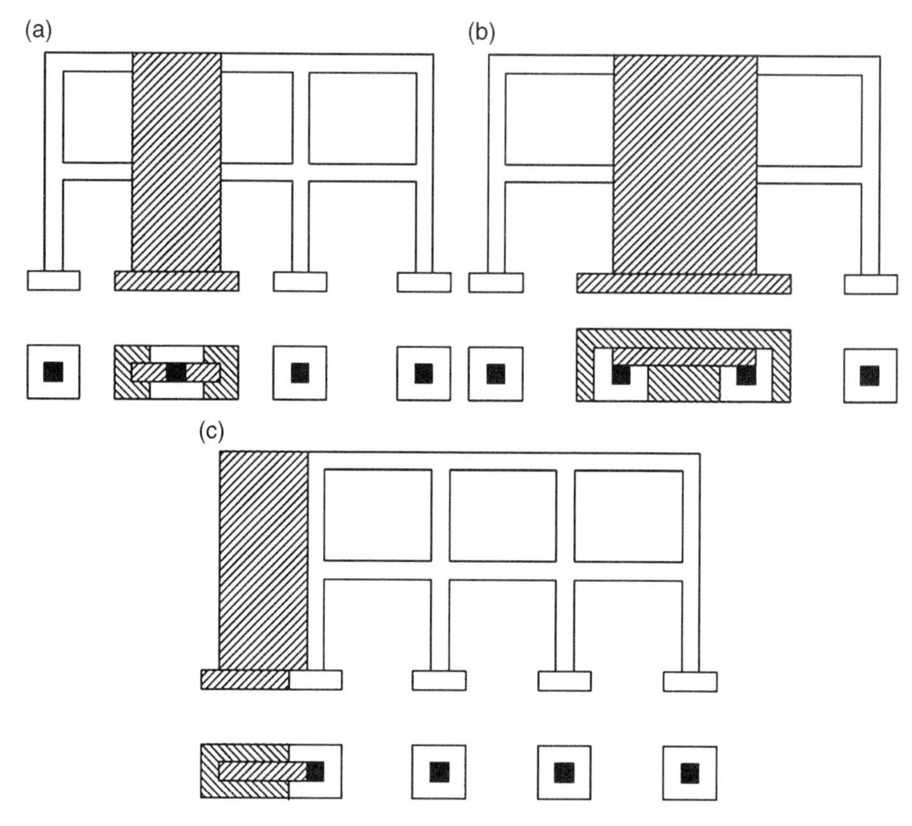


Figure 4.27 View and cross-section above the foundation of RC frames strengthened with new RC walls placed (a) internally, (b) externally, or (c) as buttress (Tsionis et al. 2014).

The first option entails higher disturbance to the occupants and increased cost due to intensive secondary interventions, i.e. repairs damage. The two others minimize disturbance, but in turn require more space outside the building, which might not be available.

Although external buttresses have all the advantages of walls constructed close to the building and they minimize the disruption, this option is rarely employed. Apart from the obvious space requirements, which are fulfilled only by a small proportion of the building stock, they have questionable aesthetic quality and pose significant architectural problems and restrictions, creating large shaded areas that reduce internal lighting. Furthermore, the connection of the buttress to the building is an intricate issue. Although such buttresses are connected to the existing building at all levels, the connection areas are subjected to very large shear force levels, even under compression, and often the available space for the placement of dowels is not enough. Buttress stability is another major issue, since buttresses are not loaded vertically but for their own weight, which increases the possibility of uplifting of the foundation and reduces their effectiveness to resist lateral loads.

The most common problem with this method is that very often the existing openings do not allow for the symmetric placing of the walls in the perimeter of the building. In the general case, there are not many walls without openings in the entire building height, or walls where the openings can be canceled without affecting the building from an architectural and operational perspective. For instance, it is not possible to add a wall in front of the entrance of a building, or a wall in a location that closes the door to a room or a balcony. Furthermore, due to the very large size of new walls with respect to the existing structural elements, this type of intervention is particularly invasive; hence, a certain sensitivity to architectural aspects is required.

Usually, in practical applications a combination of internal and external (in the perimeter) walls is employed. Where the geometry of the building allows it, the new walls are added in the perimeter. If instead this is not possible (e.g., there is not enough space, the existing openings are preventing it, or there are party walls with the adjoining properties), internal walls are chosen instead.

With the introduction of the new walls and because of their very large stiffness in the direction of their largest dimension (recall that the elastic moment of inertia is $bh^3/12$), the new walls attract a large proportion of the seismic inertia forces, which can be as much as 70% or 80% of the total, if the walls are adequately large and symmetrically placed in the perimeter of the building. This means that the seismic demand on the existing vertical members drops significantly at a similar proportion, which enables them to withstand the applied forces even with the reduced existing reinforcement. Simultaneously, there is also a considerable reduction in the fundamental period, which can be up to 50%, resulting in a completely different dynamic behavior. What is more, the new walls typically change the structural failure mechanism to a beam-sway one.

Consequently, new walls are often introduced without the strengthening of the other existing columns. This is of particular importance in buildings that are in operation for the duration of the works, because it considerably reduces the locations of intervention and the corresponding costs for non-structural damage and repair (e.g., tiles, floors, windows).

On the negative side however, these increased forces applied on the new walls leads to extreme requirements at the foundation level that often cannot be fulfilled with normal footing dimensions. Finally, increased chord rotation and shear demands are imposed on the beams that are directly connected to the wall and lie within its plane. The beams are usually unable to resist such demands, if they are not retrofitted, too.

### 4.5.2 Foundation Systems of New Shear Walls

Since so much of the base shear is undertaken by just a few vertical members of the building, it is important to consider how to transfer these forces and the corresponding overturning moments to the ground when adding large shear walls. This generally requires significant interventions on the foundation level for each wall.

A small footing of these large walls will result in rocking rotations and uplifting, overturning moments below the value calculated assuming fixed support conditions, and finally considerable increase in the lateral floor displacements and reduced effectiveness of the proposed strengthening scheme. Therefore, it is important to reduce the rocking and uplift of the wall footing, which can be achieved by the following means:

- By increasing the vertical load acting on the footing. This can be done by increasing the size and the weight of the footing, and by enclosing the footings of the adjacent columns, in order to activate a larger percentage of the gravity loads of the building.

- By connecting the new footing to the neighboring footing with large, strong, and stiff connecting beams.
- In cases of very large overturning moments, it is beneficial to use micro-piles, instead of shallow foundation elements.

All of these methods are costly and more importantly very disruptive to the building's operation, since they require serious interventions on the ground level and the uncovering of the existing foundation system in a large area. To some extent, these disruptions can be moderated with the introduction of eccentric footings, for instance in the walls at the perimeter. Nevertheless, the required size of the foundation is such that it should always be considered, when choosing the method of adding new shear walls in existing buildings.

### 4.5.3 Advantages and Disadvantages

New shear walls improve the global response of the building in terms of strength, stiffness, deformation capacity, and ductility. They significantly reduce the story drifts, they may prevent story mechanisms and, depending on the layout of the existing building and the location of the new components, they may also reduce irregularity, both height-wise and in plan. Because of their very large stiffness with respect to the existing vertical members, they can increase the lateral building strength as high as 200–300% of its initial strength. Consequently, the lateral seismic loads are mainly resisted by these new components, which are appropriately designed (reinforcement and detailing) to withstand them.

This in turn results in a very large reduction of the seismic force and drift demands to the other building members that can be as high as 70–80%. These members are expected to play a secondary role, mainly carrying gravity loads, and the need for strengthening is eliminated in all or most of them. This is of particular importance in buildings that are in operation during the duration of the works, since it considerably reduces the locations of intervention, the disturbance to the residents, and the corresponding costs for nonstructural damage and repair (e.g., tiles, floors, openings). In addition, the nonstructural building components are expected to sustain reduced damage during a large seismic event, because of the reduced deformations.

On the negative side, very often there are limited locations, where the walls can be added. Usually, the walls should be constructed throughout the entire height of the building, so as to not introduce significant irregularities in elevation, and should be placed in a symmetric fashion on plan view, otherwise significant torsional effects are induced. As a result, it is often difficult or impossible to find appropriate locations for their placement, especially in buildings that do not have a typical floor plan at the different levels.

Another problem of new shear walls is that they require the construction of very large (often gigantic) footings at the ground level, in order to avoid rocking, which is undesirable. This, apart from the obvious increased cost, leads to very disruptive works at the ground level.

### 4.5.4 Design Issues: Modeling and Analysis

New shear walls are typically designed and detailed as in new structures, taking into account all the corresponding capacity design principles. At the two edges of the cross

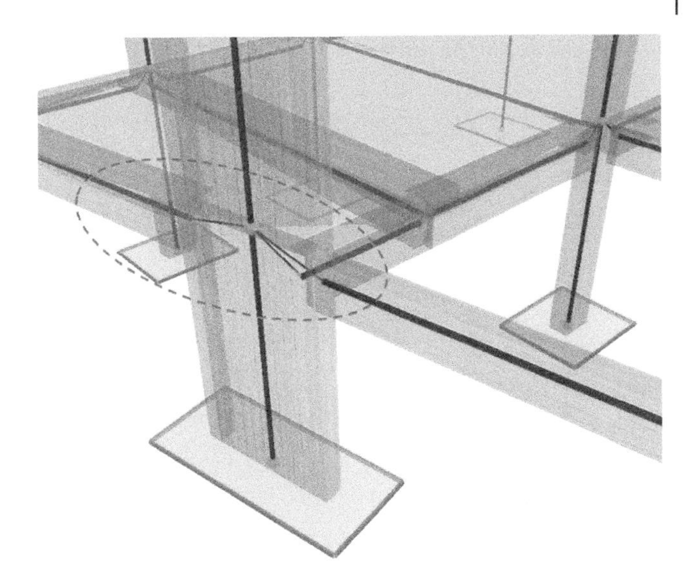

Figure 4.28 Modeling of large shear walls as one-dimensional elements with offsets. *Source:* Stelios Antoniou.

section, there are well-confined pseudo-columns with closely spaced stirrups, whereas the web remains lightly reinforced and is expected to sustain damage and to absorb energy.

The walls are over-reinforced against shear (i.e., taking into account the increased demand), in order to make sure that the wall first yields in flexure. Furthermore, plastic hinges and yielding should only be allowed at the base with the over-design (increased demand) in flexure above the plastic hinge. Plastic hinges at higher levels should be avoided to preclude the generation of unwanted higher-mode effects that could be caused by inelastic response. Steel connectors and dowels should also be calculated using capacity design provisions, thus preventing any failure of the connectors that could again cause higher-mode effects.

In linear analysis the new walls can be modeled as rectangular or triangular shell elements. On the contrary in nonlinear static (pushover) or dynamic analysis, the walls are typically modeled as one-dimensional nonlinear elements. The offset between the axis of the walls and the beam edges, which can be as high as 2.00 m or more is typically covered with rigid links (Figure 4.28).

At first the new walls do not carry any gravity loads (other than their self-weight), since they are constructed after the building has undertaken all the dead and live load from the slabs through the existing vertical members. However, for reasons of simplicity, and because the larger proportion of the demand imposed on the structural components comes from the seismic and not the gravity loads, it is assumed that the walls are constructed together with the rest of the building and participate in undertaking the initial, gravity loads.

## 4.6 RC Infilling

### 4.6.1 Application

An alternative to the construction of new shear walls that encompass the columns and beams of the existing frame is RC infilling. With this method, a reinforced concrete wall is constructed inside the RC panel and it is connected with the adjacent columns (to the left

Figure 4.29 RC infilling (Chrysostomou et al. 2014; Poljanšek et al. 2014).

and right) and beams (up and bottom) with a series of strong dowels that are designed to undertake the inertia forces developed during the earthquake and ensure a monolithic connection with the existing frame (Figure 4.29).

The reinforced concrete infills may be constructed either with cast-in-situ concrete or shotcrete. In the former case, the concrete is cast through relatively large holes opened in the beam or from one side of the panel, at the side of the beam leaving a triangular or trapezoidal-shaped concrete prism outside the plane of the infill that is demolished after the concrete has strengthened, typically one or two days after casting.

As in the case of new RC walls, special attention should be paid to the way the large overturning moments are transferred to the ground, since they are disproportionally large with respect to the vertical load of the infill. If an adequate foundation system is not ensured, this will result in rocking rotations and uplifting and in overturning moments significant well below the value calculated assuming fixed support conditions. Hence, similar measures to the case of RC walls should be considered in RC infilling as well.

As an alternative to the method, instead of demolishing existing infilled brick walls and constructing the new RC infill, the existing infill can be strengthened on both sides with shotcrete reinforced with a dense grid of rebar anchored at the surrounding columns and beams. In order to ensure adequate connection of the shotcrete-infill-shotcrete compound, a series of dowels are placed connecting the reinforcement of the two layers of shotcrete with each other.

### 4.6.2 Advantages and Disadvantages

With respect to the construction of new shear walls that encompass the existing frame members, RC infilling is significantly cheaper and much less disruptive. It only requires interventions inside isolated panels of the building frame, and it does not affect the building architecturally, since the width of the RC infill is typically similar or the same as the width of a masonry infill that lies in the same position. Furthermore, the connection of two RC infills in adjacent floors is accomplished through the existing RC frame with the placement of a series of dowels, and it is not done with lapped rebar that pass through the building slabs and beams and cause significant disruption and increased costs from the restoration of the nonstructural components (floors, tiles, plaster).

The main drawback of the method is related to the capacity of the existing members that surround the infill. In order to be able to transfer the shear forces and overturning moments from floor to floor, they need to possess a minimum capacity; otherwise, they will suffer significant local damage in the case of strong seismic events. However, the vast majority of existing buildings (at least those that require strengthening) have lightly reinforced members of relatively small dimensions. In such cases, together with the construction of the RC infill, strengthening interventions are required for the adjacent members, typically FRP wrapping, FRP laminates or steel plates. These interventions undermine the main advantage of RC infilling, which is the limited and localized damage, and make it less appealing with respect to the construction of fully reinforced shear walls that are generally much stronger.

## 4.7 Steel Bracing

### 4.7.1 Application

Steel bracing offers similar advantages to new shear walls, increasing the strength, the stiffness, and the ductility of the building. The braces are directly fitted to the concrete frame, inside the existing bays. They contribute to the lateral resistance of the structure through the axial force developing in their inclined members. The diagonals are pinned or fixed to steel plates that are anchored at the corners of each concrete bay with epoxy resins. Similarly to the case of new RC walls, the braces should be placed in symmetrical positions so as to not introduce unwanted torsion in the building, and, if possible, to reduce in-plane irregularities.

Although the construction of steel braces considerably increases the lateral capacity of the building, it only increases its stiffness moderately. Consequently, it is not as effective as other methods in stiff concrete structures, such as wall or dual systems or masonry infilled frames. But in the cases where steel bracing can be effective (relatively slender buildings), the dynamic deformations can be significantly reduced, provided that there are no early, brittle failures of the braces under buckling, of their connections to the RC frame, or of the adjacent concrete members in shear.

Since in the general case the braces are attached to existing unstrengthened concrete members of the building, the method is not suitable when the beams and columns do not possess a minimum strength; if this is not the case, the concrete elements can be strengthened with composite materials, or more often with jacketing.

Energy dissipation devices can be easily combined with steel braces, efficiently increasing the damping during the dynamic excitation. It is noted, however, that if dampers are employed, the steel braces should be designed so that they do not significantly increase the stiffness and thus compromise the efficiency of the damping mechanisms, which require large-deformations to be cost-efficient.

The construction of steel braces is not as disruptive as the construction of new RC walls, since it is a cleaner method that does not require the casting of concrete or shotcrete inside the building. Hence, it is a method better suited to buildings that do not undergo complete renovation and are in operation during the works. Furthermore, at the foundation level generally small interventions are required, since the braces do not transmit large overturning moments at the base. Typically, small footings are constructed to transfer the brace axial force to the existing footing or to the ground.

Generally, there are two types of steel bracing systems used for upgrading existing concrete frames:

1) Concentric braces, where the horizontal forces are mainly resisted by members subjected to axial loads. They are divided in three categories: diagonal-braces, X-braces, and V-braces (Figure 4.30a)
2) Eccentric braces, where the horizontal forces are again resisted by axially loaded members, but the eccentricity of the layout is such that energy can be dissipated in horizontal seismic links by means of either cyclic bending or cyclic shear (Figure 4.30b).

Concentric braces are the most widely used in practical applications of strengthening (Figure 4.31). Although eccentric bracing systems have some advantages compared to the

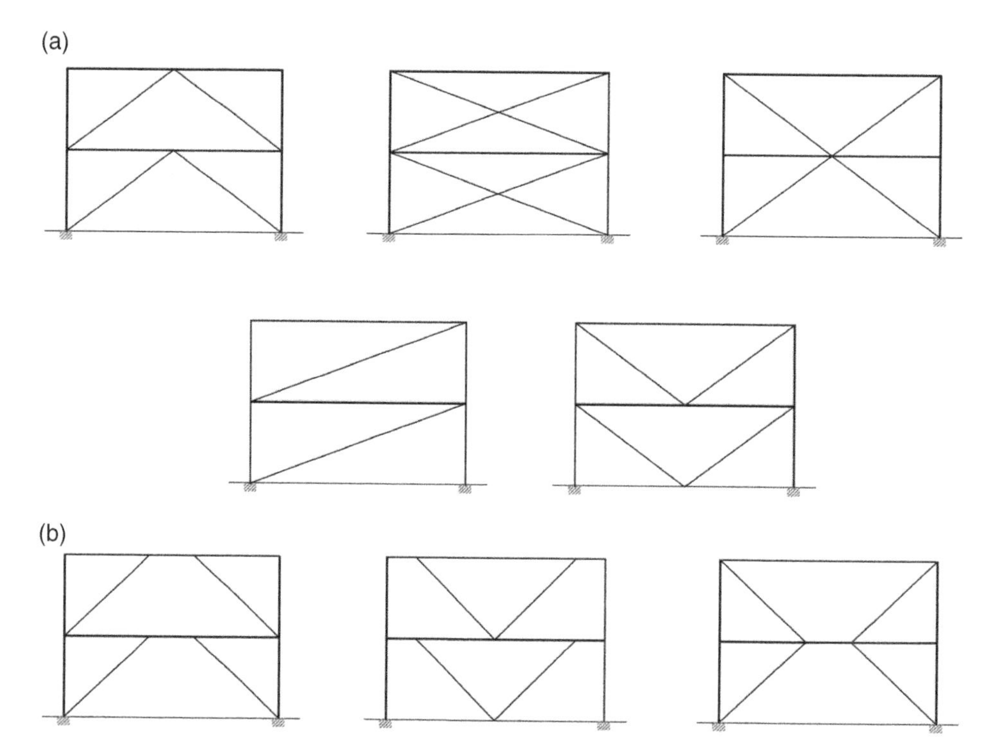


**Figure 4.30** Typical (a) Concentric and (b) Eccentric Steel Braces in retrofit. *Source:* Stelios Antoniou.

**Figure 4.31** Concentric steel braces in retrofit.

concentric ones – for instance, permitting larger energy dissipation, they are not commonly used for retrofit, because they usually require interventions to the building's concrete members. Finally, buckling-restrained bracing, in which global buckling is inhibited with the use of appropriate systems, can also be introduced.

*Concentric bracing systems* contribute to the lateral-load resistance of the structure through the horizontal projection of the axial force (mainly axial tension) developing in their inclined members. Appropriate concentric bracing systems are those with: (i) diagonal bracings, in which there is a single diagonal per braced bay of the frame, (ii) X (or cross-diagonal) bracings, with braces along both diagonals of a braced bay, which are the most common systems, and (iii) V or inverted V bracings (termed chevron bracings in the USA), in which a pair of inclined braces is connected to a point near or at the mid-span of a horizontal member (a beam, or less often a slab) of a bay of the frame.

In V-braces, special care should be taken so that the beam where the two braces connect possesses adequate strength, so that to resist the transverse (shear) load demand that develops to balance the vertical components of the forces of the two diagonals under tension and compression. Since this is rarely the case for the lightly reinforced beams of older construction, these need to be strengthened, usually through steel plates.

Sometimes the braces can be extended in more than one bays or floors of the building frame (Figure 4.32). Two V-braces, one inverted in one bay and one normal in the bay on top, can be combined to form a large X-brace between two stories. Because the braces in the two stories are effectively continuous, the shear forces from the diagonals on the intermediate beam cancel each other, and the shear demand on the beam is limited.

K bracings, in which the inclined braces are connected to a point of a column, should be avoided, in order to prevent a brittle failure in shear of the column.

*Eccentric braces* are generally an efficient technique for enhancing the seismic resistance of buildings because they provide ductility as well as strength and stiffness. Large, inelastic deformations and energy dissipation are expected to take place in a ductile horizontal (or – less often – vertical) member (link), where the forces of the brace members are transferred. The link should be designed to yield and dissipate energy, while preventing buckling of the

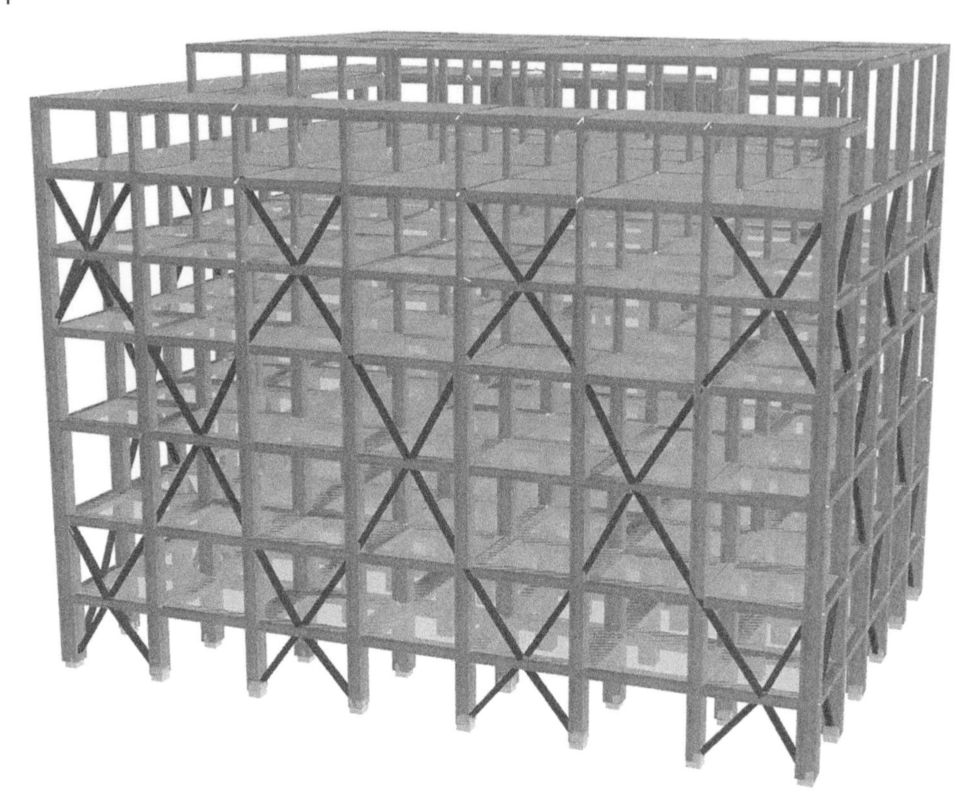

Figure 4.32 Two V-braces combined to form a larger X-brace in two stories. *Source:* Stelios Antoniou.

brace members. Different patterns are used: K, Y, and inverted Y bracing. Although good in theory, in the retrofit of existing buildings, the link is a part of a lightly reinforced concrete beam of the existing frame that cannot undertake significant nonlinear deformations. Its strengthening compromises the main advantages of the method (effectiveness, low disturbance, and works in just one bay), since, apart from the strengthening of the beam itself, significant interventions and nonstructural damage are required on the floor right above the beam.

A similar technique to the construction of steel braces inside the bays of the frame is *external trusses* (also called exoskeletons) consisting of strong steel or composite sections with diagonal members that are constructed at the perimeter of the building (Figure 4.33). The trusses are connected to the existing framing system with large steel connectors, through which they are able to undertake and transfer the seismic loading that acts on the frame to the ground. Because of the large shear demand imposed on the existing concrete beams from the connectors, these beams usually need strengthening, which is typically carried out with steel plates or shotcrete jackets. It is noted, that similarly to the RC shear walls and contrary to the internal braces, relatively large footings should be constructed to transfer the shear forces and the corresponding overturning moments of the external trusses to the ground.

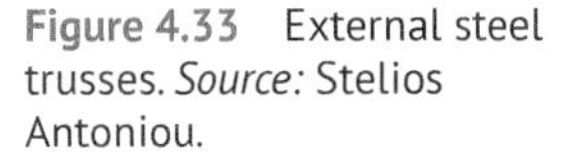
Figure 4.33 External steel trusses. *Source:* Stelios Antoniou.

Finally, there is one last bracing method that is not so common in practical applications yet, but is expected to gain significance in the future: *buckling-restrained braces*. In this method a steel core element (cross-shape or flat bar) is encased into a steel tube (called also the buckling-restraining element) and is confined by an unbonding material like concrete mortar, rubber, silicon, vinyl, etc. that prevents local buckling. In this way, the core element can resist the axial tension or compression force without the risk of local or global flexural buckling.

### 4.7.2 Advantages and Disadvantages

Steel bracing, by providing lateral strength and stiffness, significantly decreases the seismic deformations and the demand on existing members, similarly to new RC shear walls (though usually to a lesser extent with respect to the latter). What is more, the level of strength and stiffness increase can be tuned relatively easily by the choice of the number and the size of the braces; hence, it is relatively easy to avoid large irregularities in elevation, when the retrofit scheme is not applied to the entire building height. For instance, bracing is an ideal solution for the strengthening of soft ground stories, where the steel braces are applied only at the lower building level. When the bracing system is adequately designed and an early brittle failure of the braces and their connections is excluded, a satisfactory amount of ductility, hysteretic behavior, and energy dissipation can be obtained.

The added weight of the steel braces to the building is minimal, there is a minimum loss of living space by their installation, and braces possess the ability to accommodate for existing openings, as well as to achieve better architectural integration with the transparent parts of the building. Furthermore, for larger buildings, it can be easily combined with dampers, in order to increase damping during the dynamic excitation and reduce the seismic deformations and demand.

Finally, it is a rather "clean" construction method that only causes minimal disruption to the function of the buildings and its occupants, since the braces are directly fitted to the concrete frame without the introduction of concrete and the creation of dust in the building, whereas relatively small interventions to the foundation system are required.

The main disadvantage of the technique is that it is usually difficult to find suitable locations and bays of the building for the placement of the braces, considering that their distribution on plan view should be as symmetrical and uniform as possible. Furthermore, steel braces are not efficient for stiff concrete structures with large shear walls (even if these are lightly reinforced).

Moreover, it is not always easy to achieve high-quality full-penetration welds on the construction site, and the control of the interaction between new steel and the existing concrete system is also difficult. Similarly, good workmanship is also needed for the detailing of the braces to prevent local buckling and post-buckling fracture.

### 4.7.3 Design Issues: Modeling, Analysis, and Checks

Regarding the seismic analysis and the corresponding checks in the bracing system, the strength of the braces and the strength and deformations of the adjacent members and beam-column joints are evaluated using the existing design guidelines. In linear analysis, a force-based approach is employed and the checks are mainly carried out in terms of forces. In nonlinear analysis, deformation-based criteria are employed for the braces, intended to check the energy dissipation of the system. In general, it is assumed that the bracing system has a dominant role in undertaking the lateral seismic loads; hence the designation of the braces as primary members is essential.

In the braces under compression, local buckling is the most important factor limiting the ductility and the energy dissipation of the system, since it generally precipitates fracture. Regardless of whether the braces under compression are considered or neglected in the analysis, their design should consider measures against buckling, and their slenderness should be limited (e.g., Eurocode 8 requires sections of class 1 or 2 for behavior factors larger than 2.0).

## 4.8 Fiber-Reinforced Polymers (FRPs)

### 4.8.1 FRP Composite Materials

Fiber-reinforced polymer (FRP) composites comprise fibers of high-tensile strength within a polymer matrix such as epoxy, vinylester, or polyester thermosetting plastic, but most commonly epoxy resins. The polymer matrix, the original plastic that is usually stiff but relatively weak, is mixed with a reinforcing material of high-tensile capacity to yield a final product, which has the desired material or mechanical properties, i.e. large mechanical strength and elasticity. The fibers are usually made of carbon, glass, aramid, or rarely basalt, although other fibers such as paper or wood or asbestos have been used in the past.

FRP composites have evolved during the last two to three decades from being special materials used only in niche applications, to common engineering materials used in a diverse range of applications. FRP materials have a very high strength to weight ratio, and possess good fatigue, impact and compression properties. They also demonstrate impressive electrical properties and a high-grade environmental resistance and durability, along with good thermal insulation, structural integrity, UV radiation stability, and resistance to chemicals and corrosives. A key factor driving the increased number of applications of composites over the recent years is the significant drop in their price, as well as the development of new advanced forms of FRP materials, which include high-performance resin systems and new styles of reinforcement, such as carbon nanotubes and nanoparticles. The composite plastics can be tailored to suit a wide range of performance specifications, and can have a wide range of uses in many areas, including the aerospace, aviation, automotive, marine, and construction industries (Masuelli 2013).

### 4.8.2 FRP Composites in Civil Engineering and Retrofit

FRP materials in structural engineering are treated as additional reinforcement, the only difference being the initial strains that are present in the concrete and reinforcement, due to the dead load at the time of applying the FRP.

Due to their high tensile strength and low weight (compared to the conventional materials, and in particular steel), FRPs have become important structural materials for use in the construction industry as internal or more frequently external reinforcement. Other significant advantages of FRPs over steel are the ease of handling and application, the lack of requirement for heavy lifting, the minimal labor required for their installation, as well as their high resistance against corrosion, and their low thermal conductivity. Moreover, due to their exceptional formability, FRP systems provide flexibility to the practitioners and can be applied on any flat, curved, or geometrically irregular surface.

The use of FRP materials in civil engineering has increased steadily after their first appearance, four decades ago. Although FRP systems have significant potential for various civil engineering applications even in new construction, they are mostly employed in the retrofit and rehabilitation of existing reinforced concrete structures of various types, such as buildings, bridges, marine structures, and tunnels. The role of FRPs in strengthening is growing at an extremely rapid pace, owing mainly to the ease and speed of construction, and their application without significant disturbance to the functionality of the building. As a result, externally bonded FRP reinforcement has become one of the most important and most frequently used methods for enhancing the strength, energy dissipation, and stiffness characteristics of poorly detailed members.

The initial developments of the FRP-strengthening techniques took place in Germany (German General Guideline 1998) and Switzerland, where mainly the flexural strengthening of reinforced concrete members with externally epoxy-bonded FRP laminates has been thoroughly studied. Gradually, it was recognized worldwide that FRP materials can be used in a series of ways for upgrading existing members. In particular, the invention of FRP fabrics led to a series of applications that provide flexural, axial, and shear strength enhancement, especially under seismic loads, but also increased deformation capacity, confinement, and ductility. Gradually wrapping with FRP sheets has become one of the preferred methods in seismic retrofitting projects.

FRP materials in retrofit (fabrics, laminates, and less often strings and bars) are mainly used as a replacement for steel reinforcement, both longitudinal and transverse, in lightly reinforced members (which are very common in older construction), so as to increase their flexural and / or shear capacity.

The fibers are the main load-carrying element of the composite. The combination with the epoxy matrix results in a high-strength material with linear elastic behavior until failure without a yielding plateau. The complete description of the load-bearing curve of the composite is provided by the modulus of elasticity and the tensile strength of the composite material (or alternatively the maximum tensile strain). The material of the fibers usually has a tensile strength much higher than steel and is employed to mainly undertake tension, while the resins are employed to transfer and distribute these tensile stresses from the fibers to the existing member. The interface between the FRP composite and the existing member significantly affects the performance of the method, and the loss of bond between them is one of the most common failure modes. It should be noted that the strengthening of structural members with FRPs, while considerably increasing their strength, does not change their stiffness, and has no effect on the stiffness distribution of the entire structure.

Usually, the FRP materials are sold as complete systems, i.e. specific fiber types are provided with specific epoxy resins. The contribution of the FRP material to the members' capacity is taken into account, considering the properties of either the composite fiber + matrix material (called *cured laminate properties*) or the properties of the fibers (called the *dry fiber properties*). There is a direct relationship between the two sets of properties, and these can be used interchangeably through the following relationship, which is considered to be the basic "rule of mixture" in FRP materials:

$$P_{lam} = P_{fib} * V_{fib} + P_{res} * V_{res} \approx P_{fib} * V_{fib}$$

where $P_{lam}$ is the property (modulus of elasticity $E_{lam}$ or tensile strength $f_{lam,\,t}$) of the composite fiber+matrix material, $P_{fib}$ and $P_{res}$ are the corresponding properties of the fibers and the resin respectively, $V_{fib}$ is the fraction by volume of the fibers in the laminate material (which is usually of the order of 40–70%, depending on the selected system), and $V_{res} = 1 - V_{fib}$ is the fraction of the resin. Because the values of the strength and the modulus of elasticity of the resins is one or two orders of magnitude smaller than those of the fibers, the laminate properties are approximately equal to those of the fiber multiplied by the percentage of the fibers in the laminate material with a very good approximation. It is noted, however, that, when this is possible, the cured laminate properties should be established by using experimental tests, since this is considered to be the most appropriate and accurate method.

Fiber-reinforced polymers can be used in the strengthening of existing buildings mainly with two techniques, FRP fabrics and FRP laminates, however other alternatives also exist, such as FRP strings and FRP bars. The main uses of the FRP systems in structural applications are the following:

- Increase of the shear capacity of columns and beams, using FRP wraps with the fibers in the direction of the stirrups (i.e., perpendicular to the member axis).
- Enhance the confinement and ductile behavior of columns and beams by wrapping the members with fibers around the member perimeter. It is possible to both prevent rebar-buckling and improve ductile plastic-hinge behavior.
- Increase the flexural strength of columns, beams, or slabs with the use of FRP sheets or FRP laminates.
- Prevent lap-splice failure by clamping and increasing the lateral pressure using FRP jackets.
- Strengthen beam-column joints with FRP wraps, as an alternative to RC jacketing. The use of FRP wrapping can considerably increase the joint shear capacity, provided that the

sheets are adequately anchored. Yet anchoring can be very challenging and pose major practical difficulties, since significant nonstructural damage is required in each joint, both below and, more importantly, above the slab level. The strengthening of beam-column joints with FRP wraps is a subject that has not been investigated and documented adequately, thus the method should be used with caution.

The main drawbacks of FRP materials is that they generally exhibit brittle behavior with a linear elastic response in tension up to failure, and they have poor resistance to fire, as the resins quickly lose their strength when exposed to high temperatures. However, even with these disadvantages, there are many applications where the use of external FRP reinforcements is cost effective and justifiable.

### 4.8.3 FRP Composite Materials

The properties of the composite materials are mainly determined by the mechanical properties of the fibers. The FRP systems are divided into carbon (CFRP), glass (GFRP), aramid (AFRP), and basalt (BFRP) systems.

- *Carbon fiber reinforced polymers (*CFRP*).* CFRP have the best mechanical properties among other FRP composites, and have the more favorable price to properties ratio. The carbon fibers have high strength and higher modulus of elasticity with respect to the other fiber materials, which make it more appropriate for the shear strengthening of RC members. As a result, they are the most widely used FRP systems, and all the main FRP providers offer a large variety of carbon-based fabrics and laminates, in terms of size and weight.
- *Glass fiber reinforced polymers (*GFRP*).* Glass fibers have relatively lower cost with respect to the other types of FRPs, and they are the second (after carbon) most commonly used material in the construction industry. They come in three different types: (i) type E, which is the most common type. It has relatively low strength and modulus of elasticity, and its main drawbacks are that it has low humidity and alkaline resistance; (ii) type AR with increased alkaline resistance, but low strength and elasticity; and (iii) type S with high strength and elasticity modulus. Glass is more suitable for increasing the confinement of RC members, and it can also be used for flexural enhancement. Because of its low modulus, glass is seldom used for the shear capacity increase. The GFRP rebar are the most popular among other FRP rebar types, due to the combination of relatively low cost with environmental resistance.
- *Aramid fiber reinforced polymers (*AFRP*).* These fibers have high static and impact strengths, which is why they are often used for the wrapping of bridge piers, where there is high danger of car crashes. Nevertheless, their use is limited by reduced long-term strength (stress rupture) as well as high sensitivity to UV radiation. Another drawback of aramid fibers is that they are difficult to cut and process.
- *Basalt fiber reinforced polymers (*BFRP*).* Such fibers have excellent resistance to high temperatures. They possess high tensile strength as well as good durability. Other advantages are high resistance to acids, superior electro-magnetic properties, resistance to corrosion, resistance to radiation and UV light, and good resistance to vibration (Gudonis et al. 2013). All the same, basalt FRPs are seldom used in practical applications. In terms of mechanical properties and production complexity, basalt (BFRP) and aramid (AFRP) bars are somewhere in the middle.

**Table 4.2** Physical and mechanical properties of different FRP materials (Gudonis et al. 2013).

| Type of FRP | Density kg/m$^3$ | Tensile strength MPa | Deformation modulus GPa | Elongation % | Coefficient of thermal expansion 10−6 °C | Poisson ratio |
|---|---|---|---|---|---|---|
| Electrical-resistant *E*-glass | 2500 | 3450 | 72.4 | 2.4 | 5.0 | 0.22 |
| High-strength *S*-glass | 2500 | 4580 | 85.5 | 3.3 | 2.9 | 0.22 |
| Alkali-resistant *AR*-glass | 2270 | 1800–3500 | 70–76 | 2.0–3.0 | n/a | n/a |
| Carbon | 1700 | 3700 | 250 | 1.2 | −0.6 up to −0.2 | 0.20 |
| Carbon (high-modulus) | 1950 | 2500–4000 | 350–800 | 0.5 | −1.2 up to −0.1 | 0.20 |
| Carbon (high-strength) | 1750 | 4800 | 240 | 1.1 | −0.6 up to −0.2 | 0.20 |
| Aramid (Kevlar 29) | 1440 | 2760 | 62 | 4.4 | −2.0 longitudinal 59 radial | 0.35 |
| Aramid (Kevlar 49) | 1440 | 3620 | 124 | 2.2 | −2.0 longitudinal 59 radial | 0.35 |
| Aramid (Kevlar 149) | 1440 | 3450 | 175 | 1.4 | −2.0 longitudinal 59 radial | 0.35 |
| Aramid (Technora H) | 1390 | 3000 | 70 | 4.4 | −2.0 longitudinal 59 radial | 0.35 |
| Aramid (SVM) | 1430 | 3800–4200 | 130 | 3.5 | n/a | n/a |
| Bazalt (Albarrie) | 2800 | 4840 | 89 | 3.1 | 8.0 | n/a |

In Table 4.2, the most important mechanical and physical properties of different types of FRP are provided.

As mentioned above, epoxy, polyester, or vinylester resins may be employed for the polymer matrix, with epoxy resins being the most common material. In Table 4.3, the main physical and mechanical properties of these resins are given.

Fiber-reinforced polymers in civil engineering applications usually come as complete fibers + epoxy resin systems, as FRP fabrics, FRP laminates, and less often FRP strings and bars.

### 4.8.4 FRP Wrapping

In FRP wrapping, sheets of carbon, glass, aramid and basalt high strength fibers are bonded to the structural members using a two part, epoxy-based, impregnating resin. Due to the deformability of the compound fiber + resin material, it is able to conform to almost any complex or geometric shape (e.g. members' corners, beam-column joints), hence the method is versatile and can have applications in several cases. The FRP sheet usually comes in rolls (Figure 4.34a) from which the different pieces are cut to the appropriate size using fabric scissors or a sharp utility knife.

Table 4.3 Physical and mechanical properties of polyester, epoxy, and vinyl-ester resins (Gudonis et al. 2013).

| | Thermosetting resins | | |
|---|---|---|---|
| **Properties** | **Polyesters** | **Epoxy** | **Vinyl-ester** |
| Density, kg/m$^3$ | 1200–1400 | 1200–1400 | 1150–1350 |
| Tensile strength, MPa | 34.5–104 | 55–130 | 73–81 |
| Deformation modulus, GPa | 2.1–3.45 | 2.75–4.10 | 3.0–3.5 |
| Poisson ratio | 0.35–0.39 | 0.38–0.40 | 0.36–0.39 |
| Coefficient of thermal expansion, $10^{-6}$/°C | 55–100 | 45–65 | 50–75 |
| Saturation, % | 0.15–0.6 | 0.08–0.15 | 0.14–1.30 |

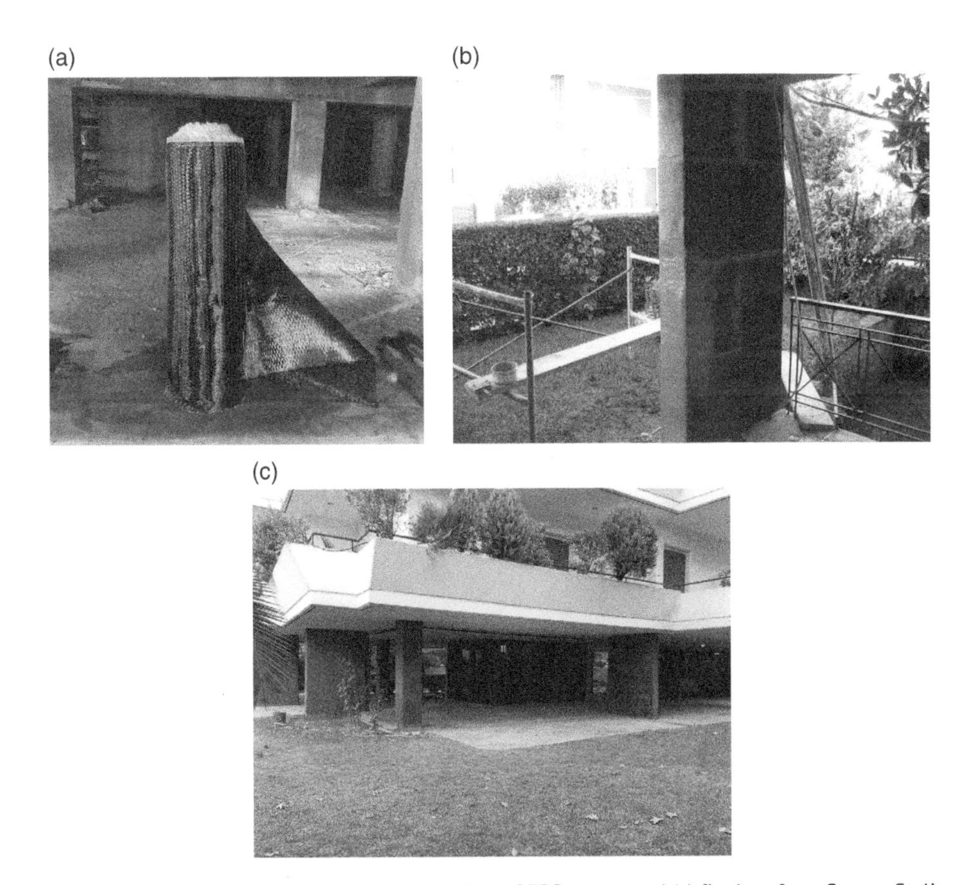


Figure 4.34 (a) FRP fabric rolls, (b) application of FRP wraps and (c) final surface. *Source:* Stelios Antoniou.

FRP wrapping is typically used in order to increase the confinement, the ductility and the shear capacity of walls, columns and beams with the fibers placed in the direction of the hoops. In certain cases, FRP wraps may be employed for the strengthening in bending of slabs, and less frequently of beams. Furthermore, they can provide significant protection against harsh environmental conditions and corrosion (e.g., in marine structures close or under the sea).

For the increase of the confinement and ductility, a complete wrapping of the member is required (Figure 4.34b,c), whereas in the strengthening in bending, the FRP wraps are usually applied on one side of the member as external longitudinal reinforcement. On the contrary, different wrapping schemes exist for shear strengthening. Whilst a complete wrapping of the structural element (column or beam) is preferred, since this is the most effective way, this may not always be possible, due to geometrical restrictions (e.g. presence of the slab). In these cases, either a U-shaped scheme (wrapping around the three sides), or very rarely side bonding (two separate FRP sheets on the two opposite faces of the beam) may be applied. The available wrapping methods for a beam are depicted in Figure 4.35. In the majority of cases, the debonding of the FRPs is the governing failure mechanism for U-shaped and side bonding, and an additional anchorage system should be provided. This can be achieved by means of FRP strings (Figure 4.36), steel or FRP anchor spikes, FRP

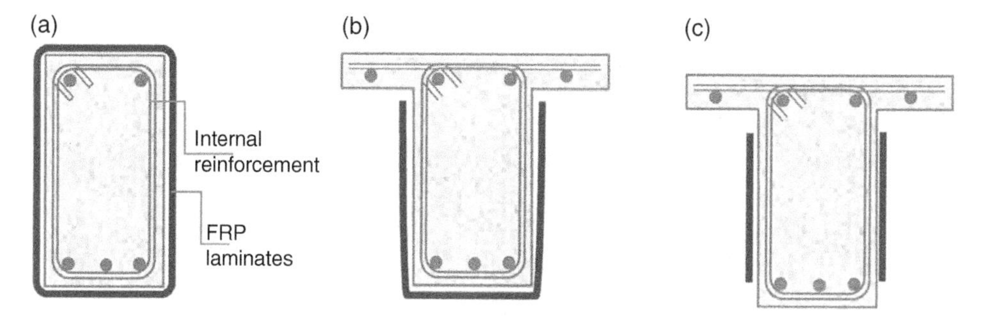


Figure 4.35 (a) Complete wrapping, (b) U-shaped wrapping and (c) side bonding (Belarbi and Acun 2013).

Figure 4.36 Anchorage of U-shaped FRP wraps using FRP strings. *Source:* Stelios Antoniou.

(a)

(b)

**Figure 4.37** Shear strengthening with FRP fabrics (a) continuous and (b) in strips. *Source:* Stelios Antoniou.

bars, or by using mechanical anchorage systems such as steel angles, steel, or FRP composite plates or anchor bolts. Finally, the application of the FRP sheets can be either continuous along the member length (i.e., covering the entire surface of the member), or in strips, perpendicular to the member axis or inclined (Figure 4.37).

The most usual FRP fabrics are unidirectional (fibers in one direction) , which is the direction perpendicular to the axis of the roll. When placed on the concrete member, the fibers are directed parallel to the direction of the reinforcement that they replace (i.e., along the member axis for flexural strengthening and perpendicular to the member axis for shear strengthening). Other fiber configurations also exist. There are unidirectional fabrics with the fibers at an angle of 45° with respect to the axis member for shear strengthening, and bidirectional fabrics at different angles, usually 0° and 90° or ±45° that can be used for the simultaneous strengthening of the member in flexure and shear (Figure 4.38).

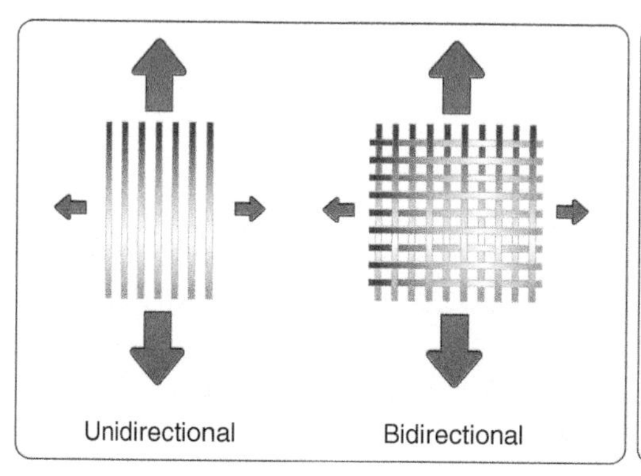


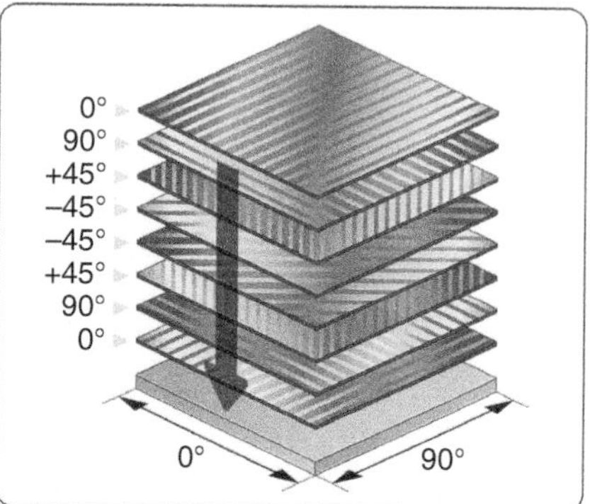


Figure 4.38 Unidirectional and bidirectional FRP fabrics, and fiber orientation types (Karataş and Gökkaya 2018).

The FRP wraps can be applied in one or more layers, but the effectiveness of each new layer decreases with the total number of layers. It is noted that all corners of the existing member should be rounded to a specific rounding radius (usually at least 30–40 mm) before the application of the FRP system, otherwise the fibers are fractured in the corners and the effectiveness of the FRP systems is significantly compromised.

A minimum concrete tensile strength should be sought, so as to achieve good bonding between the concrete surface and the epoxy resin. In general, lighter fabrics are used when the substrate strengths are low, such as stone or brick masonry buildings. For instance, bidirectional glass fabrics are very often used for increasing the shear strength of masonry walls.

Although the application of FRP wraps to concrete members is not a very difficult or daunting task, this should always be done by a skilled and experienced crew. The preparation of the concrete surface consists of the cleaning from oil, grease, dust, laitance or coatings, the removal of friable or loose particles and then grinding or sandblasting, in order to achieve a roughened but level surface with rounded edges. For some systems, the application of an epoxy primer should follow, while in others no primer is needed.

The fiber is then wrapped on the member with either the dry or the wet method. With the former the application of the resin is done on the substrate with a trowel or roller and then the dry fabric is applied to the coated substrate. With the latter, the sealing of the substrate with epoxy resin is followed by the impregnation of the fabric with the resin, manually on a table or with a saturator. The pre-wetted fabric is then applied on the sealed surface. In both methods, the fabric is carefully rolled with a plastic impregnating roller strictly in the fiber direction (Figure 4.39). One or more layers of epoxy resin and FRP wrap can be applied with the same way. In closed wrapping, a minimum overlap of the FRP sheet in the fiber direction should occur. This depends on the fabric type, but usually it is larger than 10 cm and approximately close to 20 cm. The final surface is sealed with a cementitious or epoxy coating, which is often the same resin as the one applied to the fabric, and some quartz sand is scattered on top in order to improve the adhesion of the overlay. Because of the poor resistance to fire of FRP systems, fire-resistant boards or mortars can be applied in the outer surface, in order to improve the resistance to high temperatures.

Figure 4.39 Application of FRP wraps with the dry method: (a) preparation of the substrate (b) application of the resin on the substrate, (c) FRP placement, and (d) final surface with quartz sand. *Source:* Stelios Antoniou.

It should be noted that each vendor follows its own guidelines, and variations exist between the different systems. For instance, in some systems the application of primer should precede the application of the first layer of epoxy resin, while in others no primer is required. Each company provides technical sheets with detailed and exact guidelines on the application of the FRP wraps and offers specific resin types for each fabric.

### 4.8.5 FRP Laminates

These are strips of high-strength pre-manufactured carbon + epoxy laminates, which are used as externally bonded reinforcement. For a typical FRP laminate the proportion of the fibers is between 50% and 70% by volume with the rest being filled by the epoxy matrix. Bonding with the existing concrete member is done using a two-component epoxy-based adhesive, which often has a special filler. The most common laminates are rectangular plates, although there are also L- or U-shaped special configurations. Because FRP laminates are less "bendable" with respect to FRP fabrics they are packaged in rolls of much larger diameter (Figure 4.40).

All the manufacturers provide a wide variety of strip sizes, ranging for instance from 50 mm width and 1 mm thickness to 150 mm width and 1.4 mm thickness. Different elastic moduli of the laminate material are also available, ranging from 160 to 210 GPa.

FRP plates are mainly used to increase the bending capacity of RC members, typically in beams or slabs. They can be applied at the lower member side at mid-span to undertake the positive bending moments, or at the upper member side in the supports region to undertake the negative bending moments (Figures 4.41 and 4.42).

**Figure 4.40** FRP laminates. *Source:* Stelios Antoniou.

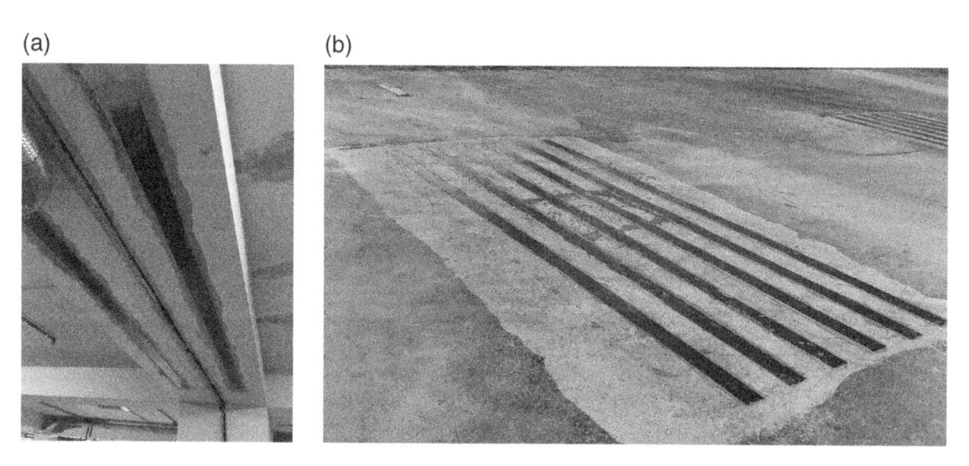


**Figure 4.41** Application of FRP laminates in beams: (a) at midspan (lower side of the beam) and (b) at the supports (upper side of the beams). *Source:* Stelios Antoniou.

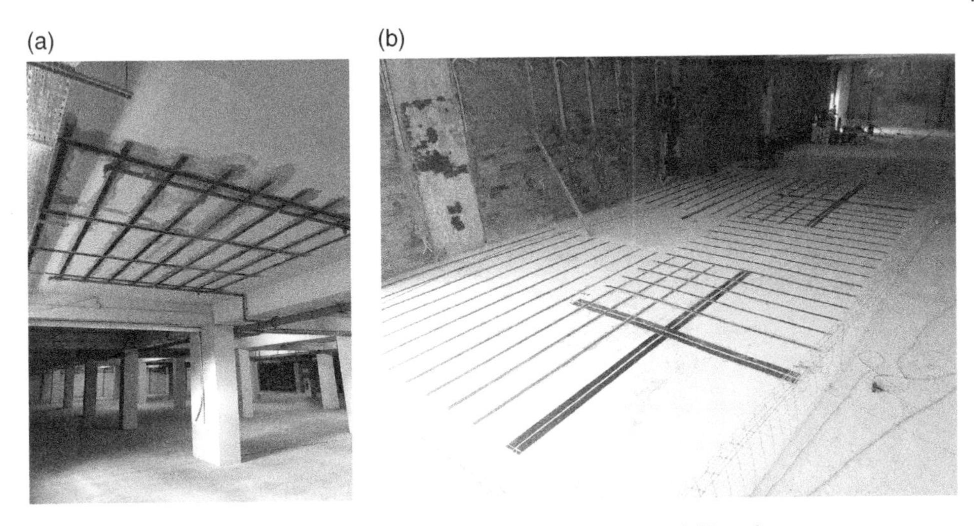


Figure 4.42 Application of FRP laminates in slabs (a) at midspan and (b) at the supports. *Source:* Stelios Antoniou.

In the case of columns, it is generally difficult to achieve the desirable upgrade in the critical regions with laminates, since this requires the lapping or anchorage of the plates inside the beam-column joint.

FRP laminates can also be used for the shear strengthening of beams with special L- or U-shaped systems. All the same, the lack of adaptability and versatility with respect to the FRP fabrics renders them less appealing for this type of intervention. Laminates can be used in conjunction with FRP wraps for the simultaneous strength increase in bending (laminates) and shear (wraps) as in Figure 4.43.

Contrary to the FRP sheets the fibers in all FRP laminates are parallel to the plate axis. Similarly to the FRP sheets they can also be applied in more than one layer. It is noted, however, that the effectiveness of each new layer decreases with their number, because of the debonding between the first layer and the concrete surface and between the adjacent layers of the laminates (Mazzotti 2011). As in the case with FRP fabrics the bond between the multiple layers and the concrete surface (which depends on the mechanical and physical properties of concrete, composite and adhesive) is very important, and bond failure is the most typical failure mode. Again, a minimum concrete tensile strength should be sought, so as to achieve good bonding conditions. When additional anchorage is needed, this is usually achieved by means of FRP strings, or special mechanical anchorage systems, usually steel or FRP composite plates placed at the edges of the laminate, in the perpendicular direction (Figure 4.44).

The application of FRP plates should always be done by a skilled and experienced crew. The preparation of the concrete surface is similar to the procedure followed for FRP wraps but generally it is done in a narrower area, due to the small width of the strips. After the substrate preparation (cleaning, removal of friable and grinding to level the surface), the laminate is thoroughly cleaned and a layer of the epoxy adhesive is applied with a spatula both on the laminate and the concrete to form a thin layer. Then the coated plate is placed onto the coated concrete surface and is pressed against it using a rubber roller, until the

Figure 4.43 Simultaneous strengthening of a beam in bending (with FRP laminates) and shear (with FRP wraps and FRP strips for anchorage). *Source:* Stelios Antoniou.

Figure 4.44 Mechanical anchorage of FRP laminates. *Source:* Stelios Antoniou.

adhesive is forced out on both sides of the laminate (Figure 4.45). The surplus adhesive is then removed. When there are multiple layers or intersections, the upper strips are applied in the same way; however, good cleaning of the lower (already placed) strips should be done beforehand. A thin layer of resin with quartz sand may be laid in order to improve the adhesion of the overlay, while measures for fire-resistance should be taken, if this is required by the design. Contrary to FRP fabrics, the systems and the installation procedures of the different manufacturers are very similar, and all systems follow the procedure described above.

Figure 4.45 Application of FRP laminates. *Source:* Stelios Antoniou.

### 4.8.6 Near Surface Mounted FRP Reinforcement

An alternative to the external application of FRP laminates is the application of NSM FRP reinforcement (Figure 4.46). In this technique, FRP plates are bonded in a groove cut in the concrete, rather than on the concrete surface. Because the application should be done inside the concrete cover, the plates are generally narrow with widths up to 20 mm, but have larger thicknesses of up to 3 mm. After the plate installation, the groove is filled with an adhesive, which is usually an epoxy-based adhesive with filler, although in some cases of lower requirements, cement based adhesive may be employed.

Usually, NSM FRP reinforcement is used for flexural strengthening, providing increased strength and stiffness, but it can also be used for shear strengthening.

The main advantage of the technique is that it does not require extensive surface preparation (other than the grooving). Unlike the external FRP application, installation of the NSM system into cracked, rough, or slightly damaged concrete is possible and independent of the surface tensile strength of the concrete. Furthermore, the procedure is generally faster and requires less installation time with respect to the externally bonded laminates. Because it has a larger bond surface, it provides better anchorage capacity and can mobilize a higher percentage of the tensile strength of the FRP plate. Finally, the system has

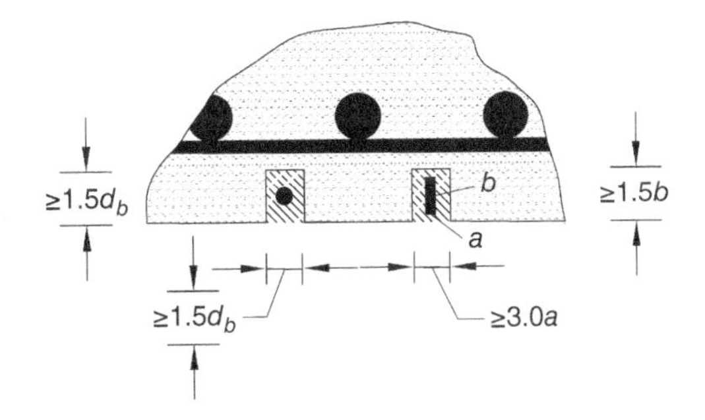


Figure 4.46 Near surface mounted (NSM) FRP reinforcement applied in the concrete cover (ACI 2017).

improved protection against freezing, elevated temperatures and fire. Since the plates are embedded inside the concrete, they are better protected against accidental debonding and vandalism. Its most important disadvantage, which is probably the reason why NSM systems are not as common as their externally bonded counterpart, is the need for a relatively large cover depth (at least 2.5–3 cm), which is not so common in older, existing structures.

Before the application, the concrete cover depth needs to be checked, so as to make sure not to cut through existing reinforcing steel, embedded ducts, or more importantly steel tendons, or other materials within the substrate. The groove is cut using a diamond blade saw or a grinder, and is thoroughly cleaned from dust and loose parts. The groove is filled halfway with the adhesive, and the FRP plate is inserted and lightly pressed to let the adhesive flow around it. Finally, the groove is filled with more paste and the surface is leveled with a trowel. In order to improve adhesion with any possible additional layers, the adhesive may be covered lightly with quartz sand.

### 4.8.7 FRP Strings

These are unidirectional FRP strings, which are used as near surface reinforcement, and more often as fiber connector and anchorage of FRP fabrics.

When installed as anchorage for the fabrics, they can be applied as (i) single connectors, whereby they are bonded into a prepared hole with an epoxy adhesive, and their outer half is usually placed into a star-shaped configuration and is attached to the FRP sheet edge; or as (ii) double connectors into a channel which goes through the concrete (for example through a beam), thus connecting the two edges of the FRP fabric – again the two edges of the FRP string are formed in a star-like shape (Figure 4.47). A very common application of the strings is to anchor the two edges of U-shaped FRP wraps that are used for the shear strengthening of beams, where the slab does not allow for complete wrapping (Figure 4.48).

When used as mounted reinforcement, this is usually done at the supports region of beams or slabs, where an FRP laminate cannot be employed, for example, because of the existence of columns or because it is the edge of the member and the laminate cannot be properly anchored. The strings are installed into small U-shaped slits that are opened in the concrete surface, which after the installation are filled with epoxy resin (Figure 4.49).

One important aspect in the application of FRP strings is that all the concrete edges should be rounded, when the string is applied at 90° angles, similarly to the case of FRP wraps. This

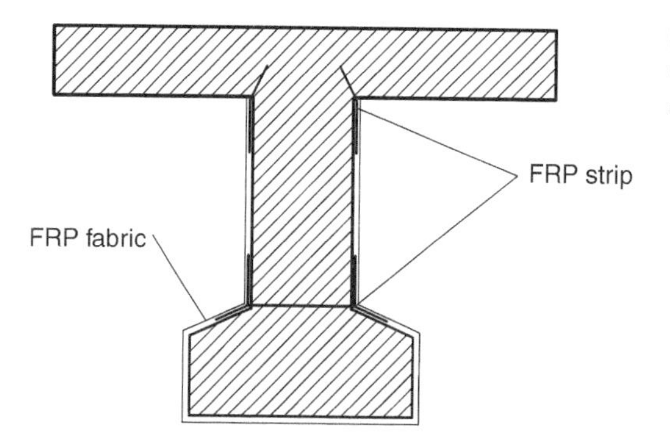


**Figure 4.47** FRP strings as single and double connectors. *Source:* Stelios Antoniou.

is done in order to prevent the fracture of the fibers (and the consequent decrease of the effectiveness of the FRP system) when bent at right angles, and usually requires the opening of a round slit at the edge of the hole, where the string is anchored (Figure 4.50).

The preparation of the concrete substrate is similar to that for the NSM FRPs, and the cutting of the string into pieces is similar to that of the application of FRP wraps. Before

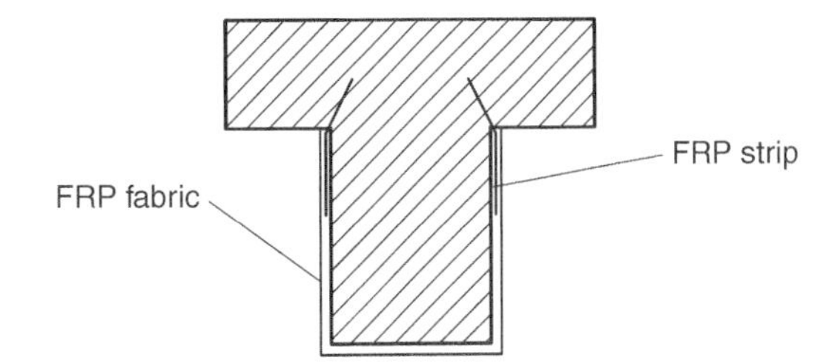


Figure 4.48 FRP strings used for the anchorage of FRP fabrics for the shear strengthening of a beam. *Source:* Stelios Antoniou.

Figure 4.49 FRP strings used as additional support reinforcement of beams and slabs at their edges. *Source:* Stelios Antoniou.

Figure 4.50 Strengthening in bending using FRP strings. *Source:* Stelios Antoniou.

installation the part of the strings that are to be anchored inside the concrete are impregnated until complete saturation, and tied with a plastic cable tie. The insertion inside the hole is done with a large spike or needle. Then the adhesive is applied to the prepared slits of the concrete surface with a brush, and the hole is filled with epoxy adhesive from the bottom up, in order to avoid air enclosures. In the case of systems for the anchorage of FRP fabrics, a dry remaining part of the string remains outside the slit. This is divided in equal parts or is opened as a tuft, and the fibers are impregnated with resin using a brush. The FRP fabric to be anchored is then installed.

### 4.8.8 Sprayed FRP

The set of FRP systems available on the market can be employed in a wide range of applications, nonetheless they also have significant drawbacks and limitations, such as the need for prior surface treatment, problems in the application of the method to strengthen beam-column joints, or in the upgrading of columns in bending. Currently under research are several new methods for the seismic strengthening of RC buildings, one of the most promising of which is the use of sprayed FRP. The method consists of the mixing of chopped glass and/or carbon fibers with epoxy and vinylester resin in open air and spraying the mix onto surfaces, similarly to using shotcrete to cast concrete jackets.

The technique provides increased shear strengths and deformation capacities and can be used for the seismic strengthening of existing RC members. This can be achieved without the need for prior surface treatment, which is important when the accessibility to the strengthened member is limited. The main disadvantages are that the method is still under research, it is not well tested, there are no guidelines available in the literature, and it requires special equipment that is not easily found on the market.

### 4.8.9 Anchoring Issues

In closed wrapping, the anchorage of the FRPs does not pose important design and construction challenges, and can be achieved with the overlapping of the FRP sheet in the fiber direction. On the contrary, in open systems, such as FRP laminates, sided-boding, or U-shaped wrapping, the debonding of the edge of the FRP fabric or plate is usually the critical failure mechanism, because of the low tensile strength of concrete. What is more debonding is a sudden and brittle type of failure. Hence, ways to improve the bond between the FRP and the concrete surface and the stress transfer between them are needed.

In some cases, this can be easily achieved by extending the plate or sheet. This is usually done in the bending strengthening of beams and slabs at mid-span. The application of the FRP is done at the lower side of the member and can be extended toward the edges for better anchorage.

In other cases, however, this type of anchorage is not possible, for example in the strengthening of slabs and beams at the supports, where the extension of the plate or the sheet is obstructed by the presence of the overlying columns or because it is the edge of the member. In such cases, the use of an effective anchorage system is required. A large variety of anchorage systems are available and have been proposed in the literature and by FRP manufacturers. These serve the purpose of increasing the total available interfacial shear stress transfer, delaying interfacial crack opening and preventing the FRP debonding failure modes, and allowing more ductile and favorable failure mechanisms to develop.

Three main anchorage categories can be distinguished:

- Direct anchorage inside the existing concrete member. This is usually accomplished with FRP strings or spikes. The main idea behind it is to open a hole in the concrete of the existing member and anchor the FRP string (or FRP fabric + spike) inside it using a large quantity of epoxy adhesive. After the hole is filled with the resin, a piece of the FRP string or fabric remains outside the concrete substrate and is attached with more adhesive to the end of the load carrying FRP fabric or laminate (Figure 4.51a).
- Increase of the ability to transfer the stresses by providing a clamping effect by means of transverse wrapping or FRP strings perpendicular to the axis of the laminate.
- Increase of the ability to transfer the stresses by providing a clamping effect by means of mechanical anchorage systems, such as steel or anchor bolts, bolted steel plate anchors (Figure 4.51b), or bolted steel angles.

A very detailed review of the available anchoring options can be found in Grelle and Sneed (2013) (Figure 4.52)

### 4.8.10 Advantages and Disadvantages of FRP Systems

The main characteristic of FRP materials is that they have a very high strength-to-weight ratio, thus allowing a significant strength increase of the reinforced concrete members without increasing the loads applied to the structure. Because they come in a variety of forms and systems, they can be tailored to meet wide-ranging performance specifications, flexural, confinement or shear only upgrade, or simultaneous flexural and shear upgrade. It is the only one, among the conventional strengthening techniques, that significantly

(a)

(b)

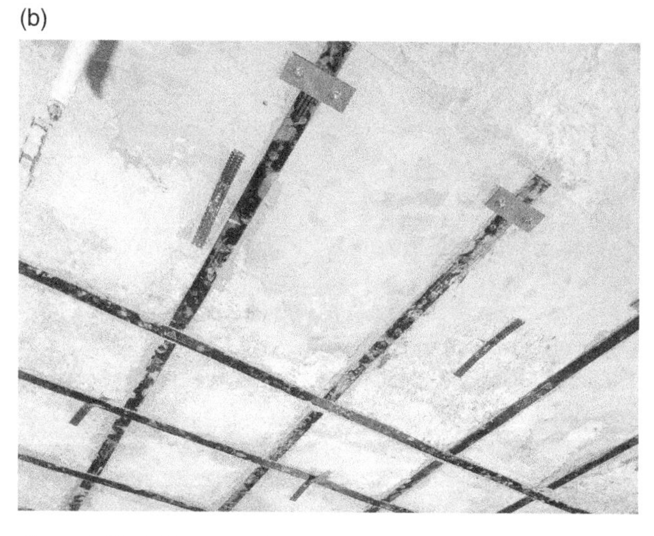

Figure 4.51 Different FRP anchorage systems: (a) anchorage with FRP strings and (b) anchorage with bolted steel plates. *Source:* Stelios Antoniou.

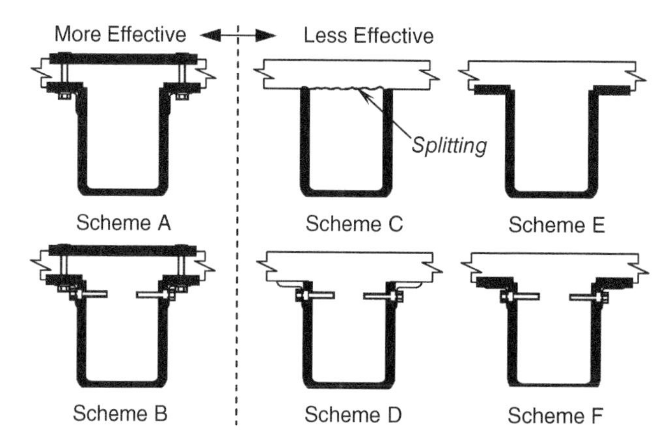


Figure 4.52 Different FRP mechanical anchorage systems (Fukuyama et al. 2001).

increases the member strengths with only a minimal change in the overall structural stiffness distribution; hence it can be readily employed in selective intervention schemes.

The FRP layers that are added with the intervention are very thin (within the order of mm in the typical applications) and do not substantially alter the dimensions of the structural elements. This make the method more attractive than other jacketing methods, in cases when the dimensions of the existing member need to remain unaltered, e.g., columns adjacent to windows or doors that we do not want to change. Compared to other methods (new shear walls, RC jacketing with shotcrete) it is a very "clean" method with minimal disturbance, and can be easily applied in buildings already in operation, without affecting the everyday life of the occupants.

A key factor driving the increased applications of composites during recent years is the significant technological progress that has been achieved, allowing for new versatile FRP systems, new styles of reinforcement, high performance resins and most importantly the significant drop in their price. This, combined with the fast application and low labor installation costs, makes them one of the most attractive strengthening solutions.

Moreover, FRP materials possess excellent environmental resistance and durability and are often employed for anti-corrosion protection even in the most adverse environments (e.g., marine and coastal environments). They possess good impact, compression, fatigue and electrical properties, good thermal insulation, UV radiation stability, and resistance to chemicals and corrosives.

On the negative side, one significant drawback of FRP materials is that they generally exhibit brittle behavior with a linear elastic response up to failure. This applies to the failure mechanism under tension (i.e., fracture of the fibers), as well as to the more common failure modes, such as debonding.

FRPs exhibit poor resistance to fire and their capacity is seriously affected, because the resins quickly lose their strength in high temperatures. Hence, protective measures such as fire-resistant boards or mortars often need to be applied. Furthermore, attention should be paid to protect FRP materials, particularly glass and aramid bars from the alkali environment in concrete, and aramid from UV radiation. All the same, protection from the alkali environment may be assured during the manufacturing stage by using proper coating materials. Finally, FRP composites are difficult to recycle, unlike other structural components such as steel.

### 4.8.11 Design Issues

Figure 4.53 depicts the most common failure modes of reinforced concrete beams strengthened with FRP composites: (i) yielding of the steel in tension preceded or followed by rupture of the FRP fabric or plate under tension; (ii) compression failure of the concrete before yielding of the reinforcing steel and without FRP damage; (iii) failure of the concrete member in shear with diagonal cracking at the end of the plate; (iv) failure of the adhesive, resulting in debonding; (v) failure of the concrete cover and separation of the cover and the FRP composite from the rest of the beam; (vi) mid-span debonding initiated by flexural cracking; and (vii) mid-span debonding initiated by flexural + shear cracks.

In most practical applications and for the typical configuration of existing members (e.g., relatively low concrete strength), the most crucial failure mode is the debonding of the FRP composite as controlled by the debonding stress. Therefore, the parameters that affect the bond and the transfer of stresses between the FRP laminate and the concrete member, such

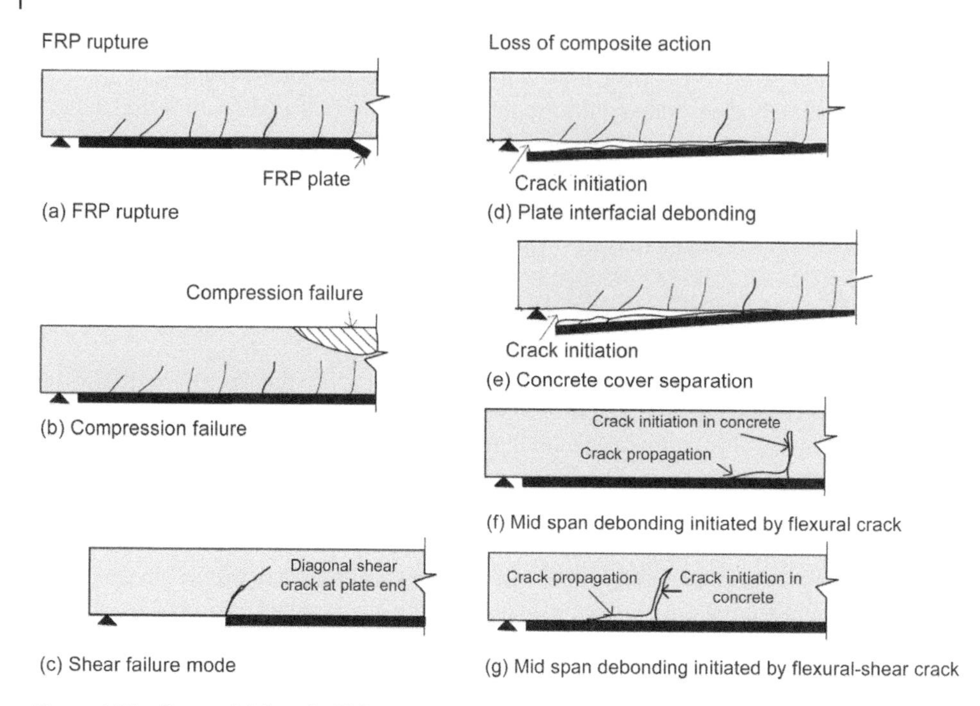


**Figure 4.53** Types of failure in FRP strengthened concrete members (Danraka et al. 2017).

as the mechanical characteristics of the epoxy adhesive, the strength of the existing concrete and the correct surface preparation of the substrate, all assume a very important role in the final capacity of the strengthened member. If adequate anchorage conditions cannot be guaranteed, additional anchoring systems should be introduced, especially in the cases when complete wrapping is not possible, due to geometrical or accessibility conditions.

Fire should also be included as a limit state, as it will influence the properties of both the FRP and the adhesive used to attach it to the concrete.

Different guidelines exist for the design of strengthening of RC members with FRP systems in shear, flexure, and confinement. The design procedure is usually based on the following assumptions: there is no slip between the FRP and the concrete, the shear deformation within the adhesive layer is neglected, the tensile strength of concrete is negligible and the FRP laminate has a linear elastic stress–strain relationship up to a brittle failure.

One important aspect of the interventions with FRP composites is that their effect in the structural stiffness distribution is negligible. Hence, it is the only method whereby the strengthening can be directly calculated based on the member demand as given from the analysis of the existing building, without the need to rerun the entire analytical procedure to make sure there is no significant increase in the member demand, as a result of the structural intervention.

Most national standards for retrofit and upgrading include relatively large sections about strengthening with FRP composites. Moreover, there are several documents that provide complete guidelines and cover the subject in more detail. Among them, the American ACI

440 document (American Concrete Institute ACI 440 (ACI) 2017) is probably the most comprehensive, influential and widely used. Other notable guidelines are the fib bulletins 90 and 14 (Fédération Internacionale du béton (FIB) 2019; FIB 2001), the Italian guidelines CNR-DT 200 R1/2013 (CNR-DT 200 2013), the CAN/CSA-S806-02 Canadian Guidelines (Canadian Standards Association 2007), the Japan Society of Civil Engineers recommendations (Japan Society of Civil Engineering Recommendations (JSCE) 2001), and Technical Report 55 from the British Concrete Society (The Concrete Society 2004).

## 4.9 Steel Plates and Steel Jackets

What can be achieved with FRP wraps and laminates can also be done using steel components. Single steel plates or straps bonded to the concrete members can improve their flexural strength, similarly to FRP laminates. Likewise, plates, straps, and angles welded together to form a jacket can increase the shear strength, improve the behavior of lap spices and provide ductility through confinement without significantly affecting the stiffness of the existing system, in a similar fashion to FRP wraps. Strengthening of reinforced concrete members using external bonding of steel plates was one of the most popular methods and very common in retrofit applications some decades ago; however, it gradually lost popularity to other more reliable and easier to use methods, in particular FRP fabrics and FRP laminates.

The primary use of steel plates is to increase the axial and bending capacity of RC members. Usually, they are attached to the side of the concrete member under tension with epoxy adhesives, often with additional fastening using bolts or dowels, applied in holes drilled in the concrete surface with epoxy resins (Figure 4.54). In a less common application, the plates are attached only with dowels, and the gap between the plate and the concrete is filled with a nonshrinking grout. The objective is to increase the effective

**Figure 4.54** Flexural strengthening with steel plates. *Source:* Stelios Antoniou.

Figure 4.54 (Continued)

reinforcement, thus increasing the bending moment capacity and stiffness. It is important to note that, since the steel plate is attached externally to the member, an additional advantage, with respect to the internal reinforcement is that it has a larger distance from the the compression zone and the neutral axis of the cross-section. The steel plates are mostly employed for the strengthening of beams (and less-often slabs), and they can be applied in single or multiple layers (similar to the multiple layers of the FRP laminates), with the former being the most common option. It is noted that special care should be taken so that the plates are not extended to the area of the concrete member under compression, since plate buckling may introduce tensile detachment stresses in the adhesive. If this cannot be avoided, additional bolts should be introduced.

Configurations of horizontal and vertical plates, straps, and angles welded together can also be used to create steel jackets that confine the reinforced concrete members, usually columns. The gap between the steel components and the concrete is again filled with epoxy resin, or less often with a nonshrinking grout. In the most typical configuration, the jacket is formed from four steel angles at the RC member corners, on which continuous steel plates or straps are welded (Figure 4.55). Often, the plates or straps are preheated prior to the application, so that when they cool down the jacket exerts some positive confinement on the member.

Confining reinforced concrete columns in steel jackets is an effective method to improve the earthquake resistant capacity by offering passive confinement of the original concrete cross section, ductility, by enhancing the lap splice conditions through clamping, and more importantly by increasing the shear strength of reinforced concrete elements. Furthermore, the axial load carrying capacity is increased and the deformation capacity of the member is slightly improved through the effect of confinement. As a result, the steel jacket contributes significantly to the prevention of these adverse failure modes, in favor of the more ductile flexural ones.

When the jacket is mainly constructed to increase the shear capacity, a gap of about 50 mm should be left between the base of the jacket and the end of the member. This gap is introduced to allow the formation of a plastic hinge at the end of the column without substantially increasing its lateral stiffness and strength. In this configuration, the steel jackets do not affect the stiffness of the existing system (similarly to FRP wrapping), since they are not somehow connected to adjacent existing members.

If the steel jacket is intended to also withstand flexure, the steel jacket should be extended outside of the length of the member, in order to achieve good anchorage and transfer forces beyond the member end. This is achieved through welding and bolting to adjacent members, or by welding the steel jackets to steel plates that are used to strengthen the adjacent beams in shear (Figure 4.56). All the same, the extension of the jacket beyond the RC member is in general difficult and error-prone. What is more, it is not very effective in resisting

**Figure 4.55** Steel jacketing in a column *Source:* Stelios Antoniou.

**Figure 4.56** A steel jacket welded to steel plates that are used for the strengthening of the adjacent beams. *Source:* Stelios Antoniou.

cyclic loading. For this reason, steel jacketing is usually not intended for flexural strengthening, since their effect on the flexural strength and stiffness is relatively small.

In circular columns, the steel jackets can be constructed by using a cylindrical steel tube that is cut in two halves and is fitted around the column and welded as closely as possible to the concrete surface. The gap between the existing column and the steel tube is again filled with nonshrinking cement-based mortar or epoxy resin.

Although steel plates can be applied to any structure, the technique should be avoided where there is evidence of ongoing deterioration due to corrosion of the reinforcement. Furthermore, even if at first sight the installation seems relatively easy (the method has been used for decades, the construction industry is familiar with it, and the technology is relatively simple), in practice things are not as straightforward. Concrete surfaces are usually dirty and have out-of-plane imperfections. The careful preparation of the concrete substrate is important, and is carried out in a similar fashion to the application of FRP fabrics. The steel surface should also be free of mill-scale, rust, grease, dirt, and other contaminants (typically, blast cleaning is used for this purpose). Moreover, the concrete surface needs to be flat to a much greater extent than when using FRP fabrics and laminates, since the steel plates are inflexible, and if they are made to follow these imperfections, additional stresses are introduced on them. Therefore, additional labor effort is required, especially in concrete surfaces that are rough and uneven. Welding of the parts of the steel jackets is also very important and requires trained and skilled crew. All these tasks are generally highly sensitive to the standards of workmanship; thus, a high level of inspection is also required. Finally, after the application of the plate, special consideration should be given to the appropriate protection of the plates against corrosion, whereas it is typical to seal the edges of the plates with a resin putty or mortar, so as to protect the surface against moisture ingress. Finally, depending on the requirements of the project, a protective coating against fire might be needed.

### 4.9.1 Advantages and Disadvantages

Retrofit with steel components is a valid and reliable method to improve the overall seismic performance of the structure and can be effectively used for the seismic upgrading of RC buildings to achieve lateral strength, axial load carrying capacity, passive confinement, ductility and shear strength. In comparison to others methods (with the exception of FRP materials), steel plates affect the stiffness of the existing system less, although this varies and depends on how the steel plates are connected to the concrete member and the members adjacent to that. Furthermore, the weight of the steel plates is very small with respect to the overall weight of the existing building; hence the increase in the structural mass of the system is also negligible.

The main disadvantage of steel plates and steel jacketing, however, is that the FRP materials, which are generally used in the same way and for the same purpose (FRP fabrics instead of steel jackets, and FRP laminates instead of steel plates), possess several advantages over steel elements. FRP materials are lighter, allow easier handling and possess high resistance to corrosion. Although the materials are more expensive, this is compensated for by faster and easier installation and considerably reduced labor costs, due to easier preparation of the concrete substrate, and no requirement for welding. What is more, the drop in the cost of FRP materials in recent years and the simultaneous increase in the cost of labor have resulted in FRP applications becoming more appealing and competitive overall in comparison to steel plates and jackets. As a result, steel plates no longer constitute a popular choice for the retrofit of existing reinforced concrete buildings.[2]

### 4.9.2 Design Issues

The design of steel plates for flexural strengthening is relatively straightforward; the plate can be considered as typical external reinforcement, provided that there is sufficient bond between the plate and the concrete substrate. Usually, the application of epoxy adhesives is not enough, and some additional fastening should be provided through bolts, in order to avoid debonding.

In steel jacketing instead the shear capacity of the strengthened member is equal to the shear strength of the existing member plus the contribution of the jacket, which can be equal to $V_j = A_j{}^*f_{yj}{}^*h$ ($A_j$ is the total area of the plates per unit length, $f_{yj}$ is the yield strength of the jacket and $h$ is the cross-sectional depth parallel to the acting shear force) times a jacket efficiency factor, which can be taken equal to 0.40 for steel plates and 0.50 or more for steel straps. Variations of the formula above can be found in the different standards; however, the philosophy is generally the same.

## 4.10 Damping Devices

Damping devices are used to reduce the amplitude of vibration, the deformations, and consequently the demand imposed on the structural members by dissipating energy during large earthquake events. There are passive, active, and hybrid dissipation systems.

The most common in practical applications are the *passive dissipation systems*. Several types of passive systems can be found on the market:

- Viscous dampers, in which the seismic energy is absorbed by silicone-based fluid passing between a piston-cylinder arrangement (Figure 4.57).
- Viscoelastic dampers, which dissipate energy by converting it into heat, through the shear deformation of a viscoelastic material.
- Friction dampers, which consists of several steel plates sliding against each other in opposite directions and dissipating energy through friction (Figure 4.58).
- Tuned mass dampers, through which a mass is placed at a specific location in a structure and reduces the amplitude of the motion during a large event.
- Metallic yield dampers, which dissipate energy through the plastic deformation (yielding) of a specially designed metallic device.

2 In fact, there is a distinct case where steel jacketing cannot be replaced by FRP fabrics. This is when a member (most often a column) needs to be strengthened both in shear and bending and the use of shotcrete is forbidden due to excessive disturbance. In such cases, the steel jacket can be welded to steel plates attached on the adjacent beams. However, this method is hardy nondisturbing. Apart from the welding itself, it also requires the demolition of the plaster coating of both the column and the beams and the demolition of the floors in order to attach plates to the beams at the bottom of the column. Also, the cases when the upgrading in bending is required and cannot be done via other means (e.g., addition of shear walls in the vicinity of the member) is indeed rare.

**Figure 4.57** Retrofit of an RC frame employing steel braces with fluid viscous dampers (Staaleson Engineering, P.C. 2021).

**Figure 4.58** A typical friction damper (Damptech A/S 2022).

Most of the systems above (the viscous, viscoelastic, friction and metallic yield dampers) can be easily combined with steel bracing members, providing increased strength, stiffness, ductility, and energy dissipation capacity. Typical bracing systems with integrated dissipative systems are depicted in Figure 4.59. It is noted that the dampers have better performance with inverted-V braces because the damping forces act in the horizontal direction of motion, which is predominant in the majority of seismic events. In X-braces, the dampers are inclined. Alternatively, the dissipative devices can be installed independently from other strengthening interventions and contribute to the structural response by dissipating energy.

The evolution in damping protection technologies for the retrofit of buildings are the *active control integrated systems*, which act by introducing forces into the structure to reduce the vibration after getting information about the building's dynamic excitation and response by special sensors. The forces are calculated instantaneously after processing the recorded data at different locations in the structure.

*Hybrid systems* also exist, which differ from active systems in that they require a much lower external power supply, without the need for a global monitoring system. The control

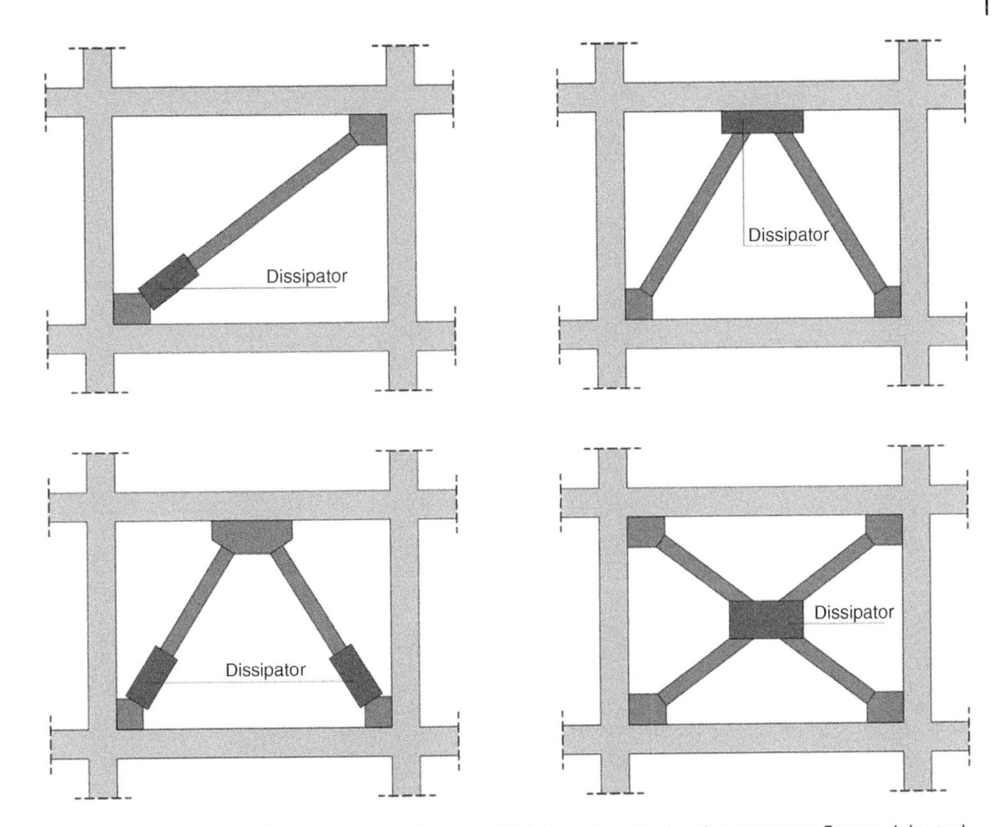


Figure 4.59 Different layouts of steel braces with integrated dissipative systems. *Source:* Adapted from Pinho et al. 2019.

is limited to the local properties of the damper and the bracing system, the rigidity of which is varied by the control computer, according to the dynamic response of the structure.

## 4.11 Seismic Isolation

Base isolation, also known as seismic isolation, is a state-of-the-art method that constitutes one of the most effective means of protecting a structure against earthquake forces. A collection of structural components, called the isolators, are used to decouple to a large extent the superstructure from the base (foundation or substructure) that rests on shaking ground, thus protecting the building's integrity (Constantinou et al. 1998; Lampropoulos et al. 2021).

When used for the seismic upgrading of existing reinforced concrete structures, seismic isolation is typically applied at the columns and walls just above the foundation level (see Figures 4.60 and 4.61). If the building has a basement, the options are to install the isolators at the top, bottom, or mid-height of the columns and walls of the basement.

In the most common configuration, a diaphragm is constructed immediately above the isolators, in order to connect the columns and prevent their independent vibration during a large seismic event. Often, a similar diaphragm is also constructed at the foundation level, right below the isolators.

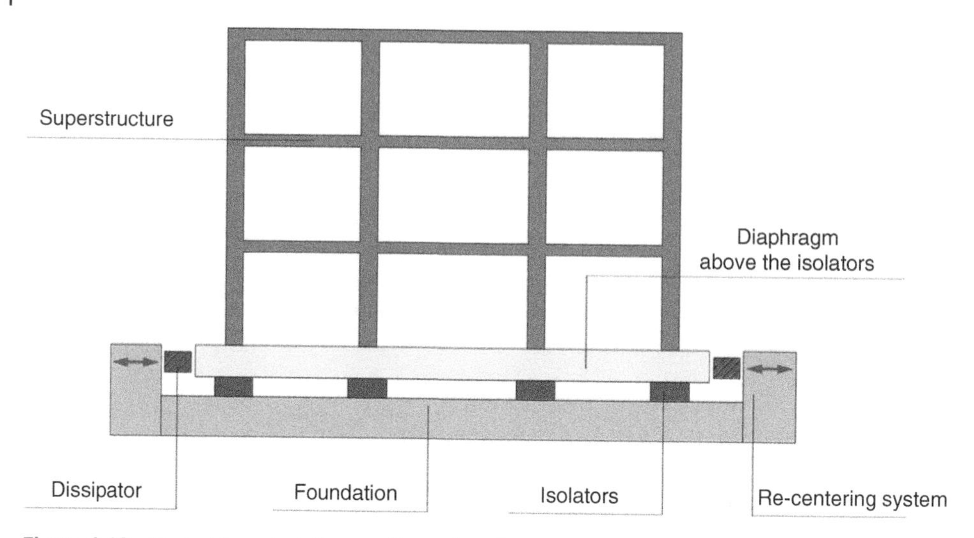


Figure 4.60 Typical base isolation configuration. *Source:* Adapted from Pinho et al. 2019.

Figure 4.61 The base isolators under the Utah State Capitol building (Wikipedia 2022c).

Before cutting the columns for the installation of the bearing, hydraulic jacks are installed in symmetrical position in the entire building plan simultaneously, or around each column separately. The superstructure is lifted with the jacks by 1–2 mm, in order to allow the decompression of the column, which is then cut with conventional methods (e.g., with a diamond saw). The bearings are installed and the gap between the concrete and the bearing is filled with mortar or epoxy resin that does not shrink.

Usually, a large wall is also constructed in the perimeter of the building at the level of the isolators in order to prevent displacements that are larger than the isolators' deformation capacity (Figure 4.60). It is also noted that allowing large relative movements of the building with respect to the ground means that even the non-structural components (e.g., partition walls), as well as the components of the electrical and the mechanical installations system (e.g., cables, pipes) that cross the plane where the isolation system is installed must be altered, in order to be able to sustain the seismic movements without interrupting their operation.

The main characteristics of a seismic isolation system are the limited stiffness at the isolators' level, which leads to the significant period elongation of the structure to fundamental periods of up to 2.5 seconds or more (see Section 4.20.8). This leads to a significant reduction in the acceleration passed on to the superstructure, the inertia forces, and the earthquake force demand. As a result, the lateral deformations, and the interstory drifts are considerably smaller, leading to light or very light damage to the structural and the non-structural components even in very large earthquake events.

The main concept behind using base isolation for retrofit is that, instead of strengthening the structural members to withstand the imposed seismic action (as is done with all the other methods), base isolation takes the opposite approach – that is, to reduce the seismic demand instead of increasing the capacity. Since it is impossible to control ground motion (e.g., earthquakes), structural protection must focus on reducing the seismic demand by preventing/reducing the motions being transferred to the superstructure from the foundation level.

The fundamental principle is to modify the response of the building, so that the ground is capable of vibrating without transmitting significant motion and inertia forces to the superstructure. A complete separation would be possible only in an ideal, fully flexible system, and no acceleration would pass to the superstructure. However, in real-world applications, it is necessary to have a system that is able to transfer the vertical loads to the base, as well as to resist the small lateral forces induced from the wind and minor seismic events.

With seismic isolation, the achieved decrease in the seismic demand is usually very large; hence, no other intervention is required in the superstructure, even if this is constructed without modern anti-seismic standards, adequate reinforcement, or good detailing. Depending on the condition of the superstructure, the design can be carried out so that it accepts limited inelastic deformations or remains totally elastic. The main drawback of the method is that the site of the building should permit horizontal displacements at the base of the order of 200 mm or more in every direction. Consequently, the method is not suitable for buildings that are not open on all sides in their perimeter.

Seismic isolation was first introduced for the design and construction of new buildings, however nowadays it is gradually gaining ground for the protection of existing structures against seismic loading. With the rapid decrease in the cost of isolators the technique is gradually changing from an "exotic" method suitable only for special applications, to one of the standard methods for seismic upgrade.

Today, seismic isolation is considered one of the preferred methods for retrofitting historical buildings that require minimum modifications, as well as for content protection, i.e. when the value of the contents of a building is greater than the value of the building itself – as, for example, in museums. Value can be financial (e.g., a very important industry that cannot afford to interrupt its operation for a large period of time), architectural (e.g., a historical building) or cultural (e.g., containing priceless works of art or history).

Moreover, seismic isolation constitutes a competitive method in purely economic terms for medium to high-rise buildings, especially when one takes into account that all the retrofit works are then carried out at a single level (typically the foundation or the ground level), which means that the disruption to the operations of the building and the cost for the business interruption are limited and considerably less with respect to other methods, such as jackets or new shear walls.

It should be noted that some prominent US monuments, such as the Pasadena City Hall, the San Francisco City Hall, the Salt Lake City and County Building, and the LA City Hall, were mounted on base isolation systems (Wikipedia 2022c). This method has also been used extensively in seismic upgrading of existing buildings in Italy and Japan, for example.

### 4.11.1 Type of Base Isolation Systems

Seismic isolation can be achieved with the use of devices like rubber bearings, friction bearings, ball bearings, and spring systems. The isolator types most used in practical applications are: (i) elastomeric rubber bearings, (ii) elastomeric lead-rubber bearings, (iii) friction-based isolators, and (iv) friction pendulum systems, FPS.

The *elastomeric isolators* (Figure 4.62a), which can be low (LDRB) or high (HDRB) damping depending on the material used, are composed of a series of horizontal layers of elastomeric material (synthetic or natural rubber) interspersed by steel plates. These bearings are very stiff in the vertical direction, being capable of supporting high vertical loads with very small deformations, but are flexible under lateral loads in both horizontal axes. The steel plates provide a significant contribution both to the vertical stiffness and to the lateral confinement, preventing excessive lateral deformations, while limiting the lateral bulging of rubber. The key parameters for the design of elastomeric isolators are the maximum vertical load that can be sustained, the horizontal stiffness, and the maximum permissible, relative horizontal displacement between the two ends of the bearing.

The *elastomeric lead-rubber bearings* (Figure 4.62b) have the same characteristics as the plain elastomeric ones, with the exception of a central cylinder at the core, which provides a source of larger damping due to the shear deformation of the lead material, since the plain bearings have limited energy dissipation capacity.

The *friction-based isolators* are sliding bearings that use sliding elements between the foundation and the superstructure. Depending on the shape of the interface between the sliding elements, they are divided into flat slider bearings and curved slider bearings. The sliding displacements are controlled by high-tension springs, laminated rubber bearings, or

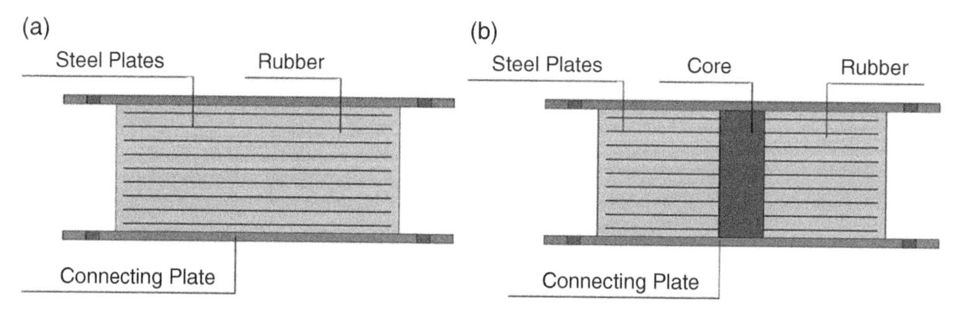


Figure 4.62 (a) A typical rubber bearing and (b) a typical lead-rubber bearing. *Source:* adapted from Pinho et al. 2019.

the curved shape of the sliding surface, and these mechanisms provide a restoring force to return the bearing and the structure to their equilibrium position. The most common types in this category are the *friction pendulum systems (FPS)*, which use the principles of a pendulum to elongate the fundamental period of the isolated structure. The FPS consists of a concave stainless steel surface covered by a Teflon-based composite material and a slider. During severe ground motion, the slider moves on the concave surface, lifting the structure and dissipating energy by friction between the spherical surface and the slider. This isolator uses its surface curvature to generate the restoring force from the pendulum action of the weight of the structure on the FPS. Figure 4.63a shows the typical cross section of a FPS base isolator, and Figure 4.63b shows this isolator installed.

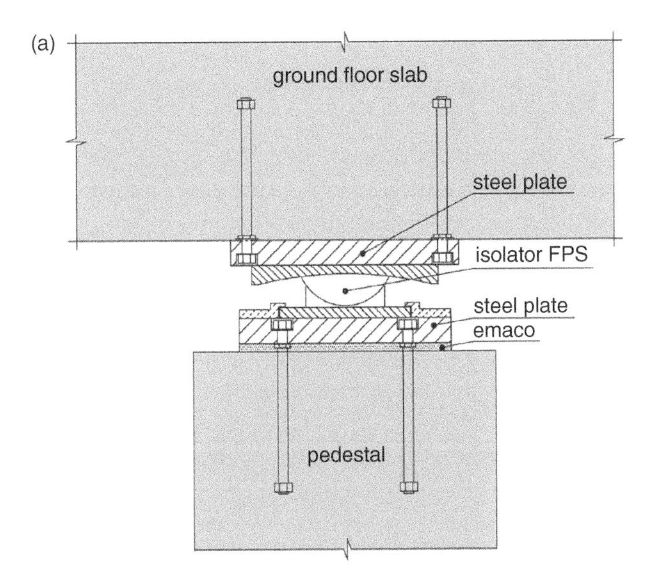

Figure 4.63 (a) and (b) A typical FPS base isolator (Giarlelis et al. 2018b).

The evolution of this system is the double-curved sliding pendulum isolator (Constantinou 2004), which combines two (rather than one) concave surfaces. It can achieve double displacement with respect to the ordinary systems, and is suitable when the aim is to limit the size of the device.

### 4.11.2 Advantages and Disadvantages

Following a different approach from all other strengthening methods, base isolation leads to a significant drop of the seismic forces applied to the structure, and is one of the safest methods for seismic upgrading. Its main advantage is that all the strengthening works are carried out at the foundation level, thereby reducing the disturbance to the residents during the construction works. This means that additional costs for nonstructural damage and business interruption are also limited.

Even in the case of a large seismic event, the damage is small and concentrated at the isolators' level; hence, the repair can be done easily and without the need for the evacuation of the building. What is more, the vibrations at the upper part of the building are very small, with limited absolute accelerations and relative displacements. This is very important in the cases, where the financial, architectural, or cultural value of the building content is very high. Base isolation is particularly suitable for historical buildings, since the method is the least invasive and does not require interventions in the building superstructure that needs to be preserved.

However, base isolation is still a very expensive solution, especially for small to moderate-sized projects (e.g., buildings up to 10 floors high), and in such cases the increased level of safety is not fully justified by the increased cost. It is noteworthy, however, that in recent years the price of isolators has dropped significantly, and they are expected to further decrease in the near future, gradually making base isolation more appealing.

Another problem with that method is that it cannot be applied to any structural configuration. It requires that the building under consideration is free to move in any direction in its perimeter, and this is not possible when the building is in contact with other properties at any of its sides. Therefore, the method cannot be applied to a large proportion of the building stock, such as buildings in city centers and other densely populated areas.

Last but not least, base isolation is a new method that is quite different in its philosophy and its details with respect to the other existing methods for retrofit. It requires specialized knowledge and expertise, both on the design and on the construction level. Because the number of projects carried out up to now employing the technique is relatively small and confined to larger, more prominent buildings and landmarks, this expertise is currently limited, and possessed only by large design or construction firms (which wrongly further corroborates the conviction that seismic isolation is currently a method for large and important projects only).

### 4.11.3 Design Issues

The main characteristics of any isolation system are the energy dissipation to control its lateral displacements, the lateral rigidity under low lateral load levels, such as wind loads or minor earthquakes, and the resistance (strength and stiffness) to vertical, gravity or live, loads. These have to be considered along with the total structural mass, and with some relatively simple and mundane calculations for SDOF oscillators the designer determines the basic parameters of the isolators and the new dynamic characteristics of the building, so as to achieve the required increase in the period of vibration and the consequent reduction of the applied inertia force. Usually, a fundamental period of 2.50 seconds or larger is sought for in both horizontal directions, so that the system is most effective.

Besides the main prerequisite that the building should be open on all sides of its perimeter (usually at least 200 mm are needed), the base isolation technique is suitable when the subsoil does not produce a predominance of long period ground motion and the lateral loads due to wind are less than 10% of the weight of structure. Ideally, isolation is most effective with buildings that lie on stiff soil or rock.

Considering that the isolation systems are almost always nonlinear and often strongly nonlinear, an equivalent linear static analysis is commonly utilized only in the preliminary design phase employing effective bearing properties, whilst the final design is usually performed with nonlinear dynamic time-history analysis using the exact dynamic and hysteretic characteristics of the isolators and the components of the superstructure.

During recent years many national standards and guidelines on base isolation have been developed and published worldwide, e.g. specific chapters in ASCE-41, in EC8, Part-1 in the Turkey Building Earthquake Regulation (TBDY 2018), and in the NZ draft guidelines (NZSEE 2019). Most of these documents are for the design of new construction, though implicitly it is assumed that they can also be employed for the upgrade of existing structures. To the best of my knowledge, only the American ASCE 41 guidelines have a dedicated chapter on the evaluation and retrofit of buildings using seismic isolation systems (ASCE 2017). Gradually, as the technology becomes more common, the methodologies and requirements for the design of seismic isolation in the seismic upgrade of structures will become better known and more standardized worldwide.

## 4.12 Selective Strengthening and Weakening Through Infills

Unreinforced masonry with clay or cement bricks is the most common way to construct infills in RC frame buildings. Their location around the building depends on architectural and operational requirements, and usually there is no consideration about safety issues related to the increased or reduced risk, due to their correct (mostly regular) or incorrect (highly irregular) placement. This is not unreasonable; even today the design of reinforced concrete buildings is carried out without consideration of the location of the infills, whereas it is very common that infills are destroyed and reconstructed in different locations throughout the lifetime of the structures.

The effect of masonry infills in the dynamic response of reinforced concrete structures subjected to earthquake loading is significant, especially in older construction with small, flexible, and lightly reinforced concrete members. Depending on the structural configuration, this effect can be either beneficial or not. Typically, in fully infilled frames, the presence of infills acts positively on the seismic behavior of the building by increasing the lateral stiffness and strength, by transferring shear forces to the foundation and by absorbing energy, thus protecting the concrete columns from increased force and deformation demands. On the contrary, in partially infilled frames, where the infill panels' placement is highly irregular, the concentration of strength and stiffness at a particular location of the building usually has considerably adverse effects on the global behavior, increasing the demand in the concrete members in the areas without infills. Typical examples are buildings with soft stories at the ground level (irregularity in elevation), buildings with many openings at one side, for example because of shop windows (irregularity in plan), or buildings with infills that do not extend to the entire floor height and the relevant restraints form short columns. It should be noted that the effect of infills in the global seismic response is more pronounced in buildings that have a flexible bare frame without large shear walls, which is the most common case in existing RC structures.

Consequently, one of the first very simple things that a designer can do in order to decrease the earthquake risk is to restore regularity by constructing infilled walls in appropriate locations. The most typical example for this is the construction of walls in the pilotis of buildings or closing or reducing the size of some openings in the weak, more flexible side. Moreover, one could also consider weakening (rather than strengthening) the building in a clever way, which renders the building more regular.

It is stressed that the vulnerability of a building with irregular distribution of infills is larger than the vulnerability of the building's bare frame. This is depicted in the succinct example of Figure 4.64, where nonlinear dynamic analysis was carried out with

**Figure 4.64** The presence of infills in the upper stories increases the interstory drift at the critical ground floor from 0.55% (1.66 cm) to 1.05% (3.15 cm). *Source:* Stelios Antoniou.

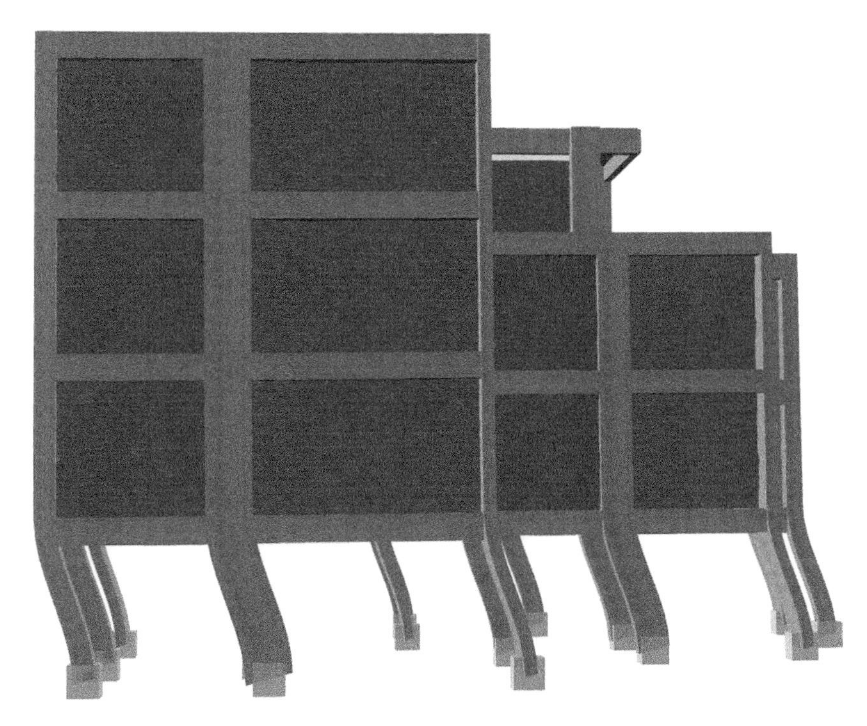

Figure 4.64 (Continued)

SeismoStruct (2023) employing the same record in the same RC frame with and without infills. The presence of infills in the upper stories increased the interstory drift at the critical ground floor from 0.55% (1.66 cm) to 1.05% (3.15 cm), although the total displacement was larger in the bare frame model. This is because the presence of infills leads to a concentration of displacements at the weak ground story, leaving the upper levels almost undeformed. This is not the case in the bare frame, where the deformations are more evenly distributed along the height.

## 4.13 Strengthening of Infills

In most practical applications, the strengthening of a building involves interventions on the reinforced concrete frame only. This is reasonable, since the RC frame is responsible for transferring the gravity loads to the ground and needs to remain intact after a large seismic event (Recall that buildings sustain damage due to lateral deformations induced from the earthquake loading, but ultimately they collapse due to the gravity loading). Furthermore, in the pre-seismic strengthening of existing buildings, it makes much more sense to carry out the retrofit at the limited locations of the columns (and possibly some of the beams), rather than at the locations of all or most of the infills.

In some cases, however, the need may arise to upgrade some of the infills of an RC building, and convert them to a more reliable source of resistance. One should recall that unreinforced masonry infill walls demonstrate poor performance even in moderate earthquakes, they lack ductility, and they suffer for premature and brittle failures (cracking, crushing and disintegration) both in the in-plane and out-of-plane directions. There are several methods available for the strengthening of infills, the most important of which are FRP fabrics, textile-reinforced mortars (TRM) and shotcrete.

### 4.13.1 Glass or Carbon FRPs

One of the most common methods for upgrading infill panels is with the use of fiber-reinforced polymers, bonded at the sides of the panel. The best scheme is to cover the whole surface of the infill on both faces with the FRP sheets, although efficient schemes can also be achieved using strips of FRP fabrics or FRP laminates, either in the orthogonal or the diagonal directions. The material usually employed is glass (GFRP) or carbon (CFRP), and similarly to the application on concrete members, the bonding to the masonry surface is done using an epoxy-based adhesive (Figure 4.65).

The FRP system is applied in a similar manner to the concrete members. Prior to the application the excess mortar is removed from the masonry surface (e.g., using a wire brush), in order to achieve a smooth flat surface (the finishing plaster has already been removed prior to the application). In most cases and for typical wall, fabric and resin properties, the main type of failure is debonding. Indeed, experimental studies have shown that FRP strengthening is not effective, unless the sheet is properly anchored to the infill and/or to the frame members (Özcebe et al. 2003); hence, improved anchoring conditions are of great importance and significantly increase the strength of the wall. If the width of the wall is equal to the surrounding RC members, the sheet can be anchored simply by its extension to the concrete surface. If instead the width is not equal, special anchorage configurations may be used, such as FRP strings and spikes or transverse wrapping. The procedures followed are similar to those described in the previous section on FRPs anchoring.

The application of FRP fabrics on the wall surface can lead to a significant improvement in its seismic response both in-plane and out-of-plane, with increased lateral strength, inelastic deformation capacity, ductility, and energy dissipation capacity. Equally important is the ability of the FRP composite materials to delay premature failures, hold the wall together and not allow a rapid and brittle disintegration once failure has occurred. These

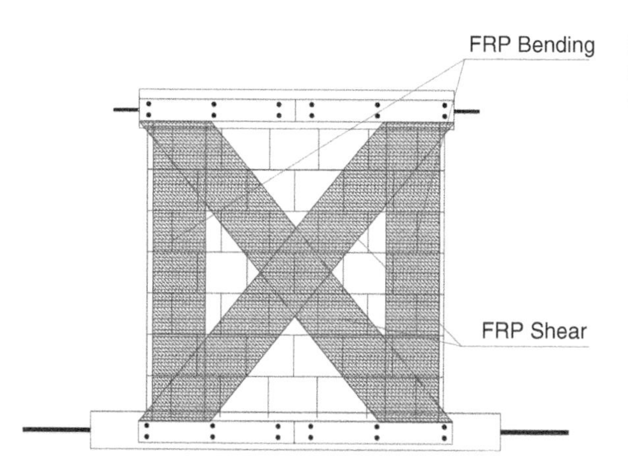


Figure 4.65 Infill panel strengthening using FRP fabrics (Wang et al. 2018).

benefits are achieved without a considerable increase in the lateral stiffness of the wall. This is essential because such an increase can lead to undesirable side effects, causing irregularities in the building's configuration or the transfer of larger lateral forces to the adjacent RC columns. Furthermore, the intervention is done with only minimal additional weight to the structure, whereas the application is as easy as in the case of concrete members.

Despite these advantages, this retrofitting technique is not problem-free. Some of the drawbacks are related to the FRP system itself and have been analyzed in a previous section (e.g., relatively high cost, poor behavior at high temperatures). The most important problem, however, seems to be the incompatibility of epoxy resins with some substrate materials such as clay (Papanicolaou et al. 2007). What is more, anchoring the fabrics to the concrete frame can transfer forces to the columns, which could move a potential failure to the structural frame.

### 4.13.2 Textile Reinforced Mortars TRM

One possible solution to the problems mentioned above regarding FRP fabrics can be the replacement of the epoxy binders with cement-based mortars, and the FRP fabrics with reinforcing meshes. The result is TRM, also known as fabric reinforced cementitious matrix (FRCM), which constitute a relatively new and promising technique.[3] TRM are applied similarly to the FRP systems as externally bonded reinforcement, and combine fibers, in the form of grids or textiles, with inorganic matrices of high strength cementitious mortar. It is most common to apply the TRMs over the entire face of the wall, preferably on both sides of it (Figure 4.66).

Similar to the use of FRPs, the application of TRM in the upgrade of infills should be supplemented with an adequate anchorage of the TRM to the surrounding reinforce concrete, in order to guarantee the transfer of tensile forces between masonry and concrete. This can be achieved with the simple extension of the TRM to the concrete surface, if this is possible, or through the use of special anchoring configurations.

The application of TRM leads to a considerable improvement of the response of the infills, both in terms of lateral resistance but also in terms of stiffness, due to the presence of the layers of high strength cementitious mortar. The progressive cracking of the cementitious mortar and the eventual activation of the textile at higher levels of deformation result in the effective dispersion of deformation demands over a broader area of the masonry infilling. This reflects the contribution of the strengthening material in absorbing energy, mainly due to the multi-cracking mechanism and the redistribution of the shear stresses on the body of the masonry infill (Koutas et al. 2014a, 2015). The gradual degradation of the TRM material

---

3 Several researchers have proposed using textile-reinforced mortars to strengthen reinforced concrete members: columns, beams, and beam-column joints. The proposed solution seems to be valid, providing good results and significant increase in the ductility, the strength and the stiffness of the members. However, the method is still under research, and the technical specifications of the TRMs, as provided by their manufacturers, propose the use of the technique for the retrofit of infills and secondarily of concrete shells. For this reason, I decided not to include TRMs in the methods for the strengthening of the RC frame.

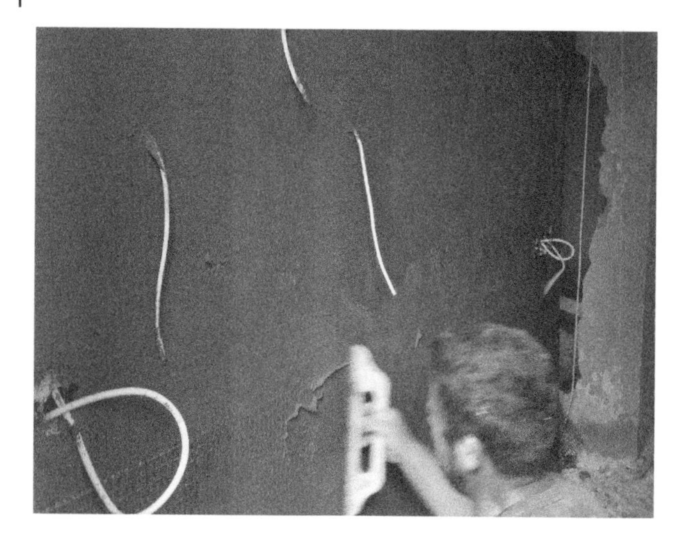

**Figure 4.66** Infill panel strengthening using Textile Reinforced Mortars TRM. *Source:* Stelios Antoniou.

ultimately leads to the progressive degradation of strength and stiffness on a global level. This constitutes a further advantageous characteristic of the TRM system: the structural integrity of the textile is maintained, rendering it capable of containing the masonry infill and reducing the risk of out-of-plane collapse. Furthermore, this more ductile response of the infills prevents their premature damage, the sudden loss of strength and the formation of a weak story at the locations of damage concentration in the building.

One important downside of TRMs is that the increased stiffness of the strengthened panel may transfer larger lateral forces to the adjacent concrete members, and cause significant localized damage to them. Hence the validity of its use should be examined, considering the global structural response – for instance, the strengthening in selected infills of the building should be done in a symmetric manner, with a philosophy similar to when adding new shear walls, or with the simultaneous upgrading of the surrounding columns.

### 4.13.3 Shotcrete

A method used quite frequently for the strengthening of infills is reinforcement concrete jackets constructed with shotcrete, similar to the strengthening of a masonry structure. The jackets are constructed preferably on both sides of the infill, connected with dowels through the panel. They are reinforced with a grid of reinforcement, and the reinforcement is anchored to the surrounding columns and beams. The behavior of an infill strengthened in this way is very similar to RC infilling, which is why the method is not developed further in the current section.

## 4.14 Connecting New and Existing Members

One of the main considerations of the designer, when introducing new members (e.g., new shear walls, RC jackets, steel braces) to existing buildings, is their connection to the rest of the building – that is, how to transfer the seismic forces that develop during the earthquake in the slabs of the existing building to the new structural components. The new members are designed to undertake a part of the inertia forces that develop during an earthquake event; however, the engineer has to make sure that these forces are indeed passed and resisted by these new members. For instance, if some large shear walls are constructed in the perimeter of an existing structure, but these walls are not well connected with the rest of the building, the building will end up vibrating unstrengthened, whereas the walls will remain undeformed, unloaded and intact.

The encapsulation of the existing members within the new members (e.g., jackets, new shear walls with pseudo-columns around the existing columns) obviously offers some type of connection, however in the majority of cases this is not enough. The effectiveness of the simple contact is generally uncertain, due to local crushing of the new and mainly the weaker existing concrete. Furthermore, with contact the forces are passed only in one direction, i.e. through compression. As a result, methods have been developed for the efficient connection of the new and the existing members (Figure 4.67):

- The roughening of the surface of the existing member with a small demolition hammer.
- The application of structural bonding agents and epoxy adhesives on the surface of the existing member (the meticulous cleaning of the surface from dust and dirt is required).
- U-shaped steel connectors that are welded to both the existing and the new reinforcement. Alternatively, steel plates may be used for the same purpose.
- Steel dowels that are anchored in the existing concrete with epoxy resins.
- Steel connectors of larger dimensions, with respect to conventional dowels that pass through the existing beams and are anchored at their back side. With this method, few connectors can be used for the transfer of the inertia floor forces.

The older method for connecting new to existing members is by welding together the old to the new reinforcement. It was a very common method a few decades ago, when the use of resins was not as widespread as today, and their technology, strength and reliability were

Figure 4.67 Connecting the existing with the new structural components (a) roughening of the existing concrete surface, ((a) *Source:* Stelios Antoniou) (b) application of epoxy adhesives (Sika 2022), (c) steel U-shaped connectors, (d) steel plate connectors (e) steel dowels (f) steel connectors. ((c)–(f). *Source:* Stelios Antoniou).

not as advanced. The problem with welding, however, is that very often the existing rebar are corroded, and the effectiveness of welding is not guaranteed. Furthermore, the process can be delayed by the existence of imperfections in the existing member (e.g., inclined longitudinal rebar), which results in different distances between the rebar to be connected along the length of the member and the need for accurate measurements on site of all the connectors, or the availability of connectors of variable sizes.

Today the standard method for connecting old and existing members is with steel dowels. These are usually L-shaped and are anchored with epoxy resins inside holes drilled in the existing concrete. The dowels have to be properly designed (both their sizes and numbers) for each new wall or jacket. The shear force through the interface that can be undertaken by each dowel depends on its diameter, the strength of the steel and the existing concrete. Because the most common failure mode for the dowels is a concrete cone failure, particular care should also be taken so that the dowels are not closely located, because in such case their capacity is seriously undermined. Typical dowel diameters for the superstructure are between ∅8 and ∅18 mm, whereas in the foundation larger diameters (e.g. ∅20 mm) are often selected, in order to minimize the effect of the corrosion due to the underground moisture.

For the case of steel members (e.g., braces, external trusses), the connection is usually accomplished with fewer larger diameter connectors. When the connectors are not protected inside the alkalinity of concrete, very often they are made of stainless steel. The connectors pass through holes in the existing members (usually the beams) and are anchored at their back side. For the effective transformation of the inertia floor forces, the strengthening of the existing member is often required; this may be achieved with steel plates that are attached with epoxy resins at the sides of its web, or through jacketing. All the same, it should be noted that the construction of few large steel connectors (with the necessary strengthening of the beams) is usually more cumbersome and expensive than the installation of a larger number of smaller dowels.

Supplementary to the two main methods described above is the roughening and the cleaning of the surface of the existing concrete, which is customarily carried out in such applications. Furthermore, this can be assisted by epoxy-based structural bonding agents laid on the interface. The problem with the use of epoxy adhesives, however, is that most of these agents require their placement right before the casting of the new concrete, so that the adhesive layer is still "tacky" during casting, and this is very often hindered by the existence of the reinforcement of the new member, which is dense in most practical applications.

Nowadays, in practical applications the most common ways of connecting the new and the existing concrete is through steel dowels, assisted by the roughening of the existing concrete surface.

### 4.14.1 Design Issues

The number and size of dowels is calculated using the shear force that is developed on the surface between the new and the existing concrete. The preferable mechanism of failure should be with the simultaneous yielding of the dowel and crushing of the existing concrete in contact with the rebar. In order to achieve this, particular consideration should be put into the following:

- The length of the dowel in the existing concrete.
- The minimum cover of the rebar (i.e., distance from the concrete edge) both in the direction of the loading and perpendicularly to it.
- The minimum distance between successive dowels.
- The interaction between dowel and pullout mechanism, if the dowels are simultaneously subjected to tension and shear.

## 4.15 Strengthening of Individual Members

During the design for seismic upgrading, a building should always be considered as a single entity and the retrofit should take into account the entire structure. As will be discussed in subsequent chapters, the general principle is that of the weakest link, especially when one considers the brittle failure types (e.g. shear) of the vertical members, which are the most common. The building is as strong as is its weakest column. There is no point in strengthening some of the columns, when the others remain unstrengthened and weak. In the cases of brittle failures types, the building will fail, when its weakest link fails.

There are cases, however, when one needs to consider the options of strengthening just one individual member. This typically can occur in the following three situations:

- When a strengthening scheme has been applied on the global level (e.g., addition of large shear walls), and the analysis has showed that there are some weak members that do not pass the checks, i.e. the demand still exceeds their capacity.
- When a failure has occurred on the member with the rest of the structure remaining largely unaffected, for instance due to a large concentration of forces locally, such as a large piece of machinery or a large load of materials during construction.
- When a member was found to have a very serious weakness due to poor workmanship, e.g. significant lack of reinforcement, because someone forgot to place it during construction.

In these cases, the designer should consider the available options for strengthening an individual member on the local level. Below, there are guidelines on how to achieve this, the possible methods that may be employed, and what the engineer should pay attention to in each particular case.

### 4.15.1 Strengthening of RC Columns or Walls

For upgrading a vertical RC member, a designer may consider either:

- RC jackets, with cast-in-place concrete or shotcrete (Figure 4.68a). The choice between the two methods of construction is related to the ease of carrying out the intervention (e.g., accessibility of the locations of the member), and to architectural issues (e.g., the larger cast-in-place jackets might lead to column sizes that are unacceptable by the architects). However, for design purposes, it is irrelevant how the jacket will be constructed.
- FRP wraps, usually for the upgrading of the shear capacity and to provide increased confinement (Figure 4.68b). Note that FRP laminates are rarely used for the strengthening of columns in bending, since it is difficult to provide sufficient anchorage inside the beam-column joints, even with special anchoring systems like FRP strings.
- Steel plates: this is rarely a choice nowadays. The use of jackets or FRP wraps is almost always a superior option.

The main difference between the first two methods (which are the most common) is that the jackets increase the strength both in shear and bending, whereas FRP wraps are usually employed for the shear strengthening of the columns and to provide confinement. Even with bidirectional fabrics, the effective strengthening in flexure is difficult for reasons related to the anchorage of the fabric (similarly to the FRP laminates). It is noted, however,

(a)

(b)

Figure 4.68 Strengthening of a column with (a) RC jackets and (b) FRP wraps. *Source:* Stelios Antoniou.

that FRP wraps are a cheaper and cleaner method, and the stiffness distribution of the structure is not affected; hence when the problem is related to reduced shear capacity, it is a more suitable option.

### 4.15.2 Strengthening of RC Beams

For the strengthening of RC beam, the engineer should consider the following options:

- RC jackets. The method again is used for the strengthening in both bending and shear. It is noted however that it also considerably affects the stiffness of the member, which might not always be desirable. The jackets can be full (four-sided) or three-sided (Figure 4.69). The full jacket is a superior option, since the stirrups are closed (usually they are welded above the beam) and longitudinal reinforcement is also placed on its

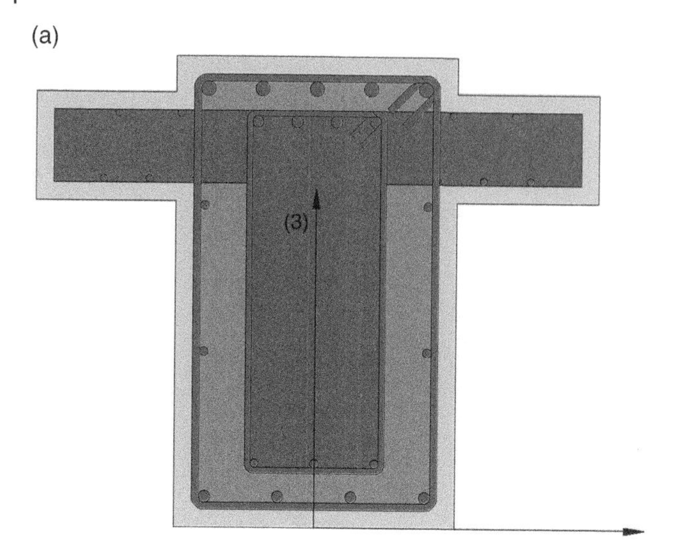


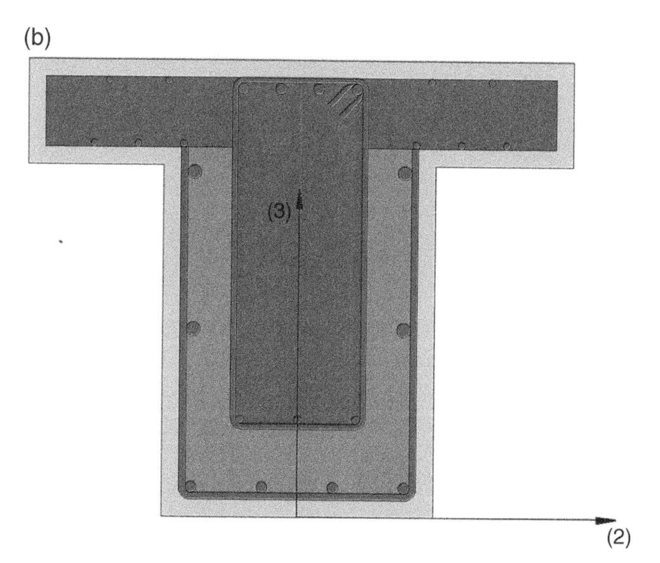


**Figure 4.69** (a) Four-sided and (b) three-sided jackets. *Source:* Stelios Antoniou.

upper side (Figure 4.70). This means that interventions are required both below and above the slab, which can cause significant additional disruption, considering that for the works on the upper side of the beam a large strip of the floor has to be demolished and reconstructed. With three-sided jackets on the other hand, special attention should be paid to the correct and effective anchorage of the stirrups inside the slab, which is not always easy, especially with older slabs of relatively small depth (11–12 cm or smaller). The part of the jacket below the slab is usually constructed with shotcrete (cast in place concrete requires the opening of large holes on the concrete of the beam every 1.0–1.5 m), whereas the part above the beam is constructed with cast-in-place concrete.

- FRP sheets can be used to upgrade beams both in bending and shear, although the former is not very common, since the use of FRP laminates is preferable due to their smaller

**Figure 4.70** RC full jackets at the lower and upper sides of the beam. *Source:* Stelios Antoniou.

width. For shear strengthening, the fabrics are applied in a U-shaped formation below the slab. Again, special attention should be paid to their efficient anchorage inside the slab. FRP strings are typically used for this purpose (Figure 4.71).

- FRP laminates is the most common option for the strengthening of beams in bending. While the application at mid-span on the lower side of the beam is straightforward, the anchorage of the laminates at the upper side of its two edges can be obstructed by the presence of the column. In such cases, the laminate can be replaced with FRP strings that are anchored inside the beam-column joint or it can bypass the column and be placed at the effective width of the beam (Figure 4.72). Note that it is rather common to combine FRP laminates for the flexural upgrade with FRP wraps to increase shear capacity (Figure 4.73).
- Steel plates: They can be used for increasing the bending capacity. Similarly to RC columns, steel plates are rarely used nowadays, since FRP laminates, which are used in the same way and for the same purposes, have significant advantages over steel plates (Figure 4.74).

Figure 4.71 Shear strengthening of a beam with FRP wraps anchored with FRP strings. *Source:* Stelios Antoniou.

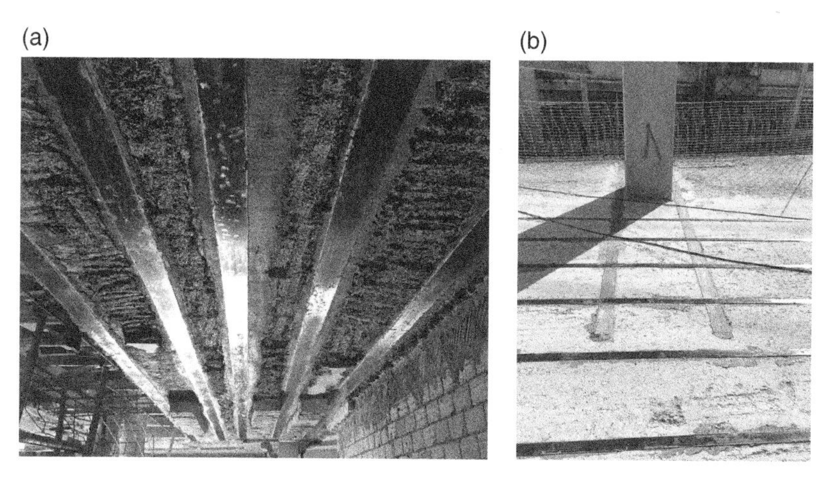


Figure 4.72 Strengthening of a beam in bending with FRP laminates (a) at mid-span and (b) at the supports. *Source:* Stelios Antoniou.

Figure 4.73 Simultaneous strengthening of a beam in bending and shear with FRP wraps and FRP laminates. *Source:* Stelios Antoniou.

Figure 4.74 Strengthening of a beam in bending with steel plates. *Source:* Stelios Antoniou.

Overall RC jackets constitute the most efficient method, providing increasing flexural and shear capacity. Yet, the method can be very disruptive and it is considerably more expensive with respect to the alternatives. As a result, very often solutions with FRPs (fabrics or laminates) are chosen, especially in occupied buildings.

### 4.15.3 Strengthening of RC Slabs

For the upgrading of RC slabs, a designer could choose one of the following options (Figure 4.75):

- A one-sided shotcrete jacket below the slab.
- A two-sided jacket with shotcrete below and cast-in-place concrete above the slab. The two parts of the jacket are connected with dowels.
- FRP laminates (or FRP fabrics); these are placed below the slab at min-span and at its upper side at the edges.
- Combinations of the methods above, e.g., a shotcrete jacket below and strengthening of the slabs' edges above with laminates.
- Construction of new reinforced concrete or (more often) steel beams below the slab, in order to reduce the span of the slab. When placing the steel beams, a small gap (of 1–2 mm) will always remain between the beam and the slab, due to the slab's imperfections. This gap should be filled with a free flowing, unshrinkable, cementitious grout. Special attention should be paid to the existing reinforcement of the slab, in the location of the new beam. If there is no negative moments' reinforcement, the slab should be strengthened in its upper side with FRP fabrics of FRP laminates (the placement of the beam will gradually change the bending moment distribution along the slab, due to additional loads and the creep of the concrete).

Figure 4.75 Different methods for the strengthening of a reinforced concrete slab (a) RC jackets, lower side, (b) RC jackets, lower side of a waffle slab, (c) RC jackets, upper side, (d) FRP fabrics, lower side, (e) FRP laminates, lower side, (f) FRP laminates, upper side. *Source:* Stelios Antoniou.

### 4.15.4 Strengthening of RC Ground Slabs

Ground slabs do not carry significant loads, since they are supported in their entire area by the ground filling, and they are typically lightly reinforced or even totally unreinforced without problems. Nevertheless, when the compaction of the ground filling is poor, these slabs can fail in bending due to the settlement of the filling, as they are not designed to sustain large bending moments. In most cases the compaction is consistently poor and the

**Figure 4.76** Grouting of the gap between the filling and the ground slab. *Source:* Stelios Antoniou.

settlement is extended throughout the entire building, resulting is a problematic slab with series of cracks throughout its entire surface. Although the cracked slab does not constitute a threat to human lives, it does cause significant disruption to the operation of the building, especially in large warehouses, where forklifts or other small vehicles operate.

In such cases, the only viable option is to fill the gap between the ground slab and the filling with grouting through a dense grid of holes on the slab surface (Figure 4.76). To this end, the surface of the slab may be mapped beforehand with the use of an ultrasonic gauge (Figure 4.77). After filling the gap below the slab, which guarantees that no or small further settlement will take place, the repair of the cracks can be done. Depending on the width of the cracks, they may be filled with low-viscosity epoxy resin (thin cracks) as in Figure 4.78a, a mix of epoxy resin and quartz sand (cracks of medium thickness) as in Figure 4.78b, or a self-leveling, cementitious or more commonly epoxy mortar after a V-shaped groove is opened along the crack (wider cracks), as in Figure 4.78c and Figure 4.78d.

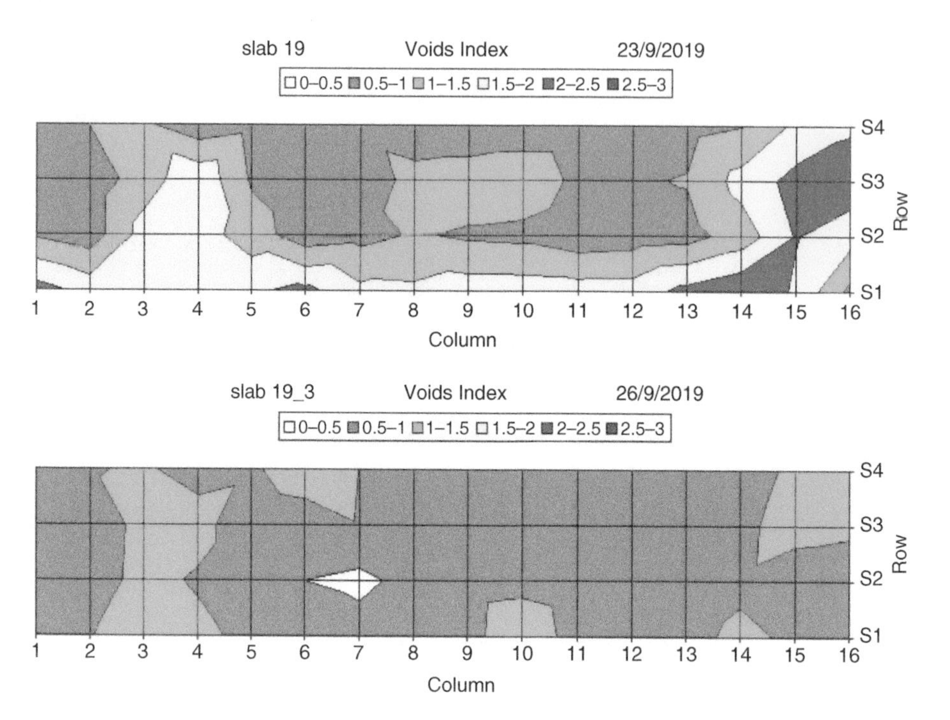


**Figure 4.77** Ultrasonic image of a RC ground slabs before and after the grouting. *Source:* Stelios Antoniou.

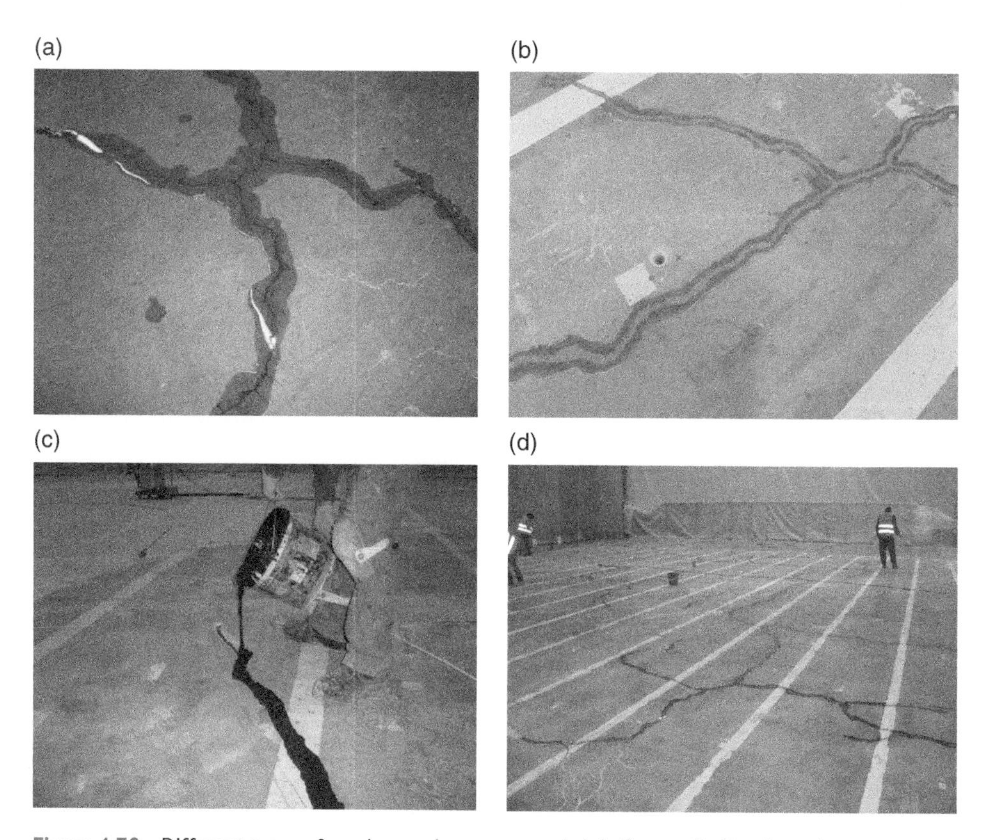


**Figure 4.78** Different types of cracks repair on a ground slab. *Source:* Stelios Antoniou.

## 4.16 Crack Repair – Epoxy Injections

The injection of low-viscosity epoxy resins is the most common method for the repair of small to moderate sized cracks in RC members, in order to restore structural integrity of damaged members, and to prevent water and chloride infiltration. Injection of cracks is normally carried out when the crack width is larger than 0.3–0.4 mm. For cracks up to 3 mm, low viscosity resins are typically employed; medium viscosity epoxies are used for widths up to 5–6 mm, whereas in the (rare) cases of even larger cracks, epoxy or cement grouts are more appropriate.

Epoxy resins are two-component thermosetting polymers with an almost instantaneous cold reaction. Compared to concrete, epoxy resins have very high compressive and tensile strength, and they are used to ensure the efficient transfer of stresses due to their strong adhesion to concrete. Their main characteristics are the ability to harden without shrinkage, low viscosity, the applicability at low temperature, and the fact that they guarantee a barrier against water infiltration that can corrode the reinforcement. On the negative side, regular epoxy resins are too sensitive to water/moisture presence during their application, and water also affects the adhesion between epoxy and concrete. Furthermore, they lose a considerable proportion of their strength at high temperatures, which makes them very sensitive to fire.

Injection of concrete cracks with epoxy resin requires a skilled and experienced crew. The main steps of its application are the following (Figure 4.79):

- For the case of very thin cracks, drill holes to permit the installation of the injection ports on the edge of the crack.
- Clean the interior of the cracks using compressed air or a vacuum system; remove oil, dust, debris, and laitance from the crack edge, where the surface seal will be applied.
- Mount the ports on the surface at appropriate intervals along the crack. The ports should be placed in locations where the crack is not too narrow to permit adequate flow of the epoxy resin.
- Seal the surface of the crack with epoxy paste.

**Figure 4.79** Cracks repair with epoxy resins. *Source:* Stelios Antoniou.

- Inject the epoxy resin with a portable, low-pressure injection machine. The injection should be done only after the sealer has hardened, and should start from the lower port for vertical cracks and from the ends for horizontal cracks.
- Once the resin appears in an adjacent port, remove the injection nozzle, seal the port, and begin injecting in the adjacent port.
- Once the operation is completed and the resin inside the crack has cured, grind away the epoxy paste from the concrete surface.

## 4.17 Protection Against Corrosion, Repair Mortars, and Cathodic Protection

In reinforced concrete members, the alkalinity (high pH) of concrete protects the reinforcement by forming a passive film on its surface that prevents or minimizes corrosion. The reduction of the concrete's pH by carbonation due to age gradually allows the ingress of air, water, or chlorides to the critical depths where the reinforcement is placed, and the steel starts to corrode in the presence of oxygen and moisture.

As steel corrodes, the rust products cause its volume to increase up to six times, and apply significant pressure to the surrounding concrete, which slowly starts to crack. Over the course of years, the cracks will appear on the surface of concrete, and result in the spalling of concrete, and the gradual degradation of the section. This, in turn, allows the ingress of more air, moisture, and chlorides and initiates a vicious cycle that quickly leads to the complete deterioration of the RC member (Figure 4.80). The lack of adequate concrete cover, which is typical in older construction, is an accelerating factor to this process.

The most common repair method in such cases is to remove the cracked and degraded concrete, clean the rust from all the rebar (with a hard brush or an angle grinder in smaller areas, with sandblasting for more extensive damage), apply an anti-corrosion coating, which also acts as a bonding agent, on the reinforcing steel, and restore the cross-section shape and size with a cementitious, nonshrinkable, repair mortar (Figure 4.81). There is a wide range of mortar types on the market, and the manufacturers typically provide different mortars for different thicknesses of the intervention.

For cases of increased deterioration of the rebar, the uncorroded area of the original bar may have been decreased up to 80% or 90%. In such cases, a designer might consider the welding of new longitudinal rebar, in order to restore the ratio of longitudinal reinforcement of the initial cross-section. The intervention involves the removal of the concrete cover along a larger length of the member, so as to allow the welding of the new rebar outside the damaged region, on the uncorroded parts of the existing reinforcement (Figure 4.82). The geometry of the section is then restored with repair mortars.

This standard repair process is sometimes supplemented by the so-called cathodic protection (CP). This is a technique through which a voltage opposite to the corrosion voltage is sent through the concrete to the steel reinforcement, thus transforming the rebar into a cathode of an electrochemical cell. Through this procedure, the rebar are connected to a

**Figure 4.80** Corrosion in RC members. *Source:* Stelios Antoniou.

more easily corroded "sacrificial metal," which acts as the anode and protects the member reinforcement (Wikipedia 2022d). Once CP is installed, the ongoing corrosion can be controlled for a long period, with future spalling and deterioration eliminated even in concrete that is severely contaminated with chlorides or carbonation. The lifetime of the anode system obviously depends on the moisture and chloride content of the concrete, as well as the air temperature, but for typical installations this lifetime ranges between 20 and 50 years.

Figure 4.81 Repair with corrosion inhibitors and repair mortars. *Source:* Stelios Antoniou.

## 4.18 Foundation Strengthening

The dynamic response of footings is a very complex problem requiring skill in soil mechanics, foundation engineering, structural dynamics, and soil structure interaction. Similarly, the accurate estimation of the seismic foundation capacity is difficult mathematically even for new footings, a task that becomes almost insurmountable when it comes to jacketed foundations. What is more, strengthening interventions at the footings is a very disruptive and expensive job, which requires excavations in the entire building at the ground level, and usually evacuation of the building for a long period of time (Figure 4.83). Moreover, in recent earthquakes there have been very few cases of failures in the foundation system, and these were mostly attributed to reasons irrelevant to the vibration and the structural response, such as soil liquefaction or slope stability.

For all these reasons, engineers usually ignore foundation strengthening, since it is difficult to justify technically and economically. What is more, most of the assessment standards largely neglect the issue: EC8 Part-3 has a simple reference to EC8 Part-1 for new structures (CEN 2005; CEN 2004), ASCE 41 and the Greek interventions Code both have very small sections (e.g. just a single paragraph in ASCE 41) on the foundation retrofit with just general discussion (ASCE 2017), and the retrofit chapter of the Italian code makes no mention to it.

**Figure 4.82** Welding of new stirrups, in order to restore the reinforcement of the RC member. *Source:* Stelios Antoniou.

**Figure 4.83** Strengthening interventions at the foundation can be very disruptive. *Source:* Stelios Antoniou.

Noticeable exceptions to the little attention that is given to the foundation system are certain cases, when very large deficiencies are found, such as the lack of reinforcement in the footings or even the total lack of footings. Unfortunately, these cases are not rare in buildings constructed before 1980, due to poor workmanship and the absence of adequate supervision (Figure 4.84). Therefore, before the structural assessment or the design of a strengthening plan for a building, the type and the state of the foundation should always be identified, with excavations and the uncovering of some of the footings (Figure 4.85).

Figure 4.84 Serious structural deficiencies at the foundation system. *Source:* Stelios Antoniou.

Figure 4.85 Uncovering of one footing in an existing building. *Source:* Stelios Antoniou.

In such cases, the most common options for strengthening the existing footings is by increasing their size with reinforced concrete jackets, usually with cast-in-place concrete (shotcrete is more expensive and has the additional disadvantage that the rebound material falls inside the jacket area). The new reinforcement forms a grid around the existing footing

Figure 4.86 Foundation strengthening with jackets and connecting beams (a) and (b), and with strip footings (c) and (d). *Source:* Stelios Antoniou.

and the connection of the new and the existing concrete is commonly accomplished with a set of dowels of relatively large diameter (e.g., ∅20 mm), in order to minimize corrosion problems (Figure 4.86). Epoxy adhesives in the concrete interface may also be applied.

Other available options are the construction of connecting beams, strip footings, or raft foundations. In certain cases, even underpinning may be considered. Finally, measures can also be taken to improve the drainage of the building area to prevent soil saturation.

## 4.19 Concluding Remarks Regarding Strengthening Techniques

In Table 4.4 the most important repair and strengthening techniques for RC buildings are presented, together with the effects that each one of them has on the global structural behavior. Most of the methods provide increase in the strength and the ductility of the members, while some of them do not affect the structural stiffness (FRPs and steel plates),

**Table 4.4** Comparison of the most important repair and strengthening techniques for RC buildings.

| | Strength | Stiffness | Ductility | Seismic demand |
|---|---|---|---|---|
| RC jackets | ✓ | ✓ | ✓ | |
| New RC walls | ✓ | ✓ | ✓ | |
| Steel bracing | ✓ | ✓ | ✓ | |
| FRP wrapping | ✓ | | ✓ | |
| FRP laminates | ✓ | | | |
| Steel plates | ✓ | | ✓ | |
| Seismic isolation | | | | ✓ |

*Source:* Stelios Antoniou.

and others (RC jackets) provide a relatively small stiffness increase. Seismic isolation is a somehow unique approach, in the sense that it operates on the demand on the structure, rather than the capacity. This table is a good reference and should be remembered in the discussion of the subsequent chapters on which strengthening method to choose and why.

## 4.20 Evaluation of Different Seismic Retrofitting Solutions: A Case Study

In this section the merits and the weaknesses of the most important retrofit techniques, which were presented in the previous sections, will be demonstrated by means of an example.

A building with a significant structural deficiency, i.e. a soft ground floor, will be analyzed with SeismoStruct (2023) and strengthened with a series of techniques. The structural response of the strengthened models will be compared with the initial construction in technical but also economical terms. The following five methods of retrofit will be applied:

- Strengthening with jacketing
- Strengthening with RC walls
- Strengthening with braces
- Strengthening with FRP wrapping
- Strengthening with base isolation

### 4.20.1 Building Configuration

This is a typical building of the late 1980s in Greece. It has four floors of approximately 200 $m^2$ each, and a soft story at the ground level. The infills of the upper levels are relatively strong with good-quality ceramic bricks and mortar of relatively high strength. The building has been modeled according to the 1984 Greek seismic code, which was the first code to impose relatively strict anti-seismic rules. The combination of the soft ground story and strong infill panels in upper floors is the most important characteristic of the building and constitutes a serious structural problem related to its seismic behavior (Figure 4.87).

The concrete grade is C20/25 ($f_{ck}$ = 20 MPa, $f_{c,mean}$ = 28 MPa) the steel grade is S400 ($f_{sk}$ = 400 MPa, $f_{s,mean}$ = 444 MPa), and in general there is adequate longitudinal reinforcement (for instance, a typical rectangular or square column has 4∅20 + 4∅16 rebar). The shear reinforcement is ∅8/20 for all the beams and ∅8/25 for the columns of the structure.

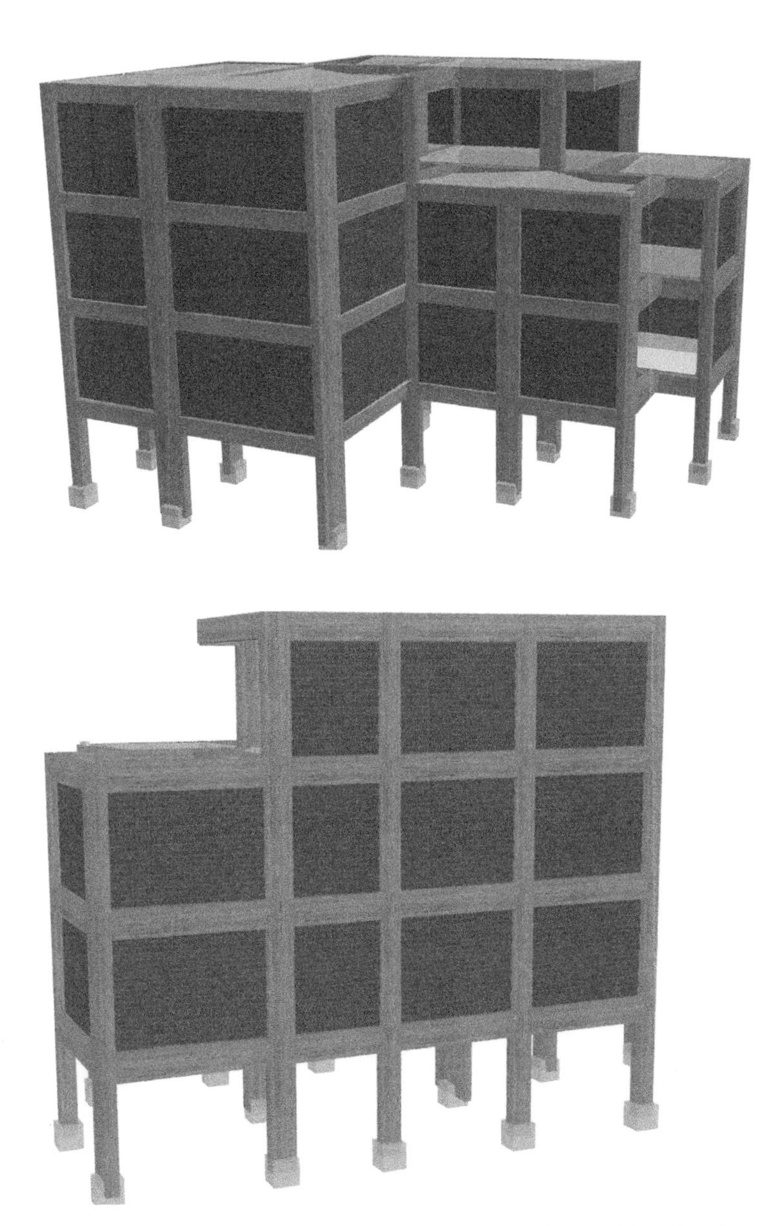

Figure 4.87 Front and back view of the building under consideration. *Source:* Stelios Antoniou.

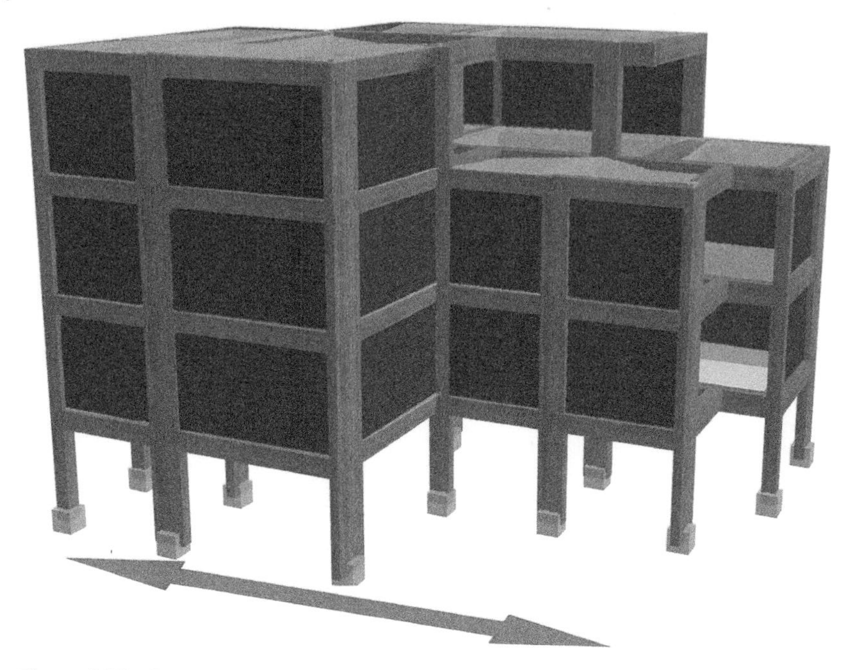

Figure 4.88 Direction of excitation in all the analyses. *Source:* Stelios Antoniou.

It should be noted that, with the exception of the soft ground floor, the overall state of the building is not bad. The reinforcement ratios are relatively large (for instance, the reinforcement of a typical column of a 1970s building could be as low as with 4Ø16 + hoops Ø6/30), and the material grades are also good.

Since the example is just for display purposes, the following simplifications have been made:

- The analyses are carried out only in one direction, as in Figure 4.88. A single ground excitation along this direction will be applied in dynamic analysis, and no accidental eccentricity is considered in the pushover analyses.
- A single lateral force distribution is used in pushover analysis: the uniform distribution.
- Checks will be carried out only in shear. This is a reasonable and realistic simplification, considering that in the vast majority of existing buildings, shear is the most critical check.
- The checks will be done according to EC8 for a single limit state, significant damage SD, which is probably the performance level mostly used for assessment purposes.
- In dynamic analysis a single earthquake ground record is applied. It is a real accelerogram recorded in the 1976 Friuli earthquake in Italy. It has a peak ground acceleration of 0.35 g and a maximum spectral acceleration of approximately 1.30 g (Figures 4.89 and 4.90, respectively). The predominant period of the record (0.27 seconds) is very close to the fundamental period of the building (0.268 seconds), hence it is expected that the excitation will be significant.

All the analyses (eigenvalue, pushover, and dynamic) and all the checks have been carried out with SeismoStruct (2023).

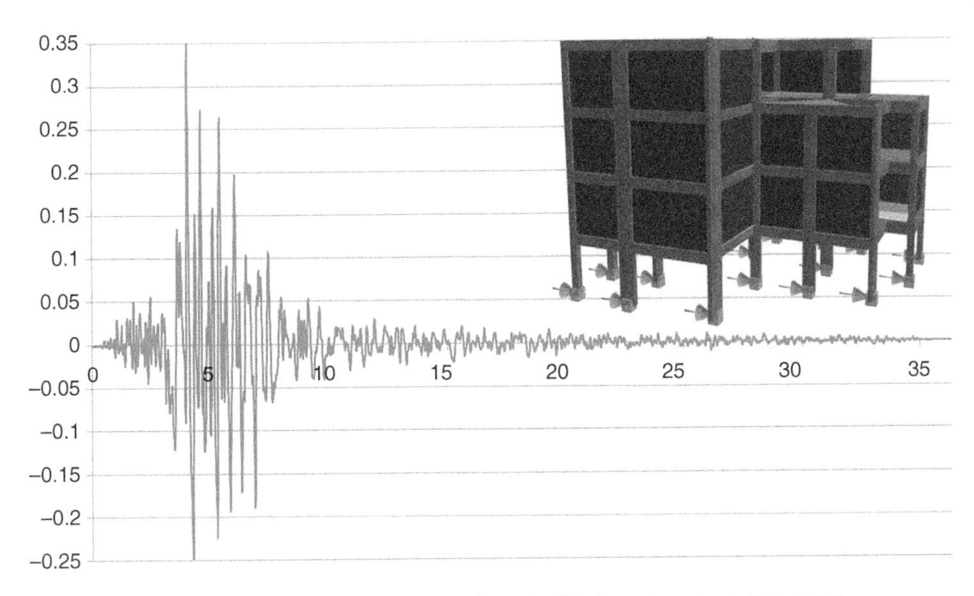


**Figure 4.89** The selected accelerogram (in g) from the Friuli earthquake in Italy, 1976. *Source:* Stelios Antoniou.

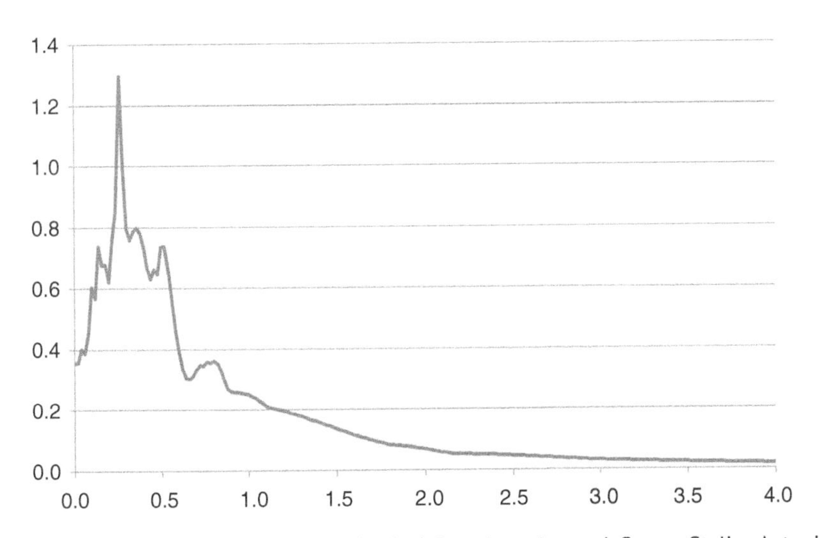


**Figure 4.90** Response spectrum (in g) of the selected record. *Source:* Stelios Antoniou.

With *eigenvalue analysis* a fundamental period of 0.268 seconds was calculated in the considered direction with an effective modal mass of 93.9%, indicating that the response is dominated by the first mode. The shape of the fundamental mode is characteristic of the weak story, with very large deformations at the ground level and negligible in upper floors (Figure 4.91).

The results of *pushover analysis* indicate a target displacement of 2.76 cm for the SD limit state, and a maximum building capacity of approximately 2.900 kN at a top displacement of 5 cm (Figure 4.92).

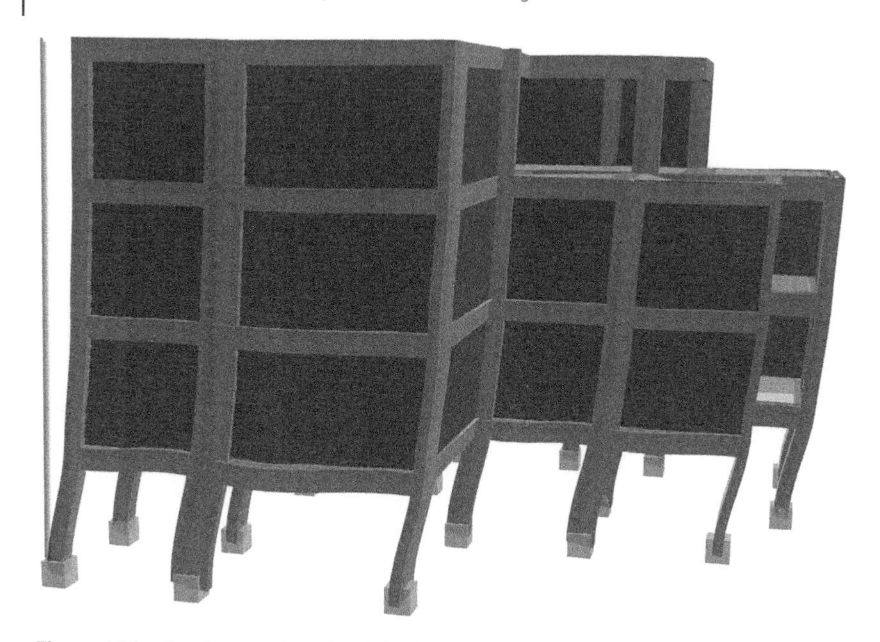

Figure 4.91 Fundamental mode of the building (T = 0.268 seconds). *Source:* Stelios Antoniou.

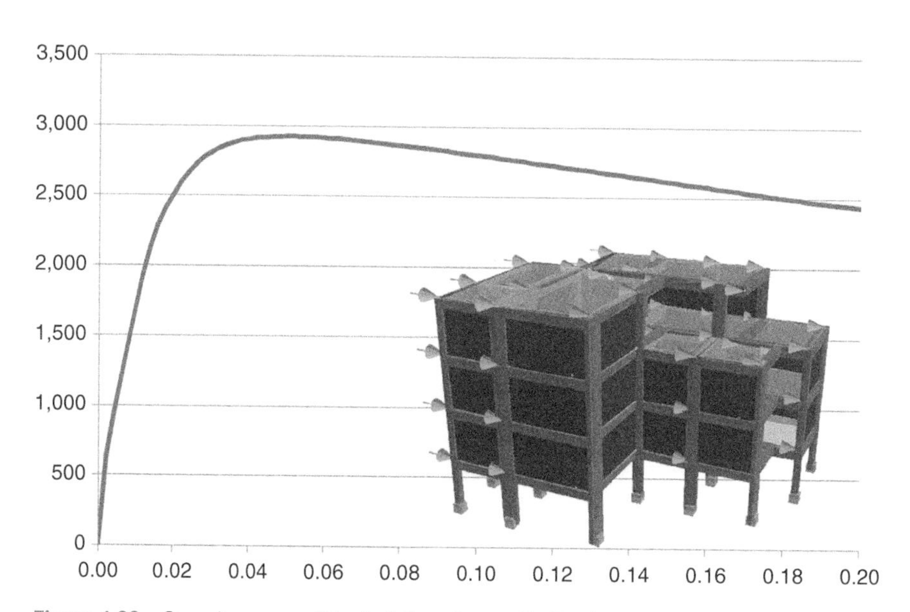


Figure 4.92 Capacity curve of the building. *Source:* Stelios Antoniou.

Finally, with *dynamic analysis* a maximum top displacement of 3.63 cm is obtained, as depicted in Figure 4.93. The top displacement indicates a total interstory drift of 0.30% for the entire building. At first sight, this is a bit surprising, considering the significant weakness of the building at the ground level. Such low drift indicates light to very light damage; in general, significant structural damage is expected for drifts close to 1.0% or more.

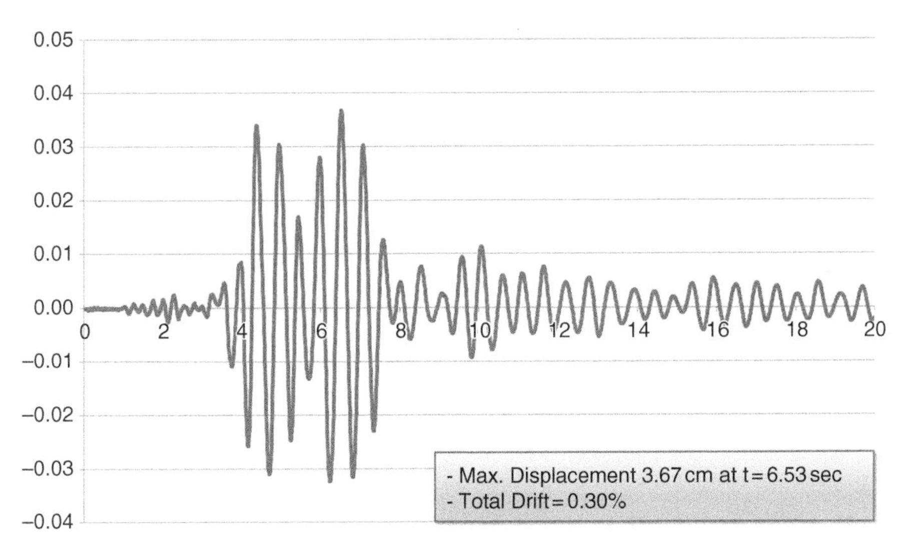


Figure 4.93 Top displacement vs. time plot. *Source:* Stelios Antoniou.

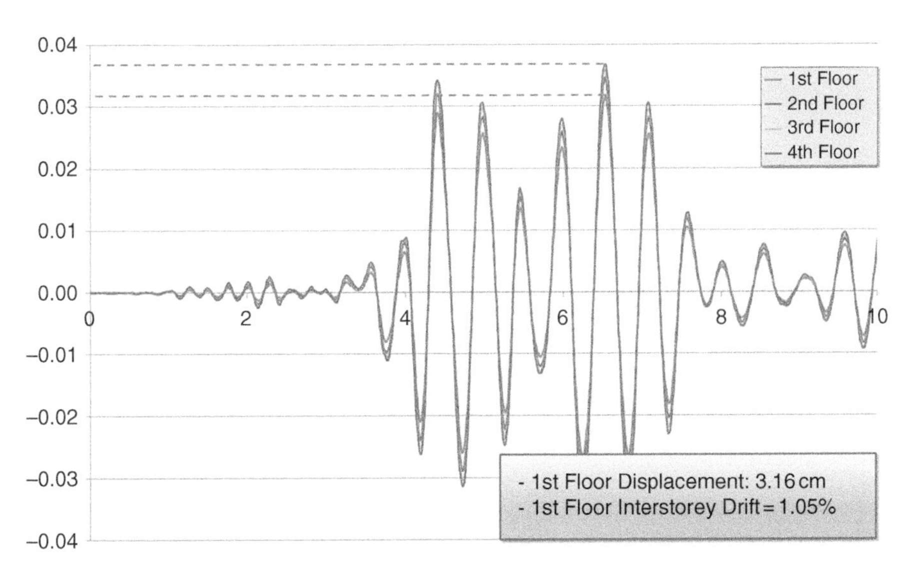


Figure 4.94 Displacement vs. time plots for all floor levels. *Source:* Stelios Antoniou.

However, a closer look at the response on the local level gives a better insight. In Figure 4.94 the response of all four stories vs. time are plotted. It can be observed that, whereas the total displacement at the top level (i.e., the total drift of all four stories) is 3.67 cm, the deformation at the ground level only is 3.16 cm, meaning that 86% of the total deformation is concentrated at the bottom! This leads to a drift of more than 1% at the ground floor, which indicates important damage in the structural members. This is better illustrated in Figure 4.95, where the deformed shape of the building at the time step of maximum displacement is shown. It can be seen that there are large deformations at the bottom, while the rest of the building remains almost undeformed.

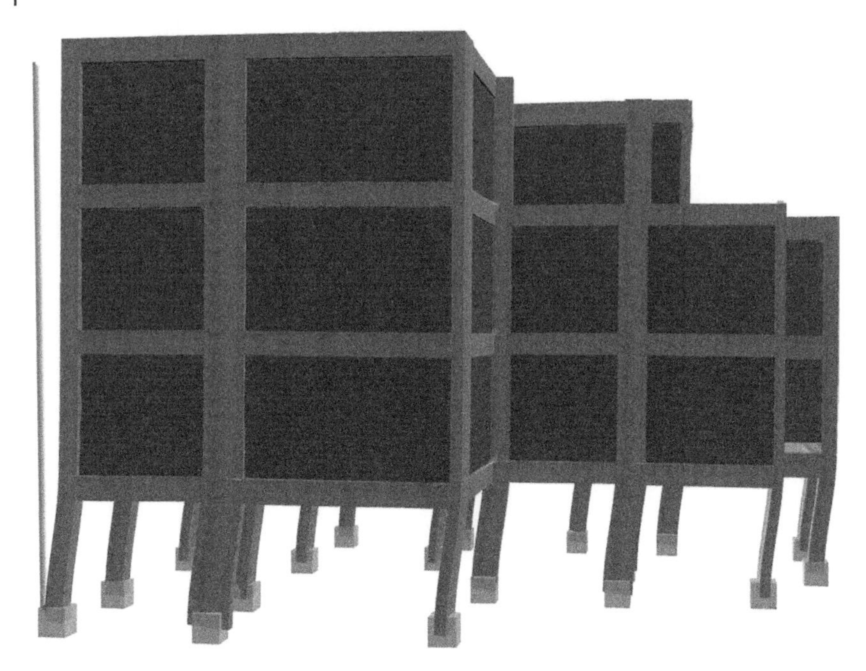

Figure 4.95 Deformed building shape at the time step of maximum displacement. *Source:* Stelios Antoniou.

If one plots the shear force diagram at the time step of maximum displacement (Figure 4.96), it is easily understood why all structural damage is to be expected at the ground level. The shear demand at the columns of the ground soft story is significantly larger than the forces at the corresponding columns of the upper floors, which indicates that the infill panels withstand a large proportion of the applied inertia forces, significantly relieving the adjacent columns. Recall that the infills are well-constructed and relatively strong; thus, they are expected to safely withstand and undertake relatively large forces.

Finally, looking at the code-based checks in Figure 4.97 (just in shear for the Significant Damage limit state, as mentioned before in this section), the maximum demand to capacity ratio DCR is as high as 1.87. The value of the DCR is an indication of whether the member can sustain the imposed demand; with DCR > 1 structural failure occurs, and with DCR < 1 the member is safe. In the plot, the structural members (columns or beams) that have failed are depicted in blue. As expected, all the observed failures are located at the ground level, confirming the fact that soft ground stories are indeed an element of increased vulnerability for existing buildings.

### 4.20.2 Effects of the Infills on the Structural Behavior

Before moving on to the next sections and investigating the effect of the different strengthening methods on the structural response, it is important to study the actual effect of the presence of the infills in upper stories. This will be done by comparing the structural response of the building under consideration with the RC frame without any infills (bare frame) and with infills in all floors (fully infilled) as in Figure 4.98.

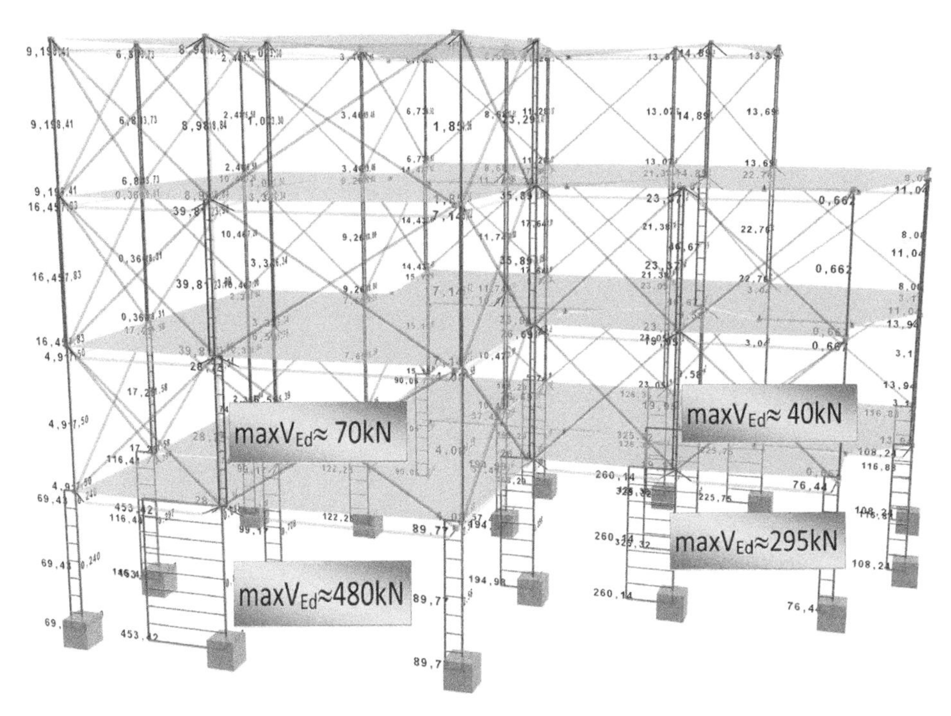


**Figure 4.96** Shear Force Diagram at the time step of maximum displacement. *Source:* Stelios Antoniou.

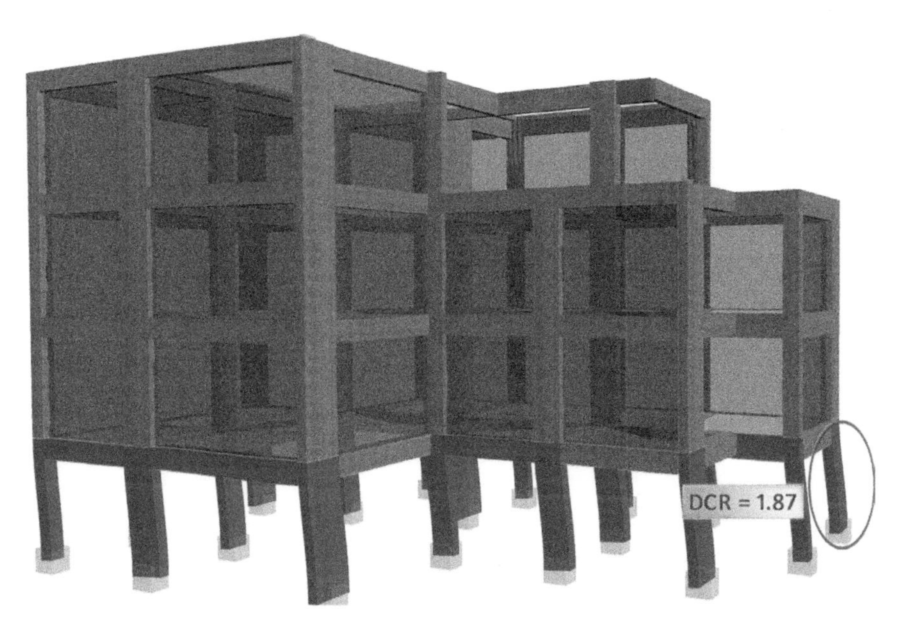


**Figure 4.97** Members failed in shear at the time step of maximum displacement. *Source:* Stelios Antoniou.

Figure 4.98 Bare and fully infilled building. *Source:* Stelios Antoniou.

*Eigenvalue analysis* indicates, as expected, that the bare frame (with a fundamental period of 0.31 seconds) is more flexible, and the fully infilled (with a fundamental period of 0.23 seconds) is less flexible than the partially infilled building. What is more, the stiffness regularity in elevation of both buildings (bare and fully infilled) results in more evenly distributed deformations along the height for all the modes of vibration (Figure 4.99).

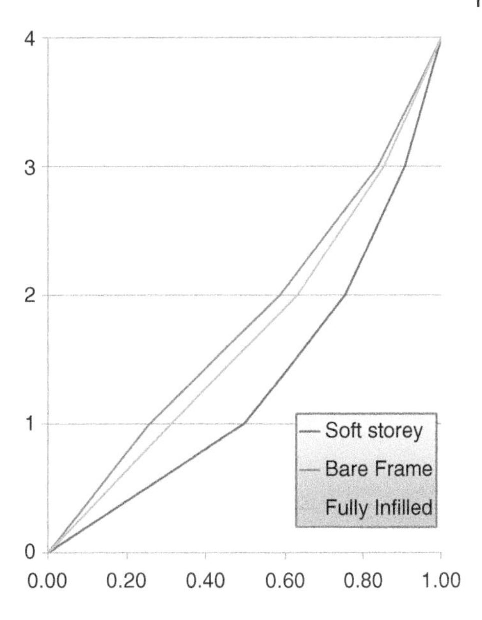


**Figure 4.99** Normalized modal shapes along the height for the 1st eigenmode of (a) the partially infilled building with the soft story, (b) the bare frame and (c) the fully infilled frame. *Source:* Stelios Antoniou.

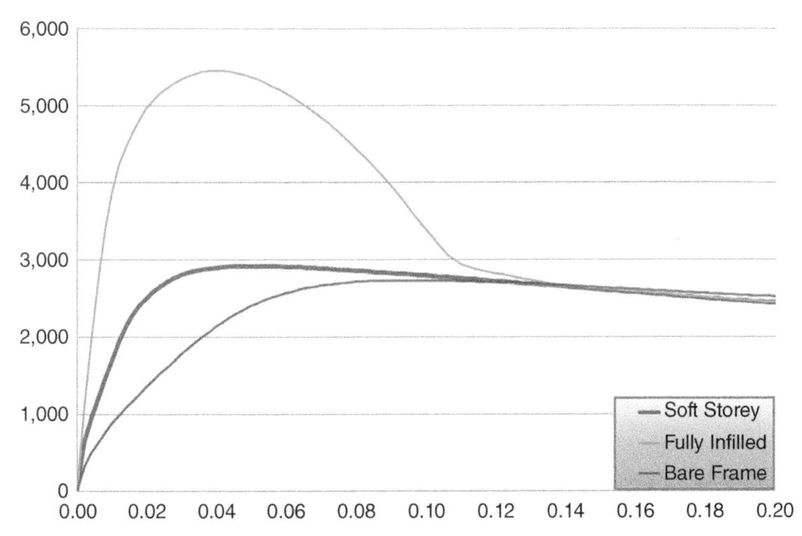


**Figure 4.100** Capacity curves of (a) the partially infilled, (b) the bare frame and (c) the fully infilled frame. *Source:* Stelios Antoniou.

The results of *pushover analysis* (Figure 4.100) show in a more obvious way the adverse effect of the infills, when these are constructed only in the upper stories. The capacity of both the partially infilled and the bare frame is approximately 3.000 kN; however, the maximum capacity of the soft story frame is reached at a considerable smaller deformation level (at approx. 3 vs. 9 cm for the bare frame). The irregularities induced by the construction of infills, despite their lateral strength and stiffness, result in a building that is much more vulnerable, highlighting the importance of maintaining (or restoring) the building's regularity, when designing retrofit interventions.

Another interesting observation from Figure 4.100 is the large drop of the capacity of the fully infilled building after the peak of the capacity curve (from 5.500 kN to approximately 2.800 kN), which is attributed to the gradual failure of the infills as the deformation level increases. Such a drop obviously does not occur for the bare frame, but it is a bit surprising at first that this drop is not present in the capacity curve of the partially infilled building. The reason for this is that the damage sustained by the columns at the weak ground level functions as a kind of seismic isolation cutting off the forces passed to the upper stories and preventing the failure of structural or non-structural components at the upper levels. Indeed, after close investigation it can be seen that even in the highly inelastic range, the infills in upper stories remain elastic, further accentuating the building's irregularity.

If the full nonlinear static procedure for structural assessment in applied, it is observed that no failures occur at the fully infilled frame, which emphasizes the fact that infills – if constructed correctly – can provide significant protection against seismic forces. On the contrary, in both the bare and the partially infilled frames, several members do not pass the code-based checks. One interesting observation though is that, whereas in the bare frame the failures are distributed along the height of the building, in the partially infilled frame failures only occur at the members of the ground story, mostly the columns (Figure 4.101).

With *dynamic analysis* similar observations with the eigenvalue and pushover analyses have been made. In the bare and the fully infilled frame, the lateral deformations are more evenly distributed. The fully infilled frame effectively resisted the excitation, but the bare frame experienced member failures in the two lower floors. Finally, in the partially infilled frame failures were observed only in the columns of the ground level.

(a)

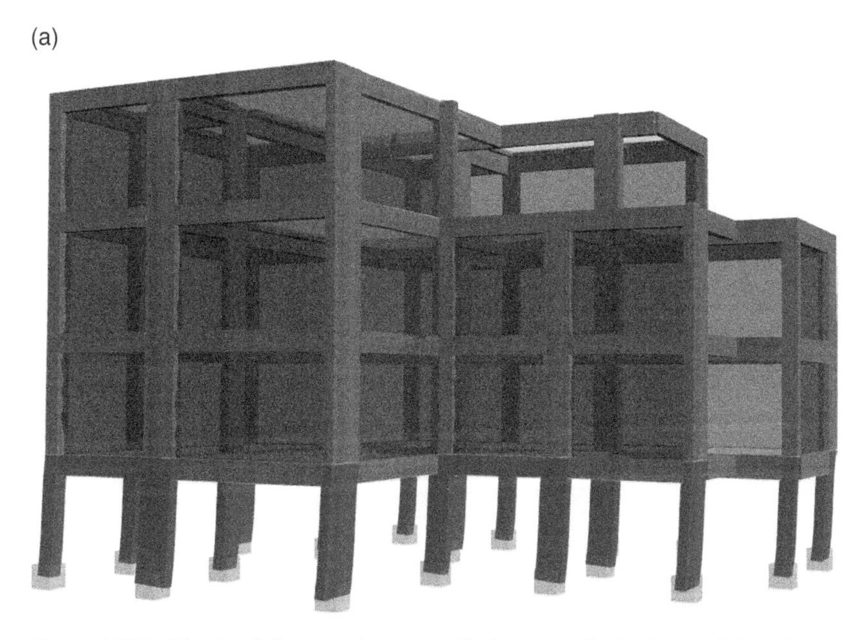

**Figure 4.101** Member failures at the target displacement for (a) the partially infilled, (b) the bare frame and (c) the fully infilled frame. *Source:* Stelios Antoniou.

(b)

(c)

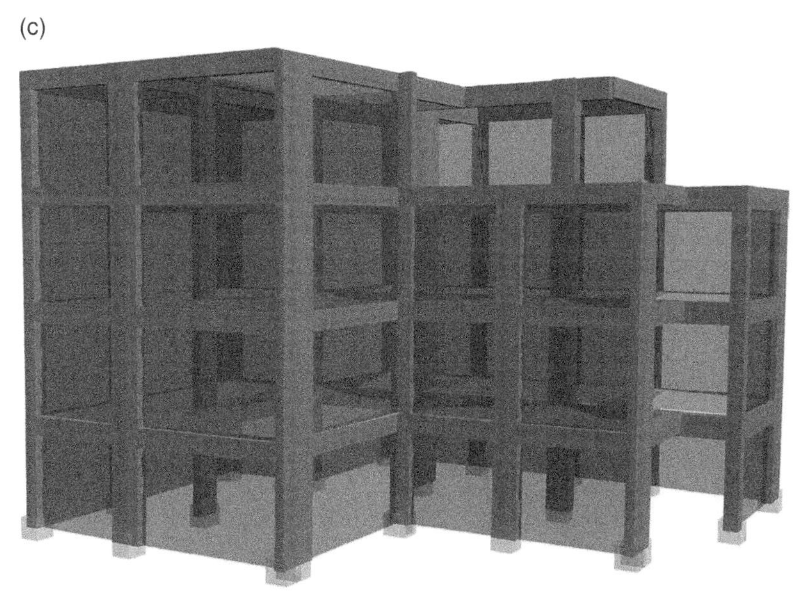

Figure 4.101 (Continued)

### 4.20.3 Strengthening with Jacketing

The retrofit scheme is applied in all the ground story columns, which were strengthened with 10 cm wide shotcrete jackets. The jackets' concrete grade is C25/30, and the steel grade is B500c. The longitudinal reinforcement of a typical rectangular jacket is 4∅20 + 4∅16, and the shear reinforcement of the members is significantly increased in all columns with stirrups of ∅10/10. Figure 4.102 shows indicative jacket layouts, as modeled in SeismoStruct.

*Eigenvalue analysis* indicates a significant increase in the stiffness of the building with a fundamental period of 0.199 seconds, i.e., a considerable decrease of more than 25% with respect to the 0.268 seconds of the initial building (Figure 4.103). What is more important

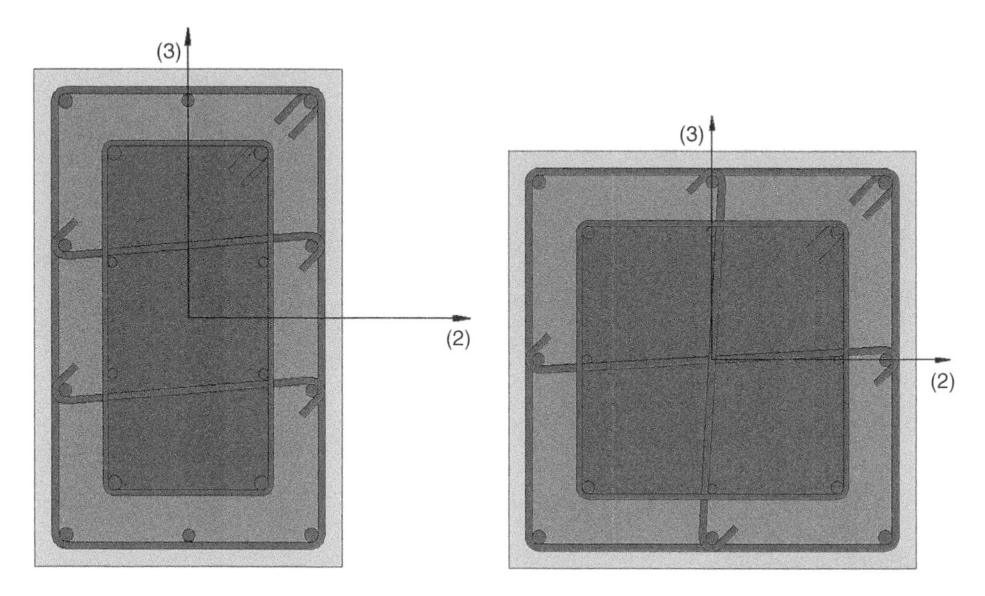


**Figure 4.102** Typical layout of the jacketed sections. *Source:* Stelios Antoniou.

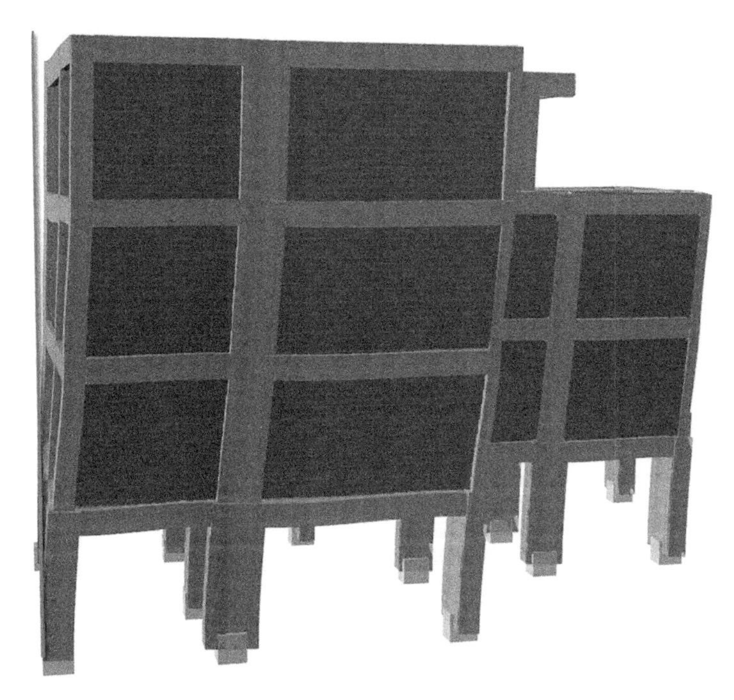

**Figure 4.103** Fundamental mode of the building strengthened with jackets (T = 0.199 seconds). *Source:* Stelios Antoniou.

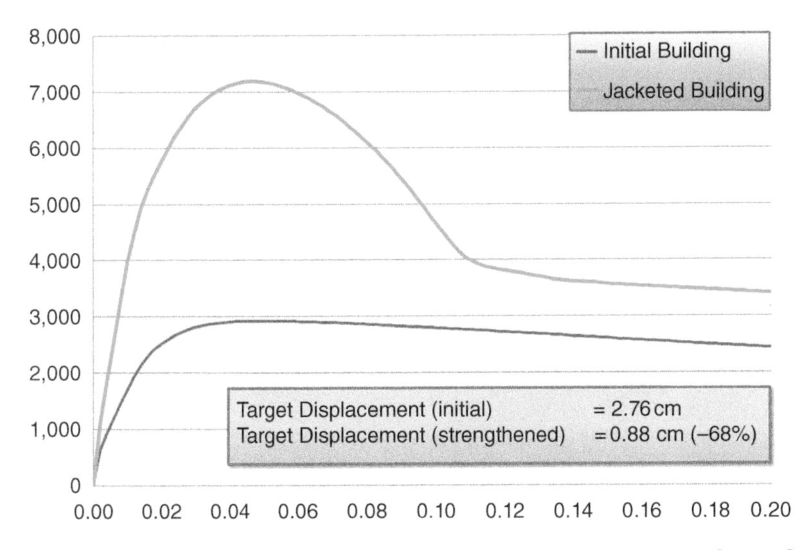


**Figure 4.104** Capacity curve of the building strengthened with jackets. *Source:* Stelios Antoniou.

is that the shape of the fundamental mode has changed significantly with larger deformations at the upper stories, rather than at the ground floor.

The results of *pushover analysis* indicate an impressive increase in the capacity of almost 150% from 2.900 kN to almost 7.200 kN (Figure 4.104). Furthermore, there is a very large decrease in the calculated target displacement of 68% from 2.76 cm to just 0.88 cm, an indication of the considerable stiffening of the building.

Another observation is the large drop of the capacity of the building after the peak of the capacity curve (from 7.200 kN to approximately 3.500 kN). This drop, as in the case of the fully infilled building of the previous section, is attributed to the gradual failure of the infills at the upper levels, and is not present in the capacity curve of the initial building. The jacketed columns at the ground level exhibit significant strength now, the ground level does not cut off the seismic forces and it does not function as seismic isolation. As a result, the infills at the upper levels sustain damage, gradually reducing the global lateral capacity.

With *dynamic analysis* (Figure 4.105) similar observations are made, i.e., significant increase of the stiffness at the ground level, larger deformations at the upper floors, increased building capacity, and fewer failures in the structural elements.

Looking at the shear code-based checks of dynamic analysis in Figure 4.106, there are very few component failures, and these have much smaller DCR values (maximum DCR = 1.07). Furthermore, the members that fail are either beams at the ground floor or columns of the second floor. The capacity is not exceeded in any of the strengthened columns at the ground level, and altogether the building has become safer.

### 4.20.4 Strengthening with New RC Walls (Entire Building Height)

The retrofit is carried out with five new reinforced concrete walls at the perimeter, which extend to the full height of the building (Figure 4.107). The reinforcement of the walls is

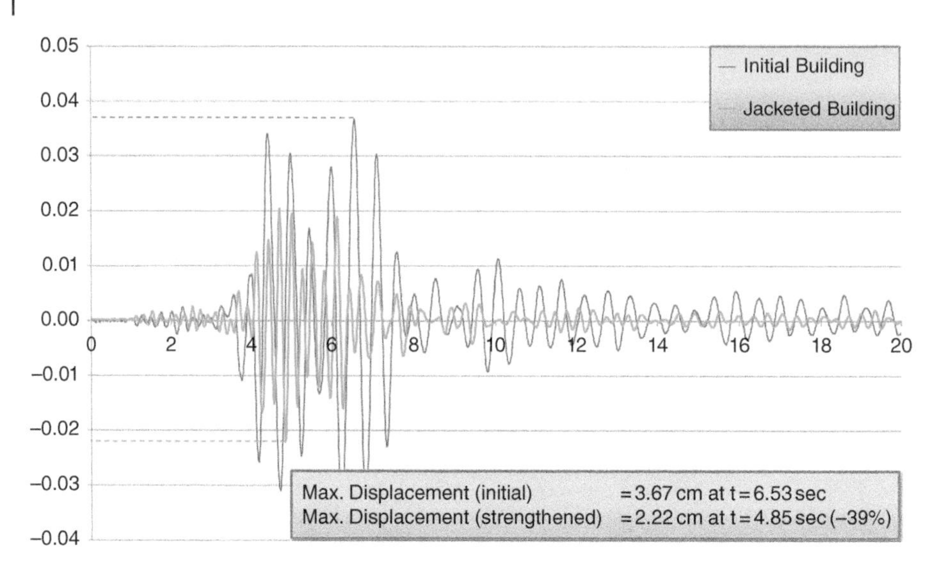


Figure 4.105 Top displacement vs. time plot of the building strengthened with jackets *Source:* Stelios Antoniou.

Figure 4.106 Shear checks at the time step of maximum displacement (building strengthened with jackets). *Source:* Stelios Antoniou.

similar to what would be found in shear walls in a new construction with pseudo-columns with a typical reinforcement of 8∅20 + 16∅16 with ∅10/10 stirrups (Figure 4.108). The walls are placed in symmetrical positions in the perimeter, in order not to introduce undesired torsional effects, and they are connected with the existing beams and the columns through a large number of dowels that are designed to transfer the seismic inertia forces from the weak building to the strong walls.

Figure 4.107 Front and back view of the building strengthened with new RC walls. *Source:* Stelios Antoniou.

*Eigenvalue analysis* indicates that the strengthened building is significantly stiffer with respect to the initial building with a fundamental period of 0.116 seconds (an impressive decrease of more than 55%). Furthermore, the eigenshape now has a cantilever-like shape, indicative of the presence of large shear walls in the construction (Figure 4.109).

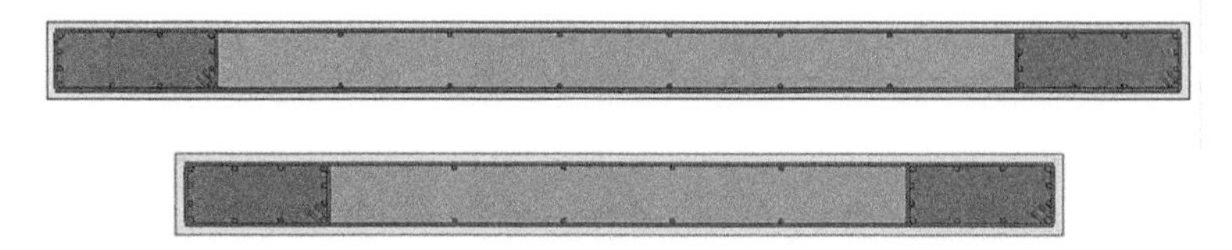

Figure 4.108 Typical configuration of new RC walls. *Source:* Stelios Antoniou.

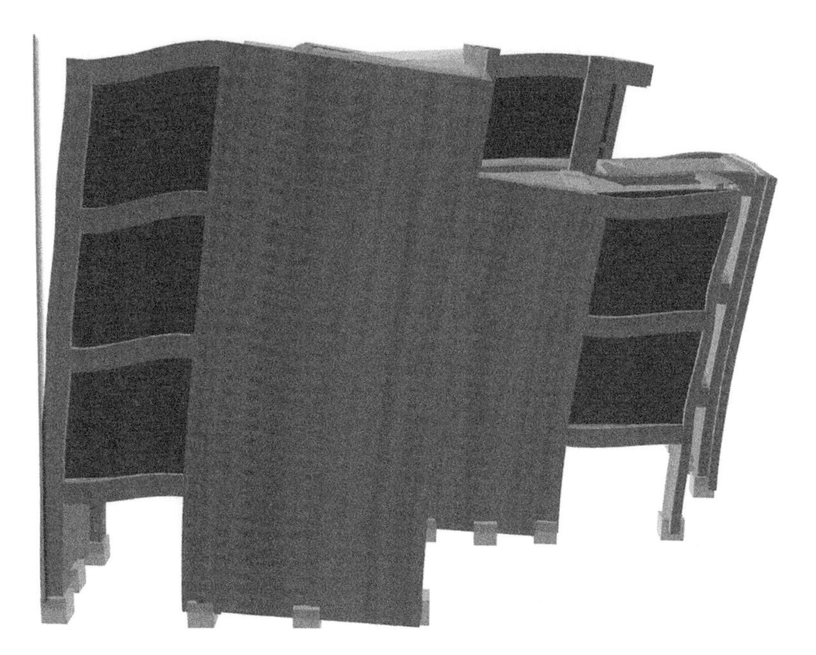

Figure 4.109 Fundamental mode of the building strengthened with new RC walls (T = 0.116 seconds). *Source:* Stelios Antoniou.

The results with *pushover analysis* show a very large increase in the building capacity of more than three times from 2.900 kN to approximately 10.000 kN (Figure 4.110), whilst the target displacement has decreased by more than 75% to 0.74 cm. Furthermore, there is no significant drop in the capacity curve after the peak, which indicates that a large proportion of the story shear is directly undertaken by the strong ductile walls, rather than the infills or the weak and without ductility existing columns, which remain mostly undamaged.

*Dynamic analysis* (Figure 4.111) shows a similarly improved performance. There is low amplitude and high frequency motion throughout the time-history, and the top maximum displacement has dropped by almost 90% to just 4 mm.

Obviously, with such low deformation level, there is no damage with the maximum DCR occurring in a beam and being equal to 0.66, i.e. much lower than 1.00 (Figure 4.112). All the vertical members are far from failure, with the most critical ones having a capacity more than twice the demand (maximum DCR = 0.47).

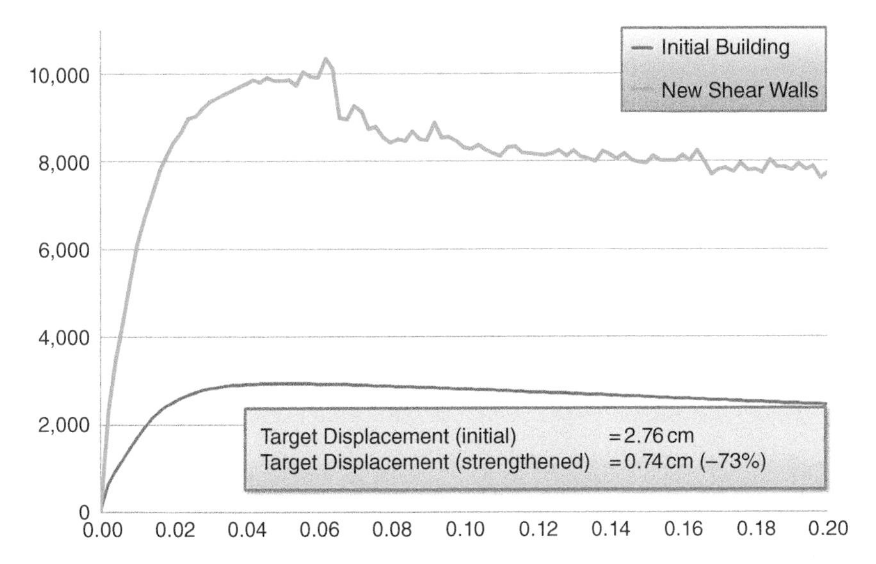


Figure 4.110 Capacity curve of the building strengthened with new RC walls. *Source:* Stelios Antoniou.

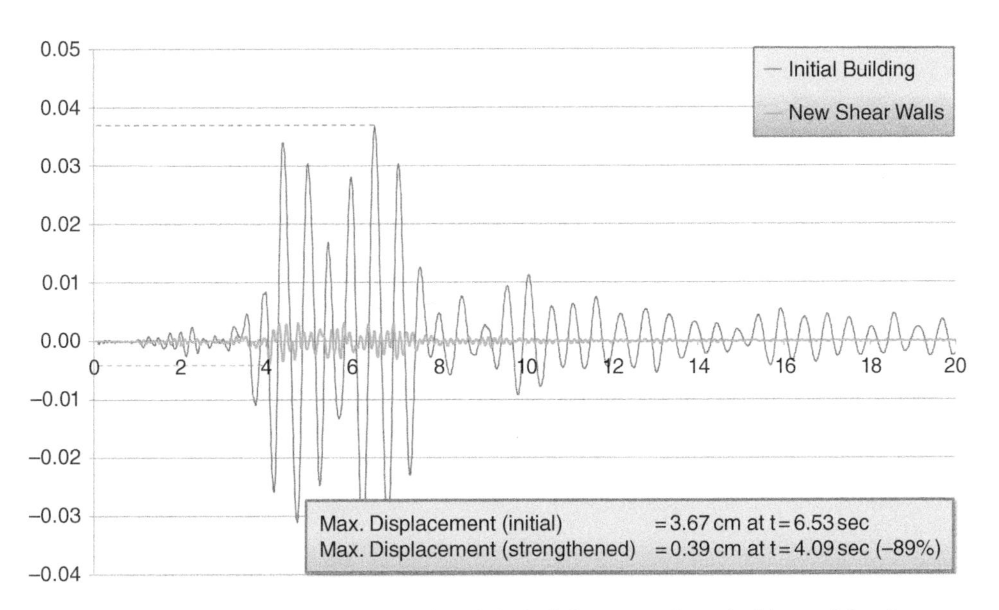


Figure 4.111 Top displacement vs. time plot of the building strengthened with new RC walls. *Source:* Stelios Antoniou.

Another interesting observation, which explains the good performance of the strengthened building, is that the walls at the ground level undertake 90% of the base shear (Figure 4.113). This means that the existing members have to withstand only 10% of the total seismic demand, and this is why they easily pass the required

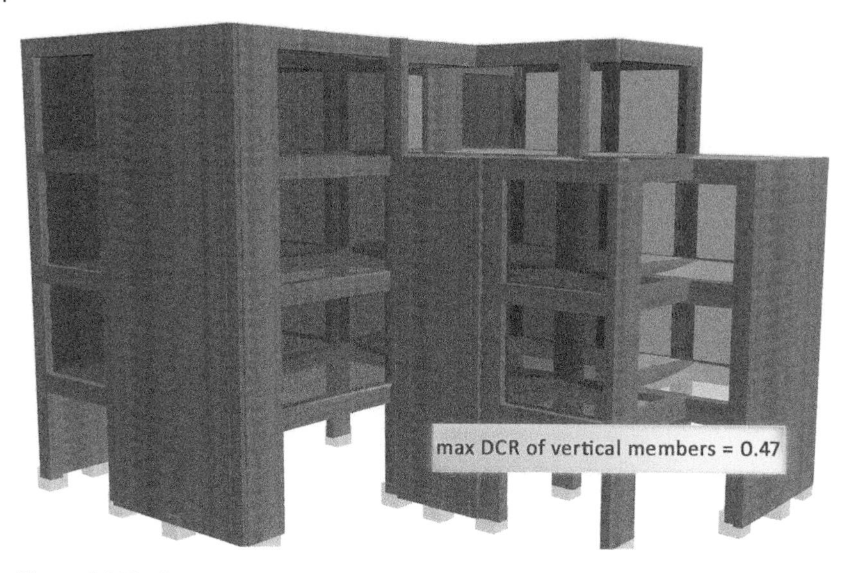


Figure 4.112 Shear checks at the time step of maximum displacement (building strengthened with new RC walls). *Source:* Stelios Antoniou.

checks.[4] This example is a good demonstration of the fact that, when large shear walls are added in an existing building, it is very common that no other intervention is required in the other structural members, provided of course that these are in relatively good condition, and possess some, non-negligible, existing reinforcement and lateral strength.

### 4.20.5 Strengthening with New RC Walls (Ground Level Only)

An alternative to the previous method is to carry out the strengthening with shear walls only at the ground level, as shown in Figure 4.114. The method has the advantage that all the interventions are carried out at the open ground level without works and disturbance to the upper stories. The walls are located in the same positions as in the case of walls in the entire building height, and their dimensions and reinforcement are the same.

*Eigenvalue analysis* shows that the strengthened building is stiffer with respect to the initial building with a fundamental period of 0.169 seconds (37% decrease), but – as expected – not as stiff as in the case with walls in the full height. The shape of the first mode indicates that all the displacements are distributed evenly in the upper stories, while the building is almost undeformed at the ground level (Figure 4.115).

4 In the example, the shear walls were assumed fully fixed to the ground – i.e., the rotations are considered negligible. This was done, in order to pronounce the contribution of large shear walls in undertaking the seismic inertia forces, and generally requires very large individual footings for the walls or a more complex foundation system with connecting beams, strip footings and/or raft foundations. If this is not possible or desirable, the elastic or inelastic behavior of the footing should be explicitly modeled, which generally decreases the demand on the walls and increases it in the other vertical members.

The results with *pushover analysis* show a very large increase in the building capacity from 2.900 to 7.250 kN (Figure 4.116), while the target displacement has again decreased significantly to 0.59 cm (78% decrease). Since the new strong walls are not constructed in the upper levels, after the maximum capacity, the pushover curve drops significantly due to the gradual failure of the infills in the upper levels.

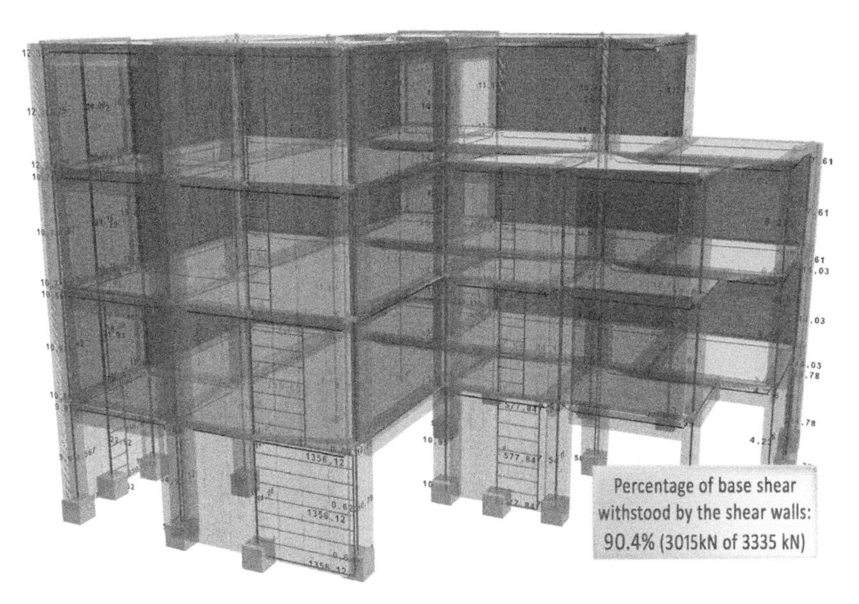

Percentage of base shear
withstood by the shear walls:
90.4% (3015kN of 3335 kN)


Figure 4.113 Percentage of base shear undertaken by the new RC walls. *Source:* Stelios Antoniou.

Figure 4.114 Front and back view of the building strengthened with new RC walls at the ground level. *Source:* Stelios Antoniou.

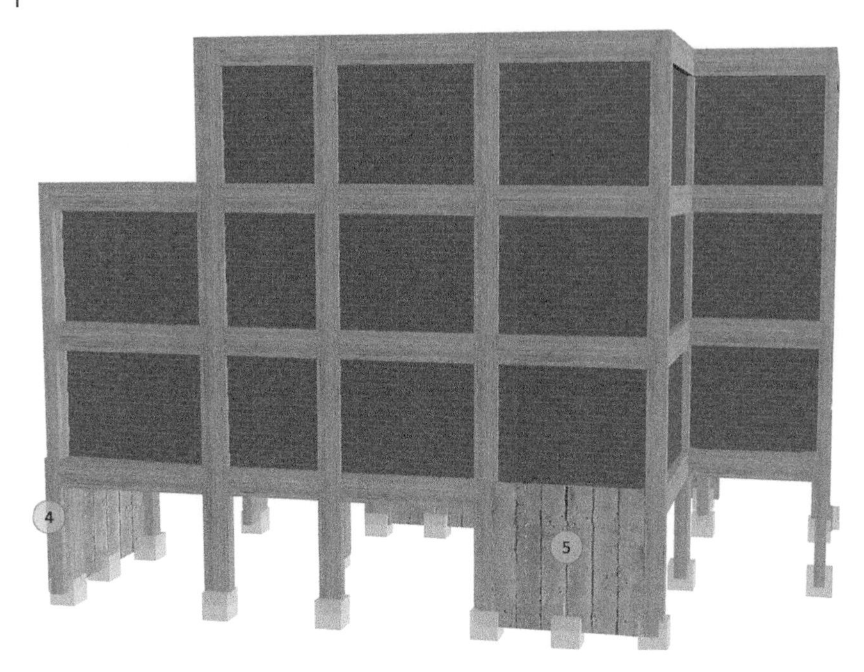


Figure 4.114 (Continued)

Figure 4.115 Fundamental mode of the building strengthened with new RC walls at the ground level (T = 0.169 seconds). *Source:* Stelios Antoniou.

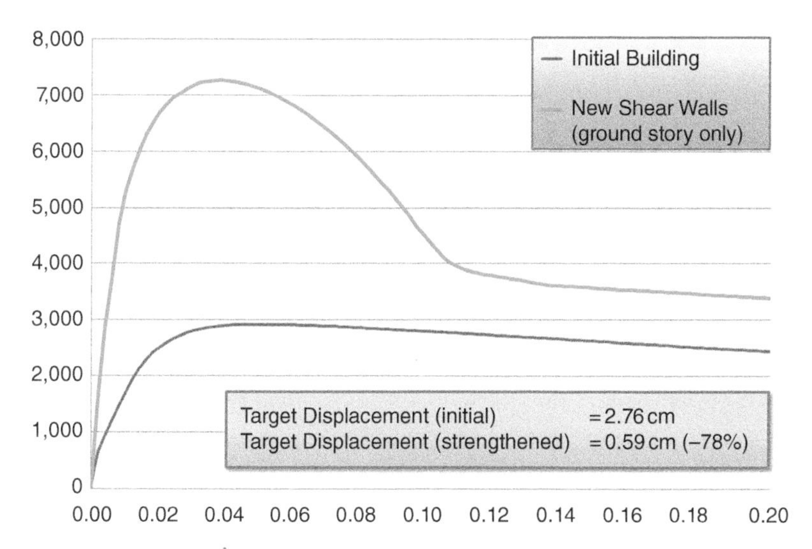


Figure 4.116 Capacity curve of the building strengthened with new RC walls at the ground level. *Source:* Stelios Antoniou.

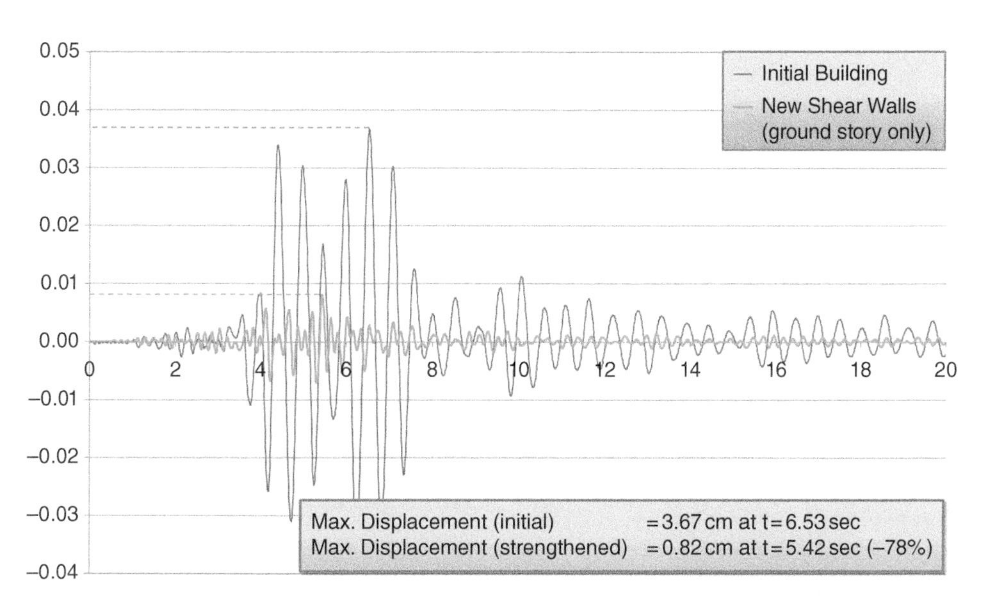


Figure 4.117 Top displacement vs. time plot of the building strengthened with new RC walls at the ground level. *Source:* Stelios Antoniou.

With *dynamic analysis* (Figure 4.117) similar observations with the case of walls in all floors are made: enhanced performance, low amplitude, and high frequency motion throughout the time-history, and significant decrease of the top maximum displacement by 78% to 8.2 mm.

Again, the maximum DCR is below unity, and none of the members fail (Figure 4.118). Furthermore, similar to the case of walls in the full building height, the new shear walls undertake a very large proportion of the base shear at the ground level, i.e. around 90%.

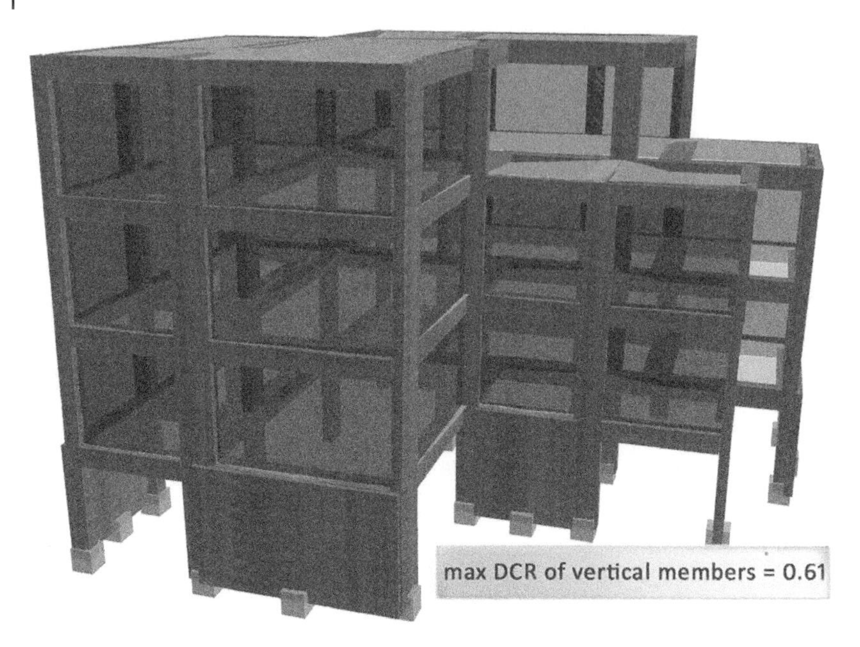


Figure 4.118 Shear checks at the time step of maximum displacement (building strengthened with new RC walls at the ground level). *Source:* Stelios Antoniou.

Again, the existing members are considerably relieved with the proposed interventions and easily pass the required code-based checks.

At first sight, the example demonstrates that the application of new shear walls is a good and sound alternative to the other strengthening methods. Indeed, in the building under consideration, the strengthened model exhibited a highly satisfactory behavior and the method can be applied.

Nonetheless, it should be recalled that in the initial unstrengthened building the infill panels were modeled to be strong on purpose, in order to emphasize the structural main weakness: the soft ground story. From my experience, this is rarely the case. Typical infilled walls in buildings of the 1980s or before are generally weak, not well-constructed, and with mortars and bricks of relatively low strength. What is more, there is no easy and inexpensive way to identify the quality of an adequate sample of the infills in a building, when carrying out the survey for structural assessment, hence in most typical cases, the strength of the infills is known with significant uncertainties.

If the infills in the building under consideration are weak, the analytical results and the conclusions are completely different. The large stiffness of the new RC walls – considerably larger with respect to any other component of the building – introduces significant irregularities in elevation, which have a very adverse effect on the structural response. This problem will be demonstrated with the next example, where the same building is analyzed with the only difference being that now the infills of the upper stories have low strengths.

*Eigenvalue analysis* gives approximately the same results, a fundamental period of 0.169 seconds with almost the same eigenshape, with small deformation at the ground level and evenly distributed displacements through the upper stories (Figure 4.119).

Figure 4.119 Fundamental mode of the building strengthened with new RC walls at the ground level – weak infills (T = 0.169 seconds). *Source:* Stelios Antoniou.

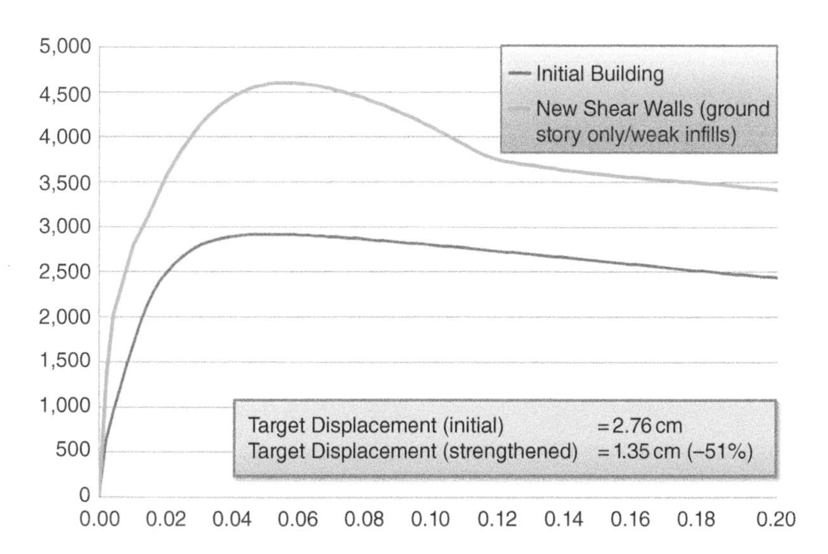


Figure 4.120 Capacity curve of the building strengthened with new RC walls at the ground level – weak infills. *Source:* Stelios Antoniou.

The maximum capacity with *pushover analysis* is now approximately equal to 4.600 kN, i.e., considerably smaller with respect to the case with strong infills. At the peak of the capacity curve a significant part of the story shear is carried by the infills, which are now much weaker. Consequently, the descending branch after the peak is also not as steep (Figure 4.120).

The problems with the structural response of the model become clear with *dynamic analysis* (Figure 4.121). The maximum displacement almost triples and increases from 8 mm to 22.3 mm, and a large number of members in the second floor fail. The maximum DCR is as high as 1.26, and the building is not safe (Figure 4.122).

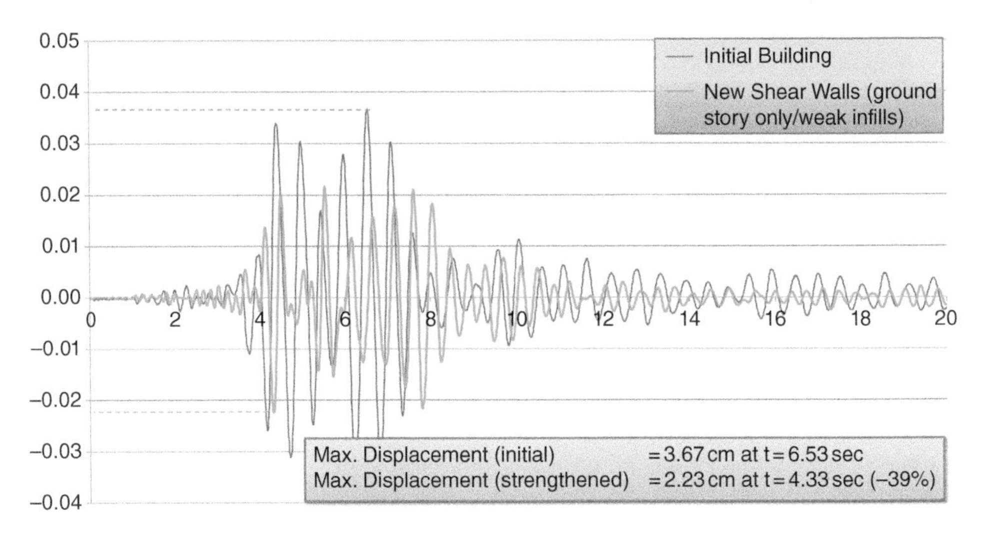


Figure 4.121 Top displacement vs. time plot of the building strengthened with new RC walls at the ground level – weak infills. *Source:* Stelios Antoniou.

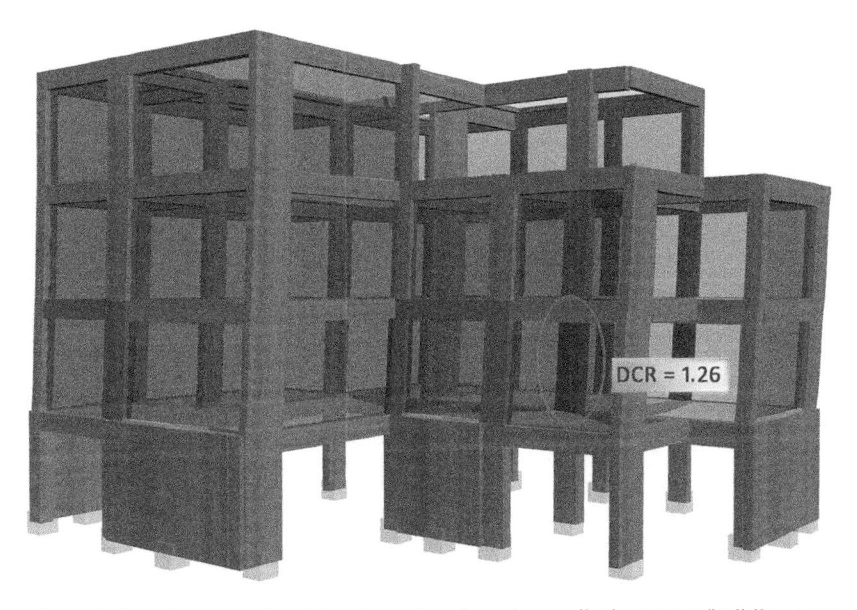


Figure 4.122 Shear checks at the time step of maximum displacement (building strengthened with new RC walls at the ground level – weak infills). *Source:* Stelios Antoniou.

Obviously, the variation in the capacity of the infills plays a very important role in the structural response. What is important to stress is that the exact strength of the infill walls is something that is not known with adequate reliability in most practical applications. Therefore, one should not rely on them to undertake a large part of the earthquake loading. When adding RC walls with very large stiffness with respect to the existing vertical members, the designer should pay particular attention so that these new walls do not cause unintentionally large stiffness irregularities that in turn give rise to undesirable higher-order effects, increasing the vulnerability of the building. This is the reason why new RC walls should generally be extended to the entire height of the building, unless the engineer has adequate data about the nonstructural components and is very certain of what he/she is doing.

### 4.20.6 Strengthening with Braces

The retrofit is carried out at the ground level only, with X-shaped steel braces at six different locations in the perimeter (Figure 4.123). For the braces, 120 × 120 × 8 hollow rectangular sections are employed, which are pinned to steel plates that are fixed at the corners of each concrete bay with steel anchors and epoxy adhesives. At the ground level special footing configurations for each brace are used. As with the case of RC walls, the braces should be placed as symmetrically as possible to avoid unwanted torsional effects.

*Eigenvalue analysis* indicates a moderate increase in the stiffness of the building with a fundamental period of 0.221 seconds (Figure 4.124). The shape of the fundamental mode has changed, and it has uniformly distributed deformations along the height, rather than large deformation concentrations at the ground level. It is noted that the increase in stiffness that can be achieved with steel braces is generally less than that with new RC walls. What is more, the stiffness of the braces can be calibrated easily (i.e., by changing the

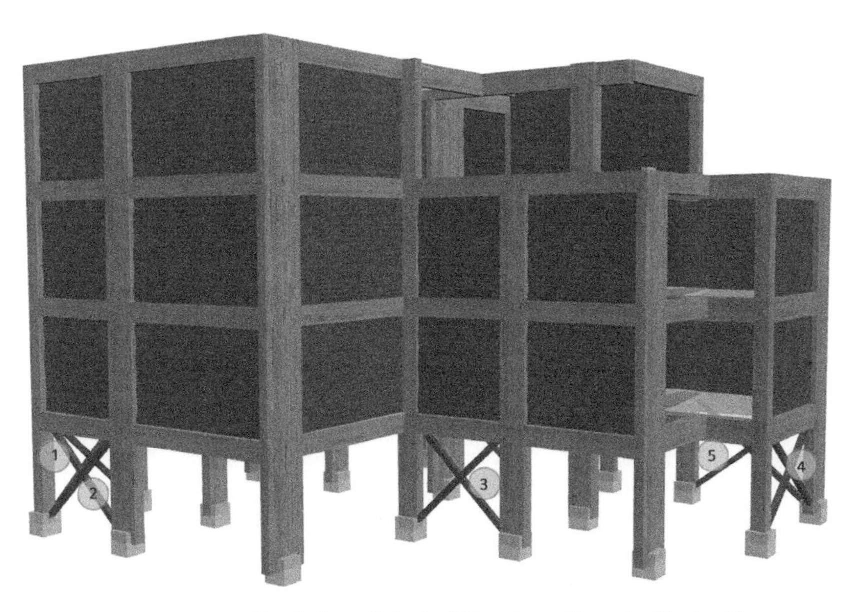


**Figure 4.123** Front and back view of the building strengthened with steel braces.

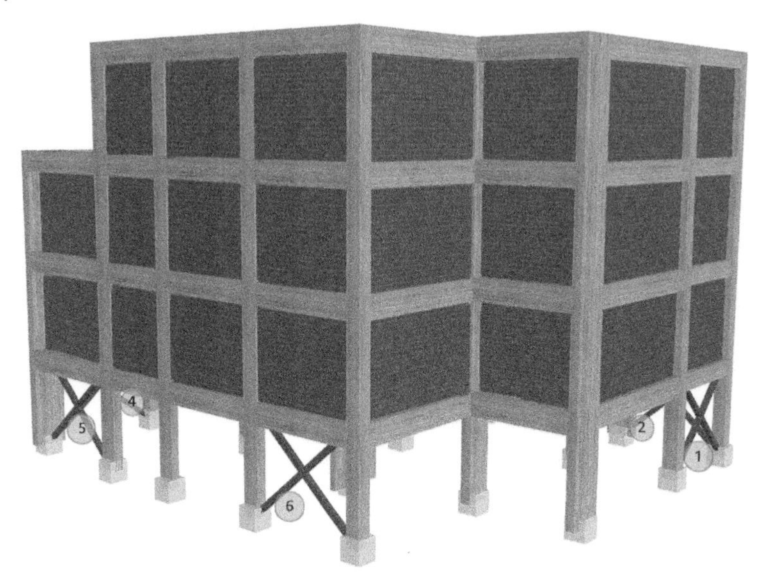


Figure 4.123 (Continued)

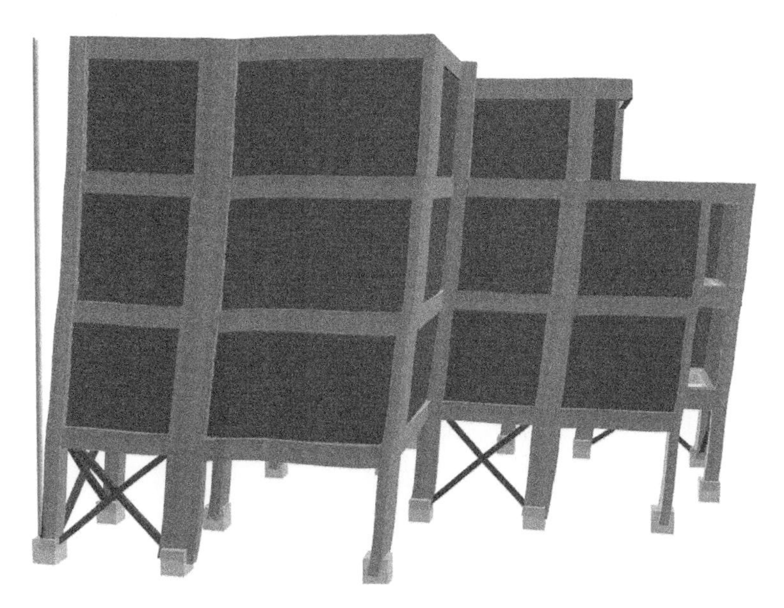

Figure 4.124 Fundamental mode of the building strengthened with steel braces (T = 0.221 seconds). *Source:* Stelios Antoniou.

cross-sections of the braces), in order to avoid irregularities in elevation, when the strengthening is not applied to the entire building height.

The *pushover analysis* results (Figure 4.125) are similar to the case of jackets, with an increased overall capacity from 2.900 kN to almost 7.200 kN, a large decrease in the calculated target displacement to 1.16 cm (58%), and a significant drop in the capacity after the peak of the curve, because of the gradual failure of infills in upper levels.

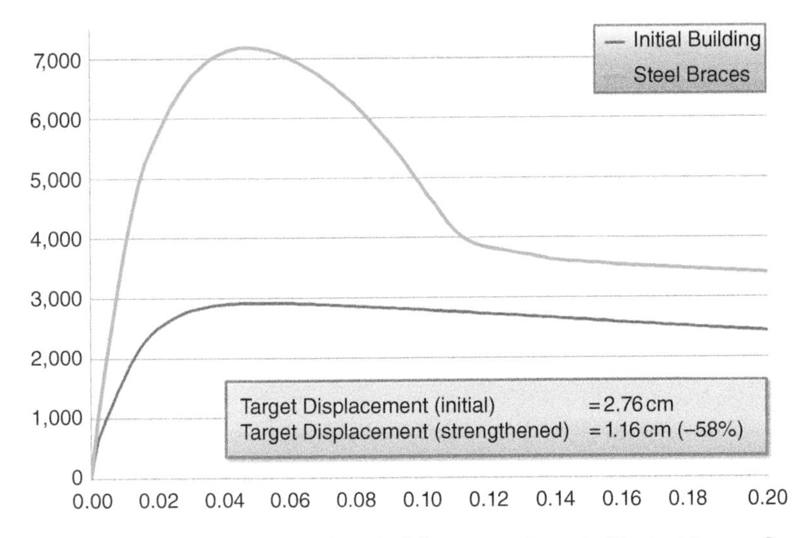


**Figure 4.125** Capacity curve of the building strengthened with steel braces. *Source:* Stelios Antoniou.

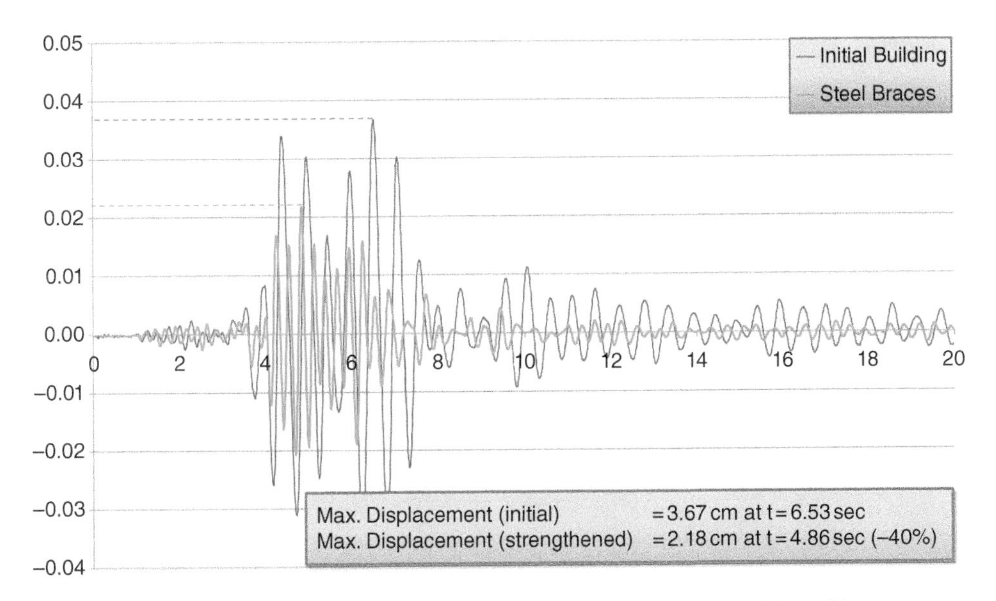


**Figure 4.126** Top displacement vs. time plot of the building strengthened with steel braces. *Source:* Stelios Antoniou.

With *dynamic analysis* (Figure 4.126) similar observations are also made: increased stiffness at the ground level, 40% decrease in the displacement demand, increased building capacity, and fewer failures in the structural elements. The latter is demonstrated in Figure 4.127: the maximum DCR is 1.18, and the maximum DCR at the previously weak ground floor is almost 1.00, indicating a much safer building with respect to the initial unstrengthened configuration.

Figure 4.127 DCRs for shear checks at the time step of maximum displacement (building strengthened with steel braces). *Source:* Stelios Antoniou.

### 4.20.7 Strengthening with FRP Wrapping

The retrofit is carried out in all the ground story columns (Figure 4.128), which are strengthened with three layers of a relatively strong FRP fabric (dry fiber thickness approx. 0.33 mm, area density of the carbon fibers approx. 600 g/m$^2$ and dry fiber tensile strength approx. 3800 N/mm$^2$).

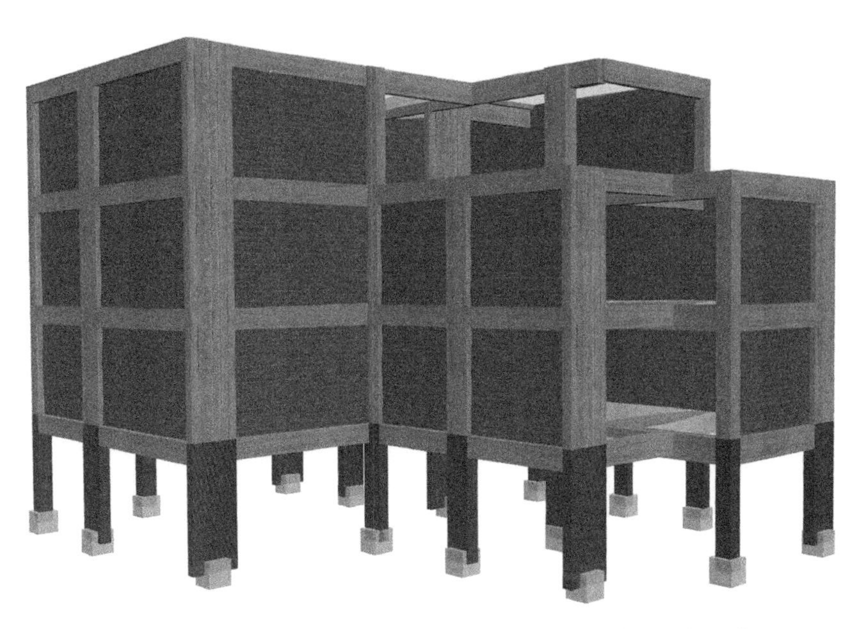

**Figure 4.128** Strengthened building with FRP wraps. *Source:* Stelios Antoniou.

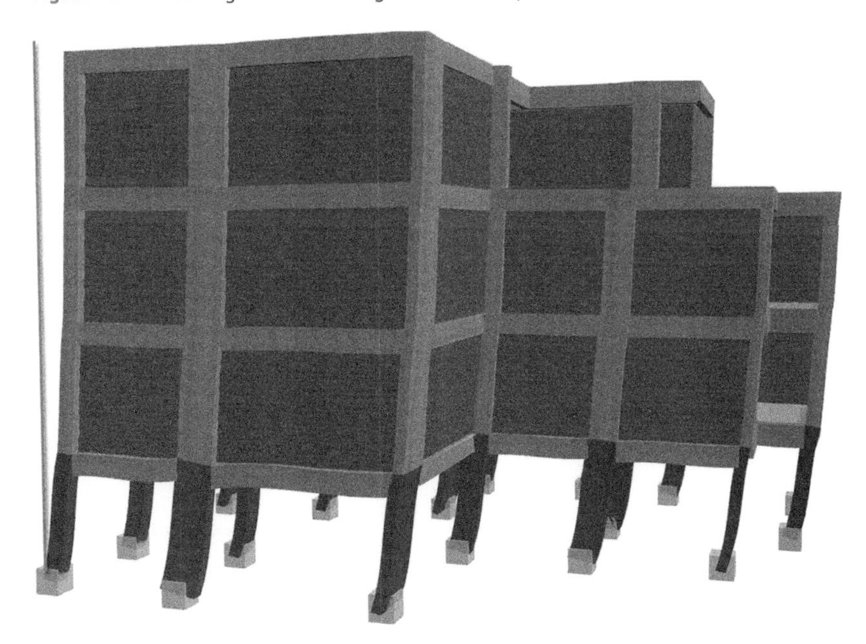

**Figure 4.129** Fundamental mode of the building strengthened with FRP wraps (T = 0.268 seconds). *Source:* Stelios Antoniou.

The results from *eigenvalue analysis* are almost identical with the results of the initial building. The fundamental period is the same, 0.268 seconds, and its shape shows large deformations at the ground level only (Figure 4.129). Similarly, the capacity curve derived from *pushover analysis* is very close to that of the initial building (Figure 4.130), with the

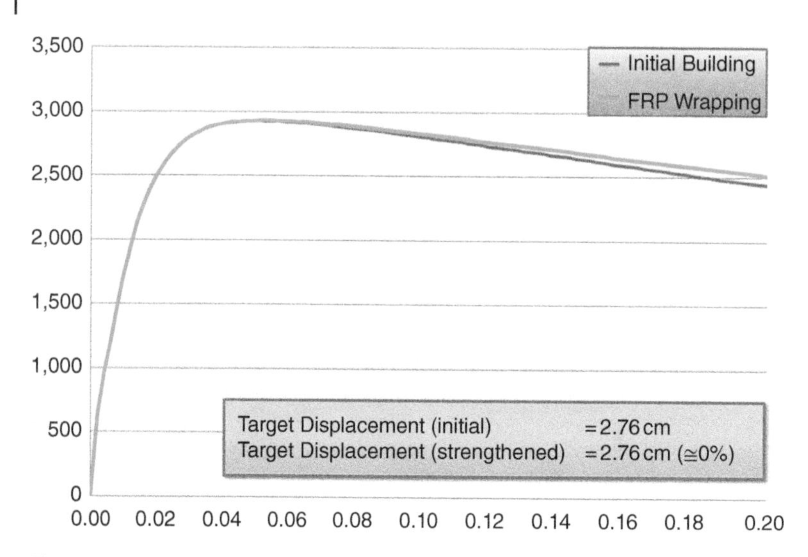


**Figure 4.130** Capacity curve of the building strengthened with FRP wraps. *Source:* Stelios Antoniou.

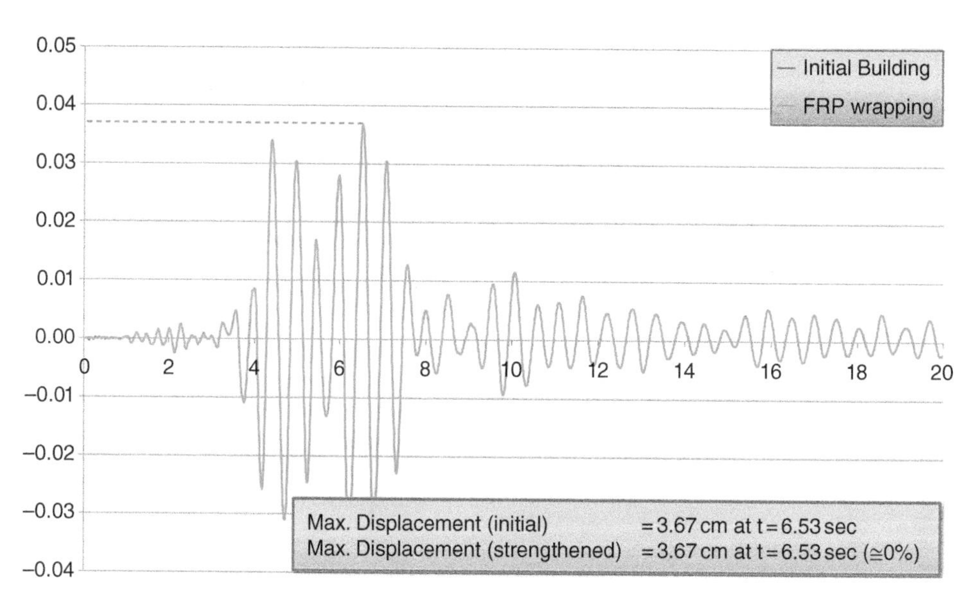


**Figure 4.131** Top displacement vs. time plot of the building strengthened with FRP wraps, the two plots coincide throughout the time-history. *Source:* Stelios Antoniou.

exception that the strengthened building exhibits larger ductility in the highly inelastic range. The target displacement is almost equal, too. The results from *dynamic analysis* are very similar as well, as shown in the plot of the time-history of the top displacement in Figure 4.131.

Despite the fact that the dynamic behavior is similar to the initial building, the shear checks are fulfilled much more easily because of the significant increase in shear capacity

Figure 4.132 DCRs for shear checks at the time step of maximum displacement (building strengthened with FRP wraps). *Source:* Stelios Antoniou.

of the strengthened columns (Figure 4.132). The larger DCR is 1.06, very close to the unity; and if one wanted to be safer, a fourth layer of the FRP wrap could have been applied to further increase the shear capacity of that column.

The example shows in a very clear way the effect that FRP wraps can have on structural behavior: one can strengthen particular members of the building without changing the load distribution in the building, and without affecting (positively or negatively) the force or deformation demand on the other components. Hence, FRPs are a good way of strengthening selected members without changing the global behavior of the entire structure.

### 4.20.8 Strengthening with Seismic Isolation

Seismic isolation is a method that was only recently introduced and employed in structural retrofit. It is relatively expensive with respect to other, more established and tested methods; however, due to the rapid decrease in the prices of isolators, it is expected to gain importance in the near future and become one of the alternative methods for the retrofit of even medium-sized constructions.

In the current example, 18 isolators are placed at the ground level, below the 18 columns of the ground floor, at the locations where the columns are connected to the footings (cutting the columns is required for this). In order not to allow large relative horizontal deformations between the isolators, two diaphragms are constructed, one at the foundation level (connecting all the individual footings) and one above the level of the isolators (connecting the 18 columns) as depicted in Figure 4.133.

With *eigenvalue analysis* a very large change in the dynamic characteristics of the structure is observed, as both the fundamental period (2.568 seconds from 0.268 seconds) and

Figure 4.133 Location of the construction of the two diaphragms (building strengthened with seismic isolation). *Source:* Stelios Antoniou.

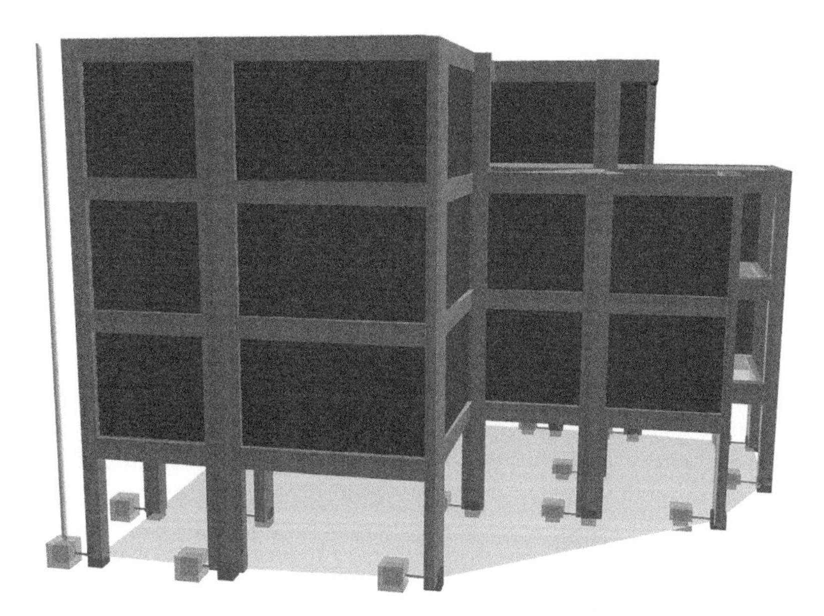

Figure 4.134 Fundamental mode of the building strengthened with base isolation (T = 2.568 seconds). *Source:* Stelios Antoniou.

the first mode shape shown in Figure 4.134 are very different. Almost all the deformations are now concentrated at the isolators with the rest of the building remaining undeformed. If the periods are plotted on the acceleration response spectrum (Figure 4.135), one can see that there is a huge drop in the demand from 1.30 g to just 0.04 g! This drop explains some

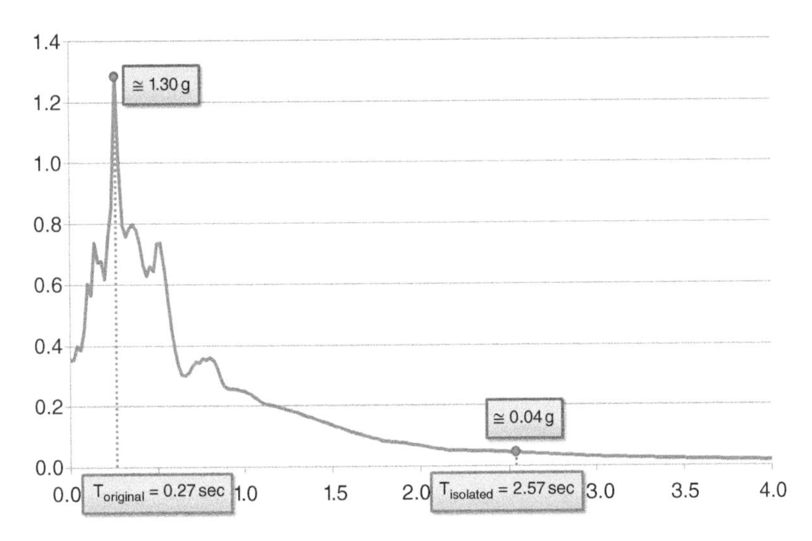


**Figure 4.135** Fundamental periods of the initial building and the building strengthened with base isolation. *Source:* Stelios Antoniou.

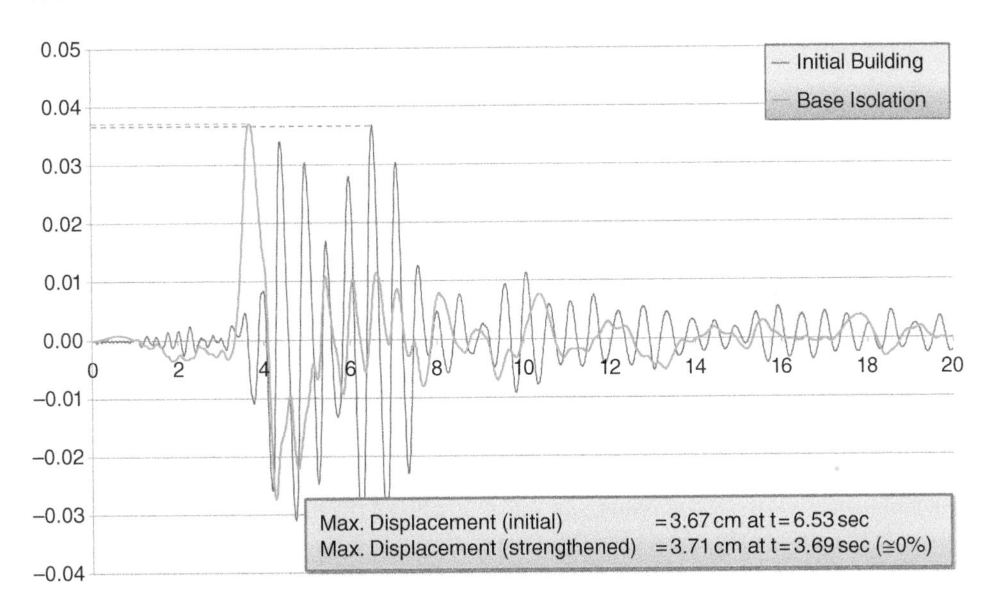


**Figure 4.136** Top displacement vs. time plot of the building strengthened with base isolation. *Source:* Stelios Antoniou.

of the observations regarding the demand on the structure from dynamic analysis that are described in this section.

The *dynamic analysis* of the isolated building shows a completely different dynamic behavior, which is to be expected due to the completely different dynamic characteristics of the new building (Figure 4.136). Although the maximum top displacement is roughly the same as the initial building, it is very impressive that 98% of the total deformations are concentrated at the isolators, leaving the upper structure almost undeformed, as shown in

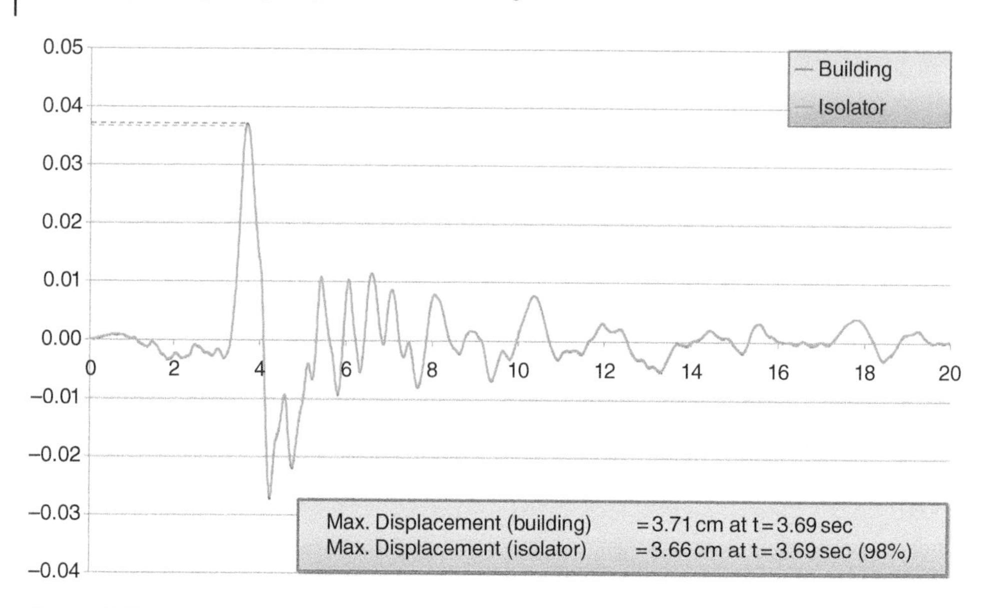


**Figure 4.137** Top displacement and isolator deformation vs. time for the building strengthened base isolation, the two plots almost coincide. *Source:* Stelios Antoniou.

Figure 4.137, where the top building displacement vs. time is plotted against the deformation of the isolators.

Looking at the shear forces diagram (Figure 4.138), we see that the shear demand at the ground story columns is just a fraction of the demand on the same columns of the original building! Consequently, the structural members are now able to effectively resist the seismic action with small DCR ratios, the maximum being equal to 0.55, significantly smaller than unity (Figure 4.139).

### 4.20.9 Comparison of the Methods

Table 4.5 depicts the most important results for the initial and the strengthened buildings. The main observation made from the table is that with all the methods the seismic performance of the building has been significantly improved, with a reduction of the demand to capacity DCR ratios from 1.87 to values between 0.47 and 1.18, which effectively means that the vulnerability of the structure has decreased considerably.

In the last two columns of Table 4.5, rough estimates of the cost of the interventions are given in € and as the ratio of the cost to the estimated building value. Obviously, these numbers can vary significantly, depending on the seismic hazard or the materials and labor costs at the location of the building. However, they give a good indication of the cost of an average retrofit scheme with respect to the value of a building: with the exception of the base isolation method, the cost of the interventions is a very small fraction of the total value.

This has serious implications for our decision on whether to upgrade a structure. Essentially, it reminds us that although a large percentage of the population of the planet lives in highly vulnerable buildings in high earthquake hazard zones, we are often

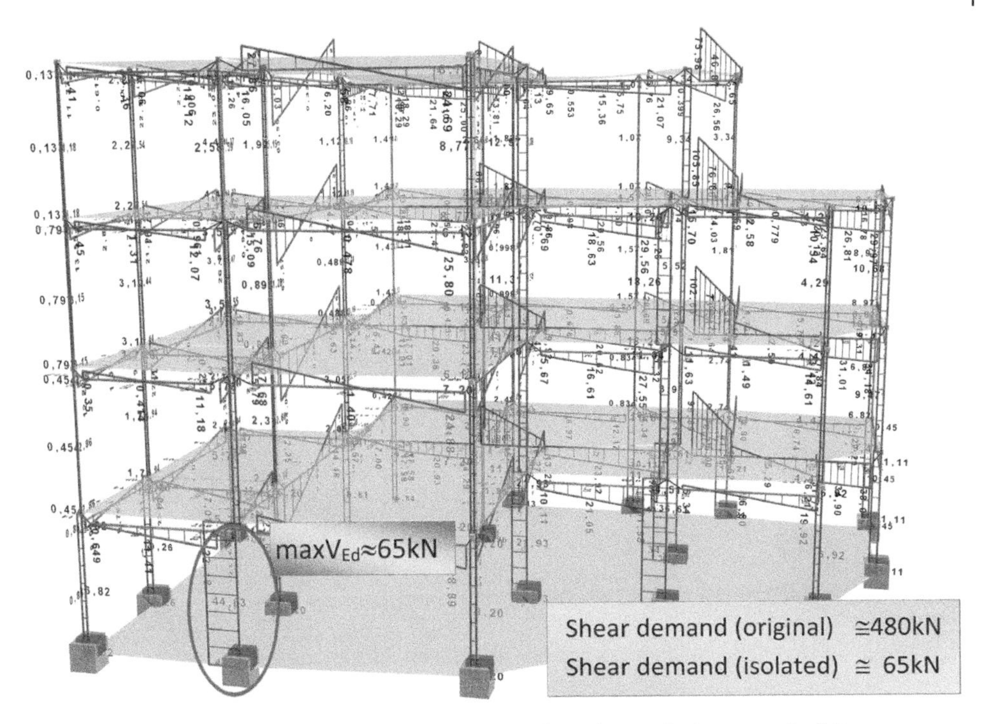


**Figure 4.138** Shear Force Diagram at the time step of maximum displacement (building strengthened with base isolation). *Source:* Stelios Antoniou.

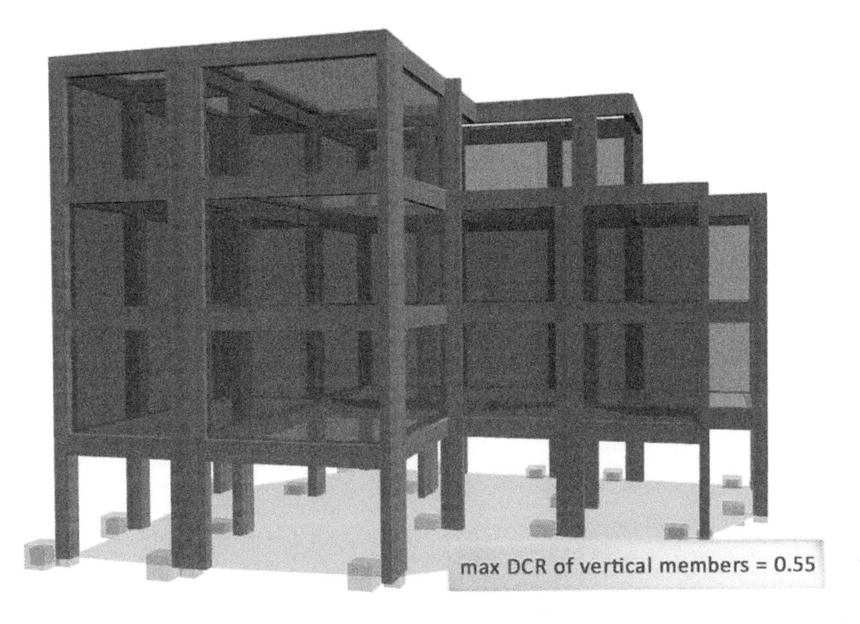


**Figure 4.139** DCRs for shear checks at the time step of maximum displacement (building strengthened with base isolation). *Source:* Stelios Antoniou.

**Table 4.5** Comparison of the most important results for the initial and the strengthened buildings.

| | Model | Period (sec) | Maximum displacement (cm) | Maximum DCR | Cost (€) | Cost/Building value (%) |
|---|---|---|---|---|---|---|
| 0 | Existing building | 0.268 | 3.67 | 1.87 | | |
| 1 | RC jackets | 0.199 | 2.22 | 1.07 | 25 000€ | 4.0% |
| 2 | New shear walls | 0.116 | 0.39 | 0.47 | 30 000€ | 5.0% |
| 3 | Steel braces | 0.221 | 2.18 | 1.18 | 18 000€ | 3.0% |
| 4 | FRP wrapping | 0.268 | 3.67 | 1.06 | 18 000€ | 3.0% |
| 5 | Base isolation | 2.57 | 3.71 | 0.55 | 90 000€ | 15.0% |

*Source:* Stelios Antoniou.

reluctant to pay just a small fraction of the total building value to upgrade them, to make them safer, and to save a significant amount of money in the event of a large seismic event that could lead to extensive damage or even to collapse (not to mention the possible loss of human lives).

## References

Abdel-Kareem, A.H., Debaiky, A.S., Makhlouf, M.H., and Badw, M. (2019). Repairing and strengthening of RC beams using thin lower concrete layer reinforced by FRP bars. *International Journal of Civil Engineering and Technology (IJCIET)* 10 (2): 1949–1966.

Aboutaha, R.S., Engelhardt, M.D., Jirsa, J.O., and Kreger, M.E. (1999, 1999). Rehabilitation of shear critical concrete columns by use of rectangular steel jackets. *ACI Structural Journal* 96 (1): 68–78.

Afshin, H., Nouri Shirazi, M.R., and Abedi, K. (2019). Experimental and numerical study about seismic retrofitting of corrosion-damaged reinforced concrete columns of bridge using combination of FRP wrapping and steel profiles. *Steel & Composite Structures* 30 (3): 231–251.

Ahmed, E., Sobuz, H.R., and Sutan, N.M. (2011). Flexural performance of CFRP strengthened RC beams with different degrees of strengthening schemes. *International Journal of the Physical Sciences* 6 (9): 2229–2238. https://doi.org/10.5897/IJPS11.304.

Akbar, J., Ahmad, N., and Alam, B. (2020). *Seismic Upgrade of RC Deficient Frames: Steel/Concrete Haunch Retrofit Solutions*. ISBN-10 : 6200787905. ISBN-13: 978-6200787903. LAP LAMBERT Academic Publishing.

Akhoundi, F., Vasconcelos, G., and Lourenço, P. (2018a). In-plane behavior of infills using glass fiber shear connectors in textile reinforced mortar (TRM) technique. *The International Journal of Structural Glass and Advanced Materials Research* 2: 1–14. https://doi.org/10.3844/sgamrsp.2018.1.14.

Akhoundi, F., Vasconcelos, G., Lourenço, P. et al. (2018b). In-plane behavior of cavity masonry infills and strengthening with textile reinforced mortar. *Engineering Structures* (156): 145–160. https://doi.org/10.1016/j.engstruct.2017.11.002.

Alashkar, Y., Nazar, S., and Ahmed, M. (2015). A comparative study of seismic strengthening of RC buildings by steel bracings and concrete shear walls. *International Journal of Civil and Structural Engineering Research* 2 (2): 24–34. https://www.researchgate.net/publication/268613512%0A.

Alferjani, M.B.S., Samad, A.A.A., Elrawaff, B.S. et al. (2013). Use of carbon fiber reinforced polymer laminate for strengthening reinforced concrete beams in shear: a review. *International Refereed Journal of Engineering and Science* 2 (2): 45–53. https://www.researchgate.net/publication/303698935_Use_of_Carbon_Fiber_Reinforced_Polymer_Laminate_for_strengthening_reinforced_concrete_beams_in_shear_A_review.

Almeida, A., Ferreira, R., Proença, J.M., and Gago, A.S. (2017). Seismic retrofit of RC building structures with buckling restrained braces. *Engineering Structures* 130: 14–22.

Almusallam, T.H. and Al-Salloum, Y.A. (2007). Behavior of FRP strengthened infill walls under in-plane seismic loading. *Journal of Composites for Construction* 11 (3): 308–318.

Al-Salloum, Y.A., Siddiqui, N.A., Elsanadedy, H.M. et al. (2011). Textile-reinforced mortar versus FRP as strengthening material for seismically deficient RC beam-column joints. *Journal of Composites for Construction* 15 (6): 920–933. https://doi.org/10.1061/(ASCE)CC.1943-5614.0000222.

Amadio, C., Clemente, I., Macorini, L., and Fragiacomo, M. (2008). Seismic behaviour of hybrid systems made of PR composite frames coupled with dissipative bracings. *Earthquake Engineering and Structural Dynamics* 37 (6): 861–879. https://doi.org/10.1002/eqe.790.

American Concrete Institute ACI 440 (ACI) (2017). *ACI PRC-440.2-17: Guide for the Design and Construction of Externally Bonded FRP Systems for Strengthening Concrete Structures*. ACI Committee 440.

American Concrete Institute ACI 506 (ACI) (2013). ACI 506.2-13: Specification for Shotcrete. Reported by ACI Committee 506.

American Society of Civil Engineers (ASCE) (2017). *Seismic Evaluation and Retrofit of Existing Buildings (ASCE/SEI 41–17)*. Reston, Virginia: ASCE.

Antonopoulos, C.P. and Triantafillou, T.C. (2003). Experimental investigation of FRP-strengthened RC beam–column joints. *ASCE Journal of Composites for Construction* 7 (1): 39–49.

Applied Technology Council (ATC) (1996). *Seismic Evaluation and Retrofit of Concrete Buildings, ATC-40 Report*. Redwood City, California: Applied Technology Council.

Aslam, Z., Riaz, W., and Khalid, A. (2016). *Seismic Retrofitting of an Existing Reinforced Concrete Structure: Because Earthquakes Do Not Kill, Unsafe Buildings Do*. ISBN-10 3659848573. ISBN-13 978-3659848575. LAP LAMBERT Academic Publishing.

ASTM International (1998). *C 1385–98, Standard Practice for Sampling Materials for Shotcrete*. West Conshohocken, PA: ASTM International.

ASTM International (2001). *ASTM C1141-01, Standard Specification for Materials for Shotcrete*. West Conshohocken, PA: ASTM International.

ASTM International (2003). *C 1116–03, Standard Specification for Fiber-Reinforced Concrete and Shotcrete*. West Conshohocken, PA: ASTM International.

ASTM International (2013). *ASTM C1436–13, Standard Specification for Admixtures for Shotcrete*. West Conshohocken, PA: ASTM International.

ASTM International (2018). *ASTM C33 / C33M-18, Standard Specification for Concrete Aggregates*. West Conshohocken, PA: ASTM International.

ASTM International (2019). *ASTM C618–19, Standard Specification for Coal Fly Ash and Raw or Calcined Natural Pozzolan for Use in Concrete*. West Conshohocken, PA: ASTM International.

ASTM International (2020a). ASTM C150 / C150M-20, Standard Specification for Portland Cement. https://doi.org/10.1520/C0150_C0150M-20.

ASTM International (2020b). *ASTM C595 / C595M-20, Standard Specification for Blended Hydraulic Cements*. West Conshohocken, PA: ASTM International.

Aydenlou, R.M. (2020). *Seismic Rehabilitation Methods for Existing Buildings*. Print Book & E-Book. ISBN 9780128199596, 9780128203842. Elsevier.

Balsamo, A., Manfredi, G., Mola, E. et al. (2005a). Seismic rehabilitation of a full-scale structure using GFRP laminates. In: *Proceedings of the 7th International Symposium on Fiber Reinforced Polymer (FRP) for Concrete Structures*, JRC32415. Washington, DC, American Concrete Institute.

Balsamo, A., Colombo, A., Manfredi, G. et al. (2005b). Seismic behavior of a full-scale RC frame repaired using CFRP laminates. *Engineering Structures* 27 (5): 769–780. https://doi.org/10.1016/j.engstruct.2005.01.002.

Barbagallo, F., Bosco, M., Ghersis, A. et al. (2014). Calibration of a design method for seismic upgrading of existing R.C. frames by BRBS. In: *2nd European Conference on Earthquake Engineering and Seismology*, (August 25–29), 1–11. Instabul: European Association for Earthquake Engineering (EAEE).

Baros, D.K. and Dritsos, S.E. (2008). A simplified procedure to select a suitable retrofit strategy for existing RC buildings using pushover analysis. *Journal of Earthquake Engineering* 12 (6): 823–848. https://doi.org/10.1080/13632460801890240.

Belal, F.M., Mohamed, H.M., and Morad, S.A. (2014). Behavior of reinforced concrete columns strengthened by steel jacket. *HBRC Journal* 33 (2): 1–12. https://doi.org/10.1016/j.hbrcj.2014.05.002.

Belarbi, A. and Acun, B. (2013). FRP Systems in Shear Strengthening of reinforced concrete structures. 11th international conference on modern building materials, structures and techniques, MBMST 2013. *Procedia Engineering* 57 (2013): 2–8. https://doi.org/10.1016/j.proeng.2013.04.004.

Bergami, A.V. and Nuti, C. (2013). A design procedure of dissipative braces for seismic upgrading structures. *Earthquakes and Structures* 4 (1): 85–108. https://doi.org/10.12989/EAS.2013.4.1.085.

Bhojkar, N. and Bagade, M. (2015). Seismic evaluation of high-rise structure by using steel bracing system. *International Journal of Innovative Science, Engineering & Technology (IJISET)* 2 (3): 264–269. http://ijiset.com/vol2/v2s3/IJISET_V2_I3_39.pdf.

Bianchi, F., Nascimbene, R., Brunesi, E. et al. (2015). Valutazione numerica del comportamento di un edificio ad uso ospedaliero in cemento armato con sistemi dissipativi aggiunti. *Progettazione Sismica* 6 (2): 35–69. IUSS Press, Pavia. In Italian.

Binici, B., Ozcebe, G., and Ozcelik, R. (2007). Analysis and design of FRP composites for seismic retrofit of infill walls in reinforced concrete frames. *Composites Part B: Engineering* 38 (5–6): 575–583.

Bournas, D.A., Pavese, A., and Tizani, W. (2015). Tensile capacity of FRP anchors in connecting FRP and TRM sheets to concrete. *Engineering Structures* 82 (1): 72–81.

Bousias, S.N., Triantafillou, T.C., Fardis, M.N. et al. (2004). Fibre-reinforced polymer retrofitting of rectangular reinforced concrete columns with or without corrosion. *ACI Structural Journal* 101 (4): 512–520.

Bousias, S.N., Biskinis, D., Fardis, M.N., and Spathis, L.-A. (2007a). Strength, stiffness, and cyclic deformation capacity of concrete jacketed members. *ACI Structural Journal; Farmington Hills* 104 (5): 521–531.

Bousias, S.N., Spathis, L.-A., and Fardis, M.N. (2007b). Seismic retrofitting of columns with lap-spliced smooth bars through FRP or concrete jackets. *Journal of Earthquake Engineering* 11: 653–674.

Braconi, A., Tremea, A., Lomiento, G. et al. (2013). Steel solutions for seismic retrofit and upgrade of existing constructions (Steelretro). In: *EU Publications*, no. EUR 25894 EN,. European Commission EC http://dx.doi.org/10.2777/7937.

Breña, S.F. and McGuirk, G.N. (2013). Advances on the behavior characterization of FRP-anchored carbon fiber-reinforced polymer (CFRP) sheets used to strengthen concrete elements. *The International Journal of Concrete Structures and Materials (IJCSM)* 7 (3–16): 2013. https://doi.org/10.1007/s40069-013-0028-1.

Briseghella, B., Zordan, T., Liu, T., and Mazzarolo, E. (2013). Friction pendulum system as a retrofit technique for existing reinforced concrete building. *Structural Engineering International* 23 (2): 219–224. https://doi.org/10.2749/101686613X13439149157759.

Bsisu, K., Hunaiti, Y., Malkavi, O., and Ynis, R. (2011). Comparing the use of CFRP laminates with light gauge galvanized steel plates in structural strengthening of beams subjected to flexural loading. *Advanced Materials Research* 163–167: 3844–3847.

C.E.B.-F.I.P. (2003). Seismic assessment and retrofit of reinforced concrete buildings - State of art report, Task Group 7.1. Bulletin n. 24, Losanna.

Canadian Standards Association (2007). CAN/CSA-S806–02 (R2007). Design and Construction of Building Components with Fibre-Reinforced Polymers.

Cardone, D. and Flora, A. (2014). Direct displacement loss assessment of existing RC buildings pre- and post-seismic retrofitting: a case study. *Soil Dynamics and Earthquake Engineering* 64: 38–49. https://doi.org/10.1016/j.soildyn.2014.03.011.

Cardone, D. and Gesualdia, G. (2014). Seismic rehabilitation of existing reinforced concrete buildings with seismic isolation: a case study. *Earthquake Spectra* 30 (4): 1619–1642. https://doi.org/10.1193/110612EQS323M.

Cardone, D., Flora, A., and Gesualdi, G. (2012). Inelastic response of RC frame buildings with seismic isolation. *Earthquake Engineering and Structural Dynamics* 42 (3): 871–889. https://doi.org/10.1002/eqe.2250.

Caterino, N., Iervolino, I., Manfredi, G., and Cosenza, E. (2008). Multi – criteria decision making for seismic retrofitting of RC structures. *Journal of Earthquake Engineering* 12 (4): 555–583. https://doi.org/10.1080/13632460701572872.

CEN (2004). *European Standard EN 1998-1: 2004. Eurocode 8: Design of Structures for Earthquake Resistance, Part 1: General Rules, Seismic Actions and Rules for Buildings*. Brussels: Comité Européen de Normalisation.

CEN (2005). *European Standard EN 1998-3: 2005. Eurocode 8: Design of Structures for Earthquake Resistance, Part 3: Assessment and Retrofitting of Buildings*. Brussels: Comité Européen de Normalisation.

Chalioris, C.E., Thermou, G.E., and Pantazopoulou, S.J. (2014). Behaviour of rehabilitated RC beams with self-compacting concrete jacketing – analytical model and test results. *Construction and Building Materials* 55: 257–273. https://doi.org/10.1016/j.conbuildmat.2014.01.031.

Chandran, H. and Subha, K. (2018). A review on strengthening techniques using NMS and FRP composites. *International Research Journal of Engineering and Technology (IRJET)* 5 (4): 3441–3445. https://www.irjet.net/archives/V5/i4/IRJET-V5I4774.pdf.

Chang, S.Y., Chen, T.W., Tran, N.C. et al. (2014). Seismic retrofitting of RC columns with RC jackets and wing walls with different structural details. *Earthquake Engineering and Engineering Vibration* 13: 279–292. https://doi.org/10.1007/s11803-014-0230-4.

Chavan, K.R. and Jadhav, H.S. (2014). Seismic response of RC building with different arrangement of steel bracing system. *International Journal of Engineering Research and Applications* 4 (7): 218–222. https://www.ijera.com/papers/Vol4_issue7/Version%203/AI04703218222.pdf.

Christopoulos, C. and Filiatrault, A. (2006). *Principles of Passive Supplemental Damping and Seismic Isolation*. Pavia: IUSS Press.

Chrysostomou, C., Kyriakides, N., Kotronis, P., and Georgiou E. (2014). RC infilling of existing RC structures for seismic retrofitting. 2nd European Conference on Earthquake Engineering and Seismology, August 2014, Istanbul, Turkey. HAL archives, HAL Id: hal-01080302.

Clemente, P. and De Stefano, A. (2011). Application of seismic isolation in the retrofit of historical buildings. *Earthquake Resistant Engineering Structures VIII* 120: 41–52.

CNR-DT 200 (2013). *Istruzioni per la Progettazione, l'Esecuzione ed il Controllo di Interventi di Consolidamento Statico mediante l'utilizzo di Compositi Fibrorinforzati*. Rome, Italy: Centro Nazionale Ricerche, Commissione di Studio per la Predisposizione e l'Analisi di Norme Tecniche relative alle costruzioni.

Colomb, F., Tobbi, H., Ferrier, E., and Hamelin, P. (2008). Seismic retrofit of reinforced concrete short columns by CFRP materials. *Composite Structures* 82 (4): 475–487.

Constantinou M.C. (2004). Friction pendulum double concave bearing. Technical Report, University of Buffalo, State University of Buffalo, NY.

Constantinou M.C., Soong T.T.; Dargush G.F. (1998). Passive Energy Dissipation System for Structural Design and Retrofit. Multidisciplinary Center for Earthquake Engineering Research MCEER, Monograph No. 1, Buffalo, NY.

Costa, A., Arêde, A., and Varum, H. (ed.) (2018). *Strengthening and Retrofitting of Existing Structures*. Hardcover ISBN 978-981-10-5857-8, eBook ISBN 978-981-10-5858-5. https://doi.org/10.1007/978-981-10-5858-5. Singapore: Springer.

Crom, T.R. (1966). *Dry-Mix Shotcrete Practice. Shotcreting, SP-14*, 15–32. Detroit, MI: American Concrete Institute.

Da Porto, F., Guidi, G., Verlato, N., and Modena, C. (2015). Effectiveness of plasters and textile reinforced mortars for strengthening clay masonry infill walls subjected to combined in-plane/out-of-plane actions. *Mauerwerk* 19: 334–354. https://doi.org/10.1002/dama.201500673.

Damptech A.S. (2022). Multi-unit friction dampers. https://www.damptech.com/dampers-for-buildings (Accessed: May 25, 2022).

Danraka, M.N., Mahmod, H.M., and Oluwatosin, O.-k.J. (2017). Strengthening of reinforced concrete beams using FRP technique: a review. *International Journal of Engineering Science* 7 (6): 13199.

Daudey, X. and Filiatrault, A. (2000). Seismic evaluation and retrofit with steel jackets of reinforced concrete bridge piers detailed with lap-splices. *Canadian Journal of Civil Engineering* 27 (1): 1–16.

Della Corte, G., D'Aniello, M., Landolfo, R., and Mazzolani, F.M. (2011). Review of steel buckling-restrained braces. *Steel Construction* 4 (2): 85–93. https://doi.org/10.1002/stco.201110012.

Dhiraj, N. and Prasad, R.V.R.K. (2016). Comparative study in the analysis of multistory RCC structure by using different types of concentric bracing system (by using software). *International Journal of Engineering Sciences & Research Technology (IJESRT)* 5 (7): 483–488. https://doi.org/10.5281/zenodo.57010.

Di Lorenzo, G., Colacurcio, E., Di Filippo, A. et al. (2020). State-of-the-art on steel exoskeletons for seismic retrofit of existing RC buildings. *Ingegneria Sismica* 37 (1): 33–50.

Di Ludovico, M., Balsamo, A., Prota, A., and Manfredi, G. (2008). Comparative assessment of seismic rehabilitation techniques on a full scale 3-story RC moment frame structure. *Structural Engineering and Mechanics* 28 (6): 727–747. https://doi.org/10.12989/SEM.2008.28.6.727.

Di Sarno, L. and Manfredi, G. (2009). Seismic Retrofitting of existing RC frames with buckling restrained braces. In: *ATC & SEI Conference on Improving the Seismic Performance of Existing Buildings and Other Structures*, 741–752. San Francisco.

Di Sarno, L. and Manfredi, G. (2010). Seismic retrofitting with buckling restrained braces: application to an existing non-ductile RC framed building. *Soil Dynamics and Earthquake Engineering* 30 (11): 1279–1297. https://doi.org/10.1016/j.soildyn.2010.06.001.

Di Sarno, L. and Manfredi, G. (2012). Experimental tests on full-scale RC unretrofitted frame and retrofitted with buckling-restrained braces. *Earthquake Engineering and Structural Dynamics* 41 (2): 315–333. https://doi.org/10.1002/eqe.1131.

Dritsos, S.E. (2005a). *Repair and Strengthening of Reinforced Concrete Structures*. Patras, Greece: University of Patras (in Greek).

Dritsos, S.E. (2005b). Seismic retrofit of buildings, a Greek perspective. *Bulletin of the New Zealand Society for Earthquake Engineering* 38 (3): 165–181. https://doi.org/10.5459/bnzsee.38.3.165-181.

Dritsos, S.E., Moseley, V.J., Lampropoulos, A. et al. (2019). Characteristic seismic failures of buildings. *IABSE Bulletins Structural Engineering Documents (SED)* 16.

Dubină, D. (ed.) (2015). Seismic retrofitting of existing structures using steel based solutions - TEHNICI DE CONSOLIDARE ANTI-SEISMICĂ A CLĂDIRILOR EXISTENTE BAZATE PE UTILIZAREA OŢELULUI. Timişoara: Orizonturi Universitare. https://www.academia.edu/32407972/SEISMIC_RETROFITTING_OF_EXISTING_STRUCTURES_USING_STEEL_BASED_SOLUTIONS_TEHNICI_DE_CONSOLIDARE_ANTI_SEISMIC%C4%82_A_CL%C4%82DIRILOR_EXISTENTE_BAZATE_PE_UTILIZAREA_O%C5%A2ELULUI_International_Workshop_organised_within_the_framework_of_RFS2_CT_2014_00022_STEELEARTH.

Durucan, C. and Dicleli, M. (2010). Analytical study on seismic retrofitting of reinforced concrete buildings using steel braces with shear link. *Engineering Structures* 32 (10): 2995–3010. https://doi.org/10.1016/j.engstruct.2010.05.019.

El-Hacha, R. and Rizkalla, S. (2004). Near-surface-mounted fiber-reinforced polymer reinforcements for flexural strengthening of concrete structures. *ACI Structures Journal* 101 (5): 717–726.

El-Hacha, R., Wight, R.G., and Green, M.F. (2001). Prestressed fibre-reinforced polymer laminates for strengthening structures. *Progress in Structural Engineering and Materials* 3 (2): 111–121. https://doi.org/10.1002/pse.76.

ElSouri, A.M. and Harajli, M.H. (2011). Seismic repair and strengthening of lap splices in RC columns: carbon fiber-reinforced polymer versus steel confinement. *Journal of Composites for Construction* 15 (5): https://doi.org/10.1061/(ASCE)CC.1943-5614.0000213.

Elwan, S.K., Elasayed, T.A., Refaat, W., and Lotfy, A.M. (2017). Experimental behavior of RC beams strengthened by externally bonded CFRP with lap splice. *International Journal of Engineering Research and Development* 13 (3): 36–47.

Eshwar, N., Nanni, A., and Ibell, T.J. (2008). Performance of two anchor systems of externally bonded fiber-reinforced polymer laminates. *ACI Structural Journal* 105 (1): 72–80.

Eskandari, R., Vafaei, D., Vafaei, J., and Shemshadian, M.E. (2017). Nonlinear static and dynamic behavior of reinforced concrete steel-braced frames. *Earthquake and Structures* 12 (2): 191–200. https://doi.org/10.12989/eas.2017.12.2.191.

Esmaeeli, E., Barros, J.A.O., Sena-Cruz, J. et al. (2015). Retrofitting of interior RC beam–column joints using CFRP strengthened SHCC: cast-in-place solution. *Composite Structures* 122: 456–467. https://doi.org/10.1016/j.compstruct.2014.12.012.

Ezz-Eldeen, H.A. (2015). An experimental study on strengthening and retrofitting of damaged reinforced concrete beams using steel wire mesh and steel angles. *International Journal of Engineering Research & Technology (IJERT)* 4 (5): 164–173.

Faella, C., Lima, C., Martinelli, E., and Realfonzo, R. (2014). Steel bracing configurations for seismic retrofitting of a reinforced concrete frame. *Proceedings of the Institution of Civil Engineers: Structures and Buildings* 167 (1): 54–65. https://doi.org/10.1680/stbu.12.00072.

Fagone, M., Ranocchiai, G., Caggegi, C. et al. (2014). The efficiency of mechanical anchors in CFRP strengthening of masonry: an experimental analysis. *Composites Part B: Engineering* 64: 1–15. ISSN 1359-8368. https://doi.org/10.1016/j.compositesb.2014.03.018.

Fahmy, M.F.M. and Wu, Z.S. (2016). Exploratory study of seismic response of deficient lap-splice columns retrofitted with near surface-mounted basalt FRP bars. *Journal of Structural Engineering* 142 (6): 04016020.

Fakharifar, M., Chen, G.D., Wu, C.L. et al. (2016). Rapid repair of earthquake-damaged RC columns with prestressed steel jackets. *Journal of Bridge Engineering* 21 (4): 04015075.

Fardis, M.N. (2009). *Seismic Design, Assessment and Retrofitting of Concrete Buildings, Based on EN-Eurocode 8*. ISBN 978-1-4020-9841-3 e-ISBN 978-1-4020-9842-0. https://doi.org/10.1007/978-1-4020-9842-0. Springer International Publishing.

Fardis, M.N., Schetakis, A., and Strepelias, E. (2013). RC buildings retrofitted by converting frame bays into RC wall. *Bulletin of Earthquake Engineering* 11 (5): 1541–1561.

Fardis, M.N., Schetakis, A., and Strepelias, E. (2014). Seismic rehabilitation of concrete buildings by converting frame bays into RC walls. In: *Seismic Evaluation and Rehabilitation of Structures. Geotechnical, Geological and Earthquake Engineering*, vol. 26 (ed. A. Ilki and M. Fardis). Cham: Springer https://doi.org/10.1007/978-3-319-00458-7_15.

Fauzan, F., Ismail, F.A., Dezardo, O., and Jauhari, Z.A. (2018). A comparison of retrofitting methods on nursing faculty building of Andalas University with concrete jacketing and shear wall systems. *MATEC Web of Conferences* 195: https://doi.org/10.1051/matecconf/201819502016.

Federal Emergency Management Agency (FEMA) (1997). *NEHRP Guidelines for the Seismic Rehabilitation of Buildings, FEMA 273 Report*. Washington, DC: Applied Technology Council and the Building Seismic Safety Council for the Federal Emergency Management Agency.

Federal Emergency Management Agency (FEMA) (2000). *Pre-standard and Commentary for the Seismic Rehabilitation of Buildings. FEMA 356 Report*. Washington, DC: American Society of Civil Engineers for the Federal Emergency Management Agency.

Federal Emergency Management Agency (FEMA) (2009). *FEMA P – 420 Risk Management Series: Engineering Guideline for Incremental Seismic Rehabilitation*. Washington, DC: FEMA.

Federal Emergency Management Agency (FEMA) (2018a). *Seismic Performance Assessment of Buildings, Volume 1 – Methodology, FEMA P-58-1. Report*. Washington, DC: Applied Technology Council for the Federal Emergency Management Agency.

Federal Emergency Management Agency (FEMA) (2018b). *Seismic Performance Assessment of Buildings, Volume 2 – Implementation Guide, FEMA P-58-2. Report*. Washington, DC: Applied Technology Council for the Federal Emergency Management Agency.

Federal Emergency Management Agency (FEMA) (2018c). *Seismic Evaluation of Older Concrete Buildings for Collapse Potential, FEMA P-2018 Report*. Washington, DC: Applied Technology Council for the Federal Emergency Management Agency.

Federal Emergency Management Agency (FEMA) (2021). Recommended Options for Improving the Built Environment for Post-Earthquake Reoccupancy and Functional Recovery Time. Special Publication FEMA P-2090/ NIST SP-1254 (January). https://doi.org/10.6028/NIST.SP.1254.

Fédération Internacionale du béton (FIB) (2001). Externally bonded FRP reinforcement for RC structures. Fib bulletin 14 Technical Report. International Federation of Structural Concrete, Lausanne. Task Group Fib TG9.

Fédération Internacionale du béton (FIB) (2019). Externally applied reinforcement for concrete structures. Fib bulletin 90. Technical report (229 pages, ISBN 978-2-88394-131-1 (July).

FEMA NEHRP (FEMA) (2006). *Techniques for the Seismic Rehabilitation of Existing Buildings, FEMA 547*. Washington, DC: Building Seismic Safety Council for the Federal Emergency Management Agency.

Ferraioli, M. and Avossa, A.M. (2012). Base isolation seismic retrofit of a hospital building in Italy. *Journal of Civil Engineering and Architecture* 6 (3): 308–321. https://doi.org/10.17265/1934-7359/2012.03.005.

Ferraioli, M., Avossa, A. M., Costanzo, R., and Lavino, A. (2011). Seismic isolation retrofit of a hospital building In: XIV convegno ANIDIS 2011, September 18–22, Bari.

[FIB] Fédération Internacionale du béton (2022). *Guide for Strengthening of Concrete Structures*. Fib bulletin 103. Technical report. 316 pages, ISBN 978-2-88394-157-1, May 2022.

Formisano, A., De Matteis, G., Panico, S., and Mazzolani, F.M. (2008). Seismic upgrading of existing RC buildings by slender steel shear panels: a full-scale experimental investigation. *Advanced Steel Construction* 4 (1): 26–45.

Formisano, A., Lombardi, L., and Mazzolani, F.M. (2016). Full and perforated metal plate shear walls as bracing systems for seismic upgrading of existing RC buildings. *Ingegneria Sismica* 33 (1–2): 16–34.

Formisano, A., Massimilla, A., Di Lorenzo, G., and Landolf, R. (2020). Seismic retrofit of gravity load designed RC buildings using external steel concentric bracing systems. *Engineering Failure Analysis* 111: 104485. https://doi.org/10.1016/j.engfailanal.2020.104485.

Fukuyama, H, Tumialan, G. and Nanni, A. (2001). Japanese design and construction guidelines for seismic retrofit of building structures with FRP composites. *Proceedings of the International Conference on FRP composites in Civil Engineering*. Hong Kong, China. December 2001.

Furtado A., Rodrigues H., Arêde A. and Varum H. (2020). Impact of the textile mesh on the efficiency of TRM strengthening solutions to improve the infill walls out-of-plane behaviour. Applied Sciences 10 (23): 8745, https://doi.org/10.3390/app10238745.

German General Guideline (1998). *Richtlinie für das Verstärken von Betonbauteilen durch Ankleben von unidirektionalen kohlenstoffaserverstärken Kunststofflamellen, Fassung*. Berlin: Deutsches Institut für Bautechnik (in German).

Ghobarah, A. and Elfath, H.A. (2001). Rehabilitation of a RC frame using eccentric steel bracing. *Engineering Structures* 23 (7): 745–755.

Ghobarah, A. and Galal, K.E. (2004). Seismic rehabilitation of short rectangular RC columns. *Journal of Earthquake Engineering* 8 (1): 45–68.

Ghosh, K.K. and Sheikh, S.A. (2007). Seismic upgrade with carbon fibre-reinforced polymer of columns containing lap-spliced reinforcing bars. *ACI Structural Journal* 104 (2): 227–236.

Giarlelis, C., Koufalis, D. and Antoniadis, P. (2018a). Seismic Rehabilitation of a RC Building through Seismic Isolation. *16th European Conference on Earthquake Engineering*, June 18–21, 2018, Thessaloniki.

Giarlelis, C., Keen, J., Lamprinou, E. et al. (2018b). The seismic isolated Stavros Niarchos Foundation Cultural Center in Athens (SNFCC). *Soil Dynamics and Earthquake Engineering* 114: 534–547.

Gkimprixis, A., Tubaldi, E., and Douglas, J. (2020). Evaluating alternative approaches for the seismic design of structures. *Bulletin of Earthquake Engineering* 18 (9): 4331–4361. https://doi.org/10.1007/s10518-020-00858-4.

Gkournelos, P.D., Bournas, D.A., and Triantafillou, T.C. (2019). Combined seismic and energy upgrading of existing reinforced concrete buildings using TRM jacketing and thermal insulation. *Earthquakes and Structures* 16 (5): 625–639.

Gkournelos, P.D., Triantafillou, T.C., and Bournas, D.A. (2021). Seismic upgrading of existing reinforced concrete buildings: A state-of-the-art review. *Engineering Structures* 240 (1 August): 112273.

Görgülü, T., Tama, Y.S., Yilmaz, S. et al. (2012). Strengthening of reinforced concrete structures with external steel shear walls. *Journal of Constructional Steel Research* 70: 226–235.

Grelle, S.V. and Sneed, L.H. (2013). Review of anchorage systems for externally bonded FRP laminates. *The International Journal of Concrete Structures and Materials (IJCSM)* 7: 17–33. https://doi.org/10.1007/s40069-013-0029-0.

Gudonis, E., Timinskas, E., Gribniak, V. et al. (2013). FRP reinforcement for concrete structures: state-of-the-art review of application and design. *Engineering Structures and Technologies* 5 (4): 147–158. https://doi.org/10.3846/2029882X.2014.889274.

Guo, T., Xu, J., Xu, W., and Di, Z. (2015). Seismic upgrade of existing buildings with fluid viscous dampers: design methodologies and case study. *Journal of Performance of Constructed Facilities* 29: 04014175.

Harajli, M.H. and Dagher, F. (2008). Seismic strengthening of bond-critical regions in rectangular reinforced concrete columns using fibre-reinforced polymer wraps. *ACI Structural Journal* 105 (1): 68–77.

Haroun, M.A. and Elsanadedy, H.M. (2005a). Behavior of cyclically loaded squat reinforced concrete bridge columns upgraded with advanced composite-material jackets. *Journal of Bridge Engineering* 10 (6): 741–748.

Haroun, M.A. and Elsanadedy, H.M. (2005b). Fibre-reinforced plastic jackets for ductility enhancement of reinforced concrete bridge columns with poor lap-splice detailing. *Journal of Bridge Engineering* 10 (6): 749–757.

Harries, K.A., Ricles, J.M., Pessiki, S., and Sause, R. (2006). Seismic retrofit of lap splices in nonductile square columns using carbon fibre-reinforced jackets. *ACI Structural Journal* 103 (6): 874–884.

Harshita, M.K. and Vasudev, M.V. (2018). Analysis of RC frames structure with structural steel braces using ETABS. *International Research Journal of Engineering and Technology (IRJET)* 5 (1): 1542–1545. https://www.irjet.net/archives/V5/i1/IRJET-V5I1336.pdf.

Hyderuddin, M., Imran, M., and Mohsin, S. (2016). Retrofitting of reinforced concrete frames using steel bracing. *International Journal for Scientific Research & Development (IJSRD)* 4 (8): 297–301. IJSRDV4I80184.

Ilki, A. and Fardis, M.N. (ed.) (2014). *Seismic Evaluation and Rehabilitation of Structures*. ISBN 978-3-319-00457-0, eBook ISBN 978-3-319-00458-7. DOI 10.1007/978-3-319-00458-7. Springer International Publishing.

Ilki A., Goksu C., Demir C. and Kumbasar N. (2007). Seismic analysis of a RC frame building with FRP-retrofitted infill walls. 6th International Conference on Fracture Mechanics of Concrete and Concrete Structures. Catania, Italy 2007.

Ilki, A., Karadogan, F., Pala, S., and Yuksel, E. (ed.) (2009). *Seismic Risk Assessment and Retrofitting with Special Emphasis on Existing Low-Rise Structures*. Hardcover ISBN 978-90-481-2680-4, eBook ISBN 978-90-481-2681-1. https://doi.org/10.1007/978-90-481-2681-1. Netherlands: Springer.

Irwin R. and Rahman A. (2002). FRP strengthening of concrete structures–design constraints and practical effects on construction detailing. *New Zealand Concrete Society Conference*, Wairakei, Oct. 2002.

Islam, A.A.K.M. (2008). Effective methods of using CFRP bars in shear strengthening of concrete girders. *Engineering Structures* 31 (3): 709–714. https://doi.org/10.1016/j.engstruct.2008.11.016.

Islam, N. and Hoque, M.M. (2015). Strengthening of reinforced concrete columns by steel jacketing: a state of review. *Asian Transactions on Engineering* 5 (3): 6–14. https://www.researchgate.net/publication/313350544_Strengthening_of_Reinforced_Concrete_Columns_by_Steel_Jackcting_A_State_of_Review.

Japan Society of Civil Engineering Recommendations (JSCE) (2001). *Recommendations for Upgrading of Concrete Structures with Use of Continuous Fiber Sheets*, Concrete Engineering Series, vol. 41 (ed. K. Maruyama). JSCE.

Jender H (2020). Gunite vs. Shotcrete: What's the Difference? https://www.riverpoolsandspas.com/blog/gunite-vs-shotcrete.

Jiang, S.F., Zeng, X.G., Shen, S., and Xu, X.C. (2016). Experimental studies on the seismic behavior of earthquake-damaged circular bridge columns repaired by using combination of near-surface-mounted BFRP bars with external BFRP sheets jacketing. *Engineering Structures* 106: 317–331.

Jumaat, M.Z., Rahman, M.M., and Alam, M.A. (2010). Flexural strengthening of RC continuous T beam using CFRP laminate: a review. *International Journal of the Physical Sciences.* 5 (6): 619–625. https://www.researchgate.net/publication/234077736_Flexural_strengthening_of_RC_continuous_T_beam_using_CFRP_laminate_A_review.

Jumaat, M.Z., Rahman, M.A., Alam, M.A., and Rahman, M.M. (2011a). Premature failures in plate bonded strengthened RC beams with an emphasis on premature shear: a review. *International Journal of Physical Sciences* 6 (2): 156–168. https://doi.org/10.5897/IJPS10.369.

Jumaat, M.Z., Rahman, M.M., and Rahman, M.A. (2011b). Review on bonding techniques of CFRP in strengthening concrete structures. *International Journal of the Physical Sciences* 6 (15): 3567–3575. https://doi.org/10.5897/IJPS10.376.

Kabeyasawa T., Kabeyasawa T., Kim Y., et al. (2010). Strength and deformability of reinforced concrete columns with wing walls. *Proceedings of the 9th U.S. National and 10th Canadian Conference on Earthquake Engineering*, July 25–29, 2010, Toronto, Ontario, Canada • Paper No 813.

Kabeyasawa, T., Kim, Y., Sato, M. et al. (2011). Tests and analysis on flexural deformability of reinforced concrete columns with wing walls. *Proceedings of the 9th Pacific Conference on Earthquake Engineering Building an Earthquake-Resilient Society*, Auckland, New Zealand.

Kaltakci, M.Y., Arslan, M.H., Yilmaz, U.S., and Arslan, H.D. (2008). A new approach on the strengthening of primary school buildings in Turkey: an application of external shear wall. *Building and Environment* 43 (6): 983–990.

Kanungo, S. and Bedi, K. (2018). Analysis of a tall structure with x-type bracing considering seismic load using analysis tool STADD.Pro. *International Journal of Engineering Sciences & Research Technology* 7 (12): 366–373. https://doi.org/10.5281/zenodo.2526223.

Kaplan, H. and Yılmaz, S. (2012). Seismic strengthening of reinforced concrete buildings. In: *Earthquake-Resistant Structures – Design, Assessment and Rehabilitation*, Chapter 16, https://doi.org/10.5772/28854 (ed. A. Moustafa). Croatia: InTech.

Karataş, M.A. and Gökkaya, H. (2018). A review on machinability of carbon fiber reinforced polymer (CFRP) and glass fiber reinforced polymer (GFRP) composite materials. *Defence Technology* 14: 318–326.

Karayannis, C.G., Chalioris, C.E., and Sirkelis, G.M. (2008). Local retrofit of exterior RC beam-column joints using thin RC jackets – an experimental study. *Earthquake Engineering and Structural Dynamics* 37 (5): 727–746. https://doi.org/10.1002/eqe.783.

Khalifa, A. and Nanni, A. (2000). Improving shear capacity of existing RC T-section beams using CFRP composites. *Cement and Concrete Composites* 22 (3): 165–174. https://doi.org/10.1016/S0958-9465(99)00051-7.

Kobraei, M., Jumaat, M.Z., and Shafigh, P. (2011). An experimental study on shear reinforcement in RC beams using CFRP-bars. *Scientific Research and Essays* 6 (16): 3447–3460. https://doi.org/10.5897/SRE11.650.

Koutas, L.N. and Bournas, D.A. (2019). Out-of-plane strengthening of masonry-infilled RC frames with textile-reinforced mortar jackets. *Journal of Composites for Construction* 23 (1): 04018079.

Koutas, L. and Triantafillou, T.C. (2013). Use of anchors in shear strengthening of reinforced concrete T-beams with FRP. *ASCE Journal of Composites for Construction* 17 (1): 101–107.

Koutas, L., Bousias, S.N., and Triantafillou, T.C. (2014a). Seismic strengthening of masonry infilled RC frames with textile-reinforced mortar: experimental study. *Journal of Composites for Construction* https://doi.org/10.1061/(ASCE)CC.1943-5614.0000507, 04014048.

Koutas, L., Pitytzogia, A., Triantafillou, T.C., and Bousias, S.N. (2014b). Strengthening of infilled reinforced concrete frames with textile-reinforced mortar (TRM): study on the development and testing of textile-based anchors. *Journal of Composites for Construction* https://doi.org/10.1061/(ASCE)CC.1943-5614.0000390.

Koutas, L., Bousias, S.N., and Triantafillou, T.C. (2015). *Textile-Reinforced Mortar as retrofitting material of masonry-infilled RC frames.* 11th International Symposium on Ferrocement and Textile Reinforced Concrete 3rd ICTRC.

Kouteva-Guentcheva, M., Stefanov, D., and Kaneva, A. (2013). Analysis of Reinforced Concrete Frame with Hysteric Seismic Protection. *12th National Congress on Theoretical and Applied Mechanics*, 23–26 September 2013, Saints Constantine and Helena, Varna, Bulgaria.

Lampropoulos, A., Apostolidi, E., Dritsos, S.E. et al. (ed.) (2021). *SED19: Seismic Isolation and Response Control.* International Association for Bridge and Structural Engineering (IABSE). https://doi.org/10.2749/sed019. ISBN: 978-3-85748-180-2.

Lazaris, A. (2019). Seismic evaluation and retrofitting of an existing building in Athens using pushover analysis. Master thesis. Karlsruhe Institute of Technology.

Lee, J. and Kim, J. (2017). Development of box-shaped steel slit dampers for seismic retrofit of building structures. *Engineering Structures* 150: 934–946.

Li, J., Gong, J., and Wang, L. (2009). Seismic behavior of corrosion-damaged reinforced concrete columns strengthened using combined carbon fiber-reinforced polymer and steel jacket. *Construction and Building Materials* 23 (7): 2653–2663.

Li, X., Lv, H.L., Zhang, G.C. et al. (2013). Seismic retrofitting of rectangular reinforced concrete columns using fibre composites for enhanced flexural strength. *Journal of Reinforced Plastics and Composites* 32 (9): 619–630.

Liu, X., Lu, Z.D., and Li, L.Z. (2018). The use of bolted side plates for shear strengthening of RC beams: a review. *Sustainability* 10 (12): 4658.

Ma, G. and Li, H. (2015). Experimental study of the seismic behavior of predamaged reinforced-concrete columns retrofitted with basalt fiber-reinforced polymer. *Journal of Composites for Construction* 19 (6): https://doi.org/10.1061/(ASCE)CC.1943-5614.0000572.

Ma, R., Xiao, Y., and Li, K.N. (2000). Full-scale testing of a parking structure column retrofitted with carbon fibre reinforced composites. *Construction and Building Materials* 14 (2): 63–71.

Mahar, J.W., Parker, H.W. and Wuellner, W.W. (1975). Shotcrete practice in underground construction. US Dept. Transportation Report FRA-OR&D 75–90. Springfield, VA: Nat. Tech. Info. Service.

Maheri, M.R., Kousari, R., and Razazan, M. (2003). Pushover tests on steel X braced and knee-braced RC frames. *Engineering Structures* 25 (13): 1697–1705.

Mahrenholtz, C., Lin, P.C., Wu, A.C. et al. (2015). Retrofit of reinforced concrete frames with buckling-restrained braces. *Earthquake Engineering and Structural Dynamics* 44 (1): 59–78. https://doi.org/10.1002/eqe2458.

Marini, A. and Meda, A. (2009). Retrofitting of R/C shear walls by means of high performance jackets. *Engineering Structures* 31 (12): 3059–3064. https://doi.org/10.1016/j.engstruct.2009.08.005.

Marneris J. and Kouskouna V. (2014). PARSANT: A 'non-destructive' method for anti-seismic strengthening of existing buildings. *14th World Conference on Earthquake Engineering,* October 12–17, Beijing, China.

Marriott D., Pampanin S., Bull D.K. and Palermo A. (2007). Improving the seismic performance of existing reinforced concrete buildings using advanced rocking wall solutions. *New Zealand Society of Earthquake Engineering Conference.*

Masuelli, M.A. (2013). Introduction of fibre-reinforced polymers – polymers and composites: concepts, properties, and processes. In: *Fiber Reinforced Polymer Composites*. Rijecka, Croatia: InTech http://dx.doi.org/10.5772/54629.

Matsagar, V.A. and Jangid, R.S. (2008). Base isolation for seismic retrofitting of structures. *Practice Periodical on Structural Design and Construction* 13 (4): 175–185. https://doi.org/10.1061/(ASCE)1084-0680(2008)13:4(175).

Mazza, F. and Mazza, M. (2019). Seismic retrofitting of gravity-loads designed r.c. framed buildings combining CFRP and hysteretic damped braces. *Bulletin of Earthquake Engineering* 17 (4): 3423–3445. https://doi.org/10.1007/s10518-019-00593-5.

Mazza, F. and Pucci, D. (2016). Static vulnerability of an existing r.c. structure and seismic retrofitting by CFRP and base-isolation: a case study. *Soil Dynamics and Earthquake Engineering* 84: 1–12. https://doi.org/10.1016/j.soildyn.2016.01.010.

Mazza, F., Mazza, M., and Vulcano, A. (2017). Nonlinear response of R.C. framed buildings retrofitted by different base-isolation systems under horizontal and vertical components of near-fault earthquakes. *Earthquakes and Structures* 12 (1): 135–144. https://doi.org/10.12989/eas.2017.12.1.135.

Mazza, F., Mazza, M., and Vulcano, A. (2018a). Base-isolation systems for the seismic retrofitting of r.c. framed buildings with soft-storey subjected to near-fault earthquakes. *Soil Dynamics and Earthquake Engineering* 109: 209–221. https://doi.org/10.1016/j.soildyn.2018.02.025.

Mazza, F., Mazza, M. and Vulcano, A. (2018b). Seismic retrofitting of in-elevation irregularly infilled R.C. framed structures by hysteric damped braces. *16th European Conference on Earthquake Engineering*, June 18–21, Thessaloniki.

Mazza, F., Mazza, M. and Vulcano, A. (2019). Design of hysteretic damped braces for the seismic retrofitting of in-elevation irregularly infilled r.c. framed structures. In: Atti del XVIII Convegno ANIDIS.

Mazzolani, F.M., Della Corte, G., and D'Aniello, M. (2009). Experimental analysis of steel dissipative bracing systems for seismic upgrading. *Journal of Civil Engineering and Management* 15 (1): 7–19. https://doi.org/10.3846/1392-3730.2009.15.7-19.

Mazzotti, C. (2011). The effect of the number of strengthening layers on the FRP-concrete bond behaviour. *European Journal of Environmental and Civil Engineering* 15 (9): 1277–1296. https://doi.org/10.1080/19648189.2011.9714855.

Megalooikonomou, K.G., Monti, G., and Santini, S. (2012). Constitutive model for fiber-reinforced polymer-and tie-confined concrete. *ACI Structural Journal* 109 (4): 569–578.

Mehta, P.K. (1987). Natural pozzolans: supplementary cementing materials in concrete. *CANMET Special Publication* 86: 1–33.

Melkumyan, M., Mihul, V. and Gevorgyan, E. (2011). Retrofitting by Base Isolation of Existing Buildings in Armenia and in Romania and Comparative Analysis of Innovative vs. Conventional Retrofitting. In: *Proceedings of the 3rd International Conference on Computational Methods in Structural Dynamics and Earthquake Engineering*, May 25–28, 2011, Corfu Island.

Memon, M.S. and Sheikh, S.A. (2005). Seismic resistance of square concrete columns retrofitted with glass fibre-reinforced polymer. *ACI Structural Journal* 102 (5): 774–783.

Mofidi, A., Thivierge, S., Chaallal, O., and Shao, Y. (2013). Behavior of reinforced concrete beams strengthened in shear using L-shaped CFRP plates: experimental investigation. *Journal of Composites for Construction* 18 (2): 04013033-1–04013033-8. https://doi.org/10.1061/(ASCE)CC.1943-5614.0000398.

Moham, A.A. and Matlab, T. (2018). Shear strengthening of RC without stirrups for deep beams with near surface mounted CFRP rods. *International Journal of Engineering Research & Technology (IJERT)* 4 (6): 545–547. https://www.researchgate.net/publication/329970278_Shear_Strengthening_of_RC_without_Stirrups_for_Deep_Beams_with_Near_Surface_Mounted_CFRP_Rods.

Molina, F.J., Sorace, S., Terenzi, G. et al. (2004). Seismic tests on reinforced concrete and steel frames retrofitted with dissipative braces. *Earthquake Engineering and Structural Dynamics* 33: 1373–1394. https://doi.org/10.1002/ eqe.408.

Mori K., Murakami K., Sakashita M., Kono S. and Tanaka H. (2008). Seismic performance of multi-storey shear wall with an adjacent frame considering uplift of foundation, 14th World Conference on Earthquake Engineering. China: Beijing.

Murali, G. and Pannirselvam, N. (2011). Flexural strengthening of reinforced concrete beams using fibre reinforced polymer laminate: a review. *Journal of Engineering and Applied Sciences* 6 (11): 41–47. https://www.researchgate.net/publication/288293590_Flexural_strengthening_of_reinforced_concrete_beams_using_fibre_reinforced_polymer_laminate_A_review.

Nagaprasad, P., Sahoo, D.R., and Rai, D.C. (2009). Seismic strengthening of RC columns using external steel cage. *Earthquake Engineering and Structural Dynamics* 38 (14): 1563–1586. https://doi.org/10.1002/eqe.917.

Nakai, T., Haruhiko, K., Tomoki, Y., and Naoki, N. (2019). Control effect of large tuned mass damper used for seismic retrofitting of existing high-rise building. *Japan Architectural Review* https://doi.org/10.1002/2475-8876.12100.

Narmashiri, K., Jumaat, Z., and Sulong, H.R. (2010). Shear strengthening of steel I-beams by using CFRP strips. *Scientific Research and Essays* 5 (16): 2155–2168. http://www.academicjournals.org/SRE.

Niemitz, C.W., James, R., and Breña, S.F. (2010). Experimental behavior of carbon fiber-reinforced polymer (CFRP) sheets attached to concrete surfaces using CFRP anchors. *Journal of Composites for Construction, ASCE* 14 (2): 185–194.

NZSEE (2019). Guideline for the Design of Seismic Isolation Systems for Buildings. Draft Report, June 2019. New Zealand Society of Earthquake Engineering, Wellington.

Oinam, R.M. and Sahoo, D.R. (2019). Using metallic dampers to improve seismic performance of soft-story RC frames: experimental and numerical study. *Journal of Performance of Constructed Facilities* 33 (1): 04018108.

Okakpu, A. and Ozay, G. (2014). A comparative study of building strengthening methods to have an efficient and economical solution. Case study in Famagusta , Cyprus. *International Journal of Civil and Structural Engineering* 5 (2): 165–176. https://doi.org/10.6088/ijcser.2014050016.

Okakpu, A. and Ozay, G. (2015). Decision selection technic for building strengthening methods. *Asian Journal of Civil Engineering (BHRC)* 16 (2): 203–218. https://www.academia.edu/10676998/DECISION_SELECTION_TECHNIC_FOR_BUILDING_STRENGTHENING_METHODS.

Okten, M.S., Ozkan, C., and Gencoglu, M. (2015). *Behavior of RC Frames with Infill Walls Strengthened by Cement Based Composites*. International Society of Offshore and Polar Engineers.

Olariu, I., Olariu, F. and Sarbu, D. (2000). Base Isolation versus Energy Dissipation for Seismic Retrofitting of Existing Structures. In: *12th World Conference on Earthquake Engineering*, January–February 2000, Auckland, New Zealand. Wellington: National Society for Earthquake Engineering.

Orton, S., Jirsa, J.O., and Bayrak, O. (2008, 2008). Design considerations of carbon fiber anchors. *Journal of Composites for Construction ASCE* 12 (6): 608–616.

Ou, Y.C. and Truong, A.N. (2018). Cyclic behavior of reinforced concrete L-and T-columns retrofitted from rectangular columns. *Engineering Structures* 177: 147–159.

Ozbakkaloglu, T. and Saatcioglu, M. (2009). Tensile behavior of FRP anchors in concrete. *Journal of Composites for Construction ASCE* 13 (2): 82–92.

Ozcan, O., Binici, B., and Özcebe, G. (2008). Improving seismic performance of deficient reinforced concrete columns using carbon fibre-reinforced polymers. *Engineering Structures* 30 (6): 1632–1646.

Özcebe G., Ersoy U., Tankut T. Erduran E., Keskin R. S. O. and Mertol H.C. (2003). Strengthening of brick-infilled RC frames with CFRP. TUBITAK Structural Engineering Research Unit Report No. 2003-1, Middle East Technical University, Ankara Turkey. March 2003.

Pantazopoulou, S.J., Tastani, S., Thermou, G. et al. (2016). Background to the European seismic design provisions for retrofitting RC elements using FRP materials. *Structural Concrete (Journal of the Fibre)* 17 (2): 194–219.

Pantelides, C.P., Okahashi, Y., and Reaveley, L.D. (2008). Seismic rehabilitation of reinforced concrete frame interior beam–column joints with FRP composites. *Journal of Composites for Construction* 12: 435–445.

Papanicolaou, C.G., Triantafillou, T.C., Karlos, K., and Papathanasiou, M. (2007). Textile-reinforced mortar (TRM) versus FRP as strengthening material of URM walls: in-plane cyclic loading. *Materials and Structures* 40: 1081–1097. https://doi.org/10.1617/s11527-006-9207-8.

Parghi, A. and Alam, M.S. (2015). Analysis of RC circular bridge columns retrofitted with fiber reinforced polymer under axial and lateral cyclic loading. In: *Canadian Society for Civil Engineering Annual Conference*, (May 27–30). Regina: Canadian Society for Civil Engineering (CSCE).

Parretti, R. and Nanni, A. (2004). Strengthening of RC members using near-surface mounted FRP composites: design overview. *Advances in Structural Engineering* 7 (5): 2004.

Phan, L.T., Lew, H.S., and Johnson, M.K. (1988). *Literature Review of Strengthening Methodologies of Existing Structures*. Gaithersburg, MD: U.S. Dept. of Commerce, National Institute of Standards and Technology NBSIR 88–3796.

Pinho, R. (2000). Selective Retrofitting of RC Structures in Seismic Areas. PhD. Imperial College London.

Pinho, R., Bianchi F. and Nascimbene R. (2019). Valutazione sismica e tecniche di intervento per edifici esistenti in c.a. Maggioli Editore (in Italian).

Pohoryles, D.A. and Bournas, D.A. (2019). A unified macro-modelling approach for masonry-infilled RC frames strengthened with composite materials. *Engineering Structures* 223: 111161.

Pohoryles, D.A. and Bournas, D.A. (2020). Seismic retrofit of infilled RC frames with textile reinforced mortars: state-of-the-art review and analytical modelling. *Composites Part B: Engineering* 183: ISSN 1359-8368. https://doi.org/10.1016/j.compositesb.2019.107702.

Poljanšek, M., Taucer, F., Ruiz, J.M. et al. (2014). Seismic Retrofitting of RC Frames with RC Infilling (SERFIN Project). JRC Science and Policy Reports, Joint Research Center.

Prasanna Kumar, M. and Vishnu, R.M. (2017). A comparative study on effect on lateral loading on steel braced reinforced concrete structure of unsymmetrical building plan. *International Journal of Civil Engineering and Technology* 8 (8): 609–616.

Pudjisuryadi, P., Tavio, and Suprobo, P. (2015). Performance of square reinforced concrete columns externally confined by steel angle collars under combined axial and lateral load. *Procedia Engineering* 125: 1043–1049.

Pugliese F., Di Sarno L. and Mannis A. (2019). Numerical evaluation of reinforced concrete frames with corroded steel reinforcement under seismic loading. A case study. *Conference ICSBS 2019, 2nd international conference on sustainable buildings and structures*, October 25–27, Suzhou (China). 10.1201/9781003000716–14.

Ravikumar, S.C. and Thandavamoorthy, T.S. (2014). Application of FRP for strengthening and retrofitting of civil engineering structures. *International Journal of Civil, Structural, Environmental and Infrastructure Engineering Research and Development (IJCSEIERD)* 4 (1): 49–60. Corpus ID: 201071944.

Realfonzo, R. and Napoli, A. (2009). Cyclic behavior of RC columns strengthened by FRP and steel devices. *Journal of Structural Engineering* 135 (10): https://doi.org/10.1061/(ASCE)ST.1943-541X.0000048.

Reggio, A., Restuccia, L., Martelli, L., and Ferro, G.A. (2019). Seismic performance of exoskeleton structures. *Engineering Structures* 198: 109459.

Sarafraz, M.E. and Danesh, F. (2012). New technique for flexural strengthening of RC columns with NSM FRP bars. *Magazine of Concrete Research* 64 (2): 151–161.

Sause, R., Harries, K.A., Walkup, S.L. et al. (2004). Flexural behavior of concrete columns retrofitted with carbon fibre-reinforced polymer jackets. *ACI Structural Journal* 101 (5): 708–716.

Sayed-Ahmed, M. S. (2012). N-M Interaction for Steel Jacket Retrofitting RC Bridge Column Due To Earthquake. https://www.academia.edu/1878194/N_M_INTERACTION_FOR_STEEL_JACKET_RETROFITTING_RC_BRIDGE_COLUMN_DUE_TO_EARTHQUAKE.

SeismoStruct (2023). SeismoStruct - A computer program for static and dynamic nonlinear analysis of framed structures. www.seismosoft.com.

Setunge, S., Kumar, A., Nezamian, A. et al. (2002). *Review of Strengthening Techniques Using Externally Bonded fiber Reinforced Polymer Composites*. Brisbane: CRC Construction Innovation.

Sevil, T., Baran, M., Bilir, T., and Canbay, E. (2011). Use of steel fiber reinforced mortar for seismic strengthening. *Construction and Building Materials* 25 (2): 892–899. https://doi.org/10.1016/j.conbuildmat.2010.06.096.

Seyhan, C.E., Goksu, C., Uzunhasanoglu, A., and Ilki, A. (2015). Seismic behavior of substandard RC columns retrofitted with embedded aramid fiber reinforced polymer (AFRP) reinforcement. *Polymers* 7 (12): 2535–2557. https://doi.org/10.3390/polym7121527.

Sharma, A., Genesio, G., Reddy, G. et al. (2010). Experimental investigations on seismic retrofitting of reinforced concrete beam-column joints. *14th Symposium on Earthquake Engineering 14SEE*, December 17–19, 2010. Paper No. A007. Roorkee: Indian Institute of Technology.

Shashikumar, N.S., Gangadharappa, B.M., Ashwini, B.T., and Chethan, V.R. (2018). Analysis of RC frame building with different types of braces in various seismic zones. *International Journal of Scientific and Engineering Research* 3 (9): 183–201. https://www.academia.edu/37611785/Analysis_of_RC_frame_building_with_different_types_of_braces_in_various_seismic_zones.

Sheikh, S.A. and Yau, G. (2002). Seismic behavior of concrete columns confined with steel and fibre-reinforced polymers. *ACI Structural Journal* 99 (1): 72–80.

Sika (2022). Sikadur®-32 EF technical specifications. https://grc.sika.com/el/45787/48448/48455/sikadur-32-ef.html (Accessed: May 15, 2022).

Skinner, R.I., Robinson, W.H., and Kelly, T.E. (2011). *Seismic Isolation for Designers and Structural Engineers*. Wellington: Robinson Seismic Ltd, Holmes Consulting Group.

Sneed, L.H. (2013). Review of Anchorage Systems for Externally Bonded FRP Laminates. *International Journal of Concrete Structures and Materials* **7**: 17–33. http://dx.doi.org/10.1007/s40069-013-0029-0.

Soong, T.T. and Dargush, G.F. (1999). *Passive Energy Dissipation and Active Control. Structural Engineering Handbook*, Chapter 27 (ed. C. Wai-Fah). Boca Raton: CRC Press LLC.

Spyrakos, C.C. (2004). *Strengthening of Structures for Seismic Loads*. Athens Greece: Technical Chamber of Greece In Greek.

Staaleson Engineering, P.C. (2021). Retrofit Steel Chevron Braced Frame with Fluid Viscous Dampers. https://www.staaleng.com/seismic_strength/default.html (Accessed: November 15, 2021).

Swetha, K.S. and James, R.M. (2018). Strengthening of RC beam with web bonded steel plates. *International Journal of Engineering and Techniques* 4 (3): 71–81. http://oaji.net/articles/2017/1992-1530709218.pdf.

Symans, M.D. and Constantinou, M.C. (1999). Semi-active control systems for seismic protection of structures: a state-of-the-art review. *Engineering Structures* 21: 469–487. https://doi.org/10.1016/S0141-0296(97)00225-3.

Symans, M.D., Charney, F.A., Whittaker, A.S. et al. (2008). Energy dissipation systems for seismic applications: current practice and recent developments. *Journal of Structural Engineering, ASCE* 134 (1): 3–21.

Szabó, K.Z. and Balázs, L.G. (2007). Near surface mounted FRP reinforcement for strengthening of concrete structures. *Periodica Polytechnica* 51 (1): 33–38. https://doi.org/10.3311/pp.ci.2007-1.05.

TahamouliRoudsari, M., Entezari, A., Hadidi, M., and Gandomian, O. (2017). Experimental assessment of retrofitted RC frames with different steel braces. *Structures* 11: 206–217.

Tarabia, A.M. and Albakry, H.F. (2014). Strengthening of RC columns by steel angles and strips. *Alexandria Engineering Journal* 53 (3): 615–626. https://doi.org/10.1016/j.aej.2014.04.005.

Tassios T.P. (2016). Theory for the Design of Repair and Strengthening. Athens, Greece (in Greek).

TBDY (2018). Türkiye Bina Deprem Yönetmeliği, Turkish Seismic Building Code. Disaster and Emergency Management Presidency, Ankara (in Turkish).

Teruna, D., Lukman, M., Majid, T. A. and Budiono, B. (2014). Evaluation of the seismic strengthening for R/C frame building with soft first story using hysteric steel damper subjected to strong earthquake. *Proceedings of Conference on the 3rd Geohazard Information Zone and the 5th Seminar and Short Course of HASTAG (GIZ 2014 & HASTAG V)*, October 20–21, Medan, Indonesia.

The Concrete Society (2004). *Design Guidance for Strengthening Concrete Structures using Fibre Composite Materials*, 2e Technical Report No. 55. The UK Concrete Society. ISBN: 1904448247.

Thermou, G.E. and Pantazopoulou, S.J. (2007). Seismic retrofit of square reinforced concrete columns using composite jacketing. In: *FRPRCS-8: 8th International Symposium on Fiber Reinforced Polymer Reinforcement for Reinforced Concrete Structures* July 16–18, 2007, 1–10. Patra: University of Patras.

Thermou, G.E. and Pantazopoulou, S.J. (2009). Fiber-reinforced polymer retrofitting of predamaged substandard RC prismatic members. *Journal of Composites for Construction* 13 (6): 535–546. https://doi.org/10.1061/(ASCE)CC.1943-5614.0000057.

Thermou, G. E. and Pantazopoulou, S. J. (2014). Criteria and methods for redesign and retrofit of old structures. In *10th U.S. National Conference on Earthquake Engineering: Frontiers of Earthquake Engineering*, 21–25 July 2014, Anchorage, Alaska. http://dx.doi.org/10.4231/D3DN3ZW75.

Thermou, G.E., Tastani, S.P., and Pantazopoulou, S.J. (2011). The effect of previous damage on the effectiveness of FRP-jacketing for seismic repairs of RC structural members. In: *American Concrete Institute*, 951–969. ACI Special Publication.

Tiwari, P. and Bhadauria, S.S. (2017). Comparative analysis of tall structure with or without bracings considering seismic load. *International Journal of Engineering Sciences & Research Technology (IJESRT)* 6 (10): 118–125. https://doi.org/10.5281/zenodo.1002663.

Triantafillou, T.C. (2001). Seismic retrofitting of structures using FRPs. *Progress in Structural Engineering and Materials* 3 (1): 57–65.

Triantafillou T.C. (2004). Strengthening of Reinforced Concrete Structures with Composite Materials (Fibre-Reinforced Polymers). Patras 2004 (in Greek).

Triantafillou, T.C., Karlos, K., Kapsalis, P., and Georgiou, L. (2018). Innovative structural and energy retrofitting system for masonry walls using textile reinforced mortars combined with thermal insulation: in-plane mechanical behaviour. *Journal of Composites for Construction* 22 (5): https://doi.org/10.1061/(ASCE)CC.1943-5614.0000869.

Tsai, K.C., Lai, J.W., Hwang, Y.C. (2004). Research and application of double-core buckling restrained braces in Taiwan. *13th World Conference of Earthquake Engineering*. Canada: Vancouver

Tsionis G., Apostolska R., Taucer F. (2014). Seismic strengthening of RC buildings. JRC Science and Policy Reports, Joint Research Center.

US Army Corps of Engineers (1993). Standard Practice for Shotcrete. Engineering Manual EM 1110-2-2005, 31 January 1993.

Valente, M. (2013). Improving the seismic performance of precast buildings using dissipative devices. *Procedia Engineering* 54: 795–804.

Vandoros, K.G. and Dritsos, S.E. (2008). Concrete jacket construction detail effectiveness when strengthening RC columns. *Construction and Building Materials* 22: 264–276.

Varum, H., Chaulagain, H., Rodrigues, H. and Spacone, E. (2013). Seismic assessment and retrofitting of existing RC buildings in Kathmandu, *9th International Congress on Pathology and Repair of Structures, CINPAR 2013*, June 2–5, JoAo Pessoa, Brazil.

Vasconcelos, G., Abreu, S., Fangueiro, R., and Cunha, F. (2012). Retrofitting masonry infill walls with textile reinforced mortar. In: *Proc. 15th World Conf. Earthq. Eng*, 10. Lisbon, Portugal.

Vrettos, I., Kefala, E., and Triantafillou, T.C. (2013, 2013). Innovative flexural strengthening of reinforced concrete columns using carbon-fibre anchors. *ACI Structural Journal* 110 (1): 63–70.

Waghmare, S.P.B. (2011). Materials and jacketing technique for retrofitting of structures. *International Journal of Advanced Engineering Research and Studies, I* I: 15–19. https://www.technicaljournalsonline.com/ijaers/VOL%20I/IJAERS%20VOL%20I%20ISSUE%20I%20%20OCTBER%20DECEMBER%202011/4%20IJAERS.pdf.

Walcoom (2022). Steel Fiber Improves Structural Strength and Integrity in Concrete. https://www.walcoom.com/pro/architecturalmesh/steel-fiber/index.html (Accessed: May 27, 2022)

Walters, M. (2003). The seismic retrofit of the Oakland City Hall. In: SMIP03 Seminar Proceedings, p. 149–163.

Wang, C., Sarhosis, V., and Nikitas, N. (2018). Strengthening/retrofitting techniques on unreinforced masonry structure/element subjected to seismic loads: a literature review. *The Open Construction & Building Technology Journal* 12: 251–268.

Wikipedia (2021). Pozzolan. https://en.wikipedia.org/wiki/Pozzolan (Accessed: November 15, 2021).

Wikipedia (2022a). Silica fume. https://en.wikipedia.org/wiki/Silica_fume (Accessed: May 5, 2022).

Wikipedia (2022b). Glass fiber reinforced concrete. https://en.wikipedia.org/wiki/Glass_fiber_reinforced_concrete (Accessed: May 15, 2022).

Wikipedia (2022c). Seismic base isolation. https://en.wikipedia.org/wiki/Seismic_base_isolation (Accessed: May 27, 2022).

Wikipedia (2022d). Cathodic protection. https://en.wikipedia.org/wiki/Cathodic_protection (Accessed: May 25, 2022).

Yamakawa, T., Rahman, M.N., and Morishita, Y. (2006). Experimental investigation and analytical approach for seismic retrofit of RC column with wing-wall. *Journal of Structural and Construction Engineering* 608: 109–117.

Yuksel, E., Ilki, A., Erol, G. et al. (2005). Seismic retrofit of infilled reinforced concrete frames with CFRP composites. In: *Advances in Earthquake Engineering for Urban Risk Reduction*.

Yuksel, E., Ozkaynak, H., Buyukozturk, C. et al. (2010, 2010). Performance of alternative CFRP retrofitting schemes used in infill RC frames. *Construction and Building Materials* 24 (4): 596–609.

Zhou, C.D., Lu, X.L., Li, H., and Tian, T. (2013, 2013). Experimental study on seismic behavior of circular RC columns strengthened with pre-stressed FRP strips. *Earthquake Engineering and Engineering Vibration* 12 (4): 625–642.

# 5

# Criteria for Selecting Strengthening Methods – Case Studies

## 5.1 Things Are Rarely Simple

In the example of Chapter 4, the focus was on a fairly simple case of retrofit. The building had a conspicuous structural problem (soft story at the ground level), it was clear that it required strengthening, it had open space on all four sides, and the interventions were applied either at the ground level only or at the perimeter of the building without the need to cause nonstructural damage in the interior of the building. Furthermore, the modeling of the strengthening did not take into account some "annoying" details of the structural configuration; the stairs, and, more importantly, the entrance of the building at the ground level that would pose some challenges in the construction of the interventions.

Outside the textbook, things can get more complicated. It is not always possible to easily identify the structural problems of the building, and more importantly, it is not always possible to carry out interventions in all the locations that are considered important. For instance, in the majority of practical applications, some of the following decisions have to be made:

- Old existing RC walls are usually very lightly reinforced and constitute the most vulnerable members of the buildings. They attract a large proportion of the lateral forces, due to their large size, but are unable to resist them, because of the lack of reinforcement. Usually, such walls require strengthening and are best strengthened with jackets. However, very frequently these walls are found in the perimeter of the elevator shafts, in which case their jacketing would require the change of the elevator since the available space gets smaller. Are the owners willing to include such a cost in the total retrofit budget?
- It is quite common that existing (weak) columns are found close to the kitchen or close to closets. The upgrading of these columns with jackets of FRP wraps would require adaptations to the shelves, disproportionally increasing the cost for the strengthening of a single column. Are there ways to avoid this?
- The jacketing or wrapping of columns in the vicinity of a bathroom imposes similar challenges, since it would be impossible to find the same tiles 30 or 40 years after the

construction of the building. Are the owners willing to change the tiles of the entire bathroom or to accept a "modern" mosaic of nonmatching tiles?

- In the general case, it is not easy to find locations in the perimeter of the building to add new walls. Will the architects accept having many locations in the perimeter without openings?
- The jacketing of columns adjacent to openings results in smaller openings, which means that the window or the door must be replaced. Is it worth the additional cost?
- It is generally difficult to add walls close to the entrance of buildings, such as in the area of hotel lobbies, so as to not limit the available space. In such cases, are there viable alternatives?

When designing strengthening interventions, the engineer should take into consideration these and similar issues, as well as the complications that these might cause during construction, and the non-negligible additional cost. All these constitute substantial factors in the decision of which technique or techniques should be employed.

Very often, hybrid solutions are favored, whereby the merits of one method are combined with the merits of another, in order to reach the target performance level. For instance, typical examples of combining more than one technique are the following:

- RC jackets to increase the strength of the vertical members, and RC walls for additional stiffness.
- RC jackets to increase the strength of the vertical members, and steel braces for additional stiffness.
- FRP wrapping to increase the shear resistance of the vertical members, and steel braces for additional stiffness.
- RC jackets in most of the vertical members to increase their strength, but FRP wrapping in certain locations, in order not to increase the size of the columns for architectural reasons (e.g., a hotel lobby).

One could even employ more complicated solutions, such as RC walls for stiffness, RC jackets to increase the capacity of the vertical and some horizontal members, and FRP wrapping for some columns, the size of which cannot be increased. The possibilities are endless, depending on the advantages and the disadvantages of each method for the building under consideration. In any case, acute engineering judgment, adequate experience, and a good understanding of the requirements of the owners and the architects are needed by the engineer who is designing the interventions.

## 5.2 Criteria for Selecting Strengthening Method

When designing new reinforced concrete buildings, there are basically two sets of criteria to be fulfilled, the *technical criteria* for the required structural safety, and the *economic criteria* that concern the relative cost of the alternative technical solutions that guarantee safety. After the architectural design is completed, the civil engineer undertakes the task of finding a design that is compatible with the architectural layout and has the lowest cost among all the different alternatives. There is always some interaction between the civil engineer and the architect (as well as the mechanical engineer), but this is generally limited; it is confined to settling issues like the exact position or the dimensions of a structural component, and it does not cause many complications in the civil engineering design. Furthermore, other issues, such as the availability of construction materials or specialized construction crews are generally known in advance and very often make the life of the engineer easier, rather than harder, since the range of technical options available is small, and the engineer is accustomed to the possible options. After the chosen solution has been outlined, it is modeled in specialized design software packages, which make the necessary calculations, estimate the required reinforcement, and determine whether a change in the concrete dimensions of the members is needed. The framework for the design of new structures is not very flexible and somehow "guides" the engineer, who can adapt the design by employing relatively simple recipes, e.g., when the calculated reinforcement of a member exceeds the maximum allowable reinforcement ratio $\rho_{max}$, the typical solution is to increase its dimensions.

When it comes to the retrofit of existing buildings, however, the criteria for strengthening a building are generally more complex. Besides the technical and the economic criteria that are standard in any structural design, other types of criteria become significant and play a considerable role in the process of selecting the most suitable technique(s).

First, many of the technical decisions have to be made on the basis of the conclusions drawn from the preliminary assessment of the structure, as well as on the structural information collected during on-site investigations. Different measures should be sought in the case of severe and distributed damage throughout the building, and in the case of light and/or concentrated damage for the selected performance objective. Furthermore, the selection of the type and the extent of the intervention also depend on the urgency, with which this intervention has to be implemented, e.g., after severe damage has been sustained in a post-earthquake scenario. In strengthening and retrofit interventions, the identification of the structural weaknesses, specific problematic members, local concentration of forces, and the consequences of the collapse of a particular member all play a major role in the selection of the intervention technique(s), as well as the locations of the interventions.

Moreover, important operational criteria should be considered, which do not constitute an important parameter in the case of new construction. These are related to the fact that the locations of interventions should not impede the functionality of the structure, for instance the location of new shear walls or steel braces should not close off the existing openings. Furthermore, the proposed works should cause the minimum possible disturbance. The obstruction of the normal operation of the building and the disruption of use or occupancy are very important parameters that should always be considered. It should not be forgotten that the buildings under consideration are existing buildings that have been used for several years, and this constitutes a significant difference (and loss of flexibility) with respect to new construction. Very often the disruption of use is a more important criterion than the direct financial cost of the intervention. The indirect loss from the suspension of operations, for rented office or residential areas might be even larger.

Note also that in the case of new buildings the architectural layout can also be adapted and the interaction with the architects usually allows for some level of flexibility in the formation of the RC bearing system. For instance, the removal of a column, because of a large opening in the facade, can be done provided that the adjacent columns and beams increase in size and strength, and the openings are changed accordingly. By contrast, in existing buildings, the size and location of all the architectural components is fixed and it is very costly or often impossible to make significant alterations. The design of the retrofit interventions should necessarily adapt to the existing architectural layout, which significantly reduces the technical options available to the engineer.

One additional consideration is that the demolition and reconstruction of nonstructural components (e.g., infilled walls) required for the strengthening of a structural member (e.g., jackets or FRP wrapping) are more expensive in certain areas of the building than others. Such locations for example are kitchens, closets or WCs, where the nonstructural costs increase considerably for the repair of the pantries, the tiles, or the drainage system.

Recognizing all these difficulties and restrictions, the authors of the different standards worldwide have introduced more performance objectives in the assessment and strengthening procedures, i.e., more seismic hazard levels and more structural performance levels. The objective is to give engineers some flexibility, which is often needed in order to overcome the complications posed by existing buildings and to enable viable technical solutions. This additional flexibility is positive overall, but it also significantly increases the options that are available to the engineer, further complicating his/her role. Note that the different options for the performance objectives correspond to different risk categories for the building; hence, close interaction with the owner is essential, so that the needs and requirements regarding the level of safety are correctly recognized, while considering the remaining life of the structure, which is generally less than new structures.

In conclusion, it should be apparent that the design of strengthening interventions poses great difficulties and challenges to the engineer, both in design and construction. The correct identification of the technical, operational, social, and performance criteria involved in the process is a prerequisite for finding the optimal solution in every case of retrofit. Very often compromises are required, making small sacrifices in certain aspects of the problem, in order to obtain greater benefits in others. Specialized knowledge, experience in both design and construction, and correct engineering judgment are needed to impartially consider all the advantages and disadvantages of the different options and to make the optimal decisions.

## 5.3 Basic Principles of Conceptual Design

The fundamental technical requirements and guiding principles for a strengthened building are similar to those for a newly constructed building (CEN 2004, Section 4.2.1, and Fardis (2009)):

- *Structural simplicity*: The existence of a clear path for the transmission of the seismic forces is important, in order to reduce uncertainty in the modeling and the analysis of the structural model, rendering the prediction of its seismic behavior and the dimensioning of new structural members much more reliable. Very often, this simplicity is not encountered in existing structures; however, the strengthening interventions that are typically designed to undertake a large proportion of the earthquake loading should be adequately simple and predictable.
- *Bidirectional resistance and stiffness:* In the vast majority of buildings the expected seismic motion is bidirectional. Very rarely is earthquake directivity so significant that the expected ground motion in one horizontal direction is much more severe than in the other direction.

Thus, any structural system shall be able to resist actions in any direction. To satisfy this requirement, the new structural components should be arranged to ensure similar stiffness and resistance characteristics in both main directions. Typically, since the new components (e.g., walls, jackets, braces) generally possess larger strength and stiffness with respect to the existing structural members, they should be placed in an orthogonal in-plan pattern.

- *Torsional resistance and stiffness:* Buildings should also possess adequate torsional resistance and stiffness, in order to limit the development of torsional deformations, which tend to stress the different structural elements in a nonuniform way. To this end, the new structural members are preferably located at the periphery, rather than the center, of the building, in order to contribute to both the lateral and the rotational stiffness.
- *Uniformity and symmetry:* Uniformity in plan requires the even distribution of the new vertical members, which ensures a more direct transmission of the inertia forces to the ground. Although this is not always feasible (e.g., it is not always easy to retrofit every column of the building), the construction of more new vertical components is preferable to the construction of fewer and larger ones. The well-distributed and symmetrical layout of the new walls and columns is necessary for the achievement of uniformity in a strengthened building, and in order to minimize the eccentricities between mass and stiffness. Moreover, uniformity in elevation is also important, since it tends to eliminate the occurrence of soft stories, and other weak zones with large force and deformation concentrations, and it prevents higher mode effects.
- *Redundancy:* The use of evenly distributed structural elements increases redundancy and allows a more favorable redistribution of action effects and spreads energy dissipation across the entire structure.
- *Diaphragmatic behavior of floors:* The floors of a building play an important role in the safe transfer of the inertia forces from the location of masses to the vertical members. They act as horizontal diaphragms and ensure that all the vertical members work together in resisting the horizontal seismic action. This diaphragmatic action is especially relevant in the strengthening of existing buildings, where some large and strong new vertical members coexist with the small and weak older members. In most existing RC buildings, the floors are concrete slabs and they are relatively thinner than modern day standards (typically 10–14 cm). All the same, they still have adequate in-plan stiffness, with respect to the lateral stiffness of the vertical members, to act as rigid or semi-rigid diaphragms, thus successfully transferring the inertia forces through them. If this is not the case, however, they need to be strengthened.

When designing interventions, the engineer should always bear these principles in mind. Respecting them will help him/her reach a satisfactory and adequate technical solution fairly early, even during the conceptual design phase. Generally, one or more of the following retrofit strategies may be adopted.

- Increase in the global lateral strength and stiffness through the addition of new large components, such as shear walls or steel braces that are designed to sustain a large proportion of the seismic action.
- Strengthening of a large number of existing structural members (usually with the simultaneous increase in their strength, stiffness and ductility), e.g., through RC jackets or FRP wrapping.

- Full replacement, through demolition and reconstruction of specific badly damaged components.
- Elimination or reduction of existing irregularities, though the introduction of new members.
- Elimination or reduction of existing irregularities, though the demolition of existing members that cause or intensify the irregular response.
- Elimination or reduction of existing irregularities, though the removal of existing expansion joints or the creation of new expansion joints.
- Mass reduction, in order to reduce the inertia earthquake forces on the structural members.
- Strengthening of the foundation system.
- Introduction of seismic isolation.
- Introduction of energy dissipation devices.
- Partial demolition of the more vulnerable parts of the building.
- Possible transformation of existing non-structural elements into structural elements.
- Local repair of slightly damaged individual structural members (e.g., epoxy injections, repair mortars).

One or more types of the aforementioned strategies may be adopted. In all cases, the effect of the proposed structural modifications should be checked employing the typical assessment procedures.

## 5.4 Some Rules of Thumb

There is a basic distinction between projects for the design of strengthening interventions for a RC building; that is whether only a structural upgrade and retrofit is to be carried out, or whether a complete renovation is being planned.

In the first case, solutions that cause minimal disturbance to the operation of the building are preferable. The strengthening interventions, when they are not combined with architectural alterations and changes in the electrical and mechanical installations, are relatively small projects that need to be carried out quickly and, if possible, without the interruption of the operations of the building. In such cases it is better to have interventions only in certain locations on the plan of the building, preferably in the perimeter, causing gradual disruption to the different levels of the building and only within specific rooms. In this way, the cost for demolition and reconstruction of the nonstructural components, required for the execution of the strengthening works, remains limited (Figure 5.1).

Methods such as the addition of new shear walls or steel braces are more suitable and are preferable to alternatives that require interventions on all vertical members (e.g., FRP wraps or RC jackets). It is stressed, however, that an important prerequisite for the selection of these techniques is that the existing columns that remain unstrengthened should possess a minimum capacity in both shear and bending. Columns with serious seismic damage, with stirrup spacing equal to 40 or 50 cm, without lap splices, or with heavily corroded steel are extremely vulnerable, cannot withstand even very small levels of lateral deformation, and need to be strengthened anyway (Figure 5.2).

On the contrary, when a radical architectural renovation is planned, especially if this includes changes in the layout of the floors with the demolition and reconstruction of partition walls, it is usually better to strengthen all the vertical members of the building, for example with jackets or FRP wraps, since there is no operational cost (the building has been evacuated anyway) and the additional cost by the repair works required by the strengthening interventions is negligible (Figure 5.3).

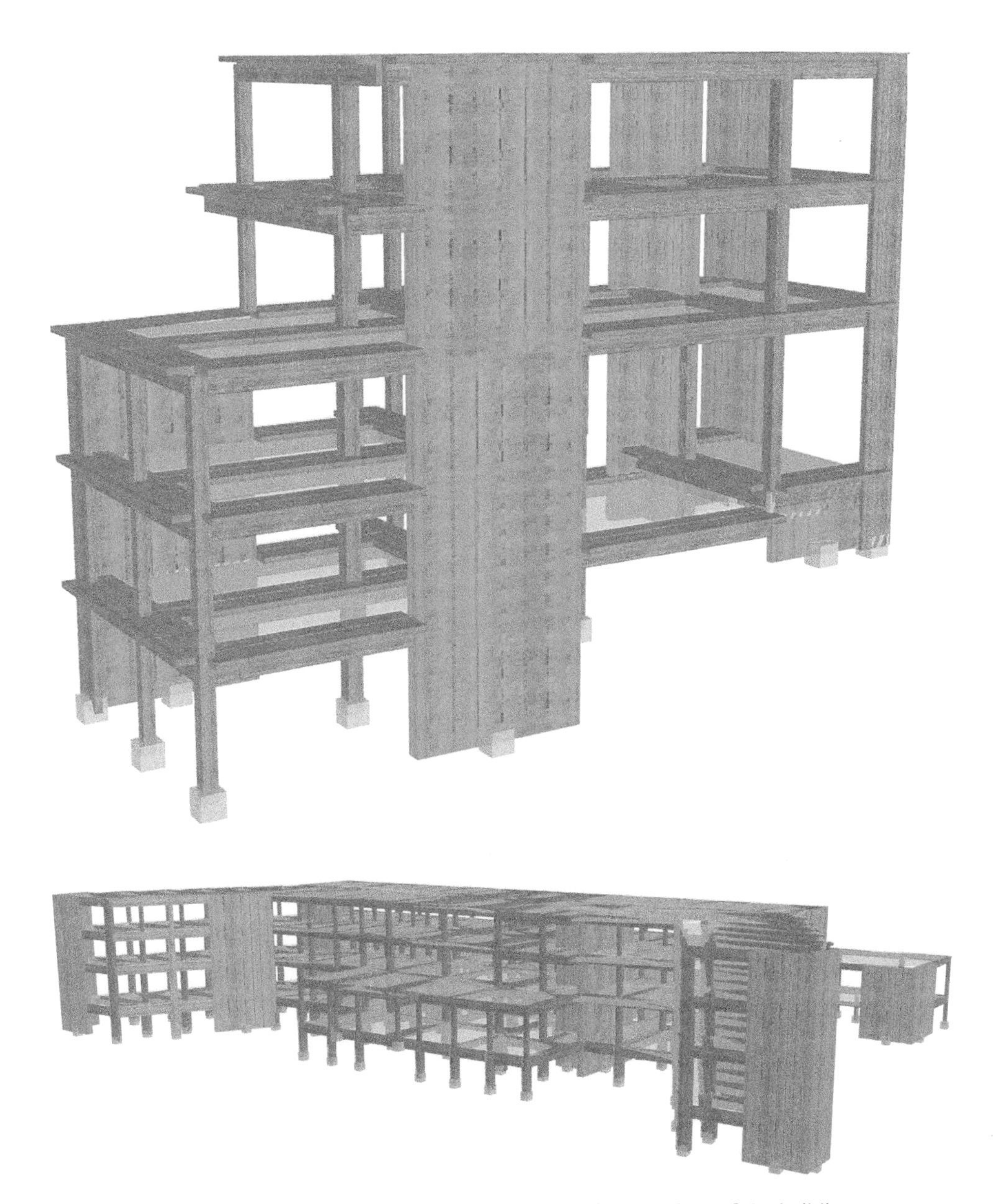

**Figure 5.1** Structural retrofit with walls or steel braces at few locations of the building. *Source:* Stelios Antoniou.

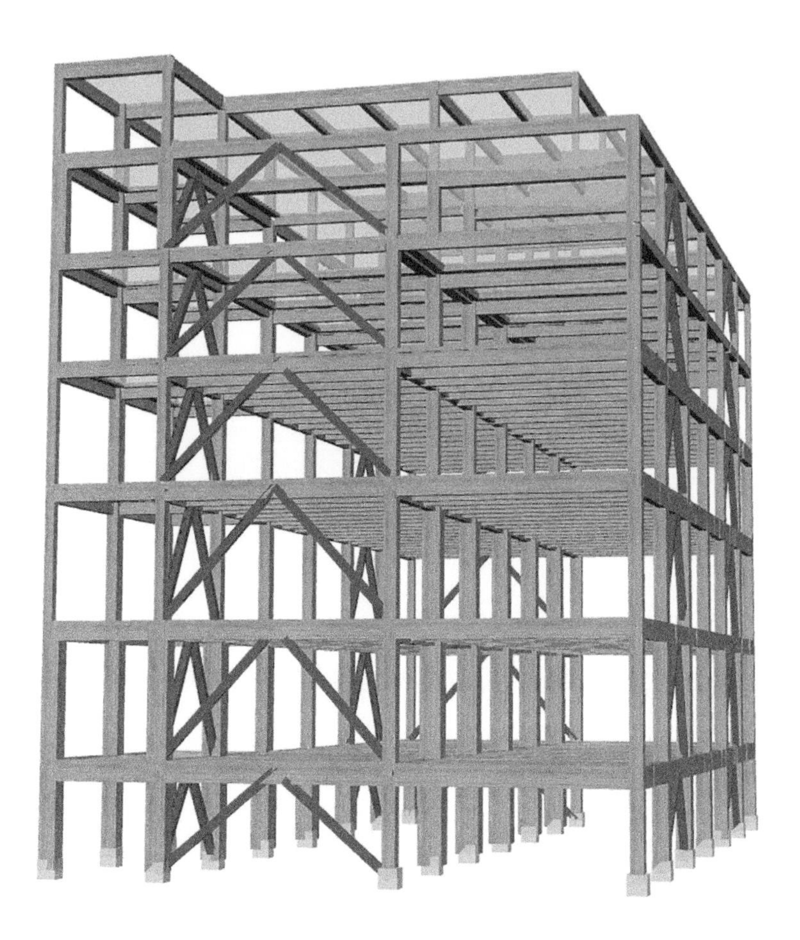

Figure 5.1 (Continued)

Figure 5.2 Columns that cannot remain without strengthening. *Source:* Stelios Antoniou.

Figure 5.3 In cases of serious renovation works the strengthening of all vertical members is advantageous. *Source:* Stelios Antoniou.

Besides this basic rule, there are certain (mostly empirical) guidelines regarding interventions on an existing building that could help readers select the most appropriate techniques for a particular building:

- Solutions that combine stiffness, strength, and ductility increase (e.g., new shear walls or RC jackets) are generally preferred to selective intervention schemes. This is because older buildings usually suffer from a lack in all these characteristics and need retrofit in all of them simultaneously.
- When strengthening the building's vertical members, the concept of the weakest link should be respected. It is generally a bad idea to leave some weak columns unstrengthened because their upgrading requires large nonstructural costs and/or causes

significant disruption. A building is as good as its weakest column (at least for the brittle failure types like shear); if the column fails, a partial or a full collapse of the building will occur, risking human lives and dooming the entire structure that would then need to be completely demolished.

- More attention is generally paid to the integrity and safety of the vertical members, which are the most important for the structural stability. The beams, at least those that do not support columns of the upper level, generally assume a secondary role in earthquake resistance, and, even if they sustain significant damage, this does not lead to collapse, even if it involves a brittle type of failure. Note that the entire philosophy of capacity design for new buildings is based on this concept of favoring failure mechanisms that absorb energy, but do not have serious consequences on the structural integrity.
- In cases of highly irregular buildings (both in terms of stiffness and strength distributions, which are usually found concurrently), the regularity in both elevation and plan should be restored. This can be achieved by the modification of the strength and the stiffness of a number of appropriately selected existing components, or by the introduction of new structural elements, such as large shear walls.
- Usually, the critical check is in shear, since older buildings feature very low transverse reinforcement ratios, contrary to the longitudinal reinforcement, which is not as low.
- When selecting the most appropriate strengthening technique, apart from the direct financial cost for the structural interventions, one should always consider the cost for the works on the nonstructural components (e.g., demolition and repair works for tiles, floors, or openings) and the indirect operational costs (e.g., suspension of operations, duration of works).
- Although building foundations are rarely the cause for severe structural damage during a seismic event, special attention should be paid in the case of very badly constructed footings. Experience has shown that the superstructure of the buildings of the 1960s and 1970s are roughly constructed as designed and shown in the drawings, possibly with some minor modifications. On the contrary, on the foundation level many cases of poor workmanship have been observed, which range from highly reduced or missing reinforcement to the absence of entire footings altogether. With a significant earthquake ground motion, these inadequacies may have serious consequences to the integrity of the entire building. Hence, it is usually necessary to uncover some of the building's footings.
- The repair and strengthening of nonstructural elements, such as infills, racks, or chimneys, should also be considered, whenever the seismic behavior of these components is a threat to human life or disproportionally affects the value of the building or the goods stored in it (e.g., racks for the storage of expensive goods in warehouses). The full or partial collapse of them should be avoided by increasing their capacity and by providing appropriate connection to the structural system.

## 5.5 Case Studies

This section discusses some characteristic case studies of strengthening interventions. These are projects that I have been somewhat involved in through either Seismosoft or Alfakat, the two companies I work for. In all the examples, after a brief description of the existing building, its deficiencies, the reasons that led to the decision for structural retrofit and the strengthening strategy are explained, together with the criteria why the specific retrofit measures were selected. All examples are slightly adapted for privacy reasons, but without altering the basic technical features in each case.

### 5.5.1 Case Study1: Seismic Upgrade of a Five-Story Hotel

This is a 5000 m$^2$ reinforced concrete hotel, which is five stories high (Figure 5.4). The strengthening was carried out during a general architectural renovation. The entire building consists

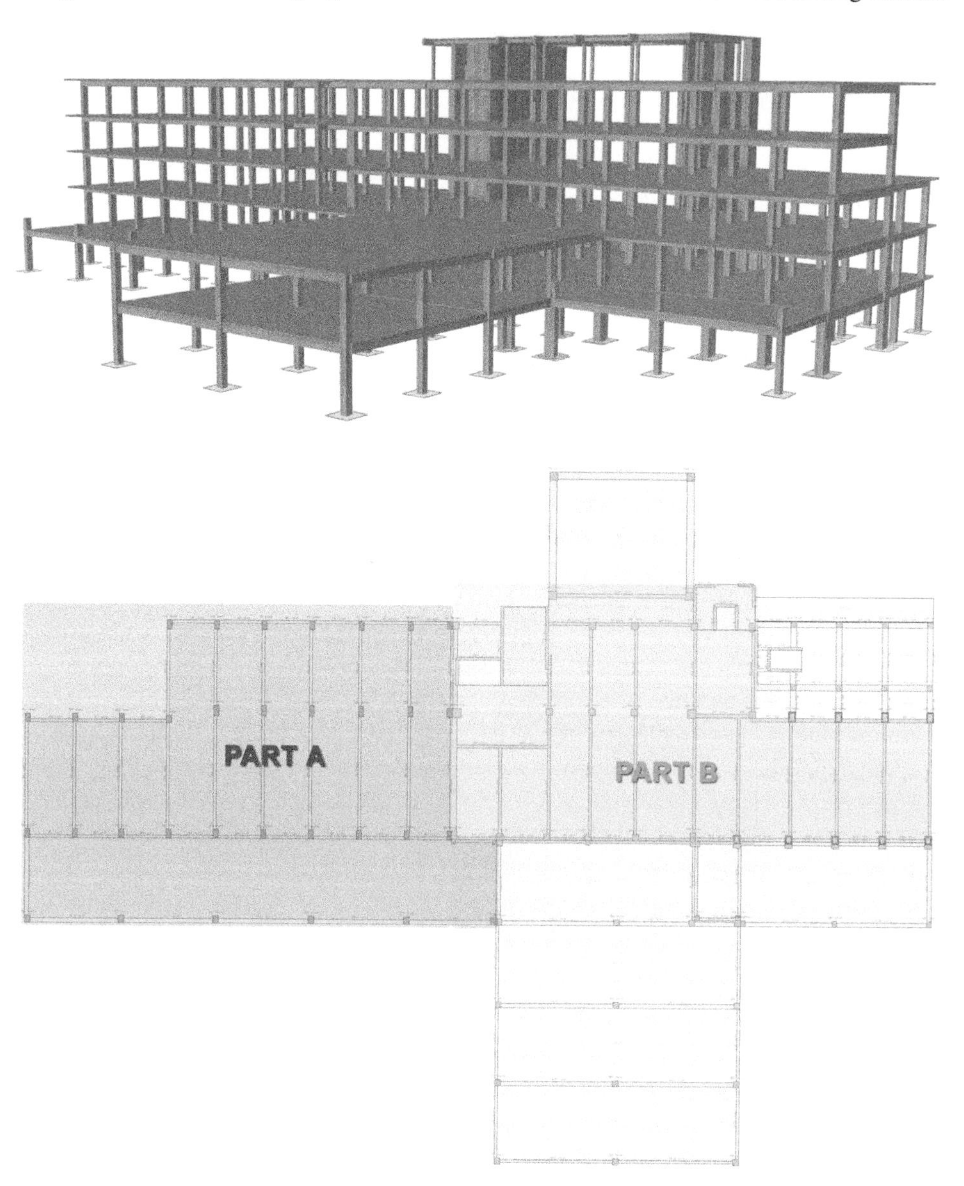


**Figure 5.4** 3D-view of the building and plan view of the ground floor, where Parts A and B are highlighted. *Source:* Stelios Antoniou.

of six statically independent parts, the largest of which, part A and part B, were constructed in the early 1970s. The other parts are smaller and were added gradually until the late 1980s.

The main challenge of the project was the strengthening of the older parts A and B, especially the former. Part A is highly irregular in elevation with a sudden and significant reduction of the size in the upper floors, with respect to the ground floor (Figure 5.5a). Part B instead is more regular, both in plan and elevation (Figure 5.5b). The structural system of both buildings includes several small and weak columns and large (up to 8 m long) beams, and both columns and beams were lightly reinforced. It is interesting that the steel class of the transverse reinforcement for the columns was S220, whereas for the beams it was S400;

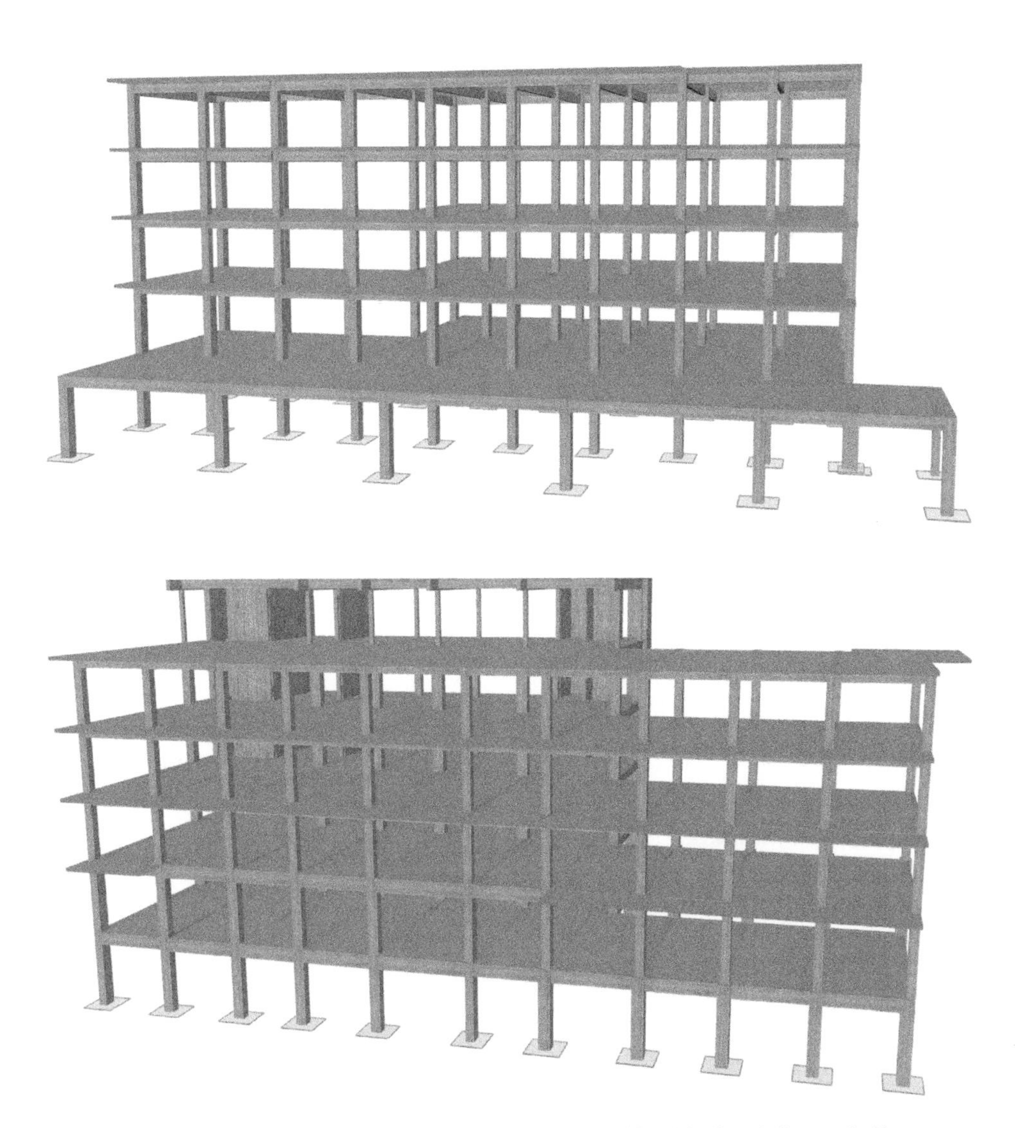

Figure 5.5 3D-view of the statically independent Parts A and B of the hotel. *Source:* Stelios Antoniou.

obviously the designers were mostly concerned about gravity loadings and the large spans of the beams, rather than any lateral, seismic force[1]. The concrete grade was measured with tests and it was relatively low ($f_{cm}$ = 14 MPa). The typical stirrup was Ø8 mm for both beams and columns, with a spacing of 20–25 cm in beams and 30–35 cm in columns (i.e., larger spacing, despite the lower steel grade).

The combination of large beam spans with columns of small dimensions resulted in a very flexible framing system (Figure 5.6), with calculated target displacements in pushover analysis of up to 15 cm, i.e., a total interstory drift of 1.00%, which is extremely large (the interstory drift at the ground level was almost double that value).

It was deemed that all columns were extremely weak to be left without strengthening, and as a first measure it was decided to upgrade all vertical members with a 10 cm RC jacket. The jackets were designed for shotcrete, but by slightly increasing their width (e.g., 12.5 cm wide) they could be constructed with cast-in-situ concrete instead. From the two buildings A and B, the latter allowed the introduction of a new, stiff RC wall in the perimeter of the building. The existing building had some slender ($d$ = 20 cm), lightly reinforced RC walls around the elevator shafts and the stairs, but because of the need to increase the size of the elevators to comply with modern standards, it had been decided to demolish the existing walls and to construct in the same locations, new RC walls with high reinforcement ratios, both longitudinal and

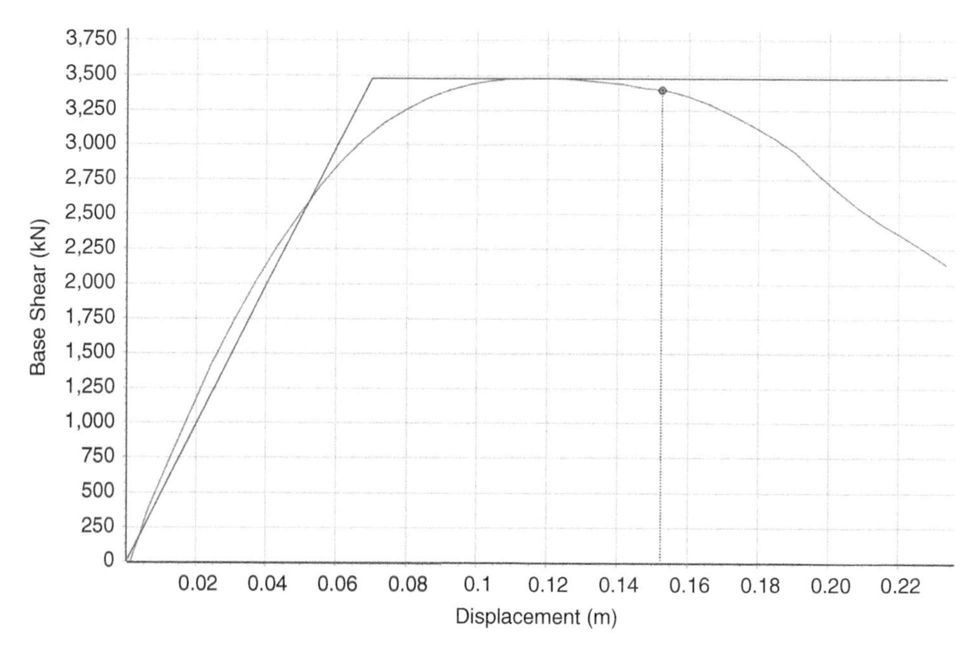


**Figure 5.6** Pushover capacity curve of the existing building of Part A; the target displacement for the Significant Damage limit state is also shown. *Source:* Stelios Antoniou.

1 It is not uncommon that in existing buildings the columns are weaker in shear than the beams; in the old days engineers very often designed only for gravity loading, which causes larger shear forces on the beams rather than on the columns.

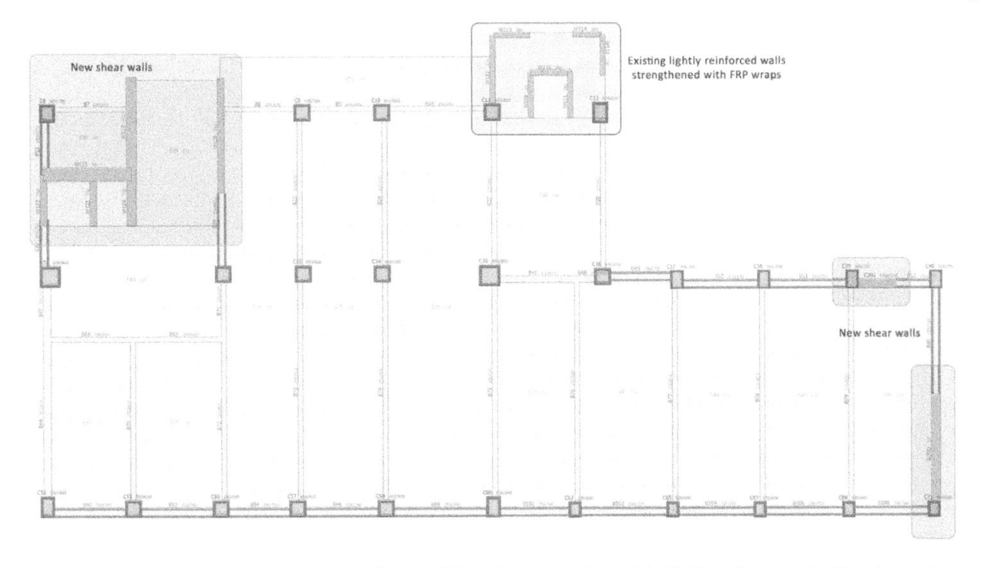


Figure 5.7 Plan view of the ground floor of Part B, strengthened building. *Source:* Stelios Antoniou.

transverse. In order not to introduce torsional effects on the building, one large and one smaller shear wall were constructed at blind walls at the opposite side of the building (Figure 5.7).

Building A instead did not have existing RC walls, nor did it have locations on plan view, where a new wall could be added in order to increase its stiffness, due to the architectural requirements (there were too many openings in the perimeter of the building, one large opening for each hotel room). There were only two blind walls that permitted the addition of a new shear wall, but these were on the same side of the building (Figure 5.8). The introduction of one or two large walls there would give rise to unwanted torsional effect and increase the vulnerability. However, leaving the building without any shear walls was not ideal, because the global slenderness of the building was still high, despite the new jackets. With several trial and error attempts, it was found that the construction of the small RC wall shown in orange in Figure 5.8 would decrease torsional effects on the one hand (the building had a tendency to rotation counterclockwise), and slightly increase the structural stiffness, mainly in the *Y–Y* axis, on the other. The size and the stiffness of the wall were adjusted after several trials.[2] It was not ideal, but it was the best that one can do, considering the structural configuration, and the architectural constraints; the addition of an RC wall in any other location but for the blind walls would abolish one hotel room, which of course the owners were very reluctant to accept.

The strengthening of the other parts of the hotel (parts C, D, E, F) was carried out with simple jackets, or was not required at all. These parts were low-rise and they had been designed with some anti-seismic provisions, which were deemed adequate even by modern standards.

2 The location and the size of the wall were selected so that the building would become more regular with its addition. A first trial was done with a 3.00 m wall located on the left, but it was causing irregularities. Then a 3.00 m wall was placed on the right, but again this caused irregularities. The final design was with a 1.80 m wall. The wall was introduced at the entire height to avoid introducing irregularities in elevation.

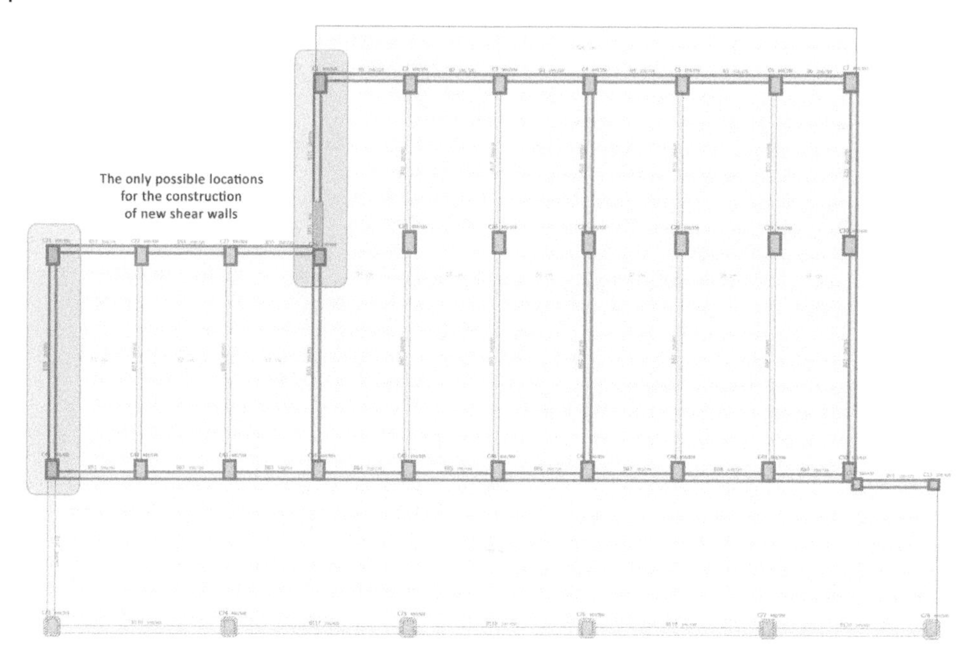


Figure 5.8 Plan view of the first and second floors of Part A. *Source:* Stelios Antoniou.

### 5.5.2 Case Study2: Seismic Upgrade of a Four-Story Hotel

This is a 3500 $m^2$ four-story reinforced concrete hotel that was built in the early 1980s. It is highly irregular in elevation with a sudden and significant reduction of the size of the upper floors, with respect to the ground floor (Figure 5.9). The structural system included several large walls, but these were lightly reinforced and constituted the most vulnerable components of the building. The capacity design approach has not been applied and generally there were large and strong beams connected to weaker and smaller columns. What is more, the hotel was in use and had been recently renovated before the intervention, hence the option of strengthening all the vertical members, e.g., with RC jackets, was immediately rejected by the owners.

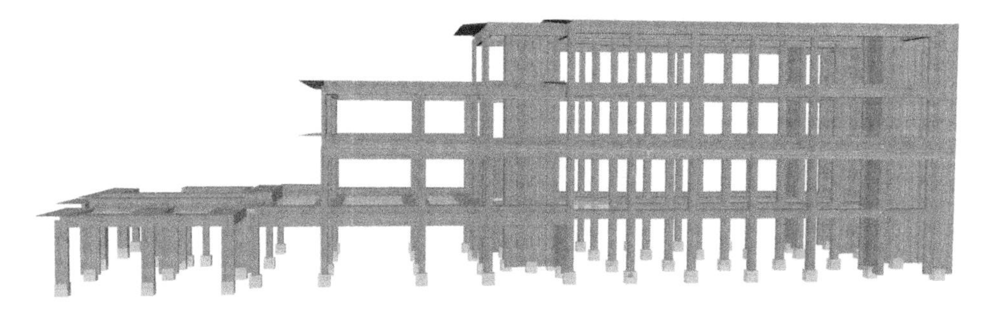

Figure 5.9 3D-view of the building, SeismoStruct (2023) screenshot. *Source:* Stelios Antoniou.

The only good pieces of information from the on-site investigation were that the concrete was of relatively high strength and that all vertical components had a rather large shear reinforcement ratio (considering the period of construction) with ∅8/25 cm stirrups consistently placed in all existing vertical members.

It was decided to strengthen the building with the construction of (i) two large RC core walls around the stairs and the elevator shafts in the central area and (ii) four RC walls (two in each horizontal direction) at the back side of the building. After the analyses were carried out, it was found that some existing, lightly reinforced walls in the entrance of the building did not pass the required checks in shear. Since these were in the reception area, the hotel owners were reluctant to allow for the construction of new shear walls that would limit the view, or the jacketing of the existing members with shotcrete, because the reception was also recently refurbished and disturbance in the hotel operations would be significant. Therefore, it was decided to strengthen the weak members with FRP wrapping, which did not alter their size and was applied quickly, efficiently, and without significant disturbance (Figures 5.10 and 5.11).

### 5.5.3 Case Study 3: Seismic Upgrade of a Four-Story Hotel

This is a four-story, 5000 m$^2$ reinforced concrete hotel. It is highly irregular, mainly in plan but also in elevation. The main part of it was designed and built in the early 1970s, whilst the two-story T-shaped extension in the front, was constructed in the early 1990s (Figures 5.12 and 5.13). The structural system of the main part consisted of many closely spaced, relatively weak columns of small dimensions that were connected with similarly small and weak beams. On the contrary the newer part featured larger structural components and the beam spans were generally larger (up to 8.00 m).

In the older part of the hotel, all members were lightly reinforced, in particular the transverse reinforcement was as low as ∅8/25 to ∅8/35 cm. Moreover, the hotel was close to the sea and many members had significant corrosion problems. One very important structural deficiency of the building was that throughout the ground floor, in the columns of the older part (approx. 80 columns in total) the transverse reinforcement was consistently placed

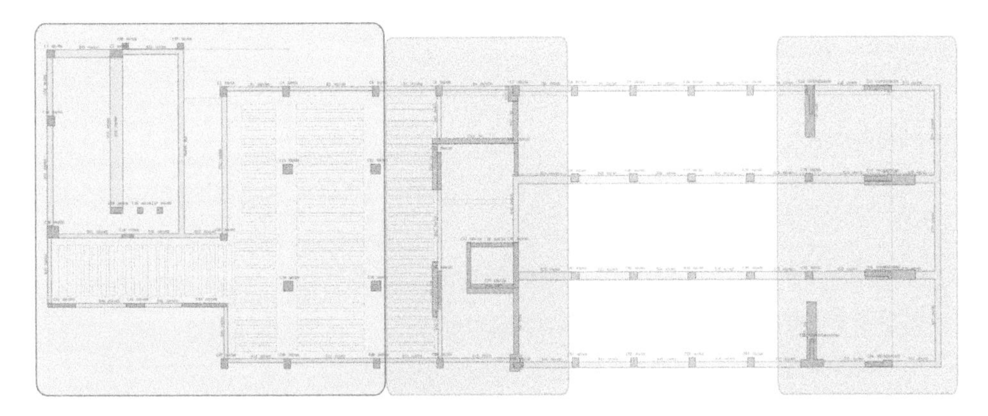

**Figure 5.10** Plan view of the ground floor with the new shear walls (in orange) and the location of application of FRP wraps (area in blue). *Source:* Stelios Antoniou.

**Figure 5.11** New shear walls and application of the FRP wraps at the reception area. *Source:* Stelios Antoniou.

**Figure 5.12** 3D-view of the building, SeismoBuild (2023) screenshot. *Source:* Stelios Antoniou.

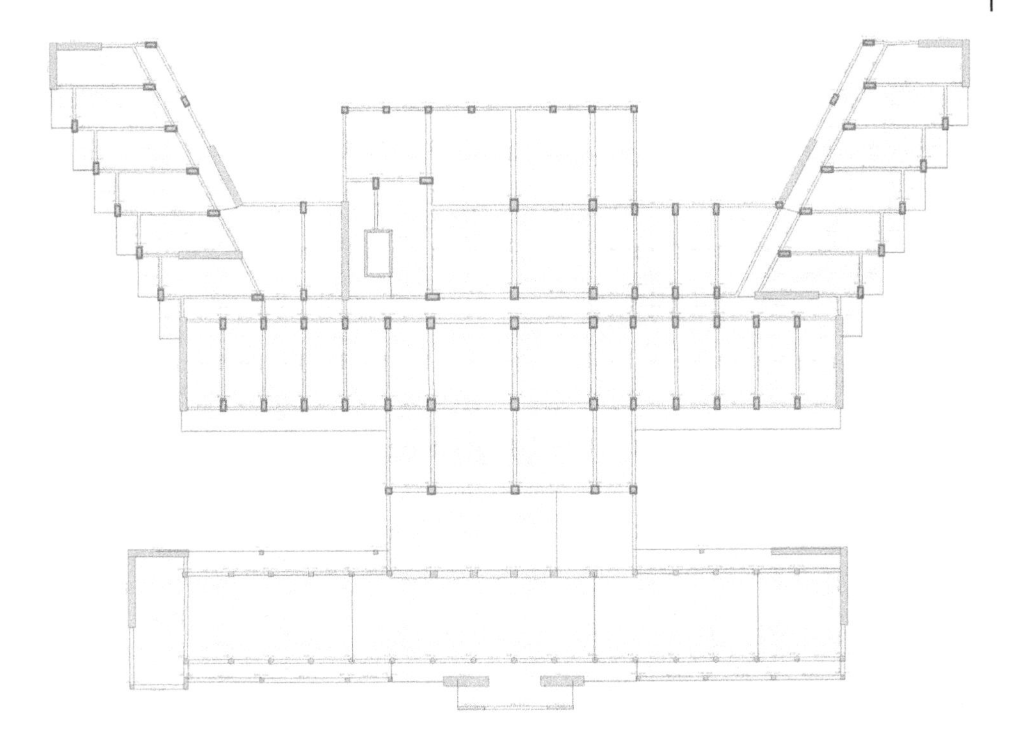

Figure 5.13 Plan view of the ground floor, where the new shear walls are shown in orange. *Source:* Stelios Antoniou.

with a spacing of 25 cm up to a height of 1.50 m from the ground floor, above which there was 1.30 m without any stirrups (Figure 5.14). It is stressed that this deficiency was consistent in all columns, which means that it cannot be attributed to poor workmanship and lack of supervision during the construction. It was probably decided by the designers of the building for unknown reasons. However, as explained in Chapter 2, this is something that should not cause too much surprise; engineers back then (1960s or early 1970s) were not really aware of the mechanisms and the risks derived from earthquake loading.

Hotel management rejected the solution of strengthening all the vertical members in all floors with shotcrete jackets because it would lead to extremely large costs to repair the nonstructural damage caused by the interventions (floors, tiles, adaptations to windows and doors). It was decided to propose the construction of many (11) very large shear walls that are designed to undertake a large proportion of the seismic force (in green in Figure 5.15). Because of the large size of the walls and the very large shear forces and bending moments transferred through them to the ground, all the walls, especially those located in the periphery of the building, required the construction of gigantic footings that enclosed the existing footings of the adjacent columns, in order to activate a larger percentage of the gravity loads of the existing building. Finally, the beams that connected the walls to the existing members were also strengthened with strong RC jackets.

Even with the construction of these new large shear walls however, it was deemed very dangerous to leave unstrengthened the problematic columns of the ground floor with 130 cm without hoops. It was decided to upgrade all of them with RC jackets (columns in blue in Figure 5.15).

**Figure 5.14** Placement of stirrups in the ground floor columns. *Source:* Stelios Antoniou.

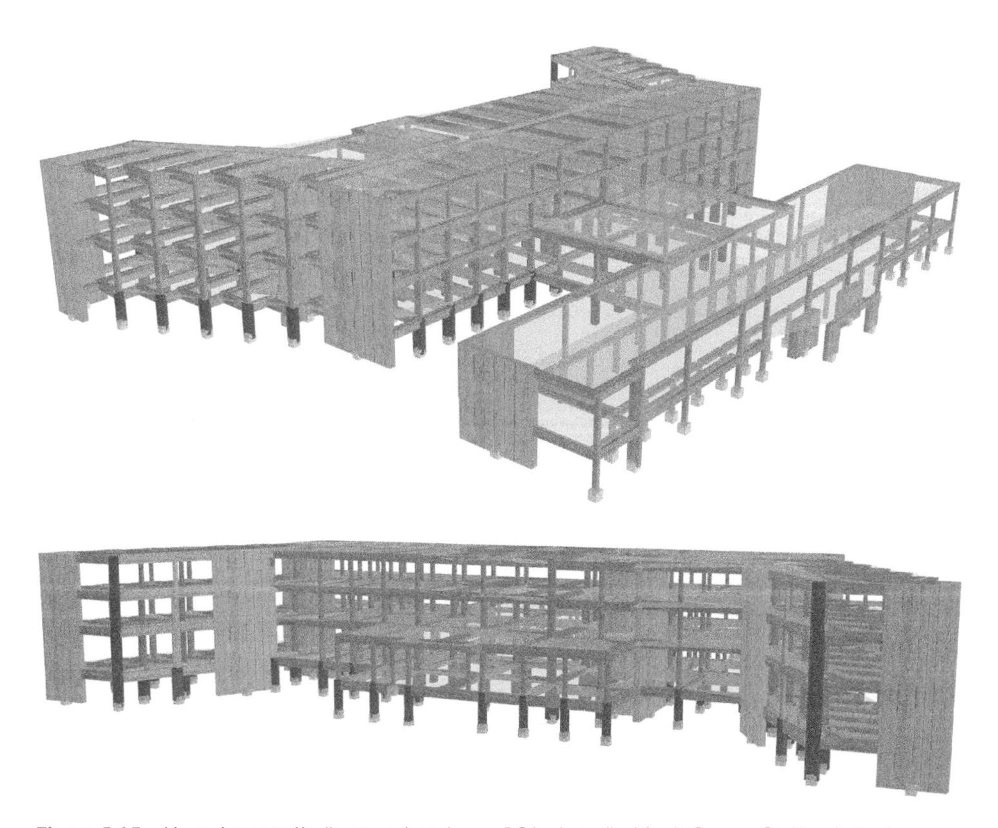

**Figure 5.15** New shear walls (in green) and new RC jackets (in blue). *Source:* Stelios Antoniou.

### 5.5.4 Case Study 4: Seismic Upgrade of a Three-Story Residential Building

This is a three-story 600 $m^2$ reinforced concrete building that was undergoing complete renovation. The structural system in all three floors was ribbed slabs supported by small lightly reinforced columns and the reinforcement in several locations was seriously corroded. The strengthening strategy was to retrofit all 66 columns (22 per story) with reinforced concrete jackets, constructed with shotcrete and to simply repair the damage from corrosion in the slabs with anti-corrosion coating and repair mortars. Simultaneously, all 22 footings were increased in size with jackets made of cast-in-place concrete.

The selection of the strengthening methods was based on two important factors: it would have been unsafe to leave any of the vertical members unstrengthened due to the lack of reinforcement, the severe corrosion problems and some cases of very poor construction. More importantly, because the building was undergoing a radical renovation (all the infilled walls and the floors had been demolished to be reconstructed) there was no additional cost for the repair of non-structural components, shifting the balance in favor of jacketing. Finally, the building was low in height, hence the construction of new shear walls was not deemed important to decrease deformations and non-structural damage in the case of a large seismic event (Figure 5.16).

### 5.5.5 Case Study 5: Seismic Upgrade of a Three-Story Residential Building for the Addition of Two New Floors

This is a relatively small three-story (basement, ground floor and first floor) residential reinforced concrete building that required seismic upgrade, in order to support the addition of two new floors (Figure 5.17). The building was constructed on a slope of strong rock on three different foundation levels, and had serious deficiencies in the foundation system with very small and very thin footings that lacked reinforcement. Because of the inclined

Figure 5.16 RC jackets constructed in all the columns (in the photographs the dowels had not been placed yet). *Source:* Stelios Antoniou.

Figure 5.16 (Continued)

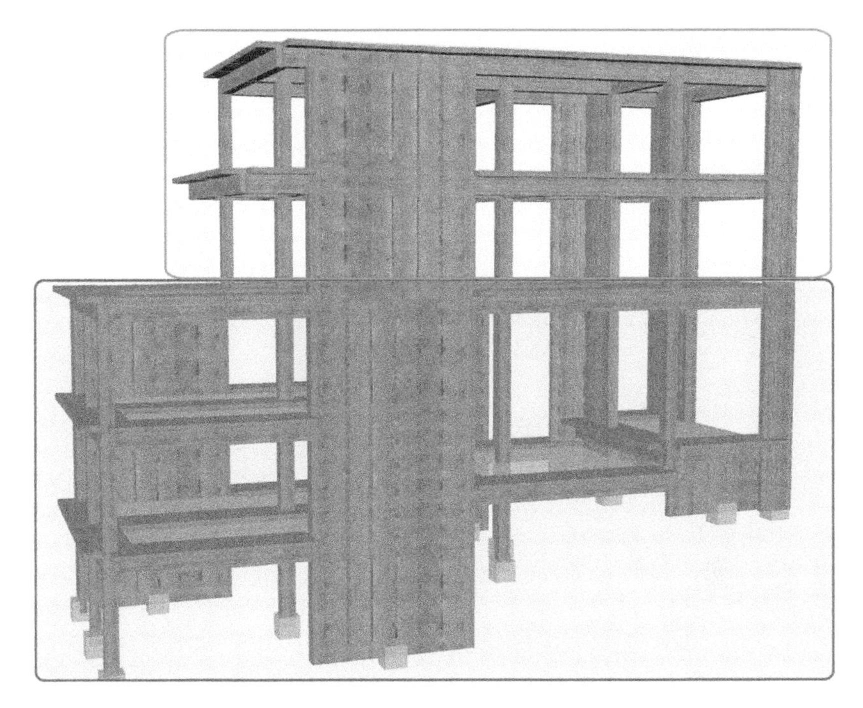

Figure 5.17 Structural model of the strengthened building, the existing floors are shown in red and the two new floors in green. *Source:* Stelios Antoniou.

ground, the slabs were located on different levels, creating a series of highly vulnerable short columns throughout the building. All the members, and in particular the columns, were small and lightly reinforced, and there were corrosion problems in many locations (the building was close to the sea).

The first and second floors were occupied and the owners were very reluctant to allow the strengthening of all the existing columns with jackets or FRP wraps. Consequently, the only way to achieve a satisfactory seismic upgrade of the building was through a series of very large shear walls in the perimeter of the building; out of the six new walls, only one L-shaped wall was located internally, around the shaft for a new elevator (Figure 5.18). In most cases, the walls enclosed the existing columns to create the pseudo-columns of the new unified component. In order not to introduce irregularities in elevation, most of the new walls (i.e., for those that it was geometrically possible) were continued to the two new floors. All members (existing ones, new shear walls, new members in the two new floors) were checked employing the analytical results for the entire building.

At the foundation level, all the vertical members in all three levels were connected to each other with a number of strip footings (Figure 5.19). In this way, a satisfactory foundation system was provided for the large new shear walls, and it was ensured that the existing columns were adequately supported. Because of the construction of the strip footings, the free height of the columns at the basement was significantly reduced creating a series of short columns that were lightly reinforced. In order to eradicate this new structural weakness, all these columns were strengthened with RC jackets constructed with shotcrete, with

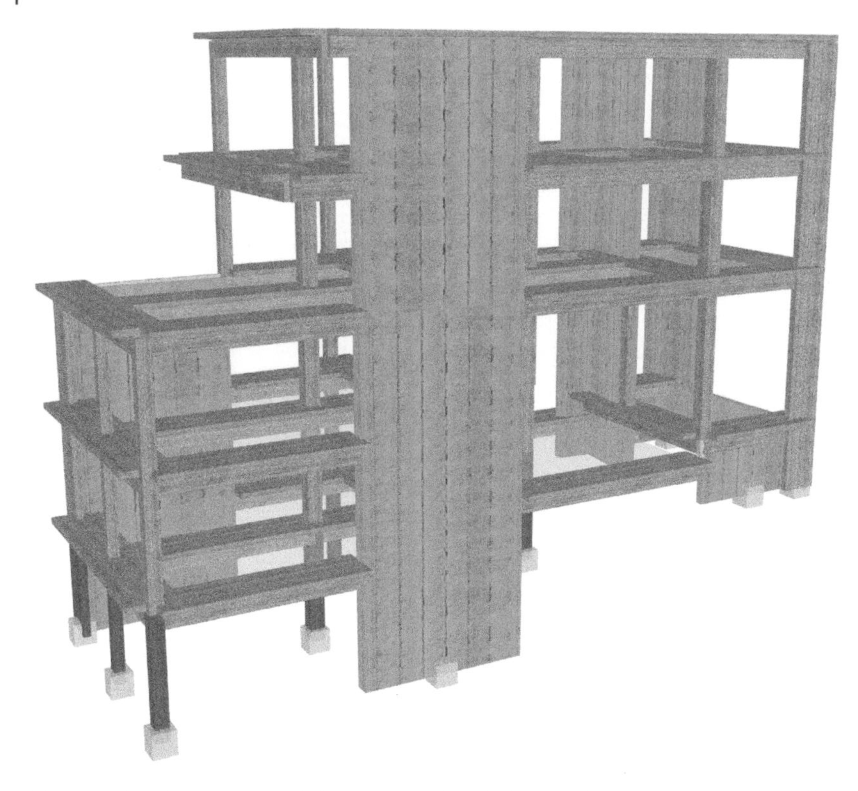

Figure 5.18 New shear walls (in green) and RC jackets at the basement (in blue). *Source:* Stelios Antoniou.

the exception of the columns at the upper part of the slope, which were enclosed in their entire height by new walls that connected the strip footings with the slabs of the upper level (Figure 5.20).

### 5.5.6 Case Study 6: Seismic Strengthening of an 11-Story Building

This is a well-designed and well-constructed building of the late 1960s. The building was found in good condition without significant problems from corrosion, despite its age. The reinforcement – both longitudinal and transverse – was way above the typical standards of the 1960s (e.g., the transverse column reinforcement was Ø8/25 cm stirrups). However, the lateral building strength was still below today's standards; the current design ground acceleration in the region is 0.16, whereas the original design was probably carried out with a value of 0.06 or 0.08.

Although the building was not occupied and it would have been relatively easy to strengthen all vertical members with jackets, it was decided to increase the lateral strength and stiffness with the construction of a number of large shear walls, since this solution was both technically satisfactory and cheaper. After the first set of analyses, a series of columns were found inadequate. They were retrofitted with shotcrete jackets (in the locations where the increase in size did not cause problems to the architectural layout) or FRP wrapping (in the locations that it did) (Figure 5.21).

Figure 5.19 Strengthening of the foundation, new shear walls and RC jackets at the basement level. *Source:* Stelios Antoniou.

Figure 5.20 Final configuration of one of the new shear walls. *Source:* Stelios Antoniou.

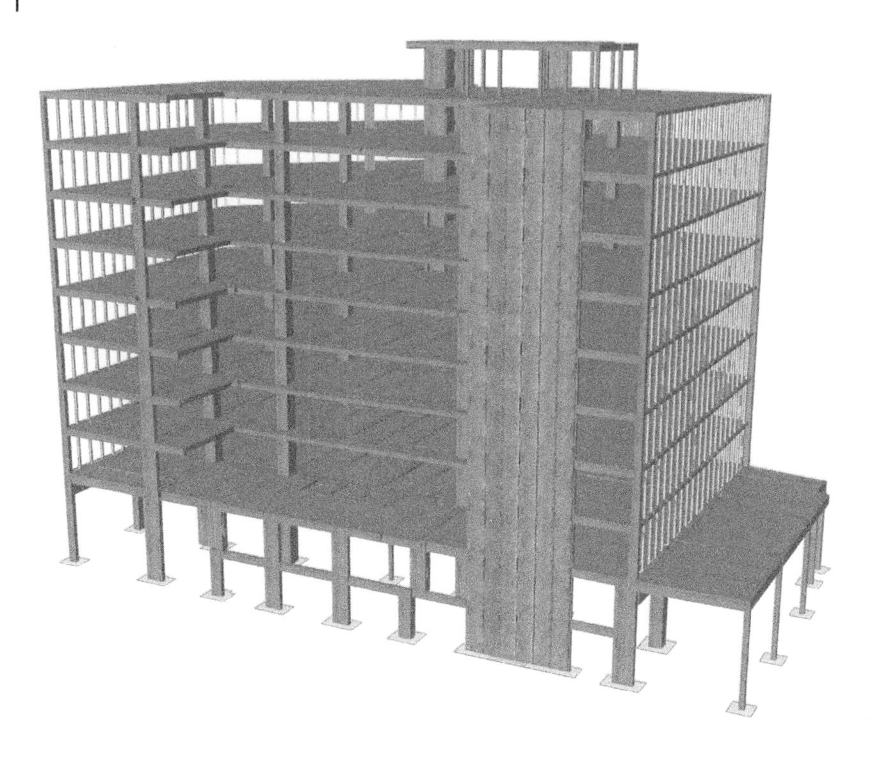

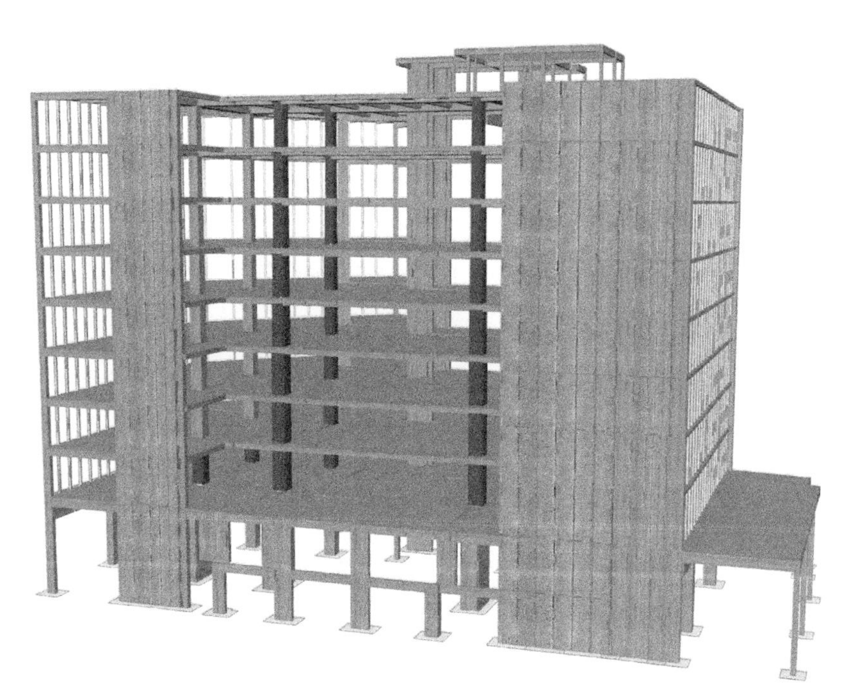

**Figure 5.21** Structural models in SeismoBuild (2023) of the existing and the strengthened building. *Source:* Stelios Antoniou.

### 5.5.7 Case Study 7: Seismic Strengthening of a Five-Story Building

This is a five-story RC building, the first three stories of which were constructed in the early 1970s and were lacking reinforcement, in particular transverse reinforcement. The upper two stories were designed and constructed in the early 1990s with increased reinforcement, which, although lower than today's standards (e.g., stirrups Ø8/20 or Ø8/25 cm), was found adequate from the analysis and the corresponding members were not strengthened.

Thus, the structural upgrade was carried out only at the three lower levels. This was done with a combination of RC jackets in the internal columns and FRP wraps in the columns in the building periphery (the architects did not allow jackets in the perimeter, because this would alter the facades).

Furthermore, the U-shaped core walls around the elevator shaft were also strengthened in shear with unidirectional FRP fabrics (with the fibers placed in the horizontal direction, as a supplement to the hoops); note that a similar result could have been accomplished with FRP laminates (again placed horizontally). Finally, because the existing foundation was quite weak, its strengthening was decided with the construction of a series of strip footings that connected all the existing footings at the ground level (Figure 5.22).

### 5.5.8 Case Study 8: Seismic Strengthening of a Three-Story Building

This is a three-story RC building constructed in two phases on inclined ground. The ground floor (which covers half of the entire building's area) and the second floor were designed in the early 1970s, whereas a third floor was added in the late 1980s. The

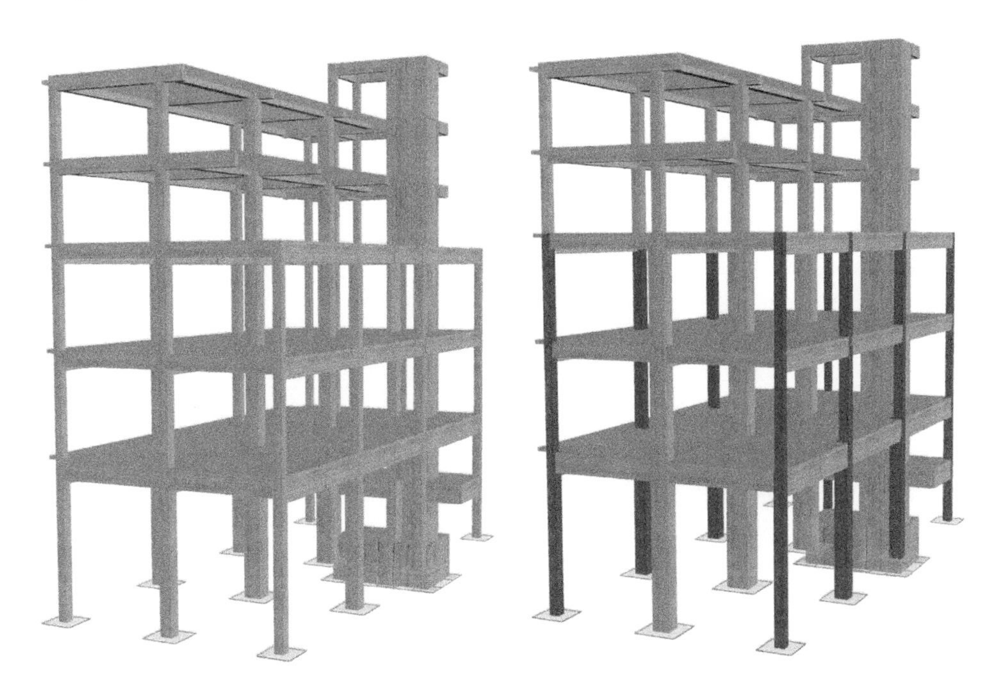

Figure 5.22 Structural models in SeismoBuild (2023) of the existing and the strengthened building. *Source:* Stelios Antoniou.

members at the two lower floors were weak and very lightly reinforced, especially in shear (stirrups Ø6/30-40 cm). Moreover, although the members in the third story were better reinforced, they were still well below current standards (e.g., stirrups Ø8/25 cm), and the preliminary analysis showed that some kind of upgrading was also required at that level. One additional complication was that, because the third floor was recently refurbished, the possible interventions were only allowed in the lower two stories and/or externally to the building.

It was decided to introduce new shear walls as the main measure for strengthening. One advantage was the fact that the building was open on all four sides; hence, it was possible to introduce external RC walls in a symmetric fashion. In total, six walls were added, four L-shaped at the corners, and two rectangular along the large side. After the analyses were carried out, 4–5 columns at the ground and the second floor were showing inadequacies in shear, and they were further strengthened with FRP wraps and RC jackets (Figure 5.23).

### 5.5.9 Case Study 9: Strengthening a Building Damaged by a Severe Earthquake

This is a small two-story building (just 18 columns per story) that had been severely damage after a large earthquake event. Apart from the age of the building (it was built in the 1970s), which meant that the members were lightly reinforced, especially in shear (stirrups Ø6/30 to Ø6/50 cm with random spacing in the columns, and Ø8/30 cm in the beams), its main structural deficiency was a combination of a weak ground story (brick infills were

(a)

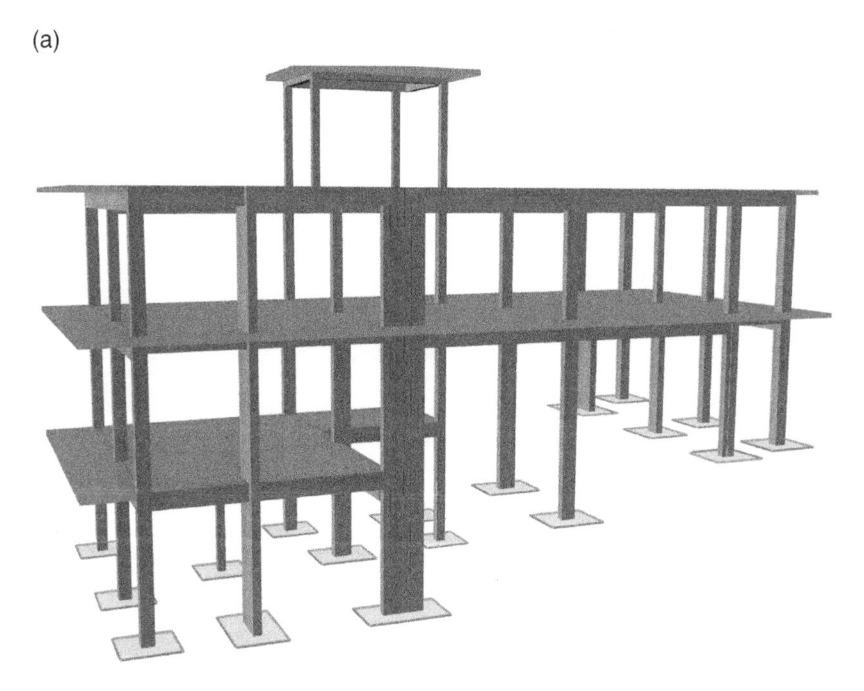

Figure 5.23 (a) SeismoBuild (2023) model of the existing building (b) SeismoBuild model of the strengthened building and (c) plan view of the second floor of the strengthened building. *Source:* Stelios Antoniou.

(b)

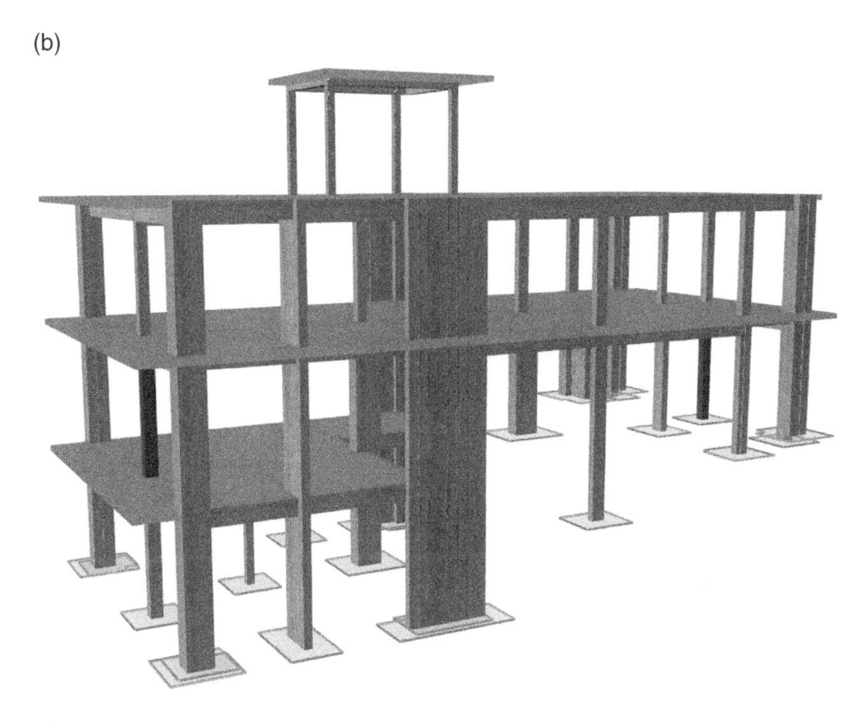

(c)

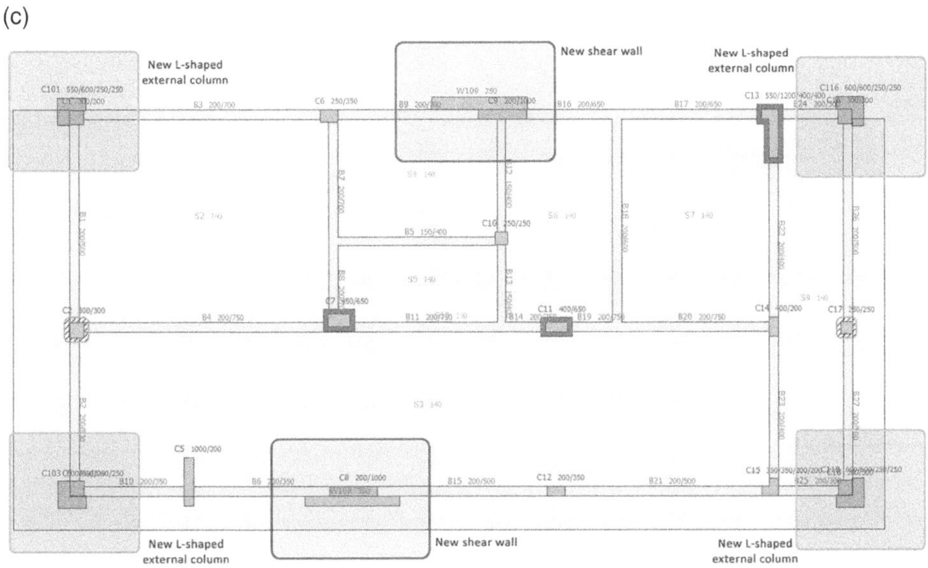


Figure 5.23 (Continued)

present mostly on the second floor), with several constraints that created a series of short columns at the ground level. All these problems resulted in a highly vulnerable structure, which expectedly sustained significant damage in all the columns of the ground floor during the earthquake. The damage was mostly concentrated in five short columns along one side of the building; all the columns failed in shear in a very brittle manner. The concrete

**Figure 5.24** Structural damage in the ground floor level. *Source:* Stelios Antoniou.

in all these columns disintegrated completely along a length of almost 1.00 m, the few stirrups fractured and the longitudinal rebar buckled (Figure 5.24).

It is noted that, although the second floor experienced significant displacements, which caused many pieces of furniture to topple over, the relative deformations at that level were negligible; not even a single crack on the walls' plaster was found (Figure 5.25), which is somehow surprising given the level of damage at the ground level. Similarly, no damage was found in the ground story beams and the foundation. The damage was concentrated solely in the columns of the soft story.

Because damage occurred in all 18 columns of the ground floor, which already had significant deficiencies and problems, it was immediately realized that jacketing all columns with shotcrete was the only viable option to repair the damage and structurally upgrade the building (Figure 5.26). It was decided not to strengthen any other member; after all, the

**Figure 5.25** The main corridor of the second floor, which remained totally undamaged. *Source:* Stelios Antoniou.

building was tested under very intense ground motion, and it was found that the only significant deficiency was the columns at the ground floor.

Of course, this retrofit strategy was supported at a later stage with all the necessary calculations from structural analysis. Recall that the strengthening of particular members of a structure might increase the demand in other structural components; the final retrofit scheme should always be checked with the detailed methodology that will be described in Chapter 6.

### 5.5.10 Case Study 10: Strengthening of an 11-Story Building

This is a rectangular 11-story RC building that was built in the late 1980s with anti-seismic standards that were, however, much lower with respect to today's modern provisions. It

**Figure 5.26** Jacketing of the columns of the ground floor. *Source:* Stelios Antoniou.

**Figure 5.26** (Continued)

had sustained slight to average damage during a recent earthquake and the owners wanted to upgrade it seismically, since the design ground acceleration in the region had been updated (increased) to 0.42 recently, which was considerably larger than the design acceleration, with which the original design had been conducted ($\cong 0.10\,g$).

The building was open in the perimeter without any buildings close to it; however, it was constructed with large cantilever beams and slabs on all four sides, which prevented the use of external shear walls for its strengthening (Figure 5.27). On the one hand, it would have been difficult to connect these walls through the slabs and the cantilever beams with the rest of the building, in order to transfer the inertia forces developed during an earthquake. On the other hand, the construction of the external walls in the building perimeter would have caused considerable esthetic disruption to its facades, since at the ground floor level the external infill walls were not constructed in the perimeter of the slabs (as in the upper stories) and the shear walls would ruin architecturally the building, which is considered a landmark in the region, where it is built.

After rejecting the external walls option, two retrofit scenarios were investigated. The first (more obvious) option was to strengthen all the columns in every level with RC jackets. This option was appealing since the building has only 16 columns per floor; however, the new strengthened structure would have relatively small stiffness, which could cause large interstory drifts and significant damage to the nonstructural elements in the case of a severe earthquake shock.

Figure 5.27 Large Cantilever beams in the perimeter of the ground story.

The second option was to try to find locations internally for the construction of new shear walls. This has proven a particularly difficult task, because the architectural layout at the different levels was not the same; hence, for instance, a location for a wall that was suitable in levels 1–3 and 7–11 was cutting a room in two in levels 4–6 (this is always the greatest complication when one has to find locations for new shear walls in an existing building). After trying several layouts, it was realized that a retrofit scheme with only RC shear walls was not possible (Figure 5.28). Consequently, some of the RC walls were replaced by steel diagonal braces that allowed doors to be placed below the inclined members.

The final scheme included the following:

- Two large shear walls parallel to the short side of the building. The walls were placed perpendicular to the floor corridors in locations where partition walls already existed on all levels. The doors on the plane of the walls were removed and installed again in special openings that were left in the walls.
- Four diagonal braces per story along the long side of the building. Again, the braces were installed on the plane of existing partition walls in most of the floors and the existing doors were slightly moved below the diagonal brace. The building model with the strengthening interventions is shown in Figure 5.29.

### 5.5.11 Case Study 11: Strengthening of a Two-Story Building with Basement

This is a new building constructed in the early 2000s with modern code standards. It has two floors and one basement, and the reason for its upgrade was the requirement of the owner for an increase in the live load sustained by the slabs from 5 to 10 kN/m$^2$. The strengthening was combined with a renovation of the building, resulting in negligible additional costs for the nonstructural damage caused by the retrofit interventions.

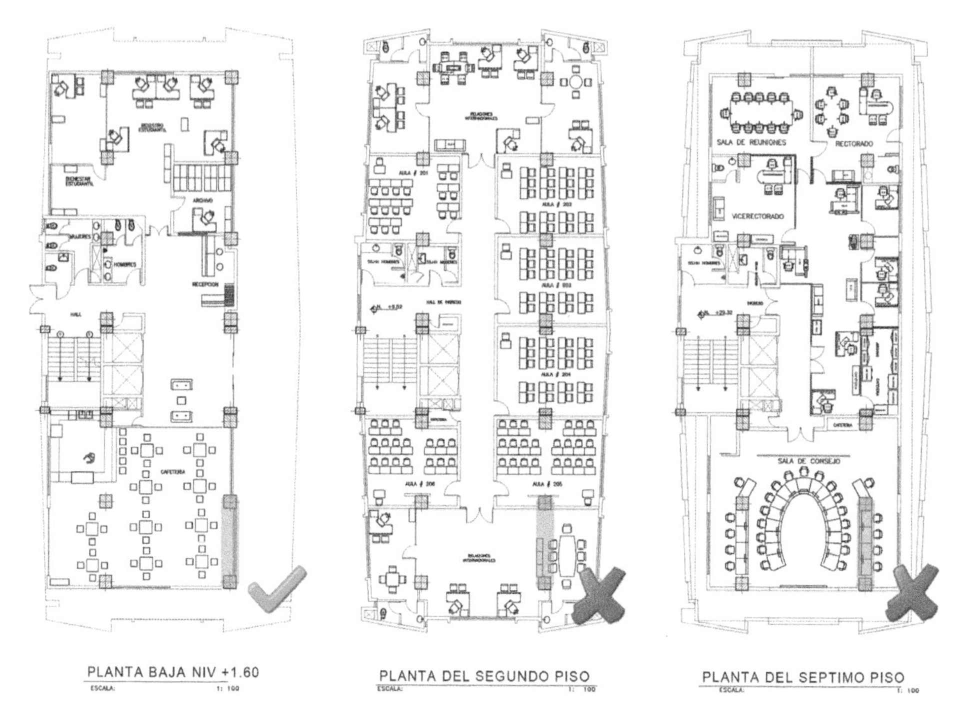


**Figure 5.28** Difficulties in finding an appropriate location of a new shear wall: a valid location in one floor blocks the operation of other floors. *Source:* Stelios Antoniou.

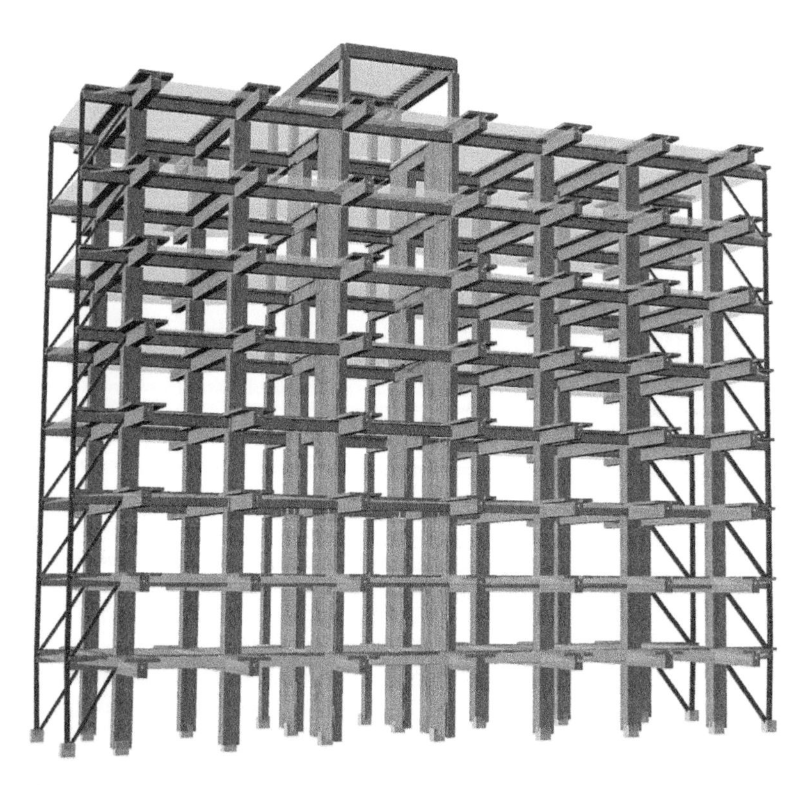

**Figure 5.29** Structural models in SeismoBuild (2023) of the building strengthened with RC walls and steel braces. *Source:* Stelios Antoniou.

Since the existing members had large reinforcement ratios and good detailing (for both the longitudinal and the transverse reinforcement), there was no point in strengthening every individual member separately, and the retrofit could be carried out with large shear walls or steel braces. Out of the two options, which were roughly equivalent in technical and economic terms, the latter was chosen. Most of the braces were installed externally, but some smaller ones were also required internally. The braces were constructed at the ground and the second story, whereas in the basement typical RC shear walls were constructed, upon which the steel trusses were installed (Figure 5.30).

### 5.5.12 Case Study 12: Strengthening of a Weak Ground Story with FRP Wraps

This case study is being presented as a bad example that should be avoided. It is a four-story RC building with a soft story at the ground level. Because the building was occupied in the

Figure 5.30 External braces, internal braces and RC walls at the basement.
*Source:* Stelios Antoniou.

upper floors, it was not possible to strengthen them, but this was not technically required; the building's important deficiency was the weak story.

It was decided to strengthen all the columns of the ground floor with FRP wrapping (Figure 5.31). For the majority of columns, this was quite easy and the works were executed without problems, since the columns had no contact with other structural or nonstructural

Figure 5.31 Strengthening of the ground floor columns with FRP wraps. *Source:* Stelios Antoniou.

components. However, that was not the case for two columns and the RC core walls at the entrance of the building. The retrofit of these members would cause both nonstructural damage and disturbance to the residents, and it was decided by the building owners not to strengthen them, and leave their upgrade for a later second phase.

Because the application of the FRP wraps does not affect the stiffness distribution in the building, the works of the first phase did not impair the capacity of the building. However, it did not improve its seismic performance, either. The building is as safe as the weakest column of the weakest floor (in this particular case, the lightly reinforced walls of the core at the ground level). Therefore, until the owners are convinced to carry out the second phase of the strengthening, the money spent during the first phase seems to be almost a complete waste of money.

### 5.5.13 Case Study 13 (Several Examples): Strengthening of RC Slabs

These are two examples of the strengthening of slabs in bending. Due to bad design, poor construction quality, and/or increased gravity loading, large slabs (usually of span widths of more than 6.00–7.00 m) may fail in bending from gravity loading resulting in increased deflections and cracking (Figure 5.32). On other occasions, thermal stresses, which were not consider in the initial design, might cause significant cracking in slabs that lack expansion joints with adequate spacing.

If the slabs are already damaged (i.e., seriously cracked), it is very important to identify the reason for this damage. Usually, the modeling of the slab with 2D finite elements and the execution of analyses with different types of loading (additional gravity loads, thermal loads or seismic loads) will provide adequate insight on the reasons for the failures. The tensile concrete stresses should match the cracking patterns with sufficient consistency; the cracks should be perpendicular to the stresses at almost all locations on the slab, as in Figure 5.33.

**Figure 5.32** Failure of a slab in bending in the negative moment region (at its support from a beam). *Source:* Stelios Antoniou.

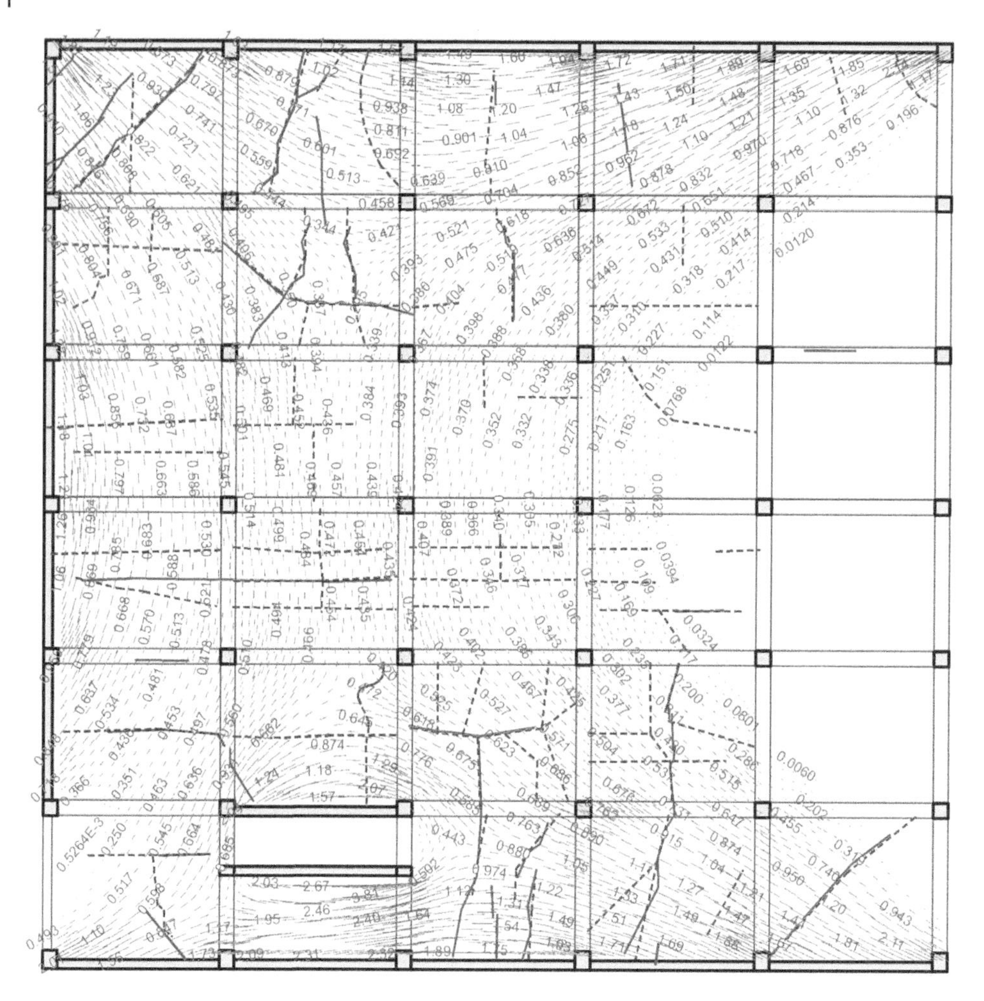

**Figure 5.33** Identification of the cause for cracks on a RC slab: existing cracks (in red and black for the upper and lower side of the slab) vs. the stress paths for tensile stresses (in blue) for the thermal load combination, which was not considered in the original design of the slab. In all locations the cracks are almost perpendicular to the stresses. *Source:* Stelios Antoniou.

Typical techniques for the strengthening of these slabs are RC jackets and FRPs. The jackets are usually made of shotcrete on the lower slab side and cast-in-place concrete on its upper side. The FRPs may be fabrics or laminates at mid-span on the lower side, and laminates or strings at the supports on the upper side (Figure 5.34). The solution of FRPs is preferable in the cases of uncracked slabs, because there is no increase in the dead load of the slab, whereas concrete jackets are more suitable for damaged slabs with large deflections. When the upper side of the slab is being used (e.g., in warehouses for the circulation of forklifts), the placement of FRP laminates or FRP fabrics is not permitted, because of the damage that the traffic might cause to them. In such cases, the strengthening with near-surface-mounted (NSM) FRP reinforcement may be considered. Combinations of FRP fabrics and RC jackets may also be applied (e.g., shotcrete at the lower surface and FRP

**Figure 5.34** Different techniques for strengthening RC slabs: (a) RC jackets, lower side, (b) RC jackets, upper side, (c) FRP laminates, lower side, (d) FRP laminates, upper side, (e) FRP fabrics, lower side, (f) FRP strings, upper side. *Source:* Stelios Antoniou.

laminates at the upper side). Note that if the slabs are cracked, they should be repaired prior to the strengthening. Usually, this is done with epoxy injections.

### 5.5.14 Case Study 14: Strengthening of a Ground Slab

This is the case of a large ground slab of a warehouse (≅30000 $m^2$), which failed in bending due to the settlement of the ground filling, caused by poor compaction. The settlement was extensive, and there was significant cracking throughout the surface of the slab, which obstructed the circulation of vehicles in the warehouse.

The retrofit of the ground slab was carried out in two phases. Initially, the gaps were mapped with ultrasonic equipment, and the gap between the ground slab and the filling was covered with grouting with the use of specialized packers through a dense grid of holes on the slab surface, so as to prevent any further settlement. In the second phase, the cracks were repaired with low-viscosity epoxy resin (thin cracks), a mix of epoxy resin and quartz sand (cracks of medium thickness), or epoxy mortar after a V-shaped groove was opened along the crack (wider cracks), as in Figure 5.35.

### 5.5.15 Case Study 15: Strengthening of Beam That Has Failed in Shear

This is a localized type of strengthening. During the renovation of the building, the beam was found to have failed in shear (Figure 5.36). It was decided to strengthen the beam both in shear, as well as in bending (to be on the safe side). The selected retrofit technique was with FRPs, because this is the only method that does not change the stiffness distribution of the entire building (Figure 5.37). The strengthening in bending was carried out only for the positive bending moments at midspan, since the supports could not be approached (it would require demolishing the upper floor, which was not possible). The strengthening in shear instead was carried out with FRP fabrics, anchored with FRP strings inside the beam. The holes for anchoring the strings were opened at 45° angles and rounded angle sides, in order to prevent the fracture of the fibers.

### 5.5.16 Case Study 16: Demolition and Reconstruction of a RC Beam

This is another case of localized intervention. During the strengthening and refurbishment project of a masonry building, the architect asked for the extension of an existing reinforced concrete beam, due to the partial demolition of the masonry wall that was supporting it. It was easy to demolish and reconstruct the beam itself; however, the connection with the two adjacent slabs was not ensured. The connection through horizontal dowels was not possible, since the depth of the slabs was just 11–12 cm, and the dowels would fail early.[3] The only solution was the construction of a beam with two side “wings” (Figure 5.38), and the connection with the slabs through a dense grid of vertical dowels.

3 Dowels would have been the best method for connecting slabs of larger depth (i.e., 25 cm or more).

Figure 5.35 Repair and strengthening of the ground slab in a warehouse: (a) placement of the packers for the cement grouting, (b) extraction of cylindrical cores from the slab after the grouting, (c) repair of the thin cracks, (d) repair of the wide cracks, and (e) final slab surface. *Source:* Stelios Antoniou.

### 5.5.17 Bonus Case Study 1: Strengthening of an Industrial Building

This is a nine-story industrial RC building. It was 34–35 m high, it was constructed in the 1960s, and it faced significant corrosion problems. Normally, in similar cases the building is demolished and reconstructed, but in this particular case the owners wanted to preserve a large rotary kiln in the center of the building, and complete the project as quickly as

**Figure 5.36** Shear failure of a beam. *Source:* Stelios Antoniou.

**Figure 5.37** Strengthening in shear and bending with FRP materials. *Source:* Stelios Antoniou.

possible (it is usually faster to strengthen than to demolish and reconstruct). The strengthening design did not pose any significant challenges, and it does not have any particularly interesting aspects; it consisted of RC jackets throughout the building, constructed with shotcrete for the beams and columns and with cast-in-place concrete in the footings and the connecting beams at the foundation. The reason why this project was included in the list of cases studies was for the range of photographs that follow, which highlight interesting details of the construction, such as dowels, welding of the stirrups, and reinforcement in the beam column joints (Figure 5.39).

### 5.5.18 Bonus Case Study 2: Strengthening of an Industrial Building

This is a 12000 $m^2$ industrial building with reinforced concrete columns and beams but steel roofs. The building was constructed in the 1970s and possessed limited seismic

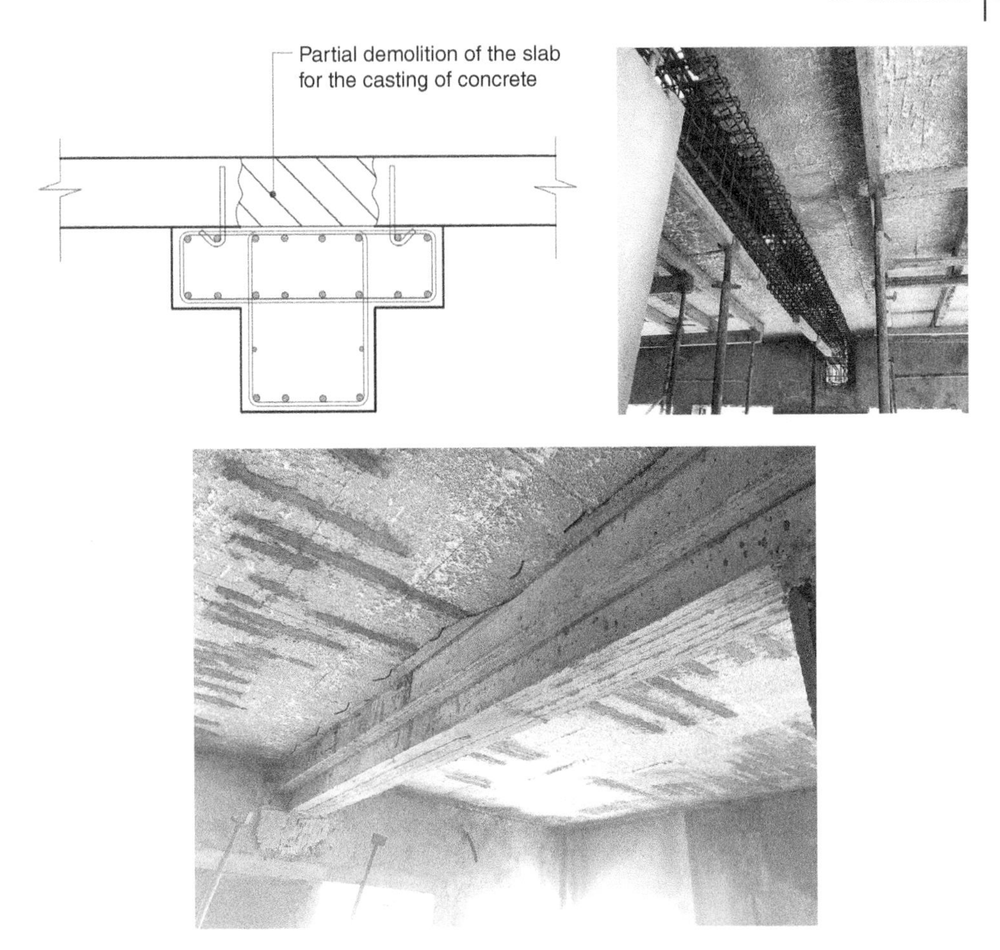


Figure 5.38 Beam with side wings for the support of the adjacent slabs. *Source:* Stelios Antoniou.

resistance; the RC members had small percentages of reinforcement (both longitudinal and transverse), and many of them faced serious problems with corrosion. The steel members were very weak with respect to modern standards, and they were also severely corroded. It was decided to retrofit the RC columns with jackets made of shotcrete and to replace the existing steel roofs, which was the only viable solution, technically and economically. The project is presented here mainly because of the photographs, which portray some interesting construction details, such as the configuration of the reinforcement in expansion joints (Figure 5.40).

### 5.5.19 Bonus Case Study 3: Strengthening of a Residential Building

Again, the design of the interventions and the retrofit strategy for this project are pretty standard. The foundation was strengthened with a series of strip footings and connecting beams and the columns of the superstructure were strengthened with shotcrete jackets. The reason for the inclusion of this project in the list of cases studies was again the photographs that highlight some aspects of the interventions at the foundation system (Figure 5.41).

**Figure 5.39** Strengthening of an industrial building. *Source:* Stelios Antoniou.

Figure 5.39 (Continued)

Figure 5.40 Strengthening of an industrial building. *Source:* Stelios Antoniou.

Figure 5.40 (Continued)

**Figure 5.41** Strengthening of a residential building. *Source:* Stelios Antoniou.

## References

CEN (2004). *European Standard EN 1998-1: 2004. Eurocode 8: Design of Structures for Earthquake Resistance, Part 1: General Rules, Seismic Actions and Rules for Buildings.* Brussels: Comité Européen de Normalisation.

Fardis, M.N. (2009). *Seismic Design, Assessment and Retrofitting of Concrete Buildings*, based on EN-Eurocode 8. ISBN 978-1-4020-9841-3 e-ISBN 978-1-4020-9842-0. DOI 10.1007/978-1-4020-9842-0. Springer International Publishing.

SeismoBuild (2023). SeismoBuild: A computer program for the linear and nonlinear analysis of reinforced concrete buildings. www.seismosoft.com.

SeismoStruct (2023). A computer program for static and dynamic nonlinear analysis of framed structures. www.seismosoft.com.

# 6

# Performance Levels and Performance Objectives

## 6.1 Introduction

Without getting into too much detail for the time being, let us simply mention that a target Building Performance Objective consists of a combination of a target Building Performance Level and a Seismic Hazard Level. The performance level is a measure of the building's performance during a seismic event (i.e. how much damage it has sustained), whereas the seismic hazard level describes the earthquake ground motion, which is expected in the region and for which the performance evaluation is carried out. In the European codes, the term performance level is replaced by the term Limit State, but the two have exactly the same meaning.

### 6.1.1 Selection of Performance Objectives in the Design of New Buildings

In the design of new reinforced concrete buildings, the selection of the target performance objectives is straightforward. Even though there are usually two limit states, for the strength and serviceability requirements, the former is dominant in the design process, and determines most of the choices the engineer has to make. For instance, in ACI 318, although it is explicitly mentioned that concrete structures should have strength, stability, serviceability, durability, and integrity, the strength requirements cover most of the document with serviceability covered only in one chapter. Similarly, in Eurocode 2 the part on the Ultimate Limit State is considerably larger with respect to that on the Serviceability Limit State.

Furthermore, for the seismic loading combinations, a single hazard level again is basically employed. This is represented by the elastic and the corresponding design acceleration response spectra, which are well documented and well known by the engineers. The spectrum represents an earthquake level with a 10% (or less often 5%) probability of exceedance in 50 years (return period equal to 475 or 975 years, respectively), and is directly given by the standard. In some cases, a second, lower hazard level is also prescribed, however again this assumes a supplementary role (e.g., an earthquake with a 10% probability of exceedance in 10 years – return period equal to 95 years – that is employed with the Damage Limitation (DL) limit state in Eurocode 8, Part-1).

Ultimately, the design process consists of a large set of checks for the strength requirements in the ultimate limit state with the "basic" earthquake spectrum for the seismic

*Seismic Retrofit of Existing Reinforced Concrete Buildings*, First Edition. Stelios Antoniou.

loading combination, and a limited number of *supplementary* checks for the serviceability (e.g., limits on the deflections or stresses on prestressed members) and/or the DL requirements. The latter could employ a smaller, supplementary seismic hazard level (i.e., a reduced spectrum). In all cases, however the role of the strength checks at the ultimate limit state is dominant.

### 6.1.2 Selection of Performance Objectives in the Assessment of Existing Buildings

For better or for worse, the process in the assessment procedures is more complicated, and combines several limit states with several earthquake hazard levels. It is less straightforward with respect to the design methodologies, but allows for much greater flexibility, which is very often needed in existing structures. The initial design and the deficiencies of the original construction impose significant restrictions for the engineer, who sometimes needs to employ a less rigid framework, when designing the strengthening interventions.

The current chapter discusses the performance levels (US nomenclature) or limit states (European nomenclature), the different seismic hazard levels, and how these two are combined to define the target performance objectives, in order to determine the design of the interventions. The main terms will be explained and a discussion will be given on how to effectively select performance levels and seismic hazard levels for the assessment or retrofit of existing buildings.

## 6.2 Seismic Assessment and Retrofit Procedures

### 6.2.1 Seismic Assessment Procedures

The seismic assessment of a building is carried out in order to demonstrate compliance with the selected target performance objective(s). The assessment is a quantitative procedure for checking whether an existing undamaged or damaged building satisfies the performance requirements for the selected limit state, were it to be exposed to the seismic action under consideration.

Within this framework, initially, a site visit is conducted and all the necessary as-built information for the building is obtained with the process described in Chapter 3. This includes the geometry, the member dimensions, the material properties, the reinforcement, the site soil conditions, the nonstructural elements and several other data points about the building and the site. The structural system is identified and understood (framing, lateral resisting system, floor and roof diaphragms, basement, and foundation system), and the nonstructural components, which affect the seismic performance and whose failure could cause serious life-threatening injuries, are detected. Subsequently, the building is modeled as in Chapter 8, and for each performance objective the force and deformation demands imposed on the structural components are estimated with the linear or nonlinear methods of structural analysis that are proposed by the standards (see Chapter 7), taking into account the corresponding seismic hazard level.

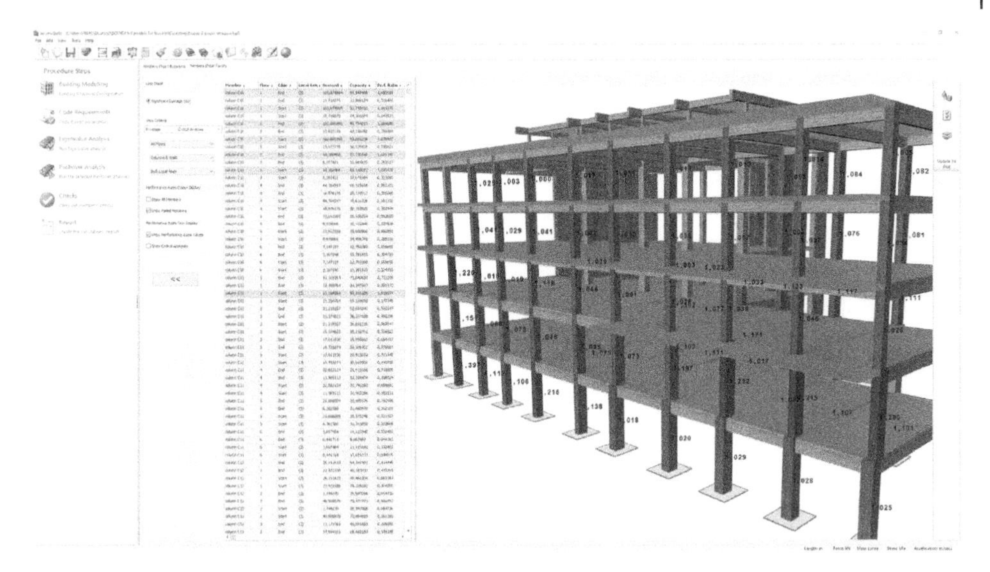

Figure 6.1 The Checks module of SeismoBuild (2023), whereby the member seismic demands (as calculated from structural analysis for the selected seismic hazard level) are compared against the member capacities (as calculated from the code-based formulae for the selected performance level). *Source:* Stelios Antoniou.

The capacity limits or acceptance criteria[1] (against which the demands are to be compared) are compatible with the selected performance level / limit state, and are calculated using code-based formulae (see Chapter 9). In general, the capacities are different for different performance levels (at least for the ductile types of failures), which means that a variable level of damage is expected, when selecting more than one performance objectives.

The seismic evaluation process concludes with a list of the seismic deficiencies, the identification of all problematic members and the consequences of their failures under the prescribed seismic load (brittle or ductile types of failures). The building is considered to satisfy the requirements of the selected limit state if the demand does not exceed the capacity for all actions of relevance and for all structural elements, including the beam-column connections (Figure 6.1).

### 6.2.2 Seismic Retrofit Procedures

The seismic retrofit of a building is performed in order to upgrade it to a state that demonstrates compliance with the selected performance objective(s), i.e., a combination of performance level and seismic hazard level. In general, it is not a direct, linear procedure, but rather it is an iterative process, in which different schemes are proposed and tested, and corrections and amendments are carried out until the building is fully compliant. The

1 According to ASCE 41, the acceptance criteria are limiting values of properties, such as drift, strength demand, and inelastic deformation, used to determine the acceptability of a component at a given performance level.

methodology followed for every intervention scheme (i.e., every step in the process) is the one described in the previous section (i.e., it is the same with the seismic assessment methodology).

Before beginning a seismic retrofit design, a detailed set of the as-built information is obtained, and a seismic evaluation of the existing building is performed to identify the seismic deficiencies relative to the selected performance objective(s). Usually, a preliminary retrofit scheme is proposed, based on the initial building information and engineering judgment. The target performance objectives are achieved by designing retrofit measures that address all the important structural deficiencies. The proposed retrofit measures need to be evaluated in conjunction with the existing structure as a whole, and their effect on strength, stiffness, and ductility should be considered and checked, employing the linear or nonlinear procedures of Chapter 7. The acceptance criteria, against which the checks are to be carried out, are established for both the existing elements and the new elements, which will be introduced as part of the retrofit.

If the proposed retrofit measures fail to comply with the acceptance criteria for the selected performance level, they are redesigned and an alternative retrofit strategy (possibly with different performance levels and/or seismic hazard levels) is adopted. This process is repeated until the design complies with the selected performance objective(s). The results and conclusions derived from each iteration of the design process are used in the following phase to introduce changes in the strengthening scheme. Once the design of retrofit measures meets the acceptance criteria for the selected performance objectives, construction documents, technical reports, and construction drawings, are prepared.

## 6.3 Understanding Performance Objectives

This section defines and details the basic concepts related to performance objectives, before moving on to a more thorough discussion on the selection of performance objectives. The concepts and naming conventions of ASCE 41 (ASCE 2017) will be employed, because a more complete framework is presented there. At the end, we will discuss how this framework is implemented in EC8, Part 3 (CEN 2005).

### 6.3.1 Target-Building Performance Levels

The building performance is expressed in terms of the target-building performance levels. In the European codes (the Eurocodes, but also the various national standards in European countries), this term is referred to as a Limit State, but note that these are two different names for the same thing. The performance levels are discrete damage states selected from among the infinite spectrum of possible damage states that buildings could experience during an earthquake. The particular damage states that are found in the different standards have been selected, because they have readily identifiable consequences associated with the post-earthquake disposition of the building, i.e. they can be described in a meaningful way to the building community, and they can be translated to quantifiable engineering quantities, such as deformations, drifts, or forces.

All the standards provide quantitative capacity limits for each performance level for all the crucial response quantities that need to be considered and checked. For the case of ASCE 41, these limits are called *acceptance criteria*. These are maximum values for the response quantities (e.g., plastic hinge rotations, chord rotations of shear forces), and a structure is considered to be in compliance with each performance level if the response of all its members is below these limits.

In ASCE 41 a target performance level consists of a combination of a Structural Performance Level, which refers to the loading bearing components of the building, and a Nonstructural Performance Level, which refers to the nonstructural components.

#### 6.3.1.1 Structural Performance Levels

The Structural Performance Level of a building describes damage patterns after an earthquake commonly associated with structural elements, and may be selected from six discrete states: Immediate Occupancy (S-1), Damage Control (S-2), Life Safety (S-3), Limited Safety (S-4), Collapse Prevention (S-5), and Not Considered (S-6).[2] The first five structural performance levels are selected to correlate with the most commonly specified and identifiable structural performance requirements. The last level is for the cases in which an evaluation or retrofit of the structural components is not needed and only the nonstructural elements are to be checked (see Table 6.1).

The performance levels are directly related to the extent of damage that would be sustained by the building and its systems in a seismic event. They provide a qualitative description of the building performance and they refer to the safety afforded to the occupants, but also to the cost and feasibility of restoring the building to its pre-earthquake condition, the length of time the building is removed from service to effect repairs, as well as the economic, architectural, or historic effects on the larger community. The damage state refers to the *worse* damage sustained by the structural members of the same type. Such damage might occur in some elements of a building at that Structural Performance Level, but it is unlikely that it will occur to all.

- **Immediate Occupancy Structural Performance Level (S-1).** Immediate Occupancy is defined as the post-earthquake damage state, in which a structure has sustained only very limited structural damage. Essentially, it retains its pre-earthquake strength and stiffness and remains safe to occupy. The risk of life-threatening injury, due to structural damage, is very low. Although some minor structural repairs might be appropriate, these repairs would generally not be required before re-occupancy, and the continued use of the building is not affected by its structural condition (though it might be limited by damage or disruption to nonstructural elements of the building, furnishings, or equipment).
- **Damage Control Structural Performance Level (S-2).** The Damage Control Structural Performance Level is set forth as a midway point between Life Safety and Immediate Occupancy. It is intended to describe a structural condition that has a greater

2 ASCE 41 also defines two intermediate Structural Performance *Ranges* (rather than discrete Levels), which allow experienced design professionals with specific requirements to create customized performance levels within these ranges: (i) the *Enhanced Safety Range*, between performance levels (S-1) and (S-3) and (ii) the *Reduced Safety Range*, between performance levels (S-3) and (S-5). For more details, refer to sections 2.3 and C2.3 of ASCE 41-17.

**Table 6.1** Structural performance levels and illustrative damage, part of Table C2-4 of ASCE 41-17 (ASCE 2017).

| Seismic-force-resisting system | Type | Structural performance levels | | |
|---|---|---|---|---|
| | | **Collapse prevention (S-5)** | **Life safety (S-3)** | **Immediate occupancy (S-1)** |
| Concrete frames | Primary elements | Extensive cracking and hinge formation in ductile elements. Limited cracking or splice failure in some nonductile columns. Severe damage in short columns. | Extensive damage to beams. Spalling of cover and shear cracking in ductile columns. Minor spalling in nonductile columns. Joint cracks. | Minor cracking. Limited yielding possible at a few locations. Minor spalling of concrete cover. |
| | Secondary elements | Extensive spalling in columns and beams. Limited column shortening. Severe joint damage. Some reinforcing buckled. | Major cracking and hinge formation in ductile elements. Limited cracking or splice failure in some nonductile columns. Severe damage in short columns. | Minor spalling in a few places in ductile columns and beams. Flexural cracking in beams and columns. Shear cracking in joints. |
| | Drift | Transient drift sufficient to cause extensive nonstructural damage. Extensive permanent drift. | Transient drift sufficient to cause nonstructural damage. Noticeable permanent drift. | Transient drift that causes minor or no nonstructural damage. Negligible permanent drift. |
| Steel moment frames | Primary elements | Extensive distortion of beams and column panels. Many fractures at moment connections, but shear connections remain intact. A few elements might experience partial fracture. | Hinges form. Local buckling of some beam elements. Severe joint distortion; isolated moment connection fractures, but shear connections remain intact. | Minor local yielding at a few places. No fractures. Minor buckling or observable permanent distortion of members. |
| | Secondary elements | Same as for primary elements. | Extensive distortion of beams and column panels. Many fractures at moment connections, but shear connections remain intact. | Same as for primary elements. |
| | Drift | Transient drift sufficient to cause extensive nonstructural damage. Extensive permanent drift. | Transient drift sufficient to cause nonstructural damage. Noticeable permanent drift. | Transient drift that causes minor or no nonstructural damage. Negligible permanent drift. |
| Braced steel frames | Primary and secondary elements | Extensive yielding and buckling of braces. Many braces and their connections might fail. | Many braces yield or buckle but do not totally fail. Many connections might fail. | Minor yielding or buckling of braces. |
| | Drift | Transient drift sufficient to cause extensive nonstructural damage. Extensive permanent drift. | Transient drift sufficient to cause nonstructural damage. Noticeable permanent drift. | Transient drift that causes minor or no nonstructural damage. Negligible permanent drift. |

| | | | | |
|---|---|---|---|---|
| Concrete walls | Primary elements | Major flexural or shear cracks and voids. Sliding at joints. Extensive crushing and buckling of reinforcement. Severe boundary element damage. Coupling beams shattered and virtually disintegrated. | Some boundary element cracking and spalling and limited buckling of reinforcement. Some sliding at joints. Damage around openings. Some crushing and flexural cracking. Coupling beams: extensive shear and flexural cracks; some crushing, but concrete generally remains in place. | Minor diagonal cracking of walls. Coupling beams experience diagonal cracking. |
| | Secondary elements | Panels shattered and virtually disintegrated. | Major flexural and shear cracks. Sliding at construction joints. Extensive crushing. Severe boundary element damage. Coupling beams shattered and virtually disintegrated. | Minor cracking of walls. Some evidence of sliding at construction joints. Coupling beams experience x-cracks. Minor spalling. |
| | Drift | Transient drift sufficient to cause extensive nonstructural damage. Extensive permanent drift. | Transient drift sufficient to cause nonstructural damage. Noticeable permanent drift. | Transient drift that causes minor or no nonstructural damage. Negligible permanent drift. |
| Unreinforced masonry infill walls | Primary and secondary | Extensive cracking and crushing; portions of outer wythe shed; some infill walls on the verge of falling out. | Extensive cracking and some crushing, but wall remains in place. No falling units. Extensive crushing and spalling of veneers at corners of openings and configuration changes. | Minor cracking of masonry infills and veneers. Minor spalling in veneers at a few comer openings. |
| | Drift | Transient drift sufficient to cause extensive nonstructural damage. Extensive permanent drift. | Transient drift sufficient to cause nonstructural damage. Noticeable permanent drift. | Transient drift that causes minor or no nonstructural damage. Negligible permanent drift. |
| Unreinforced masonry (noninfill) walls | Primary elements | Extensive cracking; face course and veneer might peel off. Noticeable in-plane and out-of-plane offsets. | Major cracking. Noticeable in-plane offsets of masonry and minor out- of-plane offsets. | Minor cracking of veneers. Minor spalling in veneers at a few comer openings. No observable out-of-plane offsets. |
| | Secondary elements | Nonbearing panels dislodge. | Same as for primary elements. | Same as for primary elements. |
| | Drift | Transient drift sufficient to cause extensive nonstructural damage. Extensive permanent drift. | Transient drift sufficient to cause nonstructural damage. Noticeable permanent drift. | Transient drift that causes minor or no nonstructural damage. Negligible permanent drift. |

margin of safety than the Life Safety Performance Level, but not to the extent required of a structure designed to meet the Immediate Occupancy Performance Level. The sustained damage enables return to function more quickly than the Life Safety Performance Level, but not as quickly as the Immediate Occupancy Performance Level does.

- **Life Safety Structural Performance Level (S-3).** Life Safety is defined as the post-earthquake damage state, in which a structure has damaged components but some margin against either partial or total structural collapse remains. Some structural elements and components are severely damaged, but this damage has not resulted in large hazards from falling debris inside or outside the building. Injuries might occur during the earthquake, however the overall risk of life-threatening injury as a result of structural damage is expected to be low. It is possible technically to repair the structure, but for economic reasons, this repair might not be practical. Although the damaged structure is not at risk of imminent collapse, it would be prudent to implement structural repairs or install temporary bracing before re-occupancy.
- **Limited Safety Structural Performance Level (S-4)**. Limited Safety is the structural performance level between the Life Safety Structural Performance Level (S-3) and the Collapse Prevention Structural Performance Level (S-5). It is intended to describe a structure with a greater reliability of resisting collapse than a structure that only meets the Collapse Prevention Performance Level, but not to the full level of safety that the Life Safety Performance Level would imply.
- **Collapse Prevention Structural Performance Level (S-5).** Collapse Prevention is defined as the post-earthquake damage state in which a structure has sustained significant damage and, although it continues to support gravity loads, it retains no margin against collapse. It is on the verge of partial or total collapse and cannot withstand any additional lateral loading. Substantial damage has occurred, including significant degradation in the stiffness and strength of the lateral-force resisting system, and there are large permanent lateral deformations. Although there can also be limited degradation in the vertical load-carrying capacity, all the significant components of the gravity-load-resisting system continue to carry their gravity loads. Significant risk of injury caused by falling hazards from structural debris might exist. The structure might not be technically practical to repair and is not safe for reoccupancy, because aftershock activity could induce collapse.
- **Structural Performance Not Considered (S-6).** This performance level is employed when the evaluation or retrofit does not address the structural components. This level was included because some owners might desire to address certain nonstructural vulnerabilities in an evaluation or retrofit program (e.g. bracing parapets or anchoring hazardous material storage containers) without addressing the performance of the structure itself. Such retrofit programs are sometimes attractive because they can permit a significant reduction in seismic risk at relatively low cost.

#### 6.3.1.2 Nonstructural Performance Levels

The Nonstructural Performance Level of a building describes damage patterns after an earthquake commonly associated with nonstructural elements, and may be selected from five discrete states: Operational (N-A), Position Retention (N-B), Life Safety (N-C), Hazards Reduced (N-D), and Not Considered (N-E) (see Table 6.2 and Table 6.3, where performance levels N-A, N-B and N-C are described.).

Table 6.2 Nonstructural performance levels and illustrative damage for architectural components, Table C2-5 of ASCE 41-17 (ASCE [illegible]).

| Component group | Nonstructural performance levels | | |
|---|---|---|---|
| | Life safety (N-C) | Position retention (N-B) | Operational (N-A) |
| Cladding panels | Distortion in connections and damage to cladding components, including loss of weather-tightness and security. Overhead panels do not fall. | Distortion in connections and damage to cladding components, including loss of weather-tightness and security. Overhead panels do not fall. | Negligible damage to panels and connections. No loss of function or weather-tightness. |
| Glazing | Some cracked panes; none broken. Limited loss of weather-tightness. | Some cracked panes; none broken. Limited loss of weather-tightness. | No cracked or broken panes. No loss of function or weather-tightness. |
| Heavy partitions (masonry and hollow clay tile or stud walls with tile or masonry veneer) | Distributed damage; cracking, crushing, and dislodging of veneer or parge coat in some areas. Damage to adjacent ceiling, but no wall failure. | Distributed damage; cracking, crushing, and dislodging of veneer or parge coat in some areas. | Minor crushing and cracking at corners. Limited dislodging of veneer or parge coat. |
| Light partitions (plaster and gypsum) | Distributed damage; some severe cracking of sheathing and racking in some areas. | Cracking at openings. Minor cracking of sheathing. | Minor cracking. |
| Ceilings | Extensive damage to suspended acoustical ceilings and grids. Plaster ceilings cracked and spalled but do not drop as a unit. Tiles in grid ceilings dislodged and falling; grids distorted and pulled apart. Plaster and gypsum board ceilings cracked and spalled but did not drop as a unit. | Limited damage. Plaster ceilings cracked and spalled but did not drop as a unit. Suspended ceiling grids largely undamaged, though individual tiles falling. | Generally negligible damage with no impact on reoccupancy or functionality. |
| Parapets and ornamentation | Minor damage; some falling of unreinforced elements in unoccupied areas. | Minor damage. | Negligible damage. |
| Canopies and marquees | Some damage to the elements, but essentially in place. | Some damage to the elements, but essentially in place. | Minor damage to the elements. |
| Chimneys and stacks | Minor damage. No collapse. | Minor damage. No collapse. | Negligible damage. |
| Stairs and fire escapes | Minor damage. Usable. | Minor damage. Usable. | Negligible damage. |

This table describes damage patterns commonly associated with nonstructural components for nonstructural performance levels. The anticipated performance of components for Hazards Reduced Performance Level are intended to be the same as for Life Safety Performance Level only for those components evaluated or retrofitted to that performance level. The damage states described in the table might occur in some elements at the nonstructural performance level, but it is unlikely that all of the damage states described will occur in all components at that nonstructural performance level. The descriptions of damage states do not replace or supplement the quantitative definitions of performance provided elsewhere in this standard and are not intended for use in postearthquake evaluation of damage or for judging the safety of, or required level of repair to, a structure after an earthquake. They are presented to assist engineers using this standard to understand the relative degrees of damage at each defined performance level.
Damage patterns in nonstructural elements depend on the modes of behavior of those elements. More complete descriptions of damage patterns and levels of damage associated with damage levels can be found in other documents, such as FEMA E-74 (2011).

**Table 6.3** Nonstructural performance levels and illustrative damage for mechanical, electrical, and plumbing systems, Table C2-6 of ASCE 41-17 (ASCE 2017).

| System or component group | Nonstructural performance levels | | |
|---|---|---|---|
| | Life safety (N-C) | Position retention (N-B) | Operational (N-A) |
| Elevators | Elevators out of service; cab and counterweights may be damaged but do not dislodge. | Elevators out of service until safety switches reset and power restored; cab and counterweight do not dislodge. | Elevators operate once safety switches are reset. |
| HVAC equipment | Units shifted on supports, rupturing attached ducting, piping, and conduit, but did not fall. Units might not operate. | Units are secure and possibly operate if power and other required utilities are available. | Units are secure and operate if emergency power and other utilities provided. |
| Manufacturing equipment | Units secure but potentially not operable. | Units secure but potentially not operable. | Units secure and operable if power and utilities available. |
| Ducts | Ducts broke loose from equipment and louvers; limited sections of ductwork dislodge | Minor damage but ducts remain serviceable. | Negligible damage. |
| Piping | Some lines rupture at joints. Some supports damaged but systems remain suspended. | Minor leaks develop at a few joints. Some supports damaged but systems remain suspended. | Negligible damage. |
| Fire suppression piping | Some sprinkler heads damaged by swaying ceilings. Minor leakage at a few heads or pipe joints. System remains operable. | Minor leakage at a few heads or pipe joints. System remains operable. | Negligible damage. System remains operable. |
| Fire alarm systems | Ceiling-mounted sensors damaged. Might not function. | System is functional. | System is functional. |
| Emergency lighting | Some lights fall. Power might be available from emergency generator or battery. | Some lights fall. Power might be available from emergency generator or battery. | System is functional. |
| Electrical distribution equipment | Units shift on supports and might not operate. Generators provided for emergency power start; utility service lost. | Units are secure and generally operable. Emergency generators start but might not be adequate to service all power requirements. | Units are functional. Emergency power is provided, as needed. |
| Light fixtures | Minor damage. Some pendant lights damaged. | Minor damage. Some pendant lights damaged. | Negligible damage. |
| Plumbing | Some fixtures broken, lines broken, but systems remain suspended. | Fixtures and lines may be damaged but serviceable; however, utility service might not be available. | System is functional if onsite water supply provided. |

This table describes damage patterns commonly associated with nonstructural components for nonstructural performance levels. The anticipated performance of components for Hazards Reduced Performance Level are intended to be the same as for Life Safety Performance Level only for those components evaluated or retrofitted to that performance level. The damage states described in the table might occur in some elements at the nonstructural performance level, but it is unlikely that all of the damage states described will occur in a component at that nonstructural performance level. The descriptions of damage states do not replace or supplement the quantitative definitions of performance provided elsewhere in this standard and are not intended for use in postearthquake evaluation of damage or for judging the safety of, or required level of repair to, a structure after an earthquake. They are presented to assist engineers using this standard to understand the relative degrees of damage at each defined performance level.

Damage patterns in nonstructural elements depend on the modes of behavior of those elements. More complete descriptions of damage patterns and levels of damage associated with damage levels can be found in other documents, such as FEMA E-74 (2011).

The last level is for the cases where an evaluation or retrofit of the nonstructural components is not needed and only the structural elements are to be checked. This decision is often made, when the retrofit of the nonstructural would significantly disrupt the normal use of the building. Since many more earthquake-related deaths result from structural collapse than from nonstructural hazards, a mitigation program that focuses on reducing casualties might reasonably require only structural evaluation and retrofit. Another possibility is to address, apart from the structural issues, only those nonstructural hazards that pose an obvious threat to human life, such as very heavy elements that could fall on the occupants.

- **Operational Nonstructural Performance Level (N-A).** This is the post-earthquake damage state, in which the nonstructural components continue to provide the functions they provided in the building before the earthquake. Most nonstructural systems required for normal use of the building are functional, although minor clean-up and repair of some items might be required.
- **Position Retention Nonstructural Performance Level (N-B).** Position Retention is the post-earthquake damage state, in which nonstructural components might be damaged, but are secured in place and damage caused by falling, toppling, or breaking of utility connections is avoided. Building access and life safety systems, including doors, stairways, elevators, emergency lighting, fire alarms, and fire suppression systems, generally remain available and operational, provided that power service is available. The occupants of the building are able to return to the building safely, though normal use might be impaired, some cleanup might be needed, and some inspection might be warranted.
- **Life Safety Nonstructural Performance Level (N-C).** Life Safety is the postearthquake damage state in which nonstructural components might have sustained significant and costly damage, but the consequential damage does not pose a threat to life safety. Nonstructural components have not been dislodged or fallen in a manner that could cause death or serious injury, either to occupants or to people in immediately adjacent areas. Egress routes within the building are not extensively blocked, but might be impaired by lightweight structural, architectural, mechanical, or furnishing debris, but life safety systems (including fire suppression systems) should be functional.
- **Hazards Reduced Nonstructural Performance Level (N-D).** Hazards Reduced is defined as the post-earthquake damage state, in which extensive damage has occurred to nonstructural components that could potentially constitute falling hazards, but large or heavy items that could pose a high risk to life safety for many people (e.g. parapets, cladding panels, heavy walls or ceilings, or storage racks) are not at risk of falling. Although isolated serious injury could occur from falling debris, failures that could injure large numbers of people – either inside or outside the structure – is avoided. The intent of the Hazards Reduced Performance Level is to address significant nonstructural hazards that pose a threat to multiple people without needing to rehabilitate all of the nonstructural components in a building. It is a nonstructural performance level that has the same life safety consequences as a partial or total collapse of a building.

#### 6.3.1.3 Target Building Performance Levels

A Target Building Performance Level is designated alphanumerically with a numeral representing the structural performance Level and a letter representing the nonstructural performance level, such as 1-B, 3-C, 5-E, or 6-C. A large number of combinations of structural and nonstructural performance levels are possible. Moving from 1-A to 5-D indicates a gradual deterioration of the building performance with increased damage to structural and nonstructural components alike. The most commonly used building performance levels are the following (Table 6.4):

- **Operational Building Performance Level (1-A)**: The building is expected to sustain minimal or no damage both to the structural and nonstructural components. The building is suitable for its normal occupancy and use, although possibly in a slightly impaired mode. All the essential utilities, such as the power and water supply, are functioning, and only some nonessential systems might not be functioning correctly. The building poses an extremely low risk to human life.
- **Immediate Occupancy Building Performance Level (1-B).** The building is expected to sustain minimal or no damage to the structural elements and only minor damage to the nonstructural components. Although it would be safe to reoccupy a building meeting this performance level immediately after a major earthquake, nonstructural systems might not be functional, either because of the lack of electrical power or because of internal damage to equipment. Even though immediate re-occupancy is possible, it might be necessary to perform some cleanup and repair and to await the restoration of utility service before the building can function in a normal mode. The risk to life safety is very low.
- **Life Safety Building Performance Level (3-C).** The performance level of Life Safety does not necessarily mean that there will be no injuries to occupants or persons in the immediate vicinity of the building. However, it means that only a few, if any, injuries are expected to be serious enough to require skilled medical attention. Buildings meeting this level may experience extensive damage to structural and nonstructural components. Repairs may be required before re-occupancy occurs, and repair, although technically feasible, may be deemed economically impractical. The risk to life safety is low.
- **Collapse Prevention Building Performance Level (5-D).** The building may experience extensive damage to structural and nonstructural components, and nonstructural falling hazards that cause serious injury or death may be. However, the total or partial building collapse has been avoided and major nonstructural falling hazards that can cause serious injury or death to large numbers of people are not likely either. Extensive repairs may be required before re-occupancy of the building occurs, and repair may be deemed economically unjustified. In the performance level of Collapse Prevention there is a significant risk to life safety.

### 6.3.2 Seismic Hazard Levels

The seismic hazard from ground shaking depends on the causative faults, the regional and site-specific geologic and geotechnical characteristics in the vicinity of the building. In theory, any level of seismic hazard may be chosen by the engineer, when defining the performance objectives for which the structure is to be assessed. Nevertheless, ASCE 41 explicitly designates specific hazard levels to be used in the assessment procedure. Each hazard level is associated with a reference probability of exceedance, $P_R$, in 50 years or equivalently

Table 6.4 Building performance levels, Table C2-3 of ASCE 41-17 (ASCE 2017).

| | Target building performance levels | | | |
|---|---|---|---|---|
| **Overall damage** | **Collapse prevention level (5-D)** | **Life safety level (3-C)** | **Immediate occupancy level(1-B)** | **Operational level (1-A)** |
| | **Severe** | **Moderate** | **Light** | **Very light** |
| Structural components | Little residual stiffness and strength to resist lateral loads, but gravity load-bearing columns and walls function. Large permanent drifts. Some exits blocked. Building is near collapse in aftershocks and should not continue to be occupied. | Some residual strength and stiffness left in all stories. Gravity-load-bearing elements function. No out-of-plane failure of walls. Some permanent drift. Damage to partitions. Continued occupancy might not be likely before repair. Building might not be economical to repair. | No permanent drift. Structure substantially retains original strength and stiffness. Continued occupancy likely. | No permanent drift. Structure substantially retains original strength and stiffness. Minor cracking of facades, partitions, and ceilings as well as structural elements. All systems important to normal operation are functional. Continued occupancy and use highly likely. |
| Nonstructural components | Extensive damage. Infills and unbraced parapets failed or at incipient failure. | Falling hazards, such as parapets, mitigated, but many architectural, mechanical, and electrical systems are damaged. | Equipment and contents are generally secure but might not operate due to mechanical failure or lack of utilities. Some cracking of facades, partitions, and ceilings as well as structural elements. Elevators can be restarted. Fire protection operable. | Negligible damage occurs. Power and other utilities are available, possibly from standby sources. |
| Comparison with performance intended for typical buildings designed to codes or standards for new buildings, for the design earthquake | Significantly more damage and greater life-safety risk. | Somewhat more damage and slightly higher life-safety risk. | Less damage and low life-safety risk. | Much less damage and very low life-safety risk. |

**Table 6.5** Probabilities of exceedance and mean return periods, according to ASCE 41-17 (ASCE 2017).

| Probability of exceedance | Mean return period (years) |
|---|---|
| 50%/30 yr | 43 |
| 50%/50 yr | 72 |
| 20%/50 yr | 225 |
| 10%/50 yr | 475 |
| 5%/50 yr | 975 |
| 2%/50 yr | 2475 |

a reference return period, $T_R$. Smaller probabilities of exceedance correspond to larger return periods of the seismic action, and larger earthquake loads (Table 6.5).[3]

The earthquake motion is usually represented by 5% damped elastic ground acceleration response spectra. Obviously, larger earthquake loading corresponds to higher spectra (Figure 6.2). Alternatively, the seismic action can be represented by ground motion acceleration time-histories that are characteristic of the seismic hazard at the site. These time-histories are compatible with the reference acceleration spectra and are used in the linear and nonlinear time-history analysis types. Typically, two-component or three-component time-histories should be introduced and these can be artificial, recorded, or simulated accelerograms. For the selection of derivation of these time-histories, specialized software is usually required (see, for instance, the SeismoArtif (2023), SeismoMatch (2023), and SeismoSelect (2023) packages).

### 6.3.3 Performance Objectives

As mentioned earlier, a Performance Objective consists of a pairing of the chosen target performance level for the building, for which the performance evaluation is made, and a seismic hazard level for the earthquake ground motion that is expected at the location of the building.

When selecting performance objectives, the prevailing philosophy is that:

- Under very low or low levels of earthquake ground motion, most buildings should be able to meet or exceed the Operational or Immediate Occupancy performance levels. Typically, it is not economically practical and justifiable to modify existing buildings to meet these performance levels for severe ground shaking, except for buildings that are very important and house essential services.
- For severe ground shaking the Life Safety performance level may be chosen, which entails relatively significant damage, which however is usually justifiable to be repaired.
- The Collapse Prevention performance level, which means very significant damage, may be chosen for the most severe ground shaking.

---

3 The value of the probability of exceedance, $P_R$, in $T_L$ years of a specific level of the seismic action is related to the mean return period $T_R$ of this level of seismic action in accordance with the expression $T_R = -T_L/\ln(1 - P_R)$. For a given $T_L$, the seismic action may equivalently be specified either via its mean return period $T_R$ or its probability of exceedance $P_R$ in $T_L$ years.

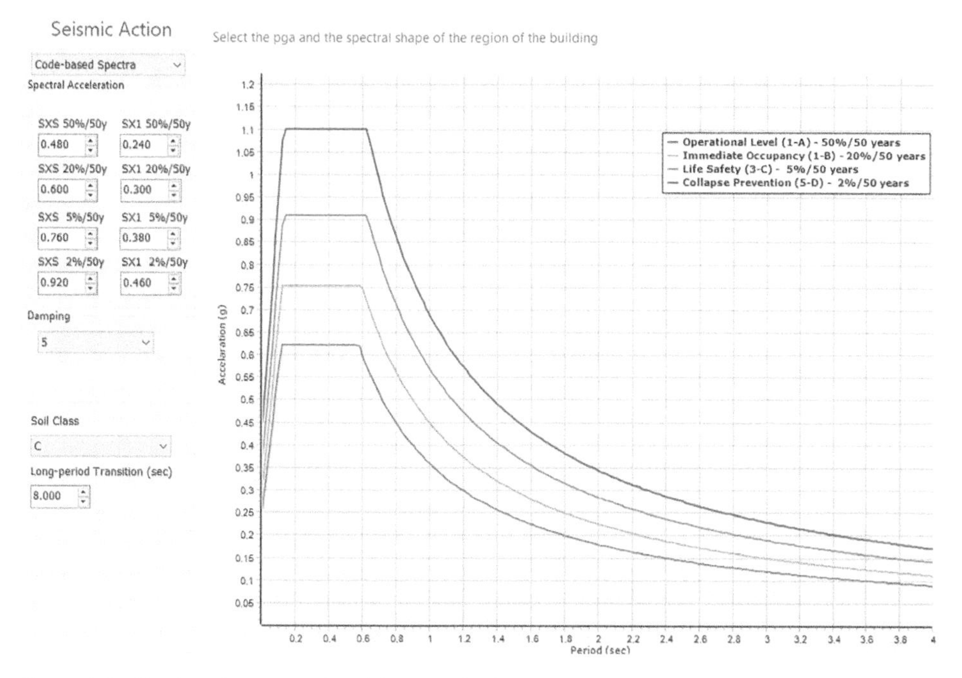


**Figure 6.2** Earthquake spectra for different seismic hazard levels, SeismoBuild screenshot (2023). *Source:* Stelios Antoniou.

**Table 6.6** Building performance levels in ASCE 41.

| | Target building performance levels | | | |
|---|---|---|---|---|
| **Seismic hazard level** | **Operational performance level (1-A)** | **Immediate occupancy performance level (1-B)** | **Life safety performance level (3-C)** | **Collapse prevention performance level (5-D)** |
| 50%/50 yr | a | b | c | d |
| BSE-1E (20%/50 yr) | e | f | g | h |
| BSE-2E (5%/50 yr) | i | j | k | l |
| BSE-2 N (ASCE 7 $MCE_R$) | m | n | o | p |

Each cell in the above matrix represents a discrete performance objective.

The pairings are made directly on a table, where the possible hazard levels are put together with the building's performance level (Table 6.6 and Figure 6.3). Each cell on the table represents a discrete performance objective.

According to the philosophy described above, within the performance objectives table it makes sense to select cells diagonally from the top-left to the bottom-right corner, i.e. with increasing ground motion, more severe damage is acccpted. It does not make sense to select cells on a horizontal line, since that would mean the assessment of the same seismic

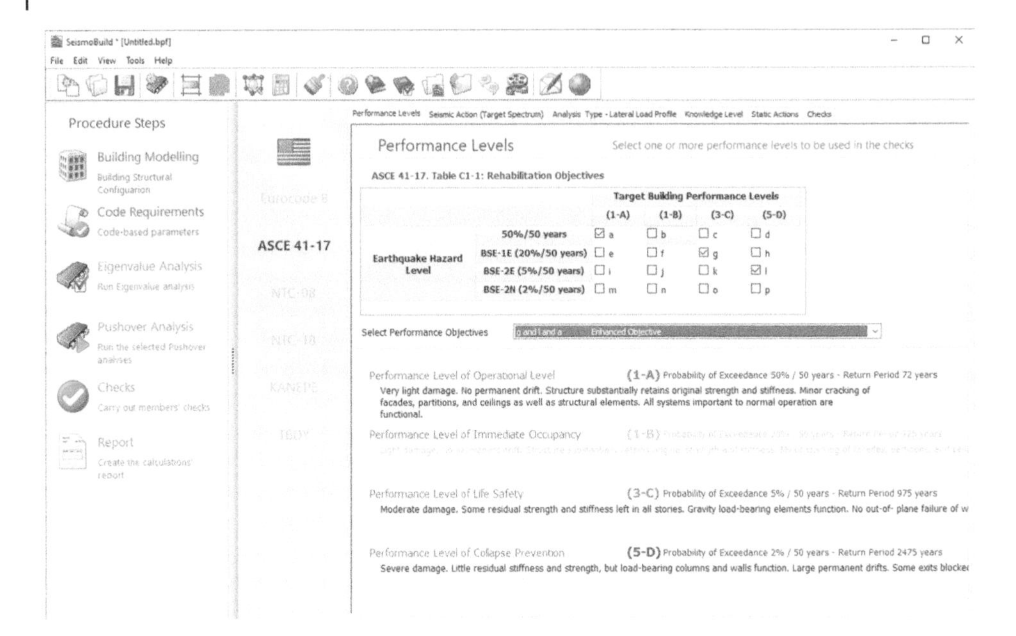


**Figure 6.3** Selection of performance objectives in SeismoBuild (2023). *Source:* Stelios Antoniou.

hazard level for different performance levels, i.e. different levels of damage; if the performance level with the lower damage is satisfied, obviously the performance level with the higher damage state is also fulfilled. Similarly, it makes little sense to select cells along a vertical line, because it would mean that the building is assessed with the same performance level (i.e. damage state) in two different hazard levels; obviously the objective with the larger seismic hazard is more critical.

### 6.3.4 Eurocode 8, Part 3, and Other Standards

Within the Eurocodes framework the performance levels are called Limit States and the damage state of the building can be described with three discrete limit states, namely:

- The *Damage Limitation (DL) Limit State*: The structure is only lightly damaged, with structural elements prevented from significant yielding and retaining their strength and stiffness properties. Nonstructural components, such as partitions and infills, may show distributed cracking, but the damage could be repaired economically. Permanent drifts are negligible. The structure does not need any immediate repair measures.
- The *Significant Damage (SD) Limit State*: The structure is significantly damaged, but some residual lateral strength and stiffness remains, and the vertical elements are capable of carrying the vertical loads. Nonstructural components are damaged, although partitions and infills have not failed out-of-plane. Moderate permanent drifts are present, and the structure can sustain after-shocks of only moderate intensity. The structure can be repaired, but it is likely that this turns out to be quite costly.
- The *Near Collapse (*NC*) Limit State*: The structure is heavily damaged, with low residual lateral strength and stiffness, although the vertical elements are still capable of sustaining the vertical loads. Most nonstructural components have collapsed. Large permanent

drifts are present. The structure would probably not survive another earthquake, even of moderate intensity.

There is a direct connection between EC8's limit states and ASCE's performance levels. The Damage Limitation limit state corresponds to the Immediate Occupancy performance level, the Significant Damage limit state to the Life Safety performance level and the Near Collapse limit state to the Collapse Prevention performance level; very similar descriptions are provided for the structural and the nonstructural damage. It is noted that an Operational level with even lighter damage than the DL limit state was not deemed important and is not available in EC8.[4]

Likewise, three hazard levels (instead of four) are specified in EC8, Part 3, with:

- A probability of exceeding 20% in 50 years (return period equal to 225 years).
- A probability of exceeding 10% in 50 years (return period equal to 475 years).
- A probability of exceeding 2% in 50 years (return period equal to 2.475 years).

Contrary to ASCE 41, the engineer is not free to select different combinations of limit states and hazard levels. Instead, a specific hazard is ascribed to each limit state: the 2.475 years level to the NC limit state, the 475 years level to the SD limit state, and the 225 years level to the DL limit state. If one created a table similar to Table C2-2 of ASCE 41 (such a table is not in fact present anywhere in Eurocode 8), the engineer would be allowed to select cells only along the large diagonal (Table 6.7).

Because each limit state is tied to a specific seismic hazard, the term Limit State is implicitly employed to describe, apart from the performance level, the corresponding performance objective, as well. This can create some confusion at first. What is more important is that this creates a rather strict framework that does not allow the designer the flexibility needed when retrofitting existing structures, in order to overcome serious restrictions imposed by the given architectural and operational constraints.

It is noted that the approach in EC8 Part 1 for the design of new structures is also similar. There are two Limit States, Near Collapse and Damage Limitation, with specific seismic hazards tied to each of them (return periods of 95 and 475 years, respectively). However, this is not an important problem with new buildings that are designed from scratch, since

**Table 6.7** Available limit states according to EC8, Part-3.

| | | Building performance levels | | |
|---|---|---|---|---|
| | | LS of DL | LS of SD | LS of NC |
| Earthquake hazard level | 20%/50y (PR = 225 yr) | ☑ a | ■ not available | ■ not available |
| | 10%/50y (PR = 475 yr) | ■ not available | ☑ e | ■ not available |
| | 2%/50y (PR = 2475 yr) | ■ not available | ■ not available | ☑ i |

*Source:* Stelios Antoniou.

4 Note that as of November 2022 a fourth limit state has been included in the draft of the new release of Eurocode 8, Part-3. The new limit state is called Fully Operational and corresponds to the Operational performance level of ASCE 41.

the civil engineer (sometimes also collaborating together with the architect and the mechanical engineer) can more easily find technical solutions (e.g., by increasing the size or the reinforcement ratio of some members).

Several other codes worldwide follow approaches for the definition of performance levels and objectives similar to either EC8 or ASCE-41. For instance, the new Italian Guidelines NTC-18 have four limit states (Operational Level, Damage Limitation, Life Safety and Collapse Prevention), that correspond to four specific seismic hazard levels (return periods 30, 50, 475, and 975 years, respectively). In NTC-18 the flexibility that is missing from EC8 is provided by the ability to change the lifetime of the structure, and the seismic force may be reduced if a smaller (e.g. 30 years instead of 50 years) remaining life is considered. The Greek Interventions Code KANEPE is very flexible as well. It has three limit states, but it also specifies (from its third revision onward) no less than nine hazard levels for the engineer to choose from. According to KANEPE, under certain conditions the proposed strengthening scheme can be accepted if the building can sustain a seismic hazard that is higher even by one level. Finally, the Turkish Code TBDY provides a 4 × 4 table, which is very similar to ASCE 41.

### 6.3.5 The Rationale for Accepting a Lower Performance Level for Existing Buildings

In the discussion of the previous section on performance objectives, the flexibility of ASCE 41 to allow for various combinations of performance levels and seismic hazards was considered a positive feature. This additional flexibility effectively means that the engineer is free to accept a somewhat higher risk with respect to new construction, and carry out the design of the interventions with lower safety requirements. In other words, the engineer accepts larger damage for the same level of seismic hazard, or equivalently the same level of damage for lower ground motion.

Is this increased risk really something that the engineers and the building owners are willing to accept? The short answer for existing buildings is yes. Often (in the majority of cases, in fact) this is something that is worth considering. Obviously, structural safety is the most important parameter when designing structural interventions. Nevertheless, many times the requirements imposed by a rigid design framework that requires "new building equivalence" are proven extremely strict and lead to very expensive interventions and serious restrictions in the operation of the building during the strengthening works. Contrary to new construction, where better building performance and higher safety can come at a marginal additional cost (e.g., by increasing the reinforcement ratio $\rho$ or increasing the cross-section dimensions by 5 or 10 cm), in existing buildings the incremental benefit of higher performance gradually comes at a disproportionate cost, financial or otherwise. Very often, these strict requirements may leave the engineer with two options, none of which is appealing: either demolish and reconstruct the building or do nothing (which, in my experience, is what is done in most cases).

Another factor that encourages lower performance objectives for existing buildings is the recognition that these have a shorter remaining life than a new building, something that is explicitly considered in the Italian NTC-18 Code through the lifetime of the structure parameter. If the traditional code-based demand for new buildings presumes a 50-year life, an existing building with, say, a 30-year life has a smaller chance of experiencing the code-level event over its remaining years.

Acknowledging all these, ASCE-41 proposes a Basic Performance Objective for Existing Buildings (BPOE) that comprises two discreet objectives g and l above the large diagonal (see Table 6.6). Seeking safety, but also taking into account the difficulties and the complications related to the retrofit of existing buildings, BPOE allows for a higher risk, with respect to new construction. This has been deemed appropriate for many mitigation programs and remains valuable for the precedent that it provides. Whereas for new buildings in the US, probabilities of exceedance of 2% in 50 years and 10% in 50 years have commonly been used (before the adoption of Risk-Targeted Maximum Considered Earthquake ground motions in ASCE 7), in the BPOE of ASCE-41 the seismic hazard levels are based on 5% in 50-year and 20% in 50-year probabilities of exceedance (for more details, refer to section C2.2.1 in ASCE 41).

## 6.4 Choosing the Correct Performance Objective

As explained in Chapter 3, in the majority of cases the significant lack of transverse reinforcement is the most important problem in existing buildings, and it is the dominant factor that determines their dynamic response and the damage they sustain during a seismic event. Ø6 mm stirrups are very common in buildings of the 1960s and 1970s, whereas, even when there are Ø8 mm stirrups, their spacing is typically very large, 25 cm or more. Corrosion also plays an important role, further decreasing the effective diameter of the hoops. In general, the total transverse reinforcement in columns of older buildings is approximately 10–15% of what would be placed nowadays, resulting in reduced shear capacity, no confinement and insignificant ductility. Consequently, in the vast majority of cases, brittle failures in shear precede the more ductile failure mechanisms. The capacity design approach simply does not apply to old buildings. Personally, I have never seen a building where the favorable ductile member failures occur before the brittle shear failures. Similar is the case with other brittle failures, such as the shear failures of the beam-column joints. This is acknowledged by the Standards worldwide, which assign extremely small behavior factors to existing buildings. For example, in the q-factor approach, EC8 adopts a behavior factor equal to 1.50 for reinforced concrete buildings (EC8, Part 3, section 4.2.(3) P), and, if the designer wants to employ larger values, a proper justification with reference to the local and global available ductility is required.

As a simple demonstration of this problem, we have carried out several pushover analyses with the sample building from the evaluation of Chapter 4, but with gradually increasing shear reinforcement in the vertical members. All the other structural parameters (geometry, member sizes, material properties, and longitudinal reinforcement) were left unchanged. It was found that in order to be able to have first a favorable, ductile failure mode in chord rotation, the columns' shear reinforcement needed to become Ø10/15 cm. This type of shear reinforcement is simply not encountered in older construction; the transverse reinforcement required in order to have capacity design behavior, is usually much higher than the actual one. What is more, the placement of the longitudinal reinforcement in existing buildings is generally much more consistent and "tidy" compared to the transverse reinforcement that might vary between the members due to poor workmanship and supervision. Therefore, it is even more likely that shear dominates the structural behavior (it is noted that each performance level is reached when a single member reaches the prescribed damage state).

Keep this in mind: shear failures will precede in the majority of cases. This is the rule, and it applies to all existing structures, except for a limited number of very special cases.

The fact that shear dominates the structural behavior has serious implications on the selection of the performance objectives. The sequence of performance levels implies a gradual increase in the sustained damage. This is translated to increasing values for the numerical acceptance criteria as we move to higher performance levels. For instance, in the nonlinear analysis types of ASCE 41, the maximum acceptable plastic hinge rotation of a beam controlled by flexure with conforming transverse reinforcement is 0.010 for the Immediate Occupancy level, 0.025 for the Life Safety level, and 0.050 for the Collapse Prevention level. Hence, it is reasonable to pair higher performance levels with higher seismic ground motions that lead to higher force and deformation demands to the members and select cells along diagonals on the performance level vs. seismic hazard table.

However, this gradual increase makes sense under one condition: the failure modes are not brittle ones with immediate consequences to structural safety. The shear failure of a column results in the immediate collapse of the building or a part of the building. For shear (and the other brittle types of failure), there is no gradation of the sustained damage, the failure is sudden and its consequences are often catastrophic. The only performance level in brittle failure modes is the Collapse Prevention level. Based on this observation, the following conclusions can be drawn:

- There is no point in specifying several performance levels with different hazard levels, as in Figure 6.4a. When shear dominates the response, the most critical performance objective will always be the one with the larger ground motion, i.e. the higher spectrum.
- Similarly, there is no point in moving horizontally on the table (Figure 6.4b). An objective with a higher performance level but the same seismic hazard, practically gives the same level of safety in shear, i.e. the same demand to capacity ratios for all members.
- On the contrary, it makes sense to move vertically on the table, i.e., changing the seismic hazard level and reducing the ground motion for the same performance level (Figure 6.4c). This will indeed lead to a reduction of the demand to capacity ratios in shear for all members.

Since ASCE-41 provides a large number of discrete hazard levels to choose from, it does indeed provide the designer the flexibility to select a lower performance objective by selecting for instance the Life Safety performance level with a hazard level with 20% probability of exceedance in 50 years (return period equal to 225 years), instead of the one with 5% that would correspond to the seismic risk of a new construction. On the contrary, the requirement of EC8 to employ a specific hazard level to each limit state is very restrictive, and often leads to all the problems mentioned above.

Obviously, this discussion does not apply to buildings that behave in a ductile manner without brittle failures. However, I reiterate that these buildings are the exception; the vast majority of existing buildings will exhibit brittle behavior failing in shear. It is also noted that, although shear is treated in the same way for vertical and horizontal members alike, the consequences of the shear failure of a beam are far less significant than that of a column, but this is not considered in the standards.[5]

5 In fact, it is indirectly considered in the Turkish code, which specifies that failures in a small proportion of the beams are acceptable depending on the performance level (e.g. 20% of the beams may experience failure in the Collapse Prevention performance level), see sections 15.8.3–15.8.5 of TBDY. Furthermore, the Greek code also accepts, under certain conditions, the exceedance of the capacity of the secondary horizontal members.

(a)

| | | Target Building Performance Levels | | | |
|---|---|---|---|---|---|
| | | (1-A) | (1-B) | (3-C) | (5-D) |
| Earthquake Hazard Level | 50%/50 years | a | b | c | d |
| | BSE-1E (20%/50 years) | e | f | g | h |
| | BSE-2E (5%/50 years) | i | j | k | l |
| | BSE-2N (2%/50 years) | m | n | o | p |

(b)

| | | Target Building Performance Levels | | | |
|---|---|---|---|---|---|
| | | (1-A) | (1-B) | (3-C) | (5-D) |
| Earthquake Hazard Level | 50%/50 years | a | b | c | d |
| | BSE-1E (20%/50 years) | e | f | g | h |
| | BSE-2E (5%/50 years) | i | j | k | l |
| | BSE-2N (2%/50 years) | m | n | o | p |

(c)

| | | Target Building Performance Levels | | | |
|---|---|---|---|---|---|
| | | (1-A) | (1-B) | (3-C) | (5-D) |
| Earthquake Hazard Level | 50%/50 years | a | b | c | d |
| | BSE-1E (20%/50 years) | e | f | g | h |
| | BSE-2E (5%/50 years) | i | j | k | l |
| | BSE-2N (2%/50 years) | m | n | o | p |

**Figure 6.4** (a) Moving along the diagonal, (b) moving horizontally, and (c) moving vertically on the performance objectives table. *Source:* Stelios Antoniou.

## References


[ASCE] American Society of Civil Engineers (2017). *Seismic Evaluation and Retrofit of Existing Buildings (ASCE/SEI 41–17)*. Reston, Virginia: ASCE.

CEN (2005). *European Standard EN 1998-3: 2005. Eurocode 8: Design of Structures for Earthquake Resistance, Part 3: Assessment and Retrofitting of Buildings*. Brussels: Comité Européen de Normalisation.

FEMA E-74 (2011). *FEMA E74: Reducing the Risks of Nonstructural Earthquake Damage – A Practical Guide*. Washington, DC: Federal Emergency Management Agency.

SeismoArtif (2023). SeismoArtif– A computer program for generating artificial earthquake accelerograms matched to a specific target response spectrum. www.seismosoft.com.

SeismoBuild (2023). SeismoBuild – A computer program for the linear and nonlinear analysis of Reinforced Concrete Buildings. www.seismosoft.com.

SeismoMatch (2023). SeismoMatch – A computer program for spectrum matching of earthquake records. www.seismosoft.com.

SeismoSelect (2023). SeismoSelect – A computer program for the selection, scaling and download of ground motion data. www.seismosoft.com.

# 7

# Linear and Nonlinear Methods of Analysis

## 7.1 Introduction

After selecting the performance objectives for the structural assessment, a series of analyses of the building that includes the retrofit measures are conducted to determine the forces and deformations imposed on the structural members by the ground motion, which corresponds to the selected seismic hazard level of that performance objectives. Within the context of all modern structural codes, four different analytical methods are proposed with small variations between the different standards:

- The Linear Static Procedure LSP, a static type of analysis with no variable load.
- The Linear Dynamic Procedure LDP, which is essentially the Response Spectrum Method (RSA). The Linear Dynamic Procedure is the method of analysis that is typically employed for the design of new structures; thus, it is the method of analysis that engineers are more familiar with.
- The Nonlinear Static Procedure NSP, which is the well-known pushover analysis, either in conventional or adaptive mode.
- The Nonlinear Dynamic Procedure NDP, which is the nonlinear dynamic time-history analysis.

The decision about the type of analysis to be used is based on several factors, including the characteristics of the structure (regularity in plan and elevation, weak story irregularities, torsional strength irregularities, fundamental period), the level of damage expected for the selected performance level and the comprehensive or limited knowledge of the structure.

The names of the methods above are the ones that are used by ASCE-41 and the different American guidelines. The corresponding names in the Eurocodes are: (i) the lateral force analysis; (ii) the modal response spectrum analysis; (iii) the nonlinear static (pushover) analysis; and (iv) the nonlinear time-history dynamic analysis. Within the context of this chapter the American nomenclature will be adopted, since it is more consistent and intuitive.

*Seismic Retrofit of Existing Reinforced Concrete Buildings*, First Edition. Stelios Antoniou.

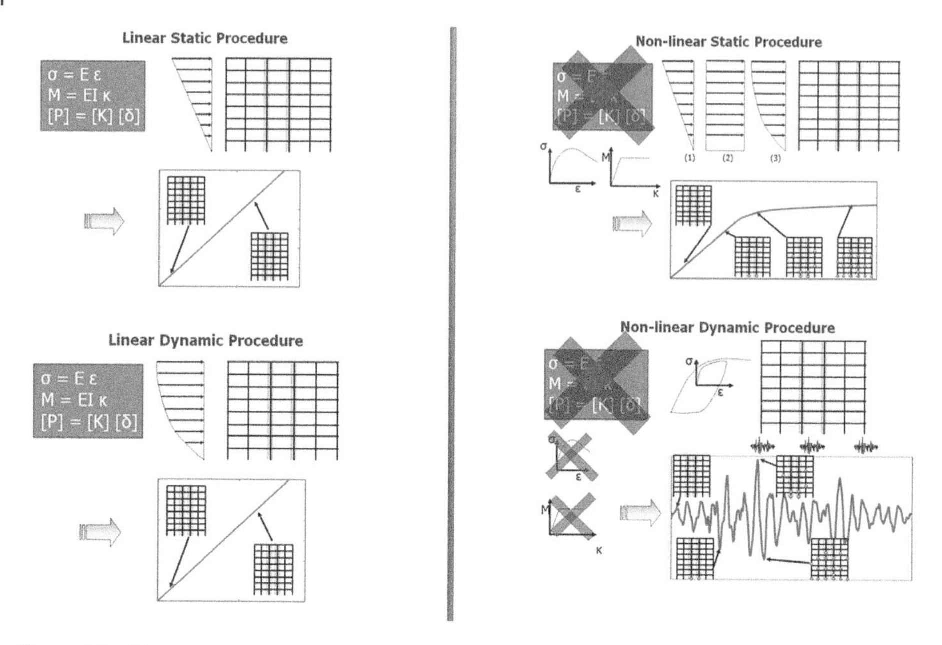


Figure 7.1 Linear and nonlinear analytical procedures. *Source:* Stelios Antoniou.

The term 'linear' in the first two procedures implies a "linearly elastic" member behavior. The basic rules of elasticity apply, the stiffness of the structure remains unchanged, and an unrealistic, linear force vs. displacement curve is obtained with the load increase (Figure 7.1). Because it is assumed that during a large seismic event the structure will sustain damage, a reduced "cracked" stiffness is employed for the analysis (in a very coarse – almost crude – manner), in order to account for the reduction of stiffness due to material inelasticity. The cracked stiffness is given as a fraction of the uncracked stiffness using factors that are given through tables, as the one from ASCE-41 shown in Table 7.1.

The term *nonlinear* in the nonlinear procedures implies explicit material nonlinearity or inelastic material response, whereas the effect of geometric nonlinearities is also included in the model. The structural behavior is no longer linear and elastic, and the rules of elasticity no longer apply. Stresses are not proportional to strains, forces are not proportional to displacements, and bending moments are not proportional to curvatures. Obviously, the structural and member stiffness is no longer constant, but rather it is updated at every step of the analysis and the structure gradually softens as plastic hinges develop at the locations of structural damage (Figure 7.1).

The nonlinear methods of analysis are numerically more advanced and are considered to be more accurate with respect to their linear counterparts, since they explicitly take into account the concentration of damage at the weakest locations of the building and the redistribution of forces upon the formation of plastic hinges. Likewise, the dynamic methods more accurately represent the dynamic nature of seismic loading with respect to the static methods. Recognizing the different levels of sophistication of the different methods, all codes impose limitations in the use of the simpler methods. The general concept is that simple methods can only represent with sufficient accuracy the structural behavior of

Table 7.1 Effective stiffness values, according to ASCE 41.

| Component | Effective stiffness values | | |
|---|---|---|---|
| | Flexural rigidity | Shear rigidity | Axial rigidity |
| Beams—nonprestressed[a] | $0.3\ E_{cE}I_g$ | $0.4\ E_{cE}A_W$ | – |
| Beams—prestressed[a] | $E_{cE}Ig$ | $0.4\ E_{cE}A_W$ | – |
| Columns with compression caused by design gravity loads $\geq 0.5A_g f'_{cE}$ [b] | $0.7\ E_{cE}I_g$ | $0.4\ E_{cE}A_W$ | $E_{cE}A_g$ |
| Columns with compression caused by design gravity loads $\leq 0.1\ A_g f'_{cE}$ or with tension[b] | $0.3\ E_{cE}I_g$ | $0.4\ E_{cE}A_W$ | $E_{cE}A_g$ (compression)<br>$E_{SE}A_S$ (tension) |
| Beam–column joints | Refer to Section 10.4.2.2.1 | | $E_{cE}A_g$ |
| Flat slabs—nonprestressed | Refer to Section 10.4.4.2 | $0.4\ E_{cE}A_g$ | – |
| Flat slabs—prestressed | Refer to Section 10.4.4.2 | $0.4\ E_{cE}A_g$ | – |
| Walls—cracked[c] | $0.35\ E_{cE}A_g$ | $0.4\ E_{cE}A_W$ | $E_{cE}A_g$ (compression)<br>$E_{SE}A_S$ (tension) |

[a] For T-beams, $I_g$ can be taken as twice the value of $I_g$ of the web alone. Otherwise, $I_g$ should be based on the effective width as defined in Section 10.3.1.3.
[b] For columns with axial compression falling between the limits provided, flexural rigidity shall be determined by linear interpolation. If interpolation is not performed, the more conservative effective stiffnesses shall be used. An imposed axial load $N_{UG}$ is permitted to be used for stiffness evaluations.
[c] See Section 10.7.2.2.

simple structural configurations under moderate loading. As shown in Table 7.2, only the nonlinear dynamic procedure (NDP) can be used for any type of building and any type of loading. By contrast, the nonlinear static procedure (NSP) cannot handle more complex seismic loading and higher mode effects, since the entire theoretical background of pushover analysis is based on the fact that the structure vibrates according to the fundamental mode only. Similarly, the linear dynamic procedure cannot handle cases where large inelasticities are expected, whilst the linear static procedure, being very simplistic, can be accepted only in cases of very regular, low-rise buildings. A more detailed discussion of the advantages and disadvantages of each method is given later in this chapter.

As explained in Chapter 2, because the linear methods are not as accurate, they are considered suitable methods for structural assessment, when the knowledge of the structural configuration is average. Both the lack of thorough knowledge of the building condition, and the lack of accuracy in the analysis are compensated by increased safety factors.

By contrast, it is not permitted to employ advanced nonlinear methods when the knowledge of the structure is incomplete. The rationale behind this limitation is that there is no point in using an advanced analytical method without adequate knowledge of the structure. What is more, advanced nonlinear methods can give a false impression of accuracy in the calculations, an accuracy that does not exist, if the knowledge of the building is

**Table 7.2** Limitations in the use of the different analysis methods.

| Analysis Procedure | Structural irregularity | High Inelastic Demand | Higher mode effects | Near source earthquakes |
|---|---|---|---|---|
| LSP | ✗ | ✗ | ✗ | ✗ |
| LDP | ✓ | ✗ | ✓ | ✗ |
| NSP | ✓ | ✓ | ✗ | ✗ |
| NDP | ✓ | ✓ | ✓ | ✓ |

*Source:* Stelios Antoniou.

limited. Consequently, advanced nonlinear methods can only be used, when there is good or preferably very good knowledge of the building, in which case the additional effort for the more advanced analytical calculations is compensated by lower safety factors.

## 7.2 General Requirements

### 7.2.1 Loading Combinations

According to ASCE 41, the deformations and the inertial forces on the components in a seismic assessment situation are evaluated by taking into account, apart from the seismic forces E, the gravity loads, the live loads and the loads from snow. For the linear types of analysis, the seismic load combination is:

$$Q_G \pm E = 1.10 \times \left(Q_D + Q_L + Q_S\right) \pm E$$

where $Q_G$ is the sum of the gravity loads, $Q_D$ is the dead loads, $Q_L$ is the actual live loads that are equal to 25% of the unreduced live load, and $Q_S$ is the snow loads.

For the nonlinear types of analysis, the 1.10 factor is omitted and the load combination becomes:

$$Q_G \pm E = 1.00 \times \left(Q_D + Q_L + Q_S\right) \pm E$$

In Eurocode 8, the snow loads are not considered and the gravity loads in the seismic load combination become:

$$\Sigma G_{K.J} \text{"}+\text{"} \Sigma \psi_{E,i} \times Q_{k,i}$$

where $\psi_{E,i}$ is the combination coefficient for variable action i, which is equal to 30% in most typical cases.

In both codes the coefficient for the live loads is relatively small, because it takes into account the likelihood of just a fraction of the loads being present over the entire structure during the earthquake, as well as a reduced participation of masses in the motion of the structure due to the non-rigid connection between them.

### 7.2.2 Multidirectional Seismic Effects

The analyses for the assessment of a building are carried out to address seismic motion in any horizontal direction. According to ASCE-41, multidirectional seismic effects should be considered for all buildings that have plan irregularities or have primary columns that form a part of two or more intersecting frames, i.e. in the vast majority of the building stock. In such cases, concurrent ground motion forces are to be applied in both horizontal directions $X$ and $Y$.

In the linear methods of ASCE 41 LSP and LDP 100% of the lateral forces will be applied in one direction and 30% in the perpendicular direction in all the combinations for the positive and negative directions in $X$ and $Y$, that is $\pm 100\%X \pm 30\%Y$ and $\pm 30\%X \pm 100\%Y$, a total of 8 combinations.

For the pushover analysis of the NSP, a similar rule is also proposed $\pm 100\%X \pm 30\%Y$ and $\pm 30\%X \pm 100\%Y$; however, it is also permitted to apply the forces in a single direction. Finally, for the NDP analysis a pair of orthogonal horizontal ground motion components should be considered, and, when the vertical earthquake effects are considered important, a vertical ground motion component should also be applied concurrently.

Contrary to ASCE 41, Eurocode 8 does not require the application of multidirectional seismic forces and the analyses are carried out in a single direction $\pm X$ and $\pm Y$.

### 7.2.3 Accidental Torsional Effects

The torsional moments determined from the building's mathematical model capture the eccentricity between the centers of mass and stiffness. Yet, an accidental torsion should also be considered in the analysis, which accounts for the additional contributing factors to torsional response that are not typically represented in the mathematical model, such as uncertainties in the location of masses or in the spatial variation of the seismic motion. In order to introduce the accidental torsion, the calculated center of mass at each floor is considered to be displaced from its nominal location in each direction by an accidental eccentricity of 5% of the floor dimension measured perpendicular to the direction of the applied load.

In the nonlinear analysis procedures, since the torsional contributions are better captured, the accidental torsion is not always required in the assessment, for instance ASCE-41 permits that it is omitted for lower seismic hazard levels, when multiple hazard levels are being considered. By contrast, accidental torsion should always be included for the highest hazard levels that are commonly associated with the life safety or collapse prevention performance levels, in which case significant asymmetric damage and large changes in the stiffness distribution are expected. Hence, there may be significant changes in the building response because of the impact of accidental torsion.

## 7.3 Linear Static Procedure

With the *linear static procedure (LSP)* (lateral force method with the EC8 naming conventions) a triangular, lateral, pseudo-seismic force distribution that is assumed to approximate the earthquake loading is applied to a linear elastic structural model in order to calculate the internal forces and the system displacements. These action effects are then compared with the members' capacities for the selected performance level, always in terms of forces, and, if the capacities are larger than the demands, the structure is considered safe.

The fundamental period of vibration of the building for lateral motion in the direction considered is calculated by eigenvalue analysis or with more approximate empirical methods, from which the ordinate of the response spectrum $S_a$ is calculated. The total lateral force is proportional to the spectral acceleration $S_a$ and the building weight $W$:

$$V = C_1 \times C_2 \times C_m \times S_a \times W \rightarrow \text{in ASCE 41 or}$$

$$V = \lambda \times S_a \times W \rightarrow \text{in Eurocode 8}$$

$C_1$, $C_2$, $C_m$ and $\lambda$ are different easily calculated modification factors that are related to higher mode effects, and parameters, such as the expected maximum inelastic displacements, the effect of pinched hysteresis shapes, the stiffness, and strength deterioration. This total force is then distributed at each floor level, according to the mass distribution of the building and the modal shape of the fundamental mode (in EC8) or an inverted triangular distribution (in both ASCE-41 and EC8).

Because of its approximate nature, the linear static procedure is permitted only in cases of very regular, low-rise constructions that sustain limited damage and do not undergo large inelastic deformations. More specifically:

- The demand to capacity ratios DCR should be small for all structural members. For the brittle failure types, they should be below unity.
- There should be no in-plane strength or stiffness discontinuities or irregularities.
- There should be no out-of-plane strength or stiffness discontinuity or irregularities.
- There should be no weak story strength or stiffness irregularities.
- There should be no torsional strength or stiffness irregularities.
- The fundamental period should not be large.

## 7.4 Linear Dynamic Procedure

The *linear dynamic procedure (LDP)* (Modal Response Spectrum Analysis, according to the EC8 naming conventions) is similar to the LSP, at least with regard to the modeling approach. The model is again elastic and there is no stiffness degradation during the analysis. However, the method is somehow more sophisticated, since the profile of the lateral forces is not arbitrary anymore but, rather is calculated as a combination of the modal contributions of the different modes of vibration of the structure. The action effects of the structural members are again compared against the capacities for the selected performance level in terms of forces, and, if the capacities are larger than the demands, the structure is considered safe. The LDP is based on the well-known response-spectrum analysis (RSA) (e.g., Rosenblueth 1951; Chopra 1995), and it is the method of analysis that is typically employed for the design of new structures.

The RSA is a pseudo-dynamic method that is capable of providing the peak values of response quantities, such as forces and deformations, of a structure under seismic excitation with a series of static analyses, rather than time-history dynamic analysis. In this context, the time–acceleration history imposed on the supports of the structure is replaced by the equivalent static forces, which are distributed to the free DOFs of the structure and represent the contribution from each natural mode of vibration. These equivalent forces are derived for each mode of vibration separately as the product of two quantities: (i) the modal inertia force distribution (hence eigenvalue analysis is needed), and (ii) the pseudo-acceleration response per mode (obtained from the 5% damped response spectrum). For each mode of interest, a static analysis is conducted, and then every final peak response quantity is derived by the superposition of the quantities corresponding to the modes.

A sufficient number of modes has to be considered, in order to capture at least 90% of the participating mass in each of two orthogonal principal horizontal directions of the building, thus neglecting only the less significant ways of vibrating in terms of participant mass. EC8 also requires that all modes with more than 5% of the participating mass in any direction should be considered.

Because the peaks in the responses of each mode generally occur at different time instants and rigorous time-history analysis has not be conducted, it is not possible to determine the exact peak values of the response quantities. Therefore, approximations need to be introduced by implementing one of the modal combination (statistical) rules, such as the absolute sum (ABSSUM), square-root-of-sum-of-squares (SRSS) and the complete quadratic combination (CQC). CQC is suggested when periods are closely spaced, with cross-correlation between the modal shapes. SRSS can be used when the periods differ by more than 10%, while ABSSUM offers a very safe, upper limit of response.

The same procedure is repeated for each desired seismic direction EX, EY, and EZ by using different or the same response spectra. It is usually requested that two or three seismic loading directions (EX, EY, EZ) be considered simultaneously, together with the gravity static loads (G + Q) of the structure (the vertical component EZ is mandatory only for the elements, where the vertical vibration is considered critical, e.g., large cantilevers). The seismic loading directions may be combined linearly ($E = \pm EX \pm EY \pm EZ$) with different factors $f_{EX}$, $f_{EY}$, $f_{EZ}$ per direction (usually $f_{EX}$, $f_{EY}$, $f_{EZ}$ = 1.00 or 0.30) or by the SRSS rule ($E = \pm\sqrt{EX^2 + EY^2 + EZ^2}$). The gravity and live loads are defined and added algebraically. Because the seismic loads are taken into account with both signs for every direction, the results of RSA loading combinations in terms of any response quantity are presented as envelopes.

Contrary to the linear static procedure, the linear dynamic procedure is suitable for buildings with larger fundamental period, where higher-mode effects are important. Apart from this, all the recommendations and limitations described for the LSP apply for the LDP as well:

- The demand to capacity ratios DCR should be small for all structural members. For the brittle failure types, they should be below unity.

- There should be no in-plane strength or stiffness discontinuities or irregularities.
- There should be no out-of-plane strength or stiffness discontinuity or irregularities.
- There should be no weak story strength or stiffness irregularities.
- There should be no torsional strength or stiffness irregularities.

## 7.5 Nonlinear Structural Analysis

Before moving on to the description of the two nonlinear methods of analysis (NSP and NDP), it is important to present the main characteristics of nonlinear analysis for those readers who are not accustomed to it. If you have used pushover or nonlinear dynamic analysis in the past and you feel confident with it, this section is probably not of much interest and you can safely move to the next sections for the detailed presentation of the code-based nonlinear analysis procedures.

### 7.5.1 Nonlinear Structural Analysis in Engineering Practice

From the onset of computer-aided structural analysis, engineers have employed linear elastic methods, implicitly assuming small deformations and limited damage of the structural members and an approximately elastic performance of all the structural components. Even in today's engineering practice, elastic methods are still vastly employed for the design of new structures. This is not unreasonable and unjustified, considering the fact that in a new structure, engineers are able to choose the strength and stiffness characteristics of the structural components, so as to have a distribution of inelasticity along the different structural members, without large concentrations of inelastic deformations. This, together with careful ductility detailing of the structural members (e.g., closely spaced stirrups in RC members, or diagonal reinforcement, where needed), provides an efficient, and reasonably accurate framework for the design of new structures with a high level of reliability.

However, true structural behavior is inevitably and inherently nonlinear, characterized by nonproportional variation of displacements with loading, especially in the presence of large deformation demands and/or significant material nonlinearities. In particular, existing buildings, especially those that have been designed only for gravity loading, frequently exhibit irregular arrangement of their structural members, and uneven distribution (in plan or elevation) of their strength, stiffness and mass, which adversely affects their behavior under earthquake loading and leads to large concentrations of the seismic forces and localized damage. Consequently, the use of elastic procedures for their seismic analysis very often leads to non-negligible inaccuracies in the estimation of both the force and the deformation demand of the structural components. What is more, in the majority of cases this approximation leads to the underestimation of the displacement demand on those members where inelastic deformations are concentrated, and which are thus the most vulnerable under seismic loading. In such cases, the use of the nonlinear methods is preferable, since it provides a more efficient and accurate framework, and allows for the determination of the realistic structural behavior, with less simplifying assumptions, more direct (high-level) design criteria, and lower safety factors.

Until relatively recently (mid-1990s), the computational power of computers was not sufficiently large to allow for the general introduction and use of inelastic analysis for structural design and assessment. However, nowadays, with the advancements in computing technologies, enhanced nonlinear solution algorithms and the increase in the available experimental data for model calibration, the use of nonlinear structural analysis is becoming increasingly widespread and common. As such, nonlinear analysis already plays an important role in the assessment of existing buildings, as well as in the design of new building, either directly through the newly developed performance based design methodologies, or indirectly by assessing the integrity of structures that have been designed with elastic methods.

The first guidelines for the application of nonlinear analysis were published in the mid-1990s, *FEMA-273: NEHRP Guidelines for the Seismic Rehabilitation of Buildings* (FEMA 1997) and *ATC-40: Seismic Evaluation and Retrofit* of Concrete Buildings (ATC 1996). Since then, improvements have been proposed in *FEMA-440: Improvement of Nonlinear Static Seismic Analysis Procedures* (FEMA 2005) and *FEMA-P440A: Effects of Strength and Stiffness Degradation on Seismic Response* (FEMA 2009a) . Nonlinear analysis methodologies have already been introduced into all modern assessment methodologies, and its concepts have also been adopted in methods for seismic risk assessment, the most widely known being HAZUS (Kircher et al. 1997a; Kircher et al. 1997b; FEMA 2009b).

Typical instances where nonlinear analysis is applied in structural earthquake engineering practice are the following (Antoniou and Pinho 2018; NIST 2010):

- *Evaluation and retrofit of existing buildings.* The majority of existing buildings fall short of meeting prescriptive detailing requirements and provisions of the relevant standards for new buildings, which presents a challenge for the evaluation and retrofit with the use of elastic analysis methods. As a result, seismic assessment and strengthening of existing buildings has been one of the primary drivers for the use of nonlinear analysis in engineering practice, since more accurate analysis may permit less conservative approaches and reduced intervention costs.
- *Verification of the design for new buildings.* More recently, the role of nonlinear analysis is being expanded beyond the analysis of existing structures to quantify building performance more completely. Nonlinear analysis is thus now also used to verify structural designs based on elastic methods, so as to allow for the better estimation of the true structural response of the building. For instance, ATC-58: Guidelines for Seismic Performance assessment of buildings (ATC 2009) employs nonlinear dynamic analyses for seismic performance assessment of new and existing buildings alike, including fragility models that relate structural demand parameters to explicit damage and loss metrics.
- *Design of new buildings that employ structural systems, materials, or other features that do not conform to current building code requirements.* Although most new buildings are designed using elastic analysis methods and prescriptive code provisions, the use of nonlinear analysis in the design of new buildings is gradually becoming increasingly common. The significant advance in engineering technology, which has taken place in recent decades with the introduction of new materials, such as Fiber-reinforced polymers (FRP) or Shape-memory alloys (SMA), or new systems, such as damping and base isolation devices, has provided engineers with a large variety of technologies and solutions that may provide enhanced structural safety with the same or reduced financial cost. However,

the use of conventional analytical methodologies is often not appropriate for such scenarios, making more advanced analysis often required. In these cases, nonlinear (static and very often dynamic) analysis is usually employed to allow for the modeling of the full hysteretic behavior of these new materials or systems.

- *Design of tall buildings in high seismicity regions.* Frequently, a building's performance must be estimated and assessed beyond the typical code recommendations. Tall buildings with nonstandard seismic-force-resisting systems are a common example of the use of nonlinear analysis in design. Toward this end, several engineering resource documents outline explicit requirements for the use of nonlinear dynamic analysis to assess the performance of tall buildings, e.g., Pacific Earthquake Engineering Research Center Guidelines for Performance-Based Seismic Design of Tall Buildings (PEER 2010), Recommendations for the Seismic Design of High-rise Buildings (Willford et al. 2008), and PEER/ATC 72-1 Modeling and acceptance criteria for seismic design and analysis of tall buildings (PEER/ATC 2010).
- *Performance-based design of new buildings with specific owner requirements.* The introduction of performance-based engineering allows for the selection of particular performance levels and objectives by the owner according to his/her specific needs. These objectives can be quantified, the performance can be predicted analytically, and the cost of improved performance can be evaluated to allow rational trade-offs based on life-cycle considerations, rather than construction costs alone. Within this framework, the use of nonlinear procedures that provide enhanced accuracy and significantly better performance predictions is in many ways advantageous and more easily accommodate the specific owner/stakeholder's requirements.
- *Seismic risk assessment.* Since the introduction of seismic risk assessment methods, such as HAZUS, in the late 1990s, the analytical procedures have been continually refined, including methods that employ building-specific analyses to improve building fragility models. HAZUS®-MH MR5 Advanced Engineering Building Module (FEMA 2009b) is an example of guidelines for using building specific nonlinear analyses in seismic risk assessment.

### 7.5.2 Challenges Associated with Nonlinear Analysis

Although undoubtedly nonlinear analysis is significantly superior, in terms of the accuracy in structural response predictions, this does not come without a cost. The computational resources required for nonlinear analysis are far from negligible, especially in the case of large models subjected to dynamic loading with acceleration time-histories applied at their base.

Further, whereas linear analysis takes into account only the mass and stiffness distribution of the structural members, nonlinear analysis involves the consideration of both the stiffness and the strength of the members, as well as their overall inelastic behavior, and limit states that depend on deformations and forces. It requires the definition of models that capture the force-deformation response of components and systems based on expected strength and stiffness properties, taking also into account large deformations. These data require a deeper knowledge of the structural configuration, and the modeling procedures followed in the analytical process, better trained and more experienced engineers, as well as significant effort and additional cost to acquire a good level of knowledge of the structure.

What is more, the results of nonlinear analyses can be very sensitive to assumed input parameters and the types of models used. Without good experience in the analytical and assessment methods that are followed, engineers can be easily misled and extract conclusions about the structural performance that are far from the true response. Generally, it is advisable to have clear expectations about those portions of the structure that are expected to undergo inelastic deformations, so as to use the analyses to confirm the locations of inelastic deformations.

Finally, in contrast to linear elastic procedures and design methods that are well established and have been used and tested extensively for decades, nonlinear inelastic analysis techniques and their application to design/assessment are relatively recent and are still evolving. In order to keep up with the latter, engineers do need to be open to seeking regular updating and training, acquiring innovative knowledge, developing new skills and gaining confidence with continuously changing tools.

## 7.5.3 Some Theoretical Background

### 7.5.3.1 Introduction

In linear structural analysis things are relatively simple and straightforward. Every element has linear cross-section properties EA, EI2, EI3, GJ, and constant stiffness. The stiffnesses of all the elements are summed up to form the global stiffness matrix [K] of the building, which is also constant, and [K] is inverted once for the solution of the well-known equation $\underline{P} = [K] \cdot \underline{u}$ and the calculation of the nodal displacements for the different load cases. What is more, because the stiffness matrix is always positive-definite and the member strengths are considered unlimited, there will always be a mathematical solution to the problem.

Because in nonlinear analysis the structural stiffness matrix is not constant anymore, but rather it is updated at every step, and because there is a strength limit in all or most of the structural members, the solution of the nonlinear equations is a much more complicated task. The next sections briefly touch on the main points of the theoretical background of nonlinear analysis.

### 7.5.3.2 Sources of Nonlinearity

The modeling of the nonlinear mechanical properties of members is a complex and wide-ranging subject. The primary source of nonlinearities in low- and medium-rise building structures is material inelasticity and plastic yielding in the locations of damage. In larger high-rise buildings, while material inelasticity still plays an important role, large deformations relative to the frame element's chord (known as P-delta effects), and geometric nonlinearities become equally important and should be taken into account.

These two aspects (material inelasticity and geometrical nonlinearity) are discussed in this section. Note that, as it will further be developed in Chapter 8, the structural analysis of buildings is typically performed through the use of one-dimensional finite elements (e.g., beams or rods). Two- and three-dimensional FE are very rarely utilized, due to the heavy computational burden and also the lack of reliable and numerically stable 2D/3D nonlinear material constitutive models.

***Material Inelasticity*** Material nonlinearities occur when the stress–strain or force-displacement law is not linear, or when material properties change with the applied loads. Contrary to linear analysis procedures, where the material stresses are always proportional to the corresponding strains and a fully elastic behavior is assumed, in nonlinear analysis the material behavior depends on current deformation state and possibly the deformation history of the cross-section. In order to estimate the stress caused by the strain at a particular location of the structure, complete expressions for the uniaxial stress–strain relationship of the material should be provided, including hysteretic rules for unloading and reloading.

***Geometric Nonlinearities*** Geometric nonlinearities involve nonlinearities in kinematic quantities, and occur due to large displacements, large rotations, and large independent deformations relative to the frame element's chord (also known as P-delta effects). Geometric nonlinearities can have very important consequences in the structural behavior, especially in large and slender structures, and they can ultimately lead to a significant decrease in the effective lateral stiffness and lateral resistance, and to dynamic instability.

#### 7.5.3.3 Solving Nonlinear Problems in Structural Analysis

Once the governing equations of geometrically nonlinear structural analysis are formed and the discretization of those equations by finite element methods is completed, a procedure is required for the solution of these equations. Because of the constant updating of the stiffness matrix, nonlinear problems in structural mechanics are solved with incremental algorithms, through a process that is considerably more articulated than common linear elastic analysis solvers.

All solution procedures of practical importance are strongly rooted on the concept of gradually advancing the solution by "continuation," that is, to follow the equilibrium response of the structure as the control and state parameters vary by small amounts. Various algorithms exist for handling such problems, but a common feature is that continuation is a multilevel process that involves a hierarchical breakdown into incremental steps, with iterative steps within. To advance the solution, the entire loading stage is broken down into incremental steps, also called increments or steps. The incremental solution methods are then divided into two broad categories: (i) Purely incremental methods, also called *predictor-only methods*, and (ii) Corrective methods, also called *predictor–corrector* or *incremental-iterative methods.*

In purely incremental methods, the iteration level is missing. In corrective methods a predictor step is followed by one or more iteration steps. The set of iterations is called the corrective phase, and its purpose is to eliminate or reduce the so-called drifting error, which is a serious problem of purely incremental methods. For this reason, the corrective methods have become the standard for the solution of the nonlinear equations in all modern finite element packages.

Solutions accepted after each increment and its corrective phase, are often of interest to users because they represent approximations to equilibrium states until the final loading state. They are therefore saved as they are computed. On the other hand, intermediate results of the iterative process are rarely of interest, since the solutions are not equilibrated and constitute "merely" an intermediate step until the next equilibrated solution, and hence most nonlinear structural analysis programs discard them.

The use of increments may seem at first sight unnecessary if one is interested primarily in the final solution. But breaking up a stage into increments serves different purposes:

- The breakdown of the entire loading stage can lead to more stable solutions and avoid convergence problems.
- The engineer can acquire a better insight into structural behavior by studying the response plot toward the final solution, which in many cases can provide more useful information than simply the structural state at the end. It is noted that in several cases failures and critical points occur before the stage end.
- The presence of path-dependent effects in nonlinear analysis problems severely restricts increment sizes because of history-tracing constraints. For example, in plasticity analysis stress states must not be allowed to stray too far outside the yield surface.

***Incremental-Iterative Algorithm*** The basic method for the solution of nonlinear equations in the majority of finite element programs is the load-control Newton-Raphson (NR) algorithm, and variations of it. The Newton-Raphson method, in its simplest form, is a numerical method for finding the roots of a function $f(x)$. Since the method is iterative, a trial guess is made at $x = x_n$. Evaluating the function at $x_n$, it is found that $f(x_n) \neq 0$, i.e. it is not a root. If $f'(x_n)$ is the tangent of the function at $x_n$, the equation of the tangent passing from $x_n$ is:

$$f(x_{n+1}) - f(x_n) = f'(x_n) \bullet (x_{n+1} - x_n) \tag{7.1}$$

With the aid of Eq. (7.1), a second trial solution is obtained at $x_{n+1}$:

$$x_{n+1} = x_n - \frac{\mathrm{f}'(x_n)}{\mathrm{f}(x_n)} \tag{7.2}$$

where $x_{n+1}$ is the point where the tangent intersects axis $X$. If $f(x_{n+1}) \approx 0$, then we have located the root, otherwise we proceed finding a new trial solution at $x_{n+2}$, until convergence to the correct solution has been reached within an acceptable convergence limit. The method is schematically shown in Figure 7.2.

In structural mechanics, the Newton-Raphson (NR) method is extended, so as to accommodate the solution of a series of nonlinear equations, arising from the general equation of (nonlinear) structural equilibrium:

$$P = k \times u \tag{7.3}$$

Because of the nonlinear nature of the problem, the stiffness matrix $\mathbf{K}$ is a function of the deformation vector $\mathbf{u}$ and is constantly updated at each iteration. As a result, Eq. (7.3) cannot be solved directly, and an incremental solution procedure needs to be followed.

The incremental-iterative Newton-Raphson method is schematically shown in Figure 7.3. The iterative procedure follows the conventional scheme, whereby the internal forces corresponding to a displacement increment are computed and the convergence of the system is checked. If no convergence is achieved, then the out-of-balance forces (difference between applied load vector and equilibrated internal forces) are applied to the structure, and the new displacement increment is computed. Such loop proceeds until convergence has been achieved or a maximum number of iterations has been reached. Load-displacement

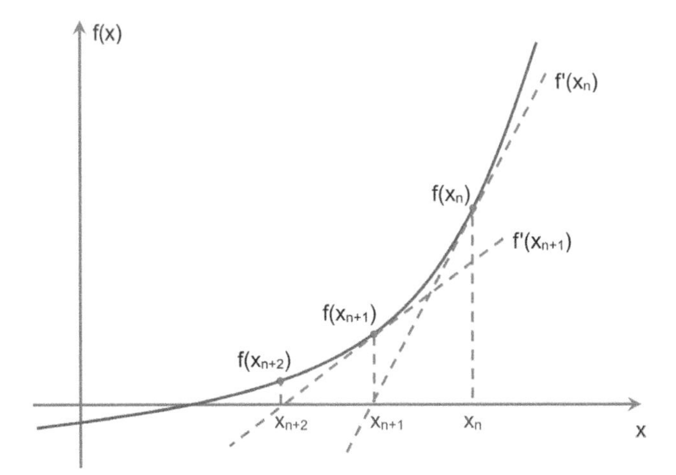


**Figure 7.2** Application of the Newton-Raphson (NR) method for finding the roots of a function $f(x)$. *Source:* Stelios Antoniou.

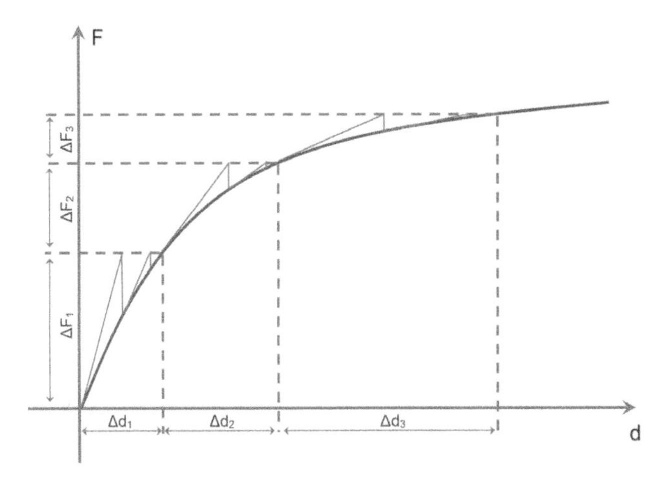


**Figure 7.3** The Newton-Raphson (NR) method in nonlinear structural analysis. *Source:* Stelios Antoniou.

plots, such as those of Figure 7.3 are exact for SDOF systems. For larger MDOF systems, they describe only schematically the structural response and the gradual convergence to the solution of the system of equations.

The full Newton-Raphson method provides a quadratic rate of convergence, meaning that it requires a small number of iterations to reach the solution. However, recalculating and inverting the stiffness matrix at every iteration requires increased computing resources. Hence, a common alternative is to recalculate and invert the stiffness matrix only at the first iteration and use it for all the corrective iterations. This approach is known as modified Newton-Raphson and is shown in Figure 7.4.

The computational savings in the formation, assembly and reduction of the stiffness matrix during the iterative process can be significant when using the mNR instead of the

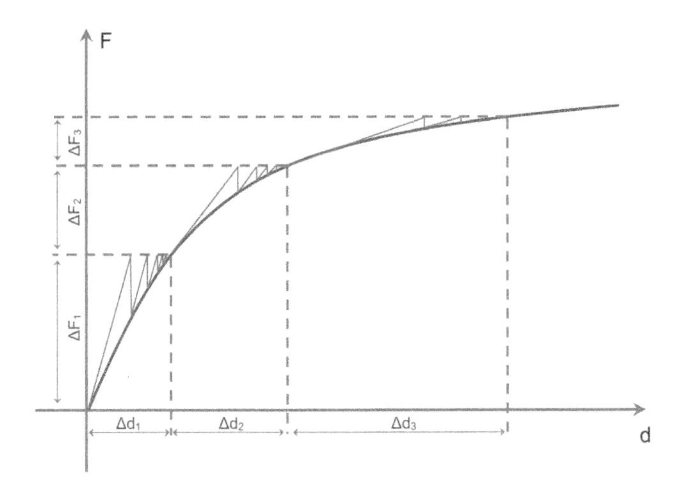


**Figure 7.4** The Modified Newton-Raphson (mNR) method in nonlinear structural analysis. *Source:* Stelios Antoniou.

NR procedures. However, more iterations are often required with the mNR, thus leading in some cases to an excessive computational effort overall. For this reason, a third NR-mNR hybrid approach is also available in the more advanced FE packages, whereby the stiffness matrix is updated only in the first few iterations of a load increment, after which a constant stiffness matrix is used. The NR-mNR solution usually leads to an optimum scenario with faster convergence (in terms of time), with respect to the other two alternatives.

For further discussion and clarifications on the algorithms described above, readers are strongly advised to refer to available literature, such as the work by Cook et al. (1989), Crisfield (1991), Zienkiewicz and Taylor (1991), Bathe (1996) and Felippa (2004), to name but a few.

#### 7.5.3.4 Convergence Criteria

As already mentioned, in iterative-incremental solution algorithms, the iterative process at each step is continued, until *convergence is achieved* – that is, the value of a norm of the out-of-balanced forces or the unbalanced deformations of the structure become smaller than given convergence criteria that have been specified by the user at the beginning of the analysis.

There are two distinct categories of convergence criteria in nonlinear analysis:

- The displacement/rotation-based criteria
- The force/moment-based criteria

Two additional convergence check schemes may arise from the combination of these distinct criteria:

- Displacement/rotation AND force/moment based scheme, where it is considered that the solution has been reached, when both the deformation and the force based criteria have been achieved.
- Displacement/rotation OR force/moment based scheme, where convergence is achieved, when either the deformation or the force based criteria has been achieved.

Usually, the *displacement/rotation* criterion consists in verifying, for each individual degree-of-freedom of the structure, that the current iterative displacement/rotation is less or equal than a user-specified tolerance. In other words, if and when all values of displacement or rotation that result from the application of the iterative (out-of-balance) load vector are less or equal to the predefined displacement/rotation tolerance factors, then the solution is deemed as having converged. This concept can be mathematically expressed in the following manner:

$$\max\left[\left|\frac{\delta d_i}{d_{tol}}\right|_{i=1}^{n_d}, \left|\frac{\delta \theta_i}{\theta_{tol}}\right|_{j=1}^{n_\theta}\right] \leq 1 \rightarrow convergence \tag{7.4}$$

where
- $\delta d_i$ is the iterative displacement at translational degree of freedom $i$
- $\delta\theta_j$ is the iterative rotation at rotational degree of freedom $j$
- $n_d$ is the number of translational degrees of freedom
- $n\theta$ is the number of rotational degrees of freedom
- $d_{\text{tol}}$ is the displacement tolerance, in the employed by the analysis length unit
- $\theta_{\text{tol}}$ is the rotation tolerance, in rad (dimensionless)

The *force/moment* criterion, on the other hand, comprises the calculation of the Euclidean norm of the iterative out-of-balance load vector, and subsequent comparison to a user-defined tolerance factor. It is therefore a global convergence check (convergence is not checked for every individual degree of freedom as is done for the displacement/rotation case) that provides an image of the overall state of convergence of the solution, and which can be mathematically described in the following manner:

$$G_{\text{norm}} = \frac{\sqrt{\sum_{i=1..n}\left[\frac{G_i}{V_{\text{REF}}}\right]^2}}{n} \leq 1 \rightarrow \textbf{convergence} \tag{7.5}$$

where

- $G_{\text{norm}}$ is the Euclidean norm of iterative out-of-balance load vector.
- $G_i$ is the iterative out-of-balance load at degree of freedom $i$.
- $V_{\text{REF}}$ is the reference "tolerance" value for forces (translational DOFs) and moments (rotational DOFs). Typically, different values are assumed for $V_{\text{REF}}$ for the translational and the rotational DOFs.
- $n$ is the number of the degrees of freedom.

#### 7.5.3.5 Numerical Instability, Divergence, and Iteration Prediction

In nonlinear analysis the maximum number of iterations is always specified, so as to avoid a situation whereby an unattainable solution is pursued infinitely. If this limit is reached and convergence has not been achieved, the analysis is either stopped or, in more advanced software packages, the load step is subdivided into smaller increments, in order to try to achieve better convergence conditions.

In addition to the convergence verification schemes described above, at the end of an iterative step three other solution checks may be carried out; numerical instability, solution divergence, and iteration prediction. These criteria serve the purpose of avoiding the computation of useless equilibrium iterations in cases, where it is apparent that convergence will not be reached, thus minimizing the duration of the analysis.

**Numerical instability.** The possibility of the solution becoming numerically unstable is checked at every iteration by comparing the Euclidean norm of out-of-balance loads, $G_{\text{norm}}$, with a predefined maximum tolerance several orders of magnitude larger than the applied load vector (e.g., 1.0E+20). If $G_{\text{norm}}$ exceeds this tolerance, then the solution is assumed as being numerically unstable and iterations within the current increment are interrupted.

**Solution divergence.** Divergence of the solution is checked by comparing the value of $G_{\text{norm}}$ obtained in the current iteration with that obtained in the previous one. If $G_{\text{norm}}$ has increased, then it is assumed that the solution is diverging from a possible solution and iterations within the current increment are interrupted.

**Iteration prediction.** Finally, a logarithmic convergence rate check is also carried out, so as to try to predict the number of iterations required for convergence to be achieved. If this estimated number of iterations is larger than the maximum number of iterations specified by the user, then it is assumed that the solution will not achieve convergence and iterations within the current increment are interrupted.

It is noted that these three additional checks described above are usually reliable and effective within the scope of applicability of nonlinear analysis, for as long as the divergence and iteration prediction check is not carried out during the first iterations of an increment when the solution might not yet be stable enough. Hence, it is advisable that these checks are carried out after an initial number of iterations.

### 7.5.4 Implications from the Basic Assumptions of Nonlinear Analysis

In the linear methods, the stiffness matrix is always positive definite and even if the loading vector is unreasonably large (e.g., because of wrong input by the user), a solution will be found. An illogical solution perhaps from the engineering point of view, of several hundreds of thousands kN or kNm for the internal element forces and some km in displacement, but a mathematical solution is always found. Unfortunately, this is not the case with nonlinear procedures, where all or most of the members have limits in their strengths, and at any step of the analysis the program carries out a series of iterations, while trying to equilibrate the out-of-balance forces in an environment of constantly updated stiffness distribution. In this framework, the achievement of convergence and the calculation of the mathematical solution is not always guaranteed.

The most typical cases occur when the external forces exceed the maximum capacity of a member, leading to structural failure, e.g., a beam cannot withstand the applied vertical loads on it. Since the member's strength is not infinite, there will not be any combination of deformations and internal forces that will be able to equilibrate the external loads, and, contrary to the linear algorithms that always provide a solution, the nonlinear algorithms will output an annoying "failed to converge" or "diverging solution" message. Other reasons for numerical instability are large load increments that the solution procedure cannot accommodate, a sudden redistribution of forces after the failure of a load-carrying member, or errors in the modeling that lead to a large concentration of forces in small and weak elements.

Remember: If at any step of the analysis the program (any FE program) cannot find a solution (i.e., cannot equilibrate the out-of-balance forces and out-of-balance displacements) in every node and for every element of the model, convergence cannot be achieved, and the analysis stops.

Following are two typical cases of this behavior. In the first (Figure 7.5), a four-story building has been strengthened with very large and stiff RC walls in all floors with the exception of the smaller penthouse at the top. Because the stiffness of the columns in the penthouse is considerably smaller with respect to the large walls, their deformations

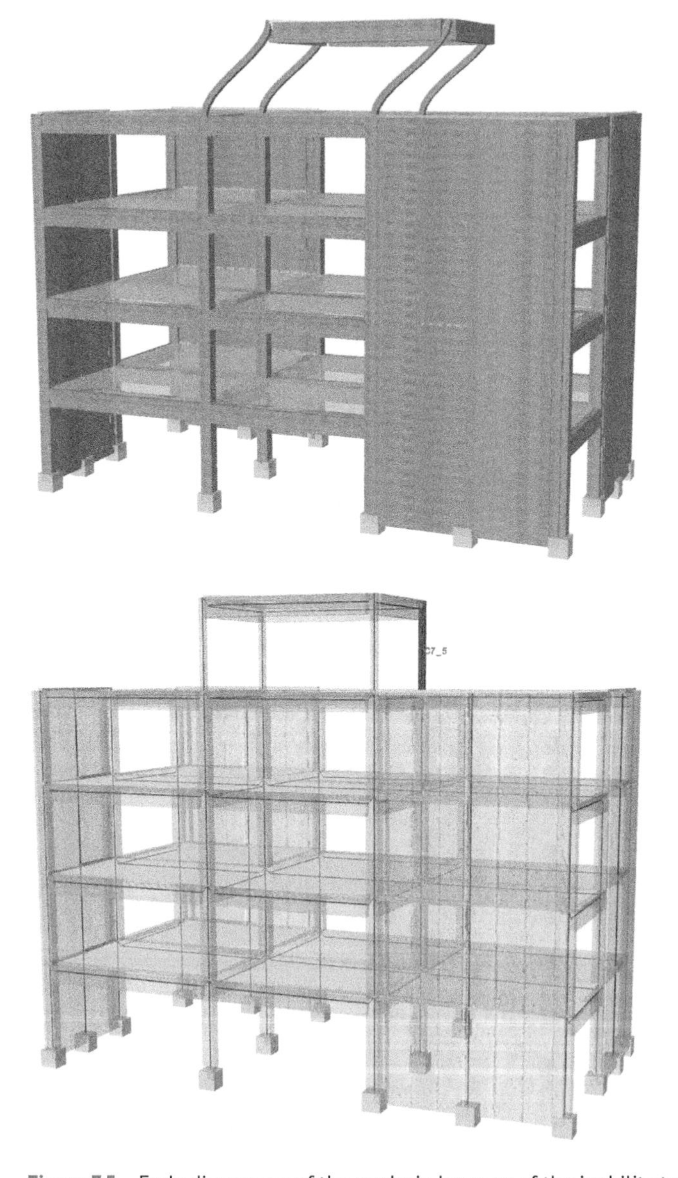

Figure 7.5 Early divergence of the analysis, because of the inability to achieve convergence locally at one or a few structural members. Case 1: a very flexible penthouse: (a) the deformed shape at the final step before divergence and (b) the diverging element as shown in the Convergence Details module of SeismoStruct. *Source:* Stelios Antoniou.

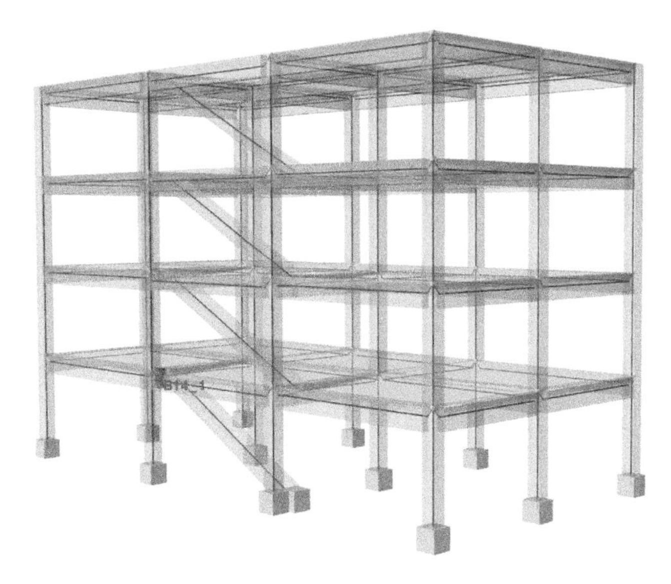

Figure 7.6 Early divergence of the analysis, because of the inability to achieve convergence locally at one or a few structural members. Case 2: a very short beam. *Source:* Stelios Antoniou.

increase disproportionally with respect to the rest of the building, leading to convergence difficulties early in the analysis. At the step of divergence, the interstory drift of the penthouse is 9.1% (at which drift the analysis cannot find a solution), and that of the "critical" ground floor just 0.03%. In the second example, a very short beam of length 0.20 m is created from SeismoStruct's (2023) Building Modeler to connect the stairs with the adjacent column. The analysis again fails prematurely due to extremely large deformation demands at that beam (Figure 7.6).

In both cases, in linear analysis such problems would not affect the structural integrity (it is elastic analysis after all) and they would pass largely unnoticed. Only a very diligent engineer might find out that the force and displacement demands on these problematic members are unreasonably high, but the analyses would run anyway. In nonlinear analysis, however, they cause the analysis to diverge prematurely, effectively preventing the structural assessment to be carried out until the target displacement.

### 7.5.5 How Reliable Are Numerical Predictions from Nonlinear Analysis Methods?

Through realistic modeling of the underlying mechanisms, nonlinear static analysis and especially nonlinear dynamic analysis, significantly reduce the uncertainty in the demand predictions, as compared to the linear methods of analysis. However, even with nonlinear dynamic analyses, it is practically impossible to always calculate accurately the demand parameters for the different structural members. Therefore, there are usually discrepancies between the analytical predictions of the response parameters and their actual values during a seismic event. Most of the time, these discrepancies are larger for structural deformations and accelerations and lower in force-controlled components of capacity-designed structures where the forces are limited by the strength of yielding members.

Since no numerical representation of the components' response is perfect, one of the sources of inaccuracy is the limitations that are related to the analytical capabilities of the selected software package, with which the analyses are carried out. Naturally, any analytical formulation that describes the hysteretic behavior of materials, sections, or members has limitations that reduce its ability to represent the structural response in a very precise manner, especially in the highly inelastic range, where the lateral stiffness is close to zero and the deformation response is thus very sensitive to small changes of the loading. Engineers should be well aware of these limitations prior to the execution of the analyses, so as to avoid modeling strategies that magnify the possible errors and affect the analytical predictions considerably.

All the same, with the continuous development of new, enhanced models to represent structural behavior, inaccuracies related to the software itself, and its analytical formulations and capabilities tend to become smaller, especially in the cases of analysis of ordinary structures with components featuring relatively predictable behavior that can be approximated reasonably well. In such cases, the main sources of inaccuracy are mostly related to human-related parameters:

- Incomplete knowledge of the structure, since it is practically impossible to measure all the important parameters of an existing structure (geometry, component dimensions, reinforcement of all members, footings dimensions and reinforcement, sustained damage by previous earthquakes or corrosion),, especially if it is in use.
- Uncertainties arising from the variability in the measured physical attributes of the structure such as material properties and structural details.
- Incomplete mathematical model representation of the actual structural behavior by the analyst.

These problems, coupled with hazard uncertainties in the ground motion intensity, and ground motion uncertainty arising from the frequency content and duration of ground motions for a given intensity, may lead to a large difference between the analytical predictions and the structure's true response during a large earthquake event.

Consequently, engineers should always bear in mind the different sources of uncertainty in the structural evaluation process, and how these may affect the predictions. The uncertainties should always be accounted for and, if possible, they should be quantified and represented in the acceptance criteria checks through the selection of appropriate values of the corresponding safety factors.

### 7.5.6 Final Remarks on Nonlinear Analysis

The use of nonlinear analysis to assess the seismic response of both existing and newly designed buildings is becoming more and more frequent, in recognition of the importance of capturing as faithfully as possible the actual dynamic response of these structures to earthquakes.

Indeed, modern earthquake-resistance design codes now explicitly stipulate the need for seismic assessment of existing buildings through the employment of nonlinear analysis. This requirement is emphasized even further in guidelines and regulations prescribed by the nuclear power industry and other energy sectors, where it has become mandatory for the safety of production plants and facilities (which include numerous types of framed structures) against earthquakes and other dynamic actions to be assessed through the employment of advanced nonlinear structural modeling.

In the past, such modeling prerequisites constituted a challenging task for practitioners, because it used to be a field very much confined to the academic research community. Nowadays, structural engineers find themselves much better equipped to tackle such regulatory requirements, thanks both to ever-increasing computational power at our disposal, as well as to the advancements in the development of practical software tools that are now capable of reproducing the response of framed structures well into the nonlinear inelastic range, as shown by verifications with experimental test results.[1]

## 7.6 Nonlinear Static Procedure

### 7.6.1 Pushover Analysis

According to ASCE 41-17, Section 7.4.3.1, the definition of the nonlinear static procedure (NSP) or pushover analysis is the following:

*A mathematical model directly incorporating the nonlinear load-deformation characteristics of individual components of the building shall be subjected to monotonically increasing lateral loads representing inertia forces in an earthquake until a target displacement is exceeded.*

In the NSP, the structural behavior is no longer elastic and linear, and the analysis accounts for the geometrical and material nonlinearities, as well as the redistribution of internal forces due to the sustained structural damage. Stresses are not proportional to strains, forces are not proportional to displacements and bending moments are not proportional to curvatures. In pushover analysis, a structural model that consists of nonlinear members (members in which material and geometric nonlinearity has been explicitly modeled) is loaded with a predefined lateral load profile and a gradually increasing load factor $\lambda$, until the displacement of a selected *control node*, typically located at the center of mass of the top story of the building, reaches the so-called *target displacement* (Figure 7.7 and Figure 7.8). The target displacement represents an approximation of the maximum displacement demand under the selected level of earthquake ground motion.

The members' stiffness is no longer constant, but rather updated at every step and the structure gradually softens as plastic hinges develop at the locations of structural damage. As a result, the force vs. deformation curve, which is the *capacity curve*, is not linear anymore, but has a parabolic shape as the structural deformations increase disproportionally with the level of lateral loading. In other words, for the same level of load increase, the increase of the deformations gets larger as we push further on in the inelastic range. At the target displacement, the demand parameters for the structural components are compared against the respective acceptance criteria for the desired performance level. System level demand parameters, such as story drifts and base shear forces, may also be checked.

1 For more information on the accuracy of nonlinear analysis, readers are advised to refer to the Verification Report of SeismoStruct that comes with the installation of the program and includes numerous validation examples.

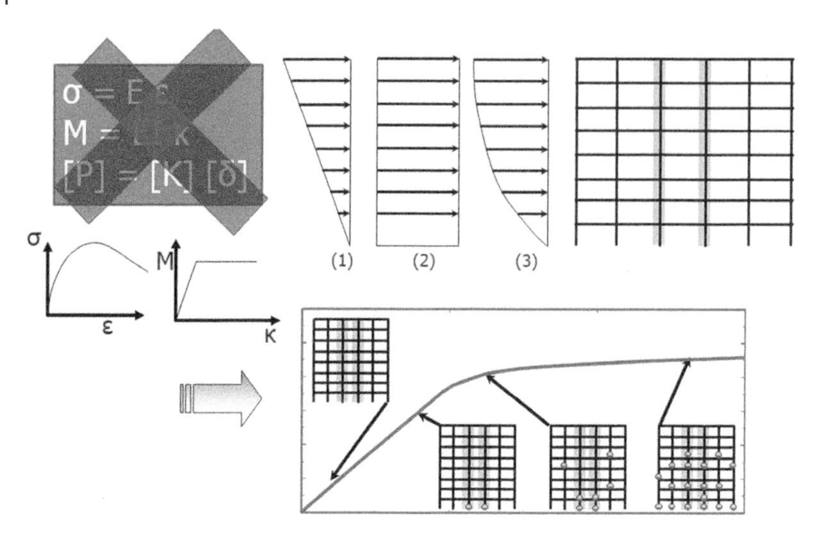


Figure 7.7 Nonlinear static procedure, NSP. *Source:* Stelios Antoniou.

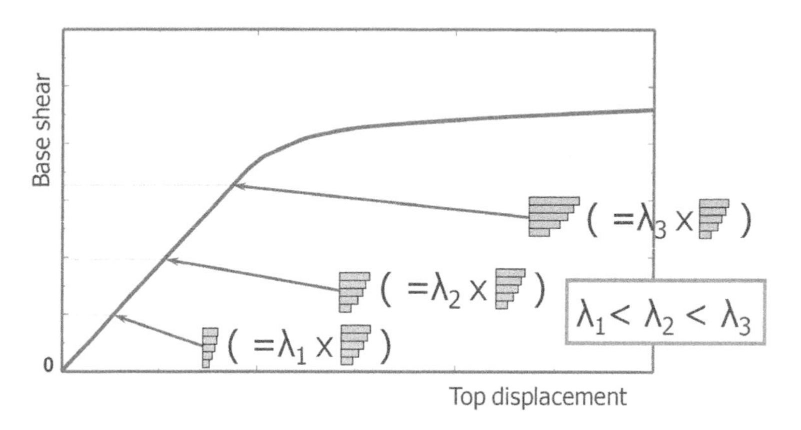


Figure 7.8 Gradual increase of the load factor λ during a pushover analysis. *Source:* Stelios Antoniou.

### 7.6.2 Information Obtained with Pushover Analysis

The purpose of pushover analysis is to assess the structural performance by considering the actual strength and deformation capacities of the structural components and comparing these capacities with the actual demands on these components. It provides crucial information on response parameters that cannot be obtained with conventional elastic methods (either static or dynamic), such as:

- The realistic force demands on potentially brittle elements, such as axial demands on columns, moment demands on beam-to-column steel connections, or shear force demands on short, shear-dominated elements.
- Estimates of the deformation demands of elements that have to deform inelastically, in order to dissipate energy.

- Consequences of the strength deterioration of particular elements on the overall structural stability.
- Identification of the critical regions, where the inelastic deformations are expected to be high.
- Identification of strength irregularities in plan or elevation that cause changes in the dynamic characteristics in the inelastic range.
- Estimates of the interstory drifts, accounting for strength and stiffness discontinuities. In this way, damage on nonstructural elements can be controlled.
- Sequence of the members' yielding and failure and the progress of the overall capacity curve of the structure.
- Verification of the adequacy of the load path, considering all the elements of the system, both structural and nonstructural.

Compared to the elastic procedures, pushover analysis treats inelasticity in a more explicit manner and as it is more "displacement-based," it is more suitable for performance-based engineering. Of course, these benefits come with the additional cost of having to accurately model the inelastic load-deformation characteristics of both structural and nonstructural members, as well as increased computational effort.

### 7.6.3 Theoretical Background on Pushover Analysis

The static pushover analysis has no robust theoretical background. It is based on the assumption that the response of the multi-degree-of-freedom (MDOF) structure is directly related to the response of an equivalent single-degree-of-freedom (SDOF) system with appropriate hysteretic characteristics. This implies that the dynamic response of the MDOF system is determined only by a single mode of vibration, and that the shape $\{\Phi\}$ of that mode is constant, throughout the time-history, regardless of the level of deformation. The reference SDOF displacement $x^*$ is related to the top displacement $x_t$ at the control node of the MDOF system through Eq. (7.6):

$$x^* = c \cdot x_t = \frac{\{\Phi\}^T \cdot M \cdot \{\Phi\}}{\{\Phi\}^T \cdot M \cdot \{1\}} \cdot x_t \tag{7.6}$$

Presuming that the vector $\{\Phi\}$ is known, the $c$ parameter can be calculated. The force deformation characteristics of the equivalent SDOF system can be determined from the results of the nonlinear pushover analysis of the MDOF system, which usually derives a base shear vs. top displacement curve that can be idealized by a bilinear curve with an effective elastic stiffness $K_e = V_y/x_{t,y}$ and a hardening stiffness $K_s = \alpha K_e$ (Figure 7.9).

The force-displacement curve of the SDOF system can be evaluated as follows:

$$x_y^* = \frac{\{\Phi\}^T M \{\Phi\}}{\{\Phi\}^T M \{1\}} x_{t,y} \quad Q_y^* = \{\Phi\}^T \{V\}_y \quad K_{SDOF} = \frac{Q_y^*}{x_y^*} \tag{7.7}$$

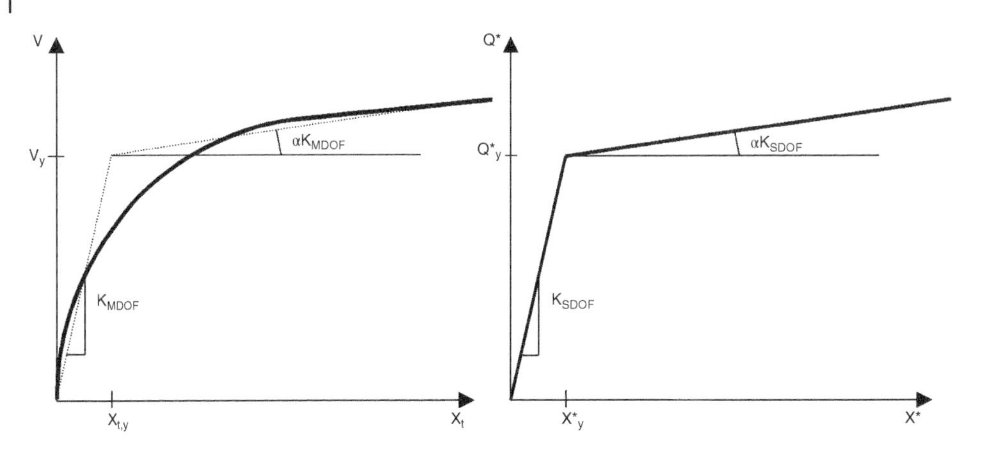


**Figure 7.9** Force-displacement characteristics of the MDOF structure and the equivalent SDOF system. *Source:* Stelios Antoniou.

where $\{V\}_y$ is the story force vector at yield for the MDOF building. The strain-hardening parameter $\alpha$ is assumed to be the same with that of the bilinear approximate curve of the MDOF system.

Although these considerations are apparently incorrect, sensitivity studies have shown that the modification factor $c$ can be considered constant for small to moderate changes in $\{\Phi\}$ and that rather accurate predictions can be attained if the structural response is dominated by the fundamental mode.

The maximum displacement of the SDOF model subjected to the expected ground motion can now be found by means of elastic or inelastic spectra or time-history analysis. The expected deformation level of the MDOF structure can then be estimated by the equation:

$$x_t = x_{\max}^{SDOF} / c = x_{\max}^{SDOF} \cdot \frac{\{\Phi\}^T \cdot M \cdot \{1\}}{\{\Phi\}^T \cdot M \cdot \{\Phi\}} \tag{7.8}$$

Several critical parameters of the procedure are worthy of consideration, namely the target displacement, the shape of the load distribution, as well as its nature (forces or displacements). These issues are discussed in the following sections.

### 7.6.4 Target Displacement

The target displacement of pushover analysis approximates the maximum level of deformation that is expected during the design earthquake. The target displacement always refers to the control node and can be calculated by any procedure that accounts for the effects of nonlinear response on displacement amplitude.

As was explained in the previous section, in pushover analysis it is assumed that the target displacement for the MDOF structure can be estimated from the displacement demand of the equivalent SDOF system, through the use of the selected shape vector $\{\Phi\}$ (that usually corresponds to that of the fundamental mode) and Eq. (7.8).

Therefore, a method is needed to determine the target displacement of the SDOF system. For increased sophistication, dynamic time-history analysis of the SDOF model can be used, assuming simple hysteretic rules. For more practical applications the so-called capacity spectrum method (CSM) can be employed. The CSM (Freeman et al. 1975; Freeman 1998) is a useful and intuitive tool, and follows a simple procedure to correlate structural damage states to amplitudes of ground motion. The method graphically pares the capacity of the lateral force-resisting system of the building (representing the strength) to response spectra values (representing the demand). The values are often plotted in an Acceleration-Displacement Response Spectrum (ADRS) format in which spectral accelerations are plotted against spectral displacements and periods $T$ are represented by radial lines. The procedure is summarized graphically in Figure 7.10.

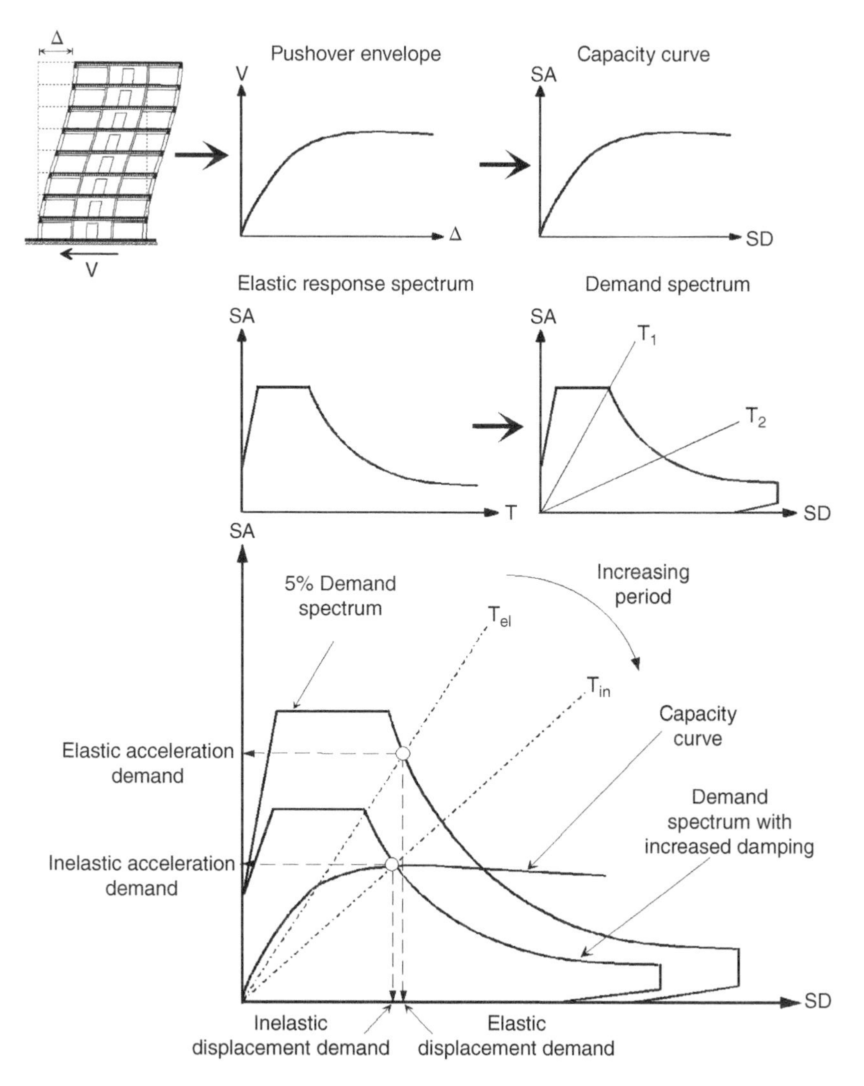


Figure 7.10 Graphical representation of the capacity-spectrum method (Mwafy 2001).

The approach proposed by Eurocode 8 for the calculation of the target displacement is a variation of the CSM. ASCE 41 instead proposes the *coefficient method*, according to which the target displacement is given with the following simplified equation (ASCE 2017):

$$\delta_t = C_0 \cdot C_1 \cdot C_2 \cdot S_a \frac{T_e^2}{4\pi^2} g \tag{7.9}$$

- $T_e$ is the effective fundamental period of the structure in the direction under consideration.
- $C_0$ is a modification factor that relates spectral displacements with the likely roof displacement.
- $C_1$ relates maximum inelastic displacements to displacements calculated for linear response. It corresponds to the ductility demand μ of the SDOF model.
- $C_2$ is a modification factor that represents the effect of the hysteresis shape on the maximum displacement response. It depends on the framing system and the selected performance level.

### 7.6.5 Applying Forces vs. Applying Displacements

Considering earthquake loading as a set of imposed energy input, ground displacements and deformations of the structural members rather than a set of lateral forces seems a much more rational approach for pushover analysis. After all, the fact that earthquake input has been modeled as forces rather than displacements can only be explained by historical reasons, related to the development of contemporary engineering methods in countries of low seismic hazards, like England and Germany, where the most significant actions are the vertical gravity loads. Had modern engineering made its initial step is earthquake-prone regions like New Zealand, California or Southern Europe, today's code provisions would probably be based on deformations. Applying displacement rather than force patterns in the pushover procedures appears to be more appropriate and theoretically correct.

However, displacement-based pushover analysis suffers from significant inherent deficiencies. Due to the constant nature of the applied patterns, it can conceal important structural characteristics, such as strength irregularities, e.g., weak ground stories. This is illustrated by means of a simple example in Figure 7.11, where a five-story simplified stick model with a soft story at the ground level has been pushed to the same target displacement with constant displacement and constant force patterns (triangular distributions). Although the interstory drift at the soft-story during an earthquake is expected to be larger than the other stories, the displacement-based pushover yields equal drifts for all the stories. By contrast, this structural deficiency becomes easily apparent when force patterns are applied.

Consequently, force-based pushover seems a far superior option, since fixing the displacements could yield seriously misleading results. Indeed, all code-based methodologies propose force lateral patterns, either conventional or adaptive. Note that there has been successful research efforts to apply adaptive displacement profiles, which do not remain constant, but rather are updated at each step of the analysis e.g., the DAP algorithm proposed by Antoniou and Pinho (2004b). All the same, such efforts have yet to be introduced into the standards worldwide.

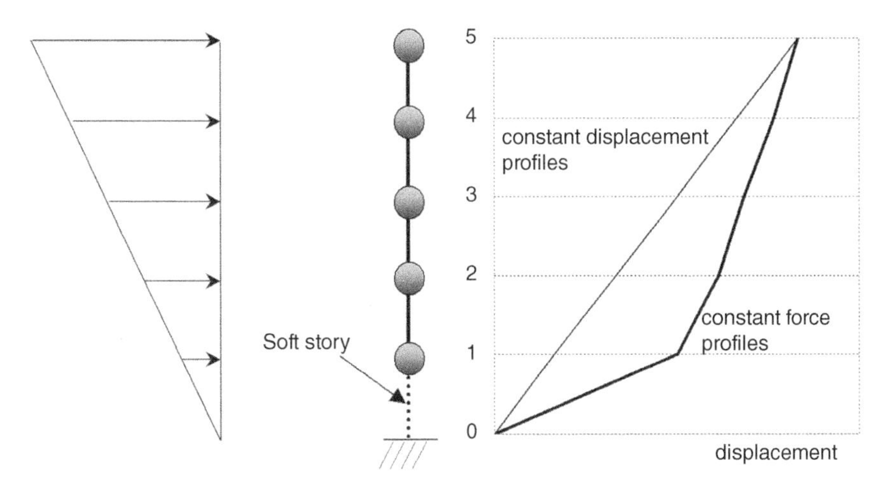


Figure 7.11 Displacement profiles of a five-story simplified model under constant-force and constant-displacement distributions. *Source:* Stelios Antoniou.

### 7.6.6 Controlling the Forces or the Displacements

In pushover analysis, the applied loading usually consists of the permanent gravity loads and the incremental lateral loads in one or both horizontal $X$ and $Y$ directions. The magnitude of the incremental loads $P_i$ at any given analysis step $i$ is given by the product of its nominal value $P_0$ and the load factor $\lambda$ at that step:

$$P_i = \lambda_i P_0 \tag{7.10}$$

The manner in which the load factor $\lambda$ is incremented throughout the pushover analysis (i.e., the loading strategy adopted) is very important. There are two main approaches in the updating of the load factor: load control and displacement control (the latter is also referred to as response control).

#### 7.6.6.1 Load Control

Load control refers to the case where the load factor is directly incremented and the global structural displacements are determined at each load factor level. In this type of loading/solution scheme, the engineer defines the target load multiplier (the factor by which all nominal loads are multiplied) and the number of increments, in which the target load vector is to be subdivided into, for incremental application. The load factor $\lambda$, therefore, varies between 0 and the target load multiplier value, with a step increment $\Delta\lambda 0$ that is equal to the ratio between the target load multiplier and the number of increments.

Although the ability to know in advance the external load at any step of the analysis is attractive, load control has one very significant disadvantage: the applied load factor can only be increased up to the peak of the curve, after which the analysis fails to converge. Hence, the derivation of the descending branch of the pushover curve and the prediction of post peak response is impossible, as depicted in Figure 7.12.

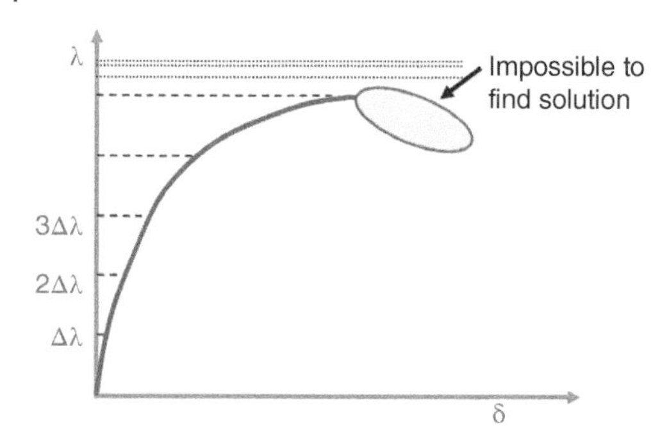


**Figure 7.12** Force controlled pushover analysis and graphical illustration of the inability of force-based schemes to derive the post-peak branch of the capacity curve. *Source:* Stelios Antoniou.

#### 7.6.6.2 Response Control

Response control refers to direct incrementation of the global displacement of one node and the calculation of the loading factor that corresponds to this displacement. In this type of loading/solution scheme, it is not the load vector that is controlled, as in the load control case, but rather the response of a particular node in the structure, commonly the control node, i.e., the center of mass of the top story of the building. The load factor $\lambda$ is not directly incremented, but is instead automatically calculated, so that the applied load vector $P_i = \lambda_i P_0$ at a particular step $i$ corresponds to the attainment of the target displacement at the controlled node at that increment. The analysis finishes when the target displacement is reached.

With this loading strategy, it is possible to (i) capture irregular response features (e.g., soft-story), (ii) capture the softening post-peak branch of the response, and (iii) obtain an even distribution of the force-displacement curve points (Figure 7.13). For these reasons, this type of scheme usually constitutes the best option for carrying out pushover analysis.

### 7.6.7 Control Node

In pushover analysis, the control node is the node, the displacement of which is controlled during the lateral push of the building. It is typically located at the center of mass of the top story or roof of the building. For buildings with a penthouse, the floor (rather than the roof) of the penthouse is regarded as the level of the control node. Mathematically, a floor could be considered a penthouse when its mass is, for instance, lower than 10% of the mass of the floor right below it.

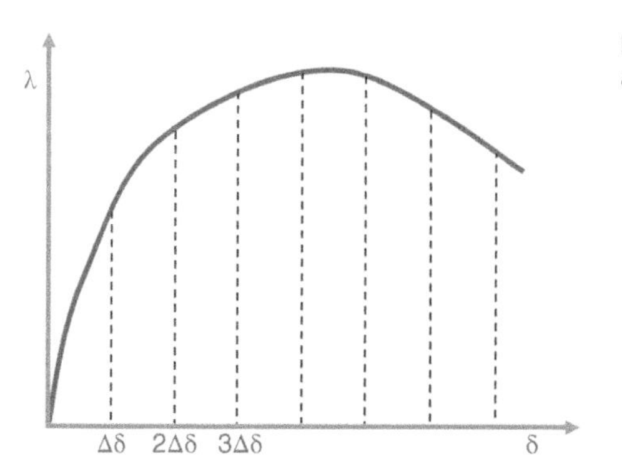


**Figure 7.13** Displacement controlled pushover analysis. *Source:* Stelios Antoniou.

### 7.6.8 Lateral Load Patterns

The lateral load patterns should approximate the inertia forces expected in the building during an earthquake. Even though it is clear that the inertia force distributions will vary with the severity of the earthquake and with time, usually an invariant load pattern is used. This approximation is likely to yield adequate predictions of the element deformation demands for low to medium-rise framed structures, where the structure behavior is dominated by a single mode. However, pushover analysis can be grossly inaccurate for structures of larger periods, where higher mode effects tend to be important.

Since the constant distributions are incapable of capturing such characteristics of the structural behavior under earthquake loading, the use of at least two different patterns is advisable. In most modern codes a uniform and a modal distribution are proposed (ASCE-41 is an exception and proposes the use of only the modal pattern). With the uniform distribution the load profile is proportional to the mass distribution of the building, whereas with the modal distribution the load profile is also proportional to the shape of the fundamental mode (or a combination of the important modes). In some standards, as an alternative to the uniform lateral pattern, a force-based adaptive pushover procedure is also proposed (Antoniou and Pinho 2004a).

Since the direction of the earthquake loading is unknown, pushover analyses at different directions should always be considered, typically loads in the positive and negative, $X$ and $Y$ horizontal directions are considered. Moreover, the effects of bidirectional loading, as well as accidental eccentricity should also be accounted for. In the general case, one could use:

- Uniform or modal distributions
- Uniaxial loading without accidental eccentricity: $\pm X \,\&\pm Y$
- Biaxial loading without accidental eccentricity: $\pm X \pm 0.3Y$ and $\pm Y \pm 0.3X$
- Uniaxial loading with single accidental eccentricity, i.e. $\pm X \pm \mathrm{ecc}Y$ and $\pm Y \pm \mathrm{ecc}X$
- Biaxial loading with double accidental eccentricity, i.e. $\pm(X \pm \mathrm{ecc}Y) \pm (0.3Y \pm \mathrm{ecc}X)$ and $\pm(0.3X \pm \mathrm{ecc}Y) \pm (Y \pm \mathrm{ecc}X)$

If all the variations are applied the number of combinations becomes too large, 168 in total. Since there is no point in running so many analyses, the different codes select just some of these variants. For instance, ASCE-41 proposes the modal distribution, biaxial loading and accidental eccentricity, i.e., 32 combinations in total, although it allows the application of uniaxial loading that leads to 8 combinations. Likewise, Eurocode 8 proposes 16 combinations (uniform + modal, uniaxial with accidental eccentricity).

### 7.6.9 Pushover Analysis Limitations

There are good reasons for using pushover analysis rather than simplified elastic methods for estimating the deformation demands, especially in existing buildings. Moreover, the simplicity of the method makes it a more attractive approach for everyday practice than time-history analysis. Pushover analysis can be an extremely useful tool, if used with caution and engineering judgment. However, it also exhibits some shortcomings and limitations, which should be considered by the engineer, who employs it as a structural assessment tool. These are summarized below:

- The theoretical background of the method is not robust and it is difficult to defend. As mentioned earlier, an important implicit assumption behind pushover analysis is that the response of a multi-degree-of-freedom structure is directly related to an equivalent single-degree-of-freedom system. Although in several cases the response is dominated by the fundamental mode, by no means can this be generalized. Moreover, in dynamic time-history analysis the shape of the fundamental mode itself may vary significantly depending on the level of inelasticity and the locations of damage.
- As a consequence of the previous point, the deformation estimates obtained from a pushover analysis may be highly inaccurate for structures where higher mode effects are significant. The method explicitly ignores the contribution of higher modes to the total response. In the cases where this contribution is significant, the pushover estimates may be quite misleading.
- The progressive stiffness degradation that occurs during the cyclic nonlinear earthquake loading of the structure is not taken into account. This degradation leads to changes in the periods and the modal characteristics of the structure that affect the inertia loading attracted during earthquake ground motion.
- As a static method, pushover analysis concentrates on the strain energy of the structure, neglecting other sources of energy dissipation, which are associated with the dynamic response, such as kinetic and viscous damping energy. Moreover, it neglects duration effects and cumulative energy dissipation.
- Only the horizontal earthquake load is considered. The vertical component of the earthquake loading, which can be of great importance in some cases, is ignored. No method has been proposed so far on how to combine pushover analysis with actions that account for the vertical ground motion.
- A separation between the supply and the demand is implicit in the method. This is clearly incorrect, as the inelastic structural response is load-path dependent and the structural capacity is always associated with the earthquake demand.

Obviously, pushover analysis lacks many important features of dynamic nonlinear analysis, and it will never substitute it as the most accurate tool for structural analysis and assessment. Nevertheless, and despite these problems, the static nonlinear procedure has become the basic method for the assessment and strengthening of existing structures, providing a very good balance between accuracy and simplicity of application.

## 7.7 Nonlinear Dynamic Procedure

According to ASCE 41–17, Section 7.4.3.1, the definition of the nonlinear dynamic procedure, NDP, or nonlinear dynamic time-history analysis is the following:

*A mathematical model directly incorporating the nonlinear load-deformation characteristics of individual components of the building shall be subjected to earthquake shaking represented by ground motion acceleration histories to obtain forces and displacements.*

The objective of the method, which is also often termed *nonlinear time-history analysis* or *nonlinear response analysis*, is to assess the capacity of the structure, considering the deformability, the strength and the hysteretic behavior of all structural members that are subjected to the specified earthquake ground motion.

The NDP constitutes a sophisticated approach for examining the inelastic demands produced on a structure by a specific seismic loading. The basis, the modeling approaches, and the acceptance criteria of the NDP are similar to those for the NSP. One additional complication with respect to the NSP is that now the monotonic force-displacement curves are not sufficient for the structural modeling, and the full hysteretic loading and unloading rules need to be introduced for all the structural members (or at least those that we expect to behave inelastically). These rules should realistically reflect the hysteretic energy dissipation in the element over the range of displacement amplitudes expected in the seismic design situation. Furthermore, the mass distribution and the equivalent viscous damping of the structure should be defined, in order to correctly model the inertia forces that are introduced in the structure from the dynamic vibrations.

Regarding the modeling of the seismic action, instead of the lateral force distributions that are used in the LSP, the LDP, and the NSP, earthquake records are now applied at the foundation level of the building in the form of acceleration time-histories. This accelerogram can be a real recorded seismic action, or an artificial or synthetic record that matches a given target (usually code-defined) spectrum (Figure 7.14).

> Nonlinear dynamic analysis may also be employed for modeling pulse loading cases (e.g., blast, impact, etc.), in which case, instead of acceleration time-histories at the foundation, force pulse functions of any given shape (rectangular, triangular, parabolic, and so on) can be employed to describe the transient loading applied to the appropriate nodes.

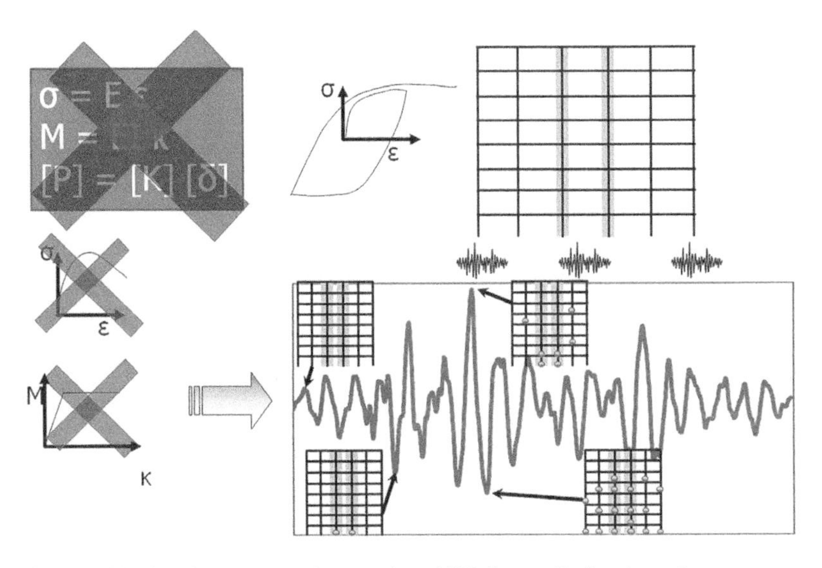


**Figure 7.14** Nonlinear dynamic procedure, NDP. *Source:* Stelios Antoniou.

The direct integration of the equations of motion is accomplished using appropriate integration algorithms, such as the numerically dissipative α-integration algorithm (Hilber-Hughes-Taylor HHT scheme (Hilber et al. 1977)) or a special case of the former, the well-known Newmark scheme (Newmark 1959).

### 7.7.1 Information Obtained with Nonlinear Dynamic Analysis

With dynamic analysis, different building response parameters at the discrete time steps are calculated:

- Deformations at the different story levels, both the absolute and the relative, the latter being of greater importance, since it is a measure of the damage caused to the structure.
- Member action effects and the base shear time-history. Particularly important are the force demands on brittle elements.
- The displacement demand on ductile components.
- The gradual formation of plastic hinges during the entire duration of the time-history.
- Stresses, strains, and curvatures of the frame members.

Contrary to pushover analysis, where the checks are carried out at the particular step of the analysis that corresponds to the target displacement, in dynamic analysis the maxima of the response parameters are checked throughout the time-history (Figure 7.15). If the capacity is exceeded, the acceptance criteria are not fulfilled; if it is not, the acceptance criteria are fulfilled.

### 7.7.2 Selecting and Scaling Accelerograms

Nonlinear dynamic analysis is performed by applying sets of acceleration time-histories at the foundation of the building. According to Section 16.2.2 of ASCE 7–22 (ASCE 2022) a suite of not less than 11 ground motions shall be selected for each target spectrum, whereas Section 4.3.3.4.3 of EC8, Part-1 specifies 7 records.[2] The ground motions consist of pairs of orthogonal horizontal ground motion components and, where vertical earthquake effects are of importance, an additional vertical ground motion component. They should be selected from events with the same tectonic regime and with consistent magnitudes and fault distances as those in the location of the building. Furthermore, they should have similar spectral shapes to the target spectrum for the selected seismic hazard level.

One of the main problems associated with the use of nonlinear dynamic analysis is related to the selection and scaling of accelerograms, which are compatible with the reference spectrum of the region of the building (see Whittaker et al. 2011). The acceleration time histories that can be used in nonlinear dynamic analyses can be natural (scaled and matched or not), simulated or artificial. The simulated records are employed when the required number of recorded ground motions is not available to the designer, and should

2 According to Eurocode 8, when at least seven records are specified, the average response is considered in the checks; when fewer records are considered, the most unfavorable value of the response quantity among the analyses should be employed.

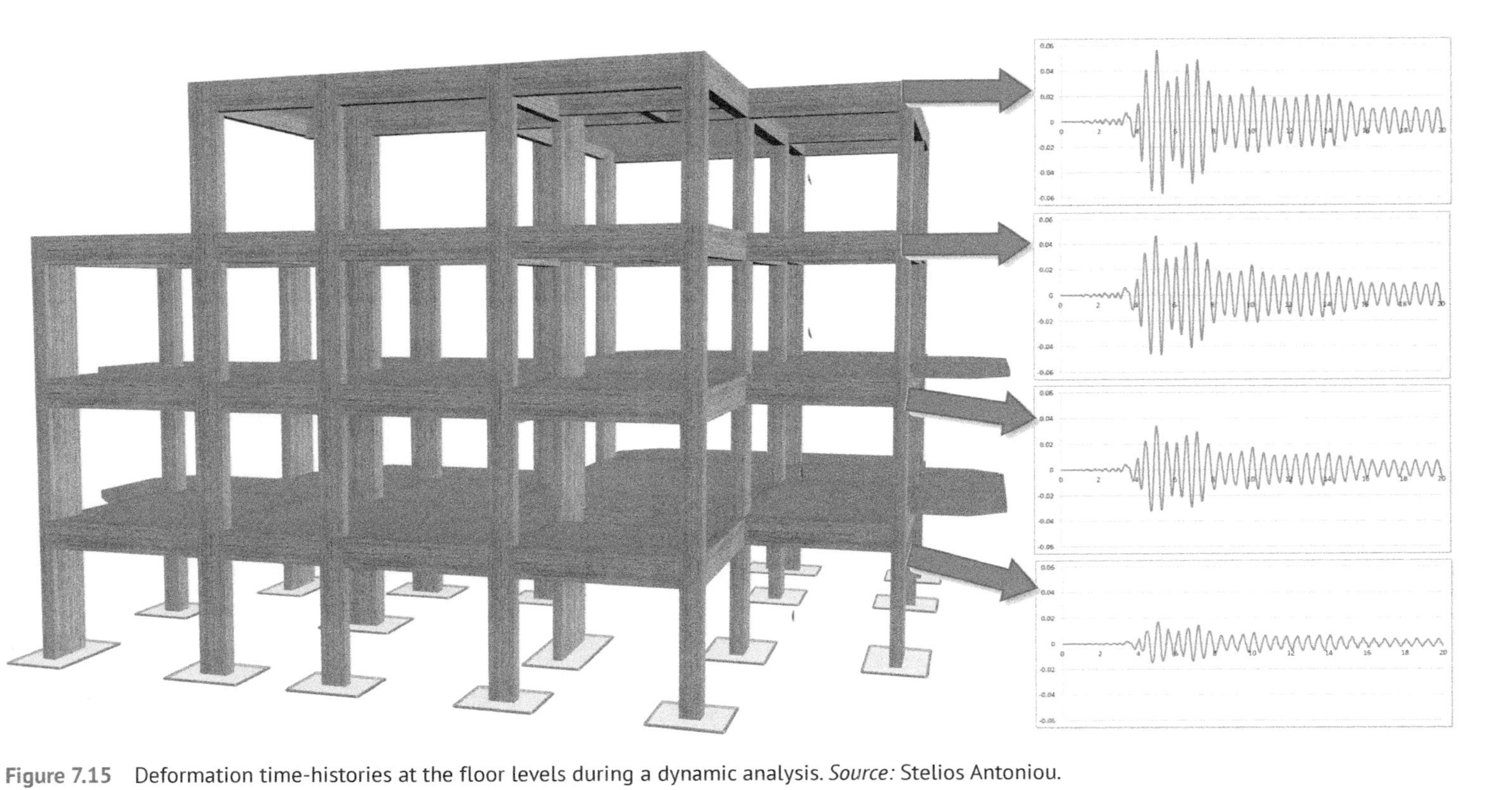

**Figure 7.15** Deformation time-histories at the floor levels during a dynamic analysis. *Source:* Stelios Antoniou.

be consistent with the magnitudes, source characteristics, fault distances, and site conditions of the target spectrum.

When scaling or matching records, the range of important periods, for which a good fit with the target response spectrum should be achieved, is between (i) 20% of the fundamental period or the period of the last mode required to achieve 90% mass participation in each principal horizontal direction, and (ii) twice the fundamental period.

#### 7.7.2.1 Natural Scaled and Matched Accelerograms

These are ground motions that have been recorded during real seismic events in the past. The use of natural accelerograms is generally preferable, compared to simulated records, as it allows for more realistic signals in terms of frequency content, duration, number of cycles, and correlation between the horizontal and vertical components of the seismic motion. Nowadays tens of thousands of records are freely available for download through a large number of natural accelerogram databases (Table 7.3).

The selection of records through one of the databases is based on a series of parameters, such as the magnitude, the epicentral distance, the peak ground acceleration, the duration of the record, the faulting style, the region and period of the seismic event, and, of course, the spectral shape. The selection can be done through the web browser directly on the website of the database (Figure 7.16) or more easily through specialized software packages, such as Seismosoft's SeismoSelect (2023), which can provide better spectrum matching much faster (Figure 7.17).

In order to achieve a better match to the target spectrum, the records are usually either amplitude-scaled or spectrally matched. When a direct amplitude scaling is applied, the same scale factor is employed for all the record components. A maximum-direction spectrum is constructed from the two horizontal components, and the scaling factor is calculated, so that the maximum-direction spectrum generally matches or exceeds the target response spectrum. Specific criteria are provided by the standards, for instance ASCE 7 requires that the average of the maximum-direction spectra from all the ground motions must not fall below 90% of the target response spectrum for any period within the period range of importance.

Alternatively, it is also possible to use specialized software, such as SeismoMatch (2023), which is capable of adjusting the records without significantly altering their natural

**Table 7.3** Some of the best-known strong motion databases worldwide.

| Database | URL |
|---|---|
| PEER Ground Motion Database (includes the NGA-West2 and NGA-East databases) | https://ngawest2.berkeley.edu |
| European Strong Motion Database ver. 2.0 | https://esm-db.eu |
| Itaca | http://itaca.mi.ingv.it |
| RESORCE | http://www.resorce-portal.eu |
| k-net | http://www.k-net.bosai.go.jp |
| Kik-net | http://www.kik.bosai.go.jp |
| COSMOS | https://www.strongmotioncenter.org |

*Source:* Stelios Antoniou.

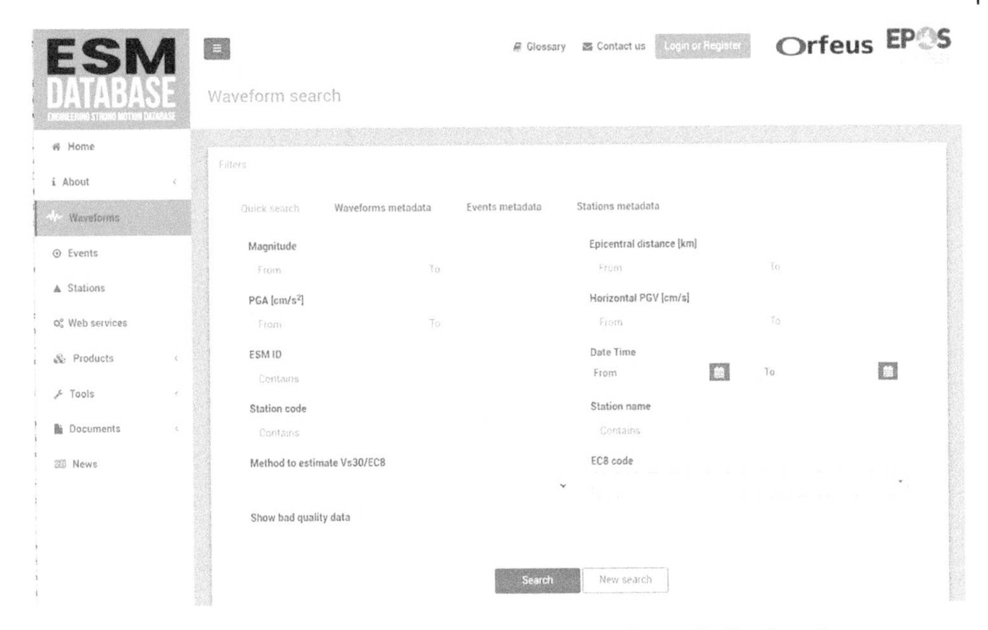


Figure 7.16 Search in the European Strong Motion Database. *Source:* Stelios Antoniou.

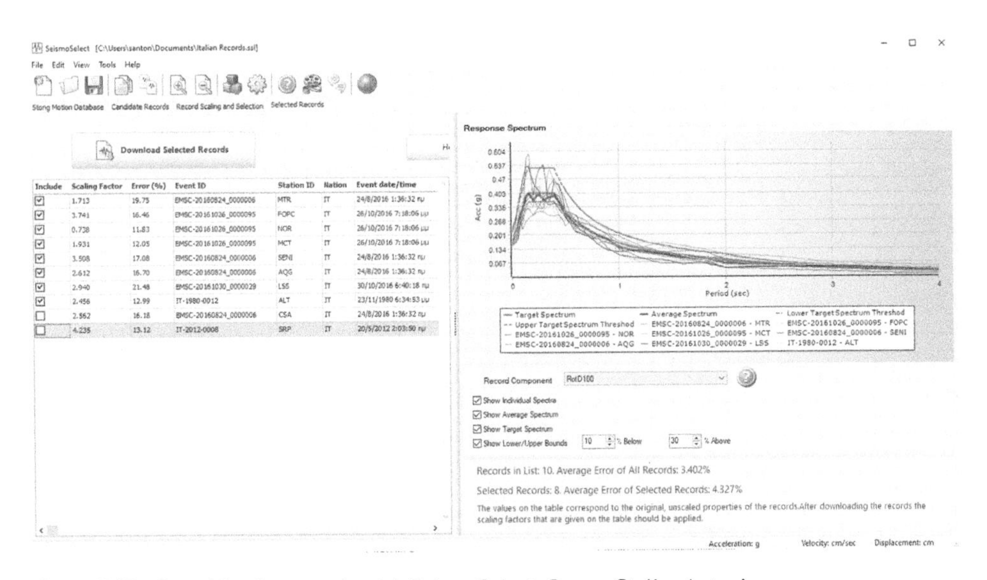


Figure 7.17 Searching for records with SeismoSelect. *Source:* Stelios Antoniou.

characteristics (Figure 7.18). This spectrum-matching process can be performed in the frequency domain, or in the time domain through the addition of "wavelets" (i.e. sinusoidal waveforms) with appropriate frequency contents (Hancock et al. 2008; Abrahamson 1992; Lilhanand and Tseng 1988). It should be noted, however, that spectral matching cannot be used for near-fault records unless the pulse characteristics of the ground motions are retained through the matching process.

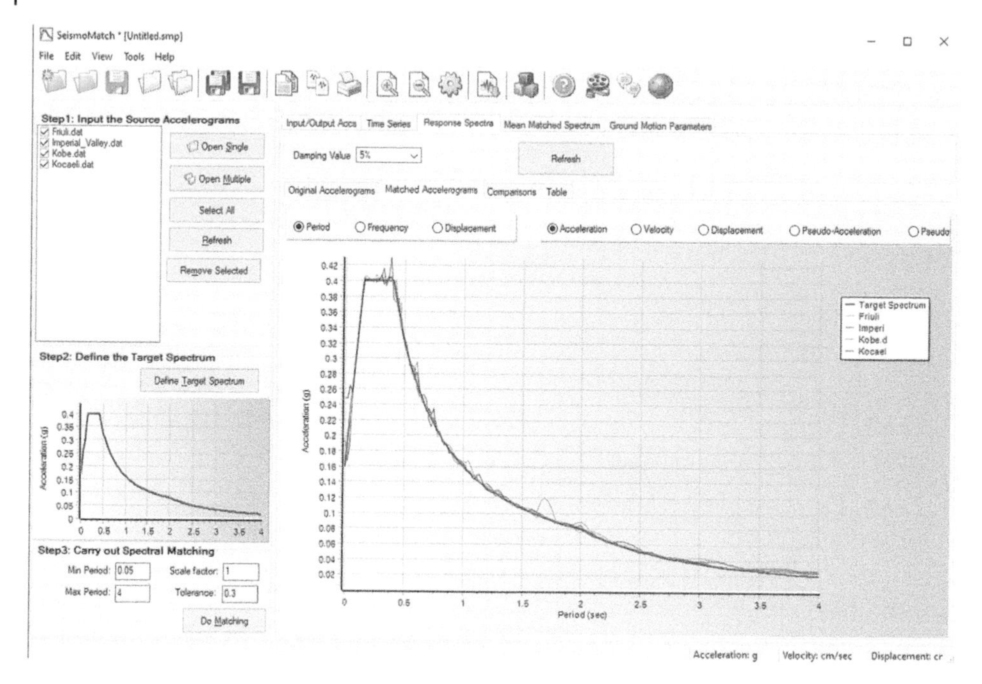


Figure 7.18 Matching records to a target spectrum with SeismoMatch. *Source:* Stelios Antoniou.

#### 7.7.2.2 Artificial and Synthetic Accelerograms

The generation of *artificial accelerograms* is based on a starting random process that is gradually adapted to a target spectrum, through the use of a selected envelope shape for the time-history and a power spectral density function.

The obvious advantage in the use of artificial accelerograms is to obtain a time series compatible with the elastic response spectrum without the need for an existing natural record, as in the case of spectral matching. The downside is that it is generally a conservative approach, and the response generated is not necessarily representative of a real seismic event, as the energy content is unreasonably high. Therefore, engineering judgment and experience is required for the assessment of the appropriateness of the generated accelerogram, since very often the generated records' characteristics differ significantly from real records.

*Synthetic accelerograms* instead are time histories that are generated with some basic (or more extended) knowledge of the earthquake history and the soil conditions relative to the region and site of interest. They are based on geophysical models that take into account the seismo-tectonic environment, the type of faulting, the generation and propagation of the seismic wave field, as well as site effects. Synthetic accelerograms tend to appear realistic (more realistic than artificial accelerograms) and are certainly a very valid tool for the derivation of records without adapting an existing one.

Different packages exist for the generation of artificial accelerograms, e.g., SeismoArtif (2023). Note that, contrary to the Eurocodes framework, ASCE 41 and ASCE 7 do not propose artificial and synthetic accelerograms as an option (Figure 7.19).

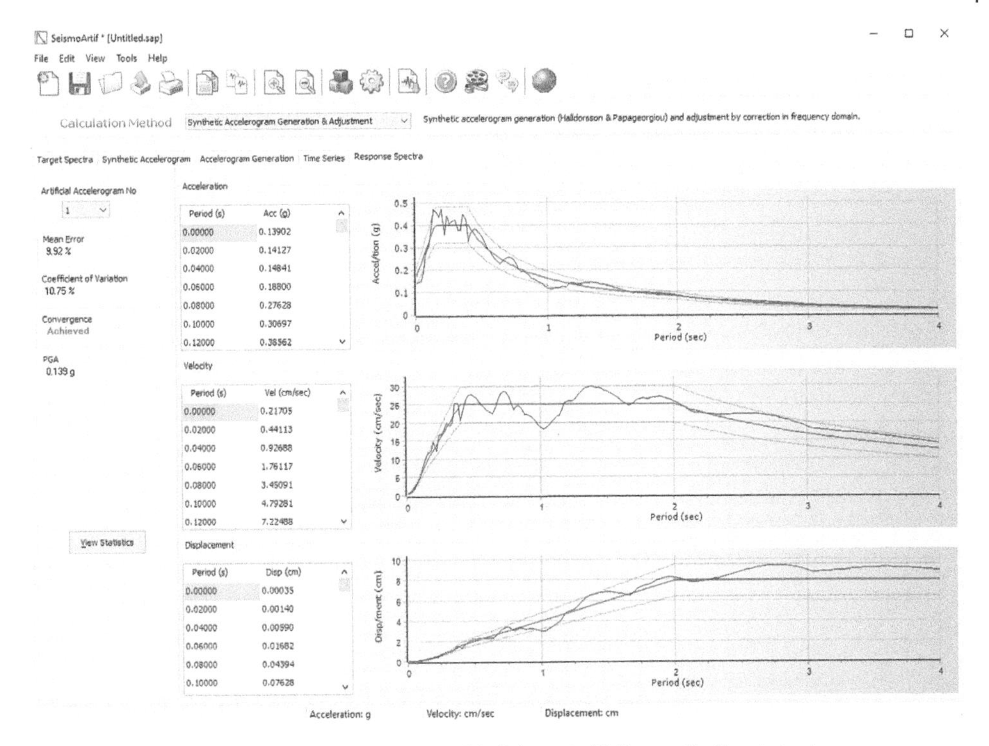


Figure 7.19 Derivation of artificial records with SeismoArtif. *Source:* Stelios Antoniou.

## 7.7.3 Advantages and Disadvantages of Nonlinear Dynamic Analysis

The NDP constitutes a sophisticated approach for examining the inelastic demands produced on a structure by a specific suite of ground motion acceleration time-histories. It is the most numerically advanced method of analysis, and it is considered the most accurate in the representation of the dynamic nature of seismic loading.

Since nonlinear dynamic analysis involves fewer assumptions than the nonlinear static procedure, it is subject to fewer limitations. It automatically accounts for higher-mode effects and shifts in inertial load patterns as structural softening occurs. In addition, it provides reliable results even for highly irregular structures, or for irregular seismic action (e.g., near-fault ground motion or loading in two or three directions simultaneously). Furthermore, for a given earthquake record, this approach directly solves for the maximum global displacement demand produced by the earthquake on the structure, eliminating the need to estimate this demand, based on general relationships and assumptions (e.g., as in the calculation of the target displacement).

As a result, the NDP is the only method that can be used for any structural configuration and any type of loading. In practice, we can analyze any building, subjected to any type of seismic action, with adequate accuracy.

Unfortunately, the increased accuracy that the NDP provides does not come without a cost, since it is time-consuming and expensive in terms of computer resources. A dynamic analysis can last up to 5 or 10 times longer than a pushover analysis for the same building, let alone the fact that the analysis needs to be carried out with multiple ground motion records, due to the inherent variability in earthquake ground motions.

Furthermore, the NDP requires non-negligible judgment and experience, and should only be used when the engineer is thoroughly familiar with nonlinear dynamic analysis techniques and limitations. For instance, there are always difficulties in the selection of the seismic motion (accelerograms), which is influenced by several factors such as the expected seismic event (magnitude, mechanism, depth), the distance from the source, the expected PGA and PGV, the geology between the source and the structure, as well as local phenomena, and the geology of the region of the structure.

Moreover, specialized knowledge is often needed for the interpretation of the results since the calculated response can be highly sensitive to characteristics of individual ground motions or small changes in the input and the initial modeling assumptions. For instance, two ground motion records enveloped by the same response spectrum can produce radically different responses both at the global and local level, with distinctive differences in the distribution and the amount of inelasticity throughout the structure.

Because of these complications, the NDP is not the first choice for the assessment and strengthening of structures in the average design office, and its use is only economically justified for large and important projects.

## 7.8 Comparative Assessment of Analytical Methods

### 7.8.1 Advantages and Disadvantages of the Analytical Methods

*Linear Static Procedure*: The LSP is the most basic method of the four with many approximations and very limited accuracy, even for relatively simple structural configurations. It is only allowed for small symmetric buildings, and it is employed in a conservative manner with large safety factors. In general, it should be avoided for everyday application, but for the most simple and regular buildings.

*Linear Dynamic Procedure*: Because the loading is calculated through the combinations of several modes (including higher ones), the LDP is suitable for tall and asymmetric buildings, where higher mode effects are of importance. However, as an elastic method, it inherently has limited accuracy in the case of large inelastic deformations, which are very common in existing buildings under large earthquake loading. Hence, the results can be highly inaccurate when applied to buildings with highly irregular structural systems, unless the building is capable of responding almost elastically at the selected seismic hazard level. Similarly to the LSP, it is employed conservatively with higher safety margins, in comparison with the nonlinear methods.

*Nonlinear Static Procedure*: Due to the explicit modeling of inelasticity, the NSP is very suitable when large inelastic deformations are expected. In such cases, the structural response can be modeled with satisfactory accuracy, allowing a less conservative approach and lower safety factors. The NSP is generally a more reliable approach for characterizing the performance of a structure than the linear procedures. However, it is not exact and cannot accurately account for changes in dynamic response as the structure degrades, and it is not suitable, when higher-mode effects are of importance, e.g., with taller buildings (more than 10–15 floors). In general, the NSP is a valid approach for the seismic assessment of existing buildings; however, it should be used with caution, when the structural response is determined by more than one mode.

*Nonlinear Dynamic Procedure*: It is by far the most sophisticated method for structural analysis. It is more accurate than the NSP in that it avoids some of the approximations made by the latter. The NDP is able to model both the inelastic material behavior and higher mode effects for a given earthquake record. It directly provides the maximum global displacement demand produced by the earthquake on the structure, eliminating the need for approximations, and it is generally suitable for any structural configuration and any earthquake loading. However, the main disadvantage of the method is a significant one: it is difficult to use, and specialized knowledge is always required (e.g., for the selection of suitable accelerograms, or the interpretation of results).

In the comparison between the methods of analysis, it is mentioned that being conservative is a negative, rather than positive, characteristic of the linear methods. At first sight, this comment seems a bit odd. Should we not be conservative when retrofitting existing buildings that are highly vulnerable?

Undoubtedly, we would all like to inhabit safer buildings, and we prefer to be on the safe side in our calculations. The problem, however, with the linear methods is that many times their requirements become very strict, and end up in interventions that are extremely expensive, and impose serious restrictions on the operation of the building. Very frequently, the engineer is left with two options, none of which is appealing: the demolition and the reconstruction of the building or to do nothing (from experience, the "do nothing" approach is what building owners choose when the required works tend to become extensive).

In such cases the use of nonlinear methods, together with a better knowledge of the building can be very advantageous, and lead to less invasive interventions.

### 7.8.2 Selection of the Best Analysis Procedure for Structural Assessment

As discussed in the previous sections, because of the inherent limitations of the linear methods, in order to be able to use them with the structural assessment methodologies one needs regularity in the geometry, both in plan and elevation, and a roughly uniform distribution of damage in the building. However, in the vast majority of existing buildings, especially those that have been built in the 1970s or before without explicit seismic provisions, it is almost impossible to fulfill these criteria. In the typical case, such buildings have significant irregularities in plan and in elevation, and they are expected to experience serious damage in the design earthquake with significant inelastic deformations at specific parts of the building (e.g., weak ground story).

Because those weak buildings are the ones that require the assessment of their seismic response and strengthening more so than others, the nonlinear methods have become more common for this purpose. In particular, the NSP is gradually becoming the "standard" methodology for assessment and retrofit, because of the simplicity in its application. In the cases of structures that have significant higher mode response, the NSP may be used together with the LDP for the verification of the adequacy of the evaluation or the retrofit.

Where this approach is taken, less restrictive criteria are permitted for the LDP, because it is recognized that improved knowledge is obtained by performing both analytical procedures.

Note that the potential of the NSP in the assessment and retrofit of existing structures is indirectly acknowledged by all the modern assessment standards (in particular ASCE-41 and EC8 Part-3), since the description of the method is covered in considerably more detail, compared to the other three methods. In particular, the NDP, despite its undeniable technical superiority, is described very briefly, as it is implicitly recognized that it still has far too many complications to be extensively used in everyday practice by smaller design companies and for small to medium-sized projects.

Last but not least, remember that it is generally preferable to carry out extensive testing in order to achieve better knowledge of the structure and to make use of the less conservative nonlinear methods together with lower safety factors. In this way, it is much easier to fulfill the acceptance criteria for all structural members.

## References

[ACI] American Concrete Institute ACI 318 (2019). *ACI CODE-318-19: Building Code Requirements for Structural Concrete and Commentary. ACI Committee 318*. American Concrete Institute.

[ASCE] American Society of Civil Engineers (2017). *Seismic Evaluation and Retrofit of Existing Buildings (ASCE/SEI 41–17)*. Reston, Virginia: ASCE.

[ASCE] American Society of Civil Engineers (2022). Minimum Design Loads and Associated Criteria for Buildings and Other Structures. (ASCE/SEI 7–22). Reston, Virginia: ASCE.

[ATC] Applied Technology Council (1996). *Seismic Evaluation and Retrofit of Concrete Buildings, ATC-40 Report*. Redwood City, California: Applied Technology Council.

[ATC] Applied Technology Council (2009). *Guidelines for Seismic Performance Assessment of Buildings, ATC 58, 50 % Draft Report*. Redwood City, CA: Applied Technology Council.

[FEMA] Federal Emergency Management Agency (1997). *NEHRP Guidelines for the Seismic Rehabilitation of Buildings, FEMA 273 Report*. Washington, D.C.: Applied Technology Council and the Building Seismic Safety Council for the Federal Emergency Management Agency.

[FEMA] Federal Emergency Management Agency (2005). *Improvement of Nonlinear Static Seismic Analysis Procedures, FEMA 440 Report*. Washington, DC: Applied Technology Council for the Federal Emergency Management Agency.

[FEMA] Federal Emergency Management Agency (2009a). *Effects of Strength and Stiffness Degradation on Seismic Response, FEMA P-440A. Report*. Washington, DC: Applied Technology Council for the Federal Emergency Management Agency.

[FEMA] Federal Emergency Management Agency (2009b). *HAZUS®-MH MR5 Advanced Engineering Building Module (AEBM) Technical and User's Manual*. Washington, DC: National Institute of Buildings Sciences for the Federal Emergency Management Agency.

[NIST] National Institute of Standards and Technology (2010). *Nonlinear Structural Analysis for Seismic Design, A Guide for Practicing Engineers, GCR 10–917-5*. Gaithersburg, Maryland: NEHRP Consultants Joint Venture, a partnership of the Applied Technology Council and the Consortium of Universities for Research in Earthquake Engineering, for the National Institute of Standards and Technology.

[PEER] Pacific Earthquake Engineering Research Centre (2010). *Tall Buildings Initiative: Guidelines for Performance-Based Seismic Design of Tall Buildings, PEER Report 2010/05*. Berkeley, California: Pacific Earthquake Engineering Research Center.

Abrahamson, N.A. (1992). Non-stationary spectral matching. *Seismological Research Letters* 63 (1): 30.

Antoniou, S. and Pinho, R. (2004a). Advantages and limitations of force-based adaptive and non-adaptive pushover procedures. *Journal of Earthquake Engineering* 8 (4): 497–522.

Antoniou, S. and Pinho, R. (2004b). Development and verification of a displacement-based adaptive pushover procedure. *Journal of Earthquake Engineering* 8 (5): 643–661, Imperial College Press.

Antoniou, S. and Pinho, R. (2018). Nonlinear seismic analysis of framed structures. In: *Engineering Dynamics and Vibrations: Recent Developments* (ed. J. Jia and J.K. Paik). Boca Raton, Florida: CRC Press.

Bathe, K.J. (1996). *Finite Element Procedures in Engineering Analysis*, 2e. Prentice Hall.

Chopra, A.K. (1995). *Dynamics of Structures: Theory and Applications to Earthquake Engineering*. New Jersey: Prentice-Hall.

Cook, R.D., Malkus, D.S., and Plesha, M.E. (1989). *Concepts and Applications of Finite Elements Analysis*. Wiley.

Crisfield, M.A. (1991). *Non-linear Finite Element Analysis of Solids and Structures*. Wiley.

Felippa, C.A. (2004). *Introduction to Finite Element Methods. Department of Aerospace Engineering Sciences and Center for Aerospace Structures*. Boulder, Colorado 80309-0429, USA: University of Colorado.

Freeman, S.A. (1998). Development and use of Capacity Spectrum Method. In: *Proceedings, Sixth U.S. National Conference on Earthquake Engineering* [computer file]. Oakland, California: Earthquake Engineering Research Inst. 12 pages.

Freeman, S.A., Nicoletti, J.P., and Tyrell, J.V. (1975). Evaluation of existing buildings for seismic risk – A case study of Puget Sound Naval Shipyard Bremerton, Washington. In: *Proceedings of the United States National Conference on Earthquake Engineering*, 113–122. Berkeley.

Hancock, J., Bommer, J.J., and Stafford, P.J. (2008). Numbers of scaled and matched accelerograms required for inelastic dynamic analyses. *Earthquake Engineering and Structural Dynamics* 37: 1585–1607.

Hilber, H.M., Hughes, T.J.R., and Taylor, R.L. (1977). Improved numerical dissipation for time integration algorithms in structural dynamics. *Earthquake Engineering and Structural Dynamics* 5 (3): 283–292.

Kircher, C.A., Nassar, A.A., Kustu, O., and Holmes, W.T. (1997a). Development of building damage functions for earthquake loss estimation. *Earthquake Spectra* 13 (4): 663–682.

Kircher, C.A., Reitherman, R.K., Whitman, R.V., and Arnold, C. (1997b). Estimation of earthquake losses to buildings. *Earthquake Spectra* 13 (4): 703–720.

Lilhanand, K. and Tseng, W.S. (1988). Development and application of realistic earthquake time histories compatible with multiple-damping design spectra. In: *Proceedings of the 9th World Conference on Earthquake Engineering*, vol. II, 819–824. Tokyo, Japan.

Mwafy, A.M. (2001). Seismic performance of code-designed RC buildings. PhD thesis. Department of Civil and Environmental Engineering, Imperial College of Science, Technology and Medicine, London.

Newmark, N.M. (1959). A method of computation for structural dynamics. *Journal of the Engineering Mechanics Division, ASCE* 85 (EM3): 67–94.
PEER/ATC (2010). Modeling and acceptance criteria for seismic design and analysis of tall buildings. PEER/ATC 72-1 Report, Applied Technology Council, Redwood City, CA.
Rosenblueth, E. (1951). A basis for a Seismic Design. PhD thesis. University of Illinois, Urbana, USA.
SeismoArtif (2023). SeismoArtif– A computer program for generating artificial earthquake accelerograms matched to a specific target response spectrum. www.seismosoft.com.
SeismoMatch (2023). SeismoMatch – A computer program for spectrum matching of earthquake records. www.seismosoft.com.
SeismoSelect (2023). SeismoSelect – A computer program for the selection, scaling and download of ground motion data. www.seismosoft.com.
SeismoStruct (2023). SeismoStruct - A computer program for static and dynamic nonlinear analysis of framed structures. www.seismosoft.com.
Whittaker, A. , Atkinson, G. , Baker, J. et al. (2011). Selecting and Scaling Earthquake Ground Motions for Performing Response-History Analyses, Grant/Contract Reports (NIST GCR 11-917-15). National Institute of Standards and Technology, Gaithersburg, MD, [online], https://tsapps.nist.gov/publication/get_pdf.cfm?pub_id=915482 (accessed May 27, 2022)
Willford, M., Whittaker, A.S., and Klemencic, R. (2008). Recommendations for the seismic design of high-rise buildings: Draft for comment 1. Council on Tall Buildings and Urban Habitat (CTBUH), Chicago, IL.
Zienkiewicz, O.C. and Taylor, R.L. (1991). *The Finite Element Method*, 4e. McGraw Hill.

# 8

# Structural Modeling in Linear and Nonlinear Analysis

## 8.1 Introduction

Linear analysis is relatively simple. The model should adequately represent the distribution of the elastic structural stiffness and the mass, in order to account for all the significant structural deformation shapes and inertia forces that are activated during the earthquake vibration. However, with nonlinear methods (which have gradually become the standard for all the assessment procedures, as explained in the previous chapter), the analytical procedure is more complicated. Apart from the mass and the stiffness, the strength and, in the case of the NDP, the nonlinear hysteretic response of the structural members should also be modeled. What is more, during nonlinear analysis, both the strength and the stiffness of the components that behave inelastically degrade, causing a gradual change in the basic structural characteristics and the global structural behavior. Nonlinear analysis has additional complications that need to be understood by engineers in order to avoid dangerous pitfalls during the modeling process.

This chapter discusses general issues regarding the modeling of reinforced concrete structures, with a particular focus on nonlinear analysis. The ways of modeling the different structural members (columns, beams, and walls) will be explained, considering also the material inelasticities and geometric nonlinearities, which constitute an integral part of the process. Advisable and avoidable practices that designers can follow will be described, and some specific subjects will be discussed, such as core walls, slabs, stairs, infills, beam-columns joints, and foundations. Furthermore, some advanced issues such as shear deformations, bar slippage, and lap splices will be presented.

## 8.2 Mathematical Modeling

In structural analysis, the model of the building should adequately represent the distribution of stiffness and mass in it, in order to correctly identify all the significant deformation shapes, and to properly estimate the inertia forces derived from the seismic action. Furthermore, in the case of nonlinear analysis, the model should also represent the distribution of strength, as well as its gradual deterioration with increased deformations.

*Seismic Retrofit of Existing Reinforced Concrete Buildings*, First Edition. Stelios Antoniou.

The structure can be considered to consist of a number of vertical, horizontal, and inclined load resisting components, connected with each other and with horizontal diaphragms that represent the in-plane rigidity of the slabs. Sometimes, the model should also account for the contribution of joint regions (end zones in beams or columns) to the deformability of the building, as well as nonstructural elements that can considerably influence the structural response (e.g., infills).

Although under certain (quite strict) prerequisites of regularity two planar structural models (one for each main direction) are allowed by most codes, nowadays the standard practice is the use of three-dimensional assemblies of all the building's components. The increase in computer power in recent years has made the use of smaller models, in order to reduce analysis time, simply pointless. It is much easier for the analyst to cope with a single model, rather than two or more. What is more, the three-dimensional model provides much better analytical predictions of the actual structural response, since it accounts for the torsional response around the vertical *Z*-axis and the interaction of frames along the two horizontal directions.

Within this 3D framework, the modeling is typically done with the use of line (one-dimensional) finite elements (e.g., beams or rods) that connect structural nodes with six degrees of freedom each. Two- and three-dimensional finite elements are rarely utilized, especially in the nonlinear case, due to the very heavy computational burden and also the lack of reliable and numerically stable constitutive models for 2D and 3D members.

Regarding the modeling of the building's mass, in the case of rigid diaphragms, the masses and the moments of inertia of each floor can be lumped at the center of gravity. All the same, usually a more accurate distribution of the structural masses is taken into account with explicit modeling of the components' masses.

## 8.3 Modeling of Beams and Columns

The accurate modeling of the mechanical properties of structural members is a complex and wide-ranging subject. In linear analysis, it is sufficient to assume that the material remains linear and elastic (i.e., that deformations are fully reversible and the stress is a unique function of strain). For the representation of the force-deformation response of the member, only the geometry and the elastic section parameters (EA, EI2, EI3, GJ) are needed.

Such a simplified assumption is appropriate only within a limited range, and is gradually being replaced by more realistic approaches. As explained in the previous chapter, in nonlinear procedures, the basic rules of elasticity do not apply anymore. The primary source of nonlinearities in low- and medium-rise building structures is material inelasticity and plastic yielding in the locations of damage. In larger high-rise buildings, while material inelasticity still plays an important role, large deformations relative to the frame element's chord (known as P-delta effects), and geometric nonlinearities become equally important and should be taken into account.

### 8.3.1 Material Inelasticity

Material nonlinearities occur when the stress–strain or force-displacement law is not linear, or when material properties change with the applied loads. Contrary to linear analysis procedures, where the material stresses are always proportional to the corresponding strains and fully elastic behavior is assumed, in nonlinear analysis the material behavior depends on the current deformation state and possibly on the past history of the deformation. In order to estimate the stress from the strain of a particular location of the structure, complete expressions for the uniaxial stress–strain relationship of the material should be provided, including hysteretic rules for unloading and reloading.

The source of such inelasticity is defined at the sectional level, through the creation of a fiber model for the section. A fiber section consists in the subdivision of the area in $n$ smaller areas, each of which is attributed a material stress–strain relationship, i.e., for reinforcing steel and concrete, confined and unconfined. After defining the material laws of every material of the section, and calculating the stresses at the fibers, the sectional moment-curvature state of the element is then obtained through the integration of the nonlinear uniaxial stress–strain response of the individual fibers. The discretization of a typical reinforced concrete cross-section is depicted in Figure 8.1.

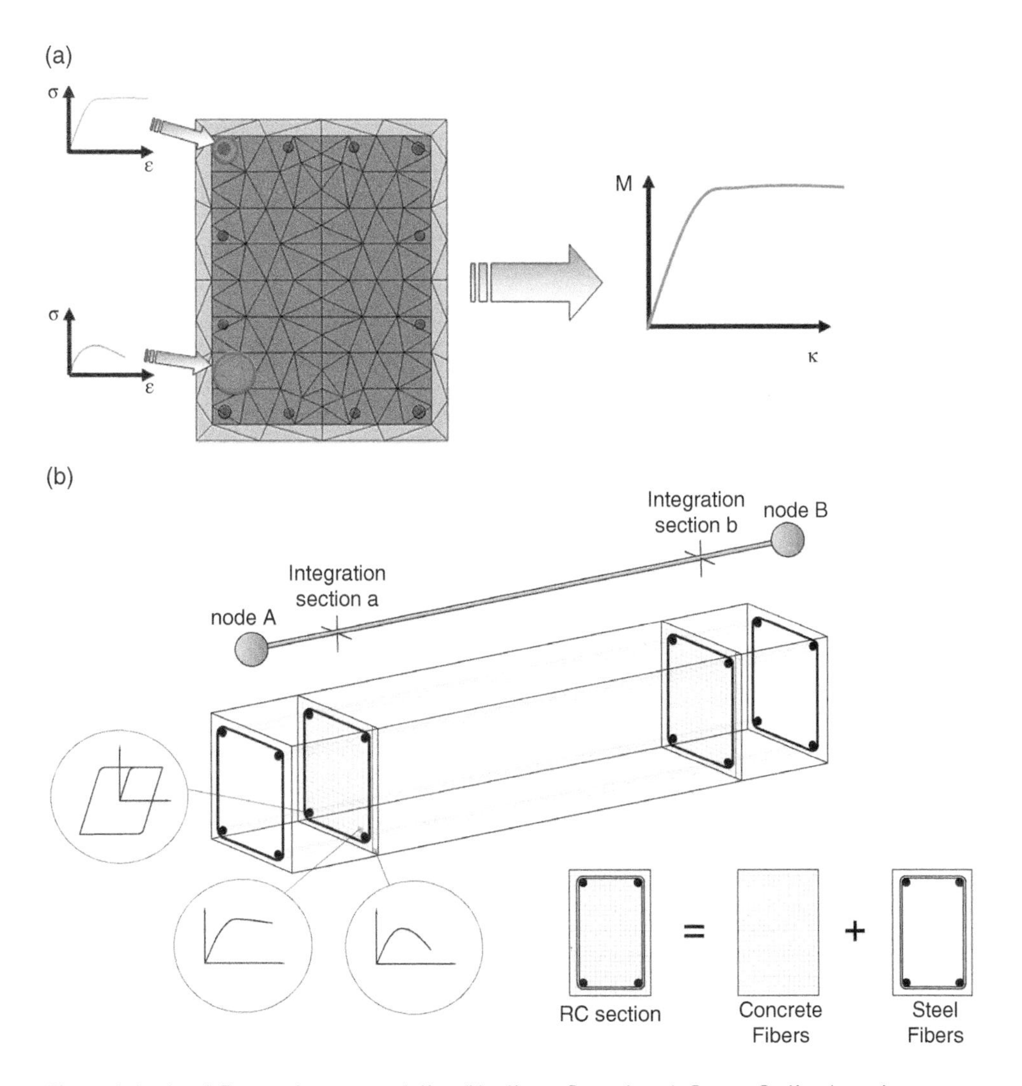


Figure 8.1 (a–c) Frame element modeling (Nonlinear Procedures). *Source:* Stelios Antoniou.

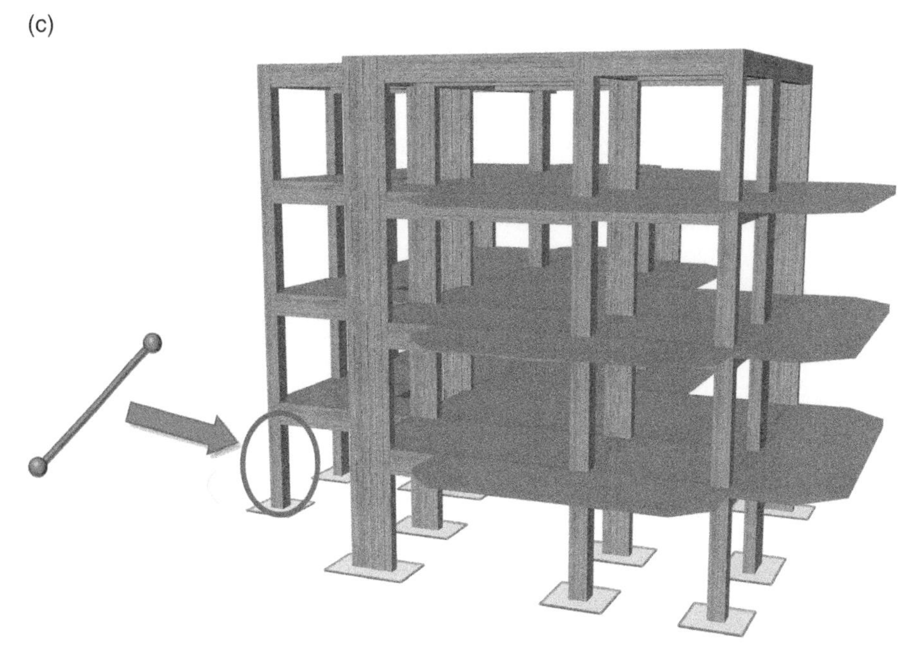


Figure 8.1 (Continued)

Estimating the inelastic response of the structural member requires the integration of the stresses that have been calculated at appropriately selected integration cross-sections along the member (called Gauss sections a and b, in Figure 8.1). The integration provides the tangent stiffness matrix of the element at the current analysis step, as well as its inelastic internal forces, which are then included in the global stiffness matrix and the global forces vector for the solution of the equations of the system.

### 8.3.2 Geometric Nonlinearities

Geometric nonlinearities involve nonlinearities in kinematic quantities, and they occur due to large displacements, large rotations, and large independent deformations relative to the frame element's chord (also known as P-delta effects).

The effect of geometric nonlinearities on the response of structures can range from negligible, in cases where large deformations are not expected, to extreme, in large and slender structures. In the general case, geometric nonlinearities must be modeled as they can ultimately lead to loss of lateral resistance, ratcheting (a gradual buildup of residual deformations under cyclic loading), and dynamic instability. Large lateral deflections magnify the internal force and moment demands, causing a decrease in the effective lateral stiffness. With the increase of internal forces, a smaller proportion of the structure's capacity remains available to sustain lateral loads, leading to a reduction in the effective lateral strength.

For the numerical simulation of geometric nonlinearities and the inclusion of its effects in the analysis, the most advanced formulation is a total co-rotational formulation (Correia and Virtuoso 2006), which is based on an exact description of the kinematic transformations

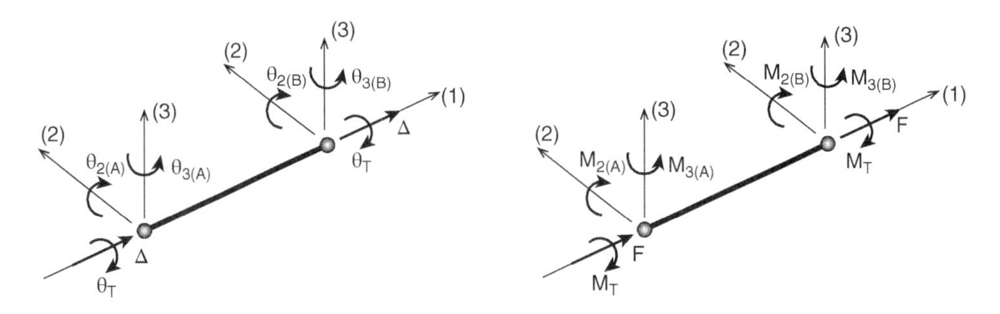


**Figure 8.2** Local chord system of a beam-column frame element. *Source:* Stelios Antoniou.

associated with large displacements and three-dimensional rotations of the frame member. This leads to the correct definition of the element's independent deformations and forces, as well as to the natural definition of the effects of geometrical nonlinearities on the stiffness matrix.

In the local chord system of the frame element, six basic displacement degrees-of-freedom ($\theta 2(A)$, $\theta 3(A)$, $\theta 2(B)$, $\theta 3(B)$, $\Delta$, $\theta T$) and corresponding element internal forces ($M2(A)$, $M3(A)$, $M2(B)$, $M3(B)$, $F$, $MT$) are defined, as shown in Figure 8.2.

According to all modern codes, the geometric nonlinearities should be considered in the analysis, directly in the structural model (in nonlinear analysis), or via simplified assumptions (in linear analysis). The inclusion of geometric nonlinearities in the structural model is the default option in most finite element packages, and from the analyst's perspective it is easy to activate them (in most cases, the only thing to do is to select this option from the program settings).

### 8.3.3 Modeling of Structural Frame Elements

Since the introduction of nonlinear analysis, various types of structural component models have been introduced to simulate beam-column behavior. The models vary in sophistication, computational resources required, and accuracy in the representation of the nonlinear response. The main ways to model the nonlinear response of a structural member are depicted in Figure 8.3.

The most basic models are shown on the left, and are the so-called plastic hinge models. They are concentrated plasticity models, in which all of the nonlinear effects are lumped into an inelastic spring, whereas the rest of the element remains linear and elastic. On the right of Figure 8.3 is a detailed continuum finite element model with explicit nonlinear rules for the representation of the response of its three-dimensional components. Between the two extremes, there are various types of distributed plasticity fiber elements that provide hybrid representations of the structural behavior.

As explained in NIST (2013):

> To some extent, all of the models have a phenomenological basis, because they all ultimately rely on mathematical models that are calibrated to simulate nonlinear phenomena observed in tests. However, the concentrated plasticity models rely

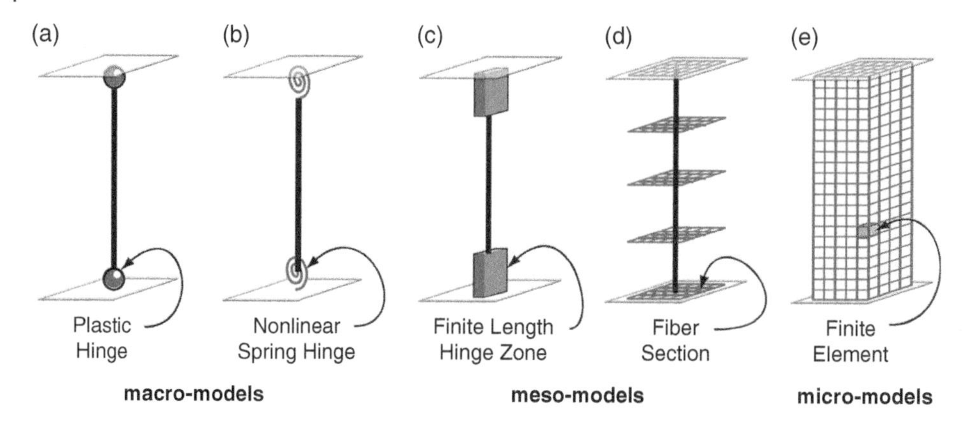


Figure 8.3 Types of structural component models: (a) Plastic hinge with zero length, (b) Nonlinear spring hinge with zero length, (c) Plastic hinge with finite length, (d) Fiber modelling and (e) Finite element modelling (NIST 2010).

almost exclusively on phenomenological representation of the overall component behavior, while the continuum finite element models include a more fundamental representation of the response, where only the most basic aspects (e.g., material constitutive relationships) rely on empirical data.

The use of the more refined and complex FE models cannot be justified but for the modeling of single members or very small buildings, due to the considerable computational resources that are required. Consequently, two main different strategies are generally employed in the reproduction of the inelastic response of structures, the *concentrated plasticity* and the *distributed inelasticity* models.

In concentrated (or lumped) plasticity, the plastic deformations are lumped at the ends of a linear elastic element and are based on the moment-rotation relationships of the end sections for a given axial force. On the other hand, distributed plasticity elements allow for the formation of plastic hinges at any location along the member length, while inelasticity is represented in terms of stresses and strains at the fibers of the integration sections, thus explicitly accounting for axial-moment and biaxial interaction. There are different variations and implementations of these two modeling philosophies, as well as hybrid approaches that employ features of both strategies. Each FE software package follows its own formulations, although the main principles for the modeling of frame members remain the same.

#### 8.3.3.1 Concentrated Plasticity Elements

Historically, concentrated plasticity models were the first to consider the effect of nonlinearity. In the concentrated plasticity philosophy, nonlinear behavior is assumed at the extremities of the structural element, while the body is modeled as an elastic part, and the nonlinear behavior is represented through nonlinear rotational springs or plastic hinges. The first concentrated plasticity elements were presented by Clough and Johnston (1966), a model consisting of two elements in parallel, and by Giberson (1967), with a slightly different model consisting of two elements in series.

The concentrated plasticity modeling assumptions lead to simple models with reduced computational cost, but the simplification of the concentrated plasticity may fail to correctly describe the hysteretic behavior of some reinforced concrete members; hence, it cannot be applied to all structural configurations without careful consideration. Furthermore, in order to obtain accurate results, this type of element requires adequate knowledge by the user on the calibration of the inelastic parameters, such as the exact hysteretic rules of the stress–strain relationship of the nonlinear parts, or the location of the plastic hinges in the structure.

#### 8.3.3.2 Advantages and Disadvantages of Concentrated Plasticity Models

The main advantage of concentrated plasticity models is the significant reduction in computational cost and requirements for data storage in three-dimensional finite element models. Despite their simplicity, concentrated plasticity models can generally capture most of the important behavioral effects of the response of beams or columns in steel or reinforced concrete moment frames, from the onset of yielding through to the highly inelastic range. Strength and stiffness degradation, associated with spalling, steel yielding, concrete crushing, and reinforcing bar buckling (concrete members) or local flange and web buckling (steel members) can be modeled effectively. Under certain conditions, these nonlinear effects can be captured with simple hinge models as reliably as with other finite element or distributed plasticity models.

However, effects such as the interaction between axial and flexural failure in concrete members, or the interaction of local and torsional-flexural buckling in steel members, are difficult to capture using concentrated plasticity models. Furthermore, concentrated plasticity models fail to adequately approximate the behavior of members, where large bending moments are not concentrated at the two ends of the members. This inadequacy can be very significant in certain cases, such as large shear walls in buildings that feature large bending moments throughout the height of the lower floors. The limitations of the concentrated plasticity modeling strategy in such cases has serious implications in the accuracy of the predictions, considering that such members usually dominate the structural response, due to their size.

What is more, significant knowledge in advanced nonlinear modeling is usually necessary from the engineer since some type of calibration of the moment-rotation curves is usually required. An accurate prediction of the structural response cannot be obtained, when users do not possess adequate knowledge on issues related to the definition of the inelastic element parameters. As a result, the plastic hinge length and the characterization of the inelastic section parameters should be given additional attention (and time), during the structural modeling process.

#### 8.3.3.3 Distributed Plasticity Elements – Fiber Modeling

Contrary to the concentrated plasticity models, distributed inelasticity elements allow inelastic deformations to be developed anywhere within the member. They have gained increased popularity recently, and they are becoming widely employed in earthquake engineering applications, either for research or professional engineering purposes. The fiber approach is used to represent the cross-section behavior. The sectional moment-curvature curve is obtained through the integration of the nonlinear uniaxial stress–strain response of the individual fibers, and the inelastic response of the structural member is calculated from the integration of the forces at the integration cross-sections along the member, as in Figure 8.3d.

In fiber modeling, users are asked to define the number of section fibers employed. The ideal number of section fibers sufficient to guarantee an adequate reproduction of the stress–strain distribution across the element's cross-section varies with the shape and material characteristics of the latter, depending also on the degree of inelasticity the element will face.

One of the main advantages of the distributed plasticity models, with respect to the concentrated plasticity models, is the nonexistence of a predetermined location or length where inelasticity occurs, which allows for a much closer approximation of the response. On the other hand, however, distributed plasticity modeling requires additional computational capacity, in terms of increased analysis time, but also in memory and CPU consumption.

#### 8.3.3.4 Types of Distributed Plasticity Elements

There are two formulations for the distributed inelasticity frame elements: the classical displacement-based (DB) (e.g., Hellesland and Scordelis 1981; Mari and Scordelis 1984), and the more recent force-based (FB) formulation (e.g., Spacone et al. 1996; Neuenhofer and Filippou 1997).

The *displacement-based formulation* follows a standard FE approach, where the element deformations are interpolated from an approximate displacement field. In order to approximate the nonlinear element response, constant axial deformation and linear curvature distribution are enforced along the element length, which is exact only for prismatic linear elastic elements, and relatively accurate with short, inelastic elements. Consequently, a refined discretization (meshing) of the structural element (typically four to five elements per structural member) is required for the accurate representation of the structural response, especially deep in the inelastic range.

By contrast, the *force-based formulation* does not depend on the assumed sectional constitutive behavior, and it does not restrain the displacement field of the element in any way. In this sense, this formulation can be regarded as always "exact," the only approximation being introduced by the discrete number of the controlling sections along the element that are used for the numerical integration. A minimum of three Gauss-Lobatto integration sections are required to avoid under-integration; however, this option in general will not simulate the spread of inelasticity in an acceptable way. The suggested minimum number of integration points is four, although five to seven integration sections are typically used. This feature enables the modeling of each structural member with a single FE element. Therefore, it allows a one-to-one correspondence between structural members (beams and columns) and model elements, and leads to considerably smaller models, with respect to assemblies of the DB elements, and much faster analyses, notwithstanding the heavier element equilibrium calculations.

The use of a single element per structural member gives users the possibility of readily employing element chord-rotation output for seismic code verifications. Instead, when the structural member has to be discretized in two or more frame elements (necessarily in the case of DB elements), then users need to post-process nodal displacements/

rotation in order to estimate the members chord-rotations. On the negative side, FB formulations suffer from so-called localization issues – that is, the loss of objectivity of the solution, depending on the location of the integration sections along the length of the member.

*Force-based plastic-hinge models*: A combination of distributed inelasticity models with concentrated plasticity elements, featuring a similar distributed inelasticity formulation, but concentrating such inelasticity within fixed lengths at the ends of the element, was proposed by Scott and Fenves (2006). The advantages of this formulation are not only reduced analysis time (since fiber integration is carried out for the two end sections of the member only) and increased stability of the nonlinear solutions, but also full control/calibration of the plastic hinge length (or spread of inelasticity), which remedies localization issues, as discussed in, for example, Calabrese et al. (2010).

#### 8.3.3.5 Advantages and Disadvantages of Distributed Plasticity Models

The main advantage of distributed inelasticity elements is that they do not require any calibration of the inelastic response parameters, as is the case with concentrated-plasticity phenomenological models. As a result, they do not need advanced knowledge from the side of the engineer, since all that is required is the introduction of the geometrical and material characteristics of structural members. Furthermore, there is no requirement for a prior moment-curvature analysis of the members, and there is no need to introduce any element hysteretic response, since this is directly obtained from the material constitutive models programmed in the software.

In addition, because of their ability to allow for the development of nonlinearities anywhere along the member length, distributed plasticity models allow yielding to occur at any location along the element, which is especially important in the case of large shear walls, or in the presence of distributed element loads (girders with high gravity loads). They possess the ability to track gradual inelasticity (e.g., steel yielding and concrete cracking) over the cross section and along the member length, to directly model the axial load-bending moment interaction, in terms of both strength and stiffness, and to allow the straightforward representation of biaxial loading, and interaction between the flexural strength in orthogonal directions. Because of these significant advantages of distributed plasticity models, in general their predictions are more accurate and closer to reality, with respect to their concentrated plasticity counterparts.

However, these advantages come with the cost of increased computational requirements. Furthermore, the ability of these models is limited with regard to capturing degradation associated with bond slip in concrete joints, local buckling and fracture of steel reinforcing bars and steel members, although specialized material models have been developed to represent these phenomena.

Finally, the calculation of curvatures (and stresses and strains) along the member length can be sensitive to the specified hardening (or softening) modulus of the materials, the assumed displacement (or force) interpolation functions along the member length, and the type of numerical integration and discretization of integration points along the member. This can lead to considerable errors and inconsistencies in the curvature and strain demands calculated in the analysis, along with the associated stress resultants (i.e., forces) and members' stiffness.

#### 8.3.3.6 Considerations Regarding the Best Frame Model for Structural Members

Many models are available in the literature, and they have been implemented in existing finite element packages for the representation of the inelastic behavior of frame elements. These nonlinear structural analysis models can vary significantly and lead to very different predictions with different modeling assumptions. Although it is highly desirable to directly simulate response at a fundamental level, ultimately response must be validated with empirical test data at the component and the global levels. The choice between the phenomenological or fundamental models is not always straightforward and it depends on several factors that need to balance practical design requirements with available modeling capabilities and computational resources.

Some of the points that should be considered, when trying to determine the most appropriate model for each case, are:

- The balance between the desire for accuracy in the analytical model and other unknowns since the overall uncertainty is not necessarily reduced when a more sophisticated model is used (e.g., due to the lack of reliable data for the geometric or material characteristics of the building under consideration).
- The ability of the engineer to use and calibrate the hysteretic curves and the plastic hinge length of concentrated plasticity models.
- The availability of resources in terms of human capital (time and effort) and computational tools (analysis software and computing capabilities).
- The need (or not) for the exact simulation of all modes of behavior, based on the expected structural response.
- The need (or not) to reliably simulate the response of buildings over their full range of response, up to collapse.
- The analysis objectives, the required demand parameters, and the need for increased accuracy at the different levels of deformation (elastic or inelastic range).

During the years of structural analysis after the introduction of the fiber elements, there has been a continuous discussion (if not open confrontation) in the engineering community about which of the two basic modeling approaches is better and for which applications (research, practical applications in a standard design office, larger projects). The heated discussion is still raging on, sometimes with too much fervor and passion.[1] During this debate many bold statements have been made from both sides, many of which are to a great extent prejudices, fallacies and public myths. Below I have included a list of some of these statements and my personal opinion about them. Since SeismoStruct (2023) is one of the few programs worldwide with reliable models for both approaches, I believe that, modesty aside, I have a relatively experienced perspective on the subject.

- *Lumped plasticity models are very inaccurate and they are not suitable for practical applications.* The vast majority of the building stock, especially older construction, has significant irregularities in plan, not just in the geometry of the floors' layout, but also irregular

1 This is not unreasonable, the success (or not) of an FE package depends largely on whether it is accepted as a reliable and accurate tool by the engineering community.

stiffness distribution with the vertical members unevenly located on the building's plan, according to the architectural needs. Furthermore, structures are very rarely hit along their two main horizontal directions X or Y. Hence, inevitably it is very likely that during a seismic event there will be torsional effects and biaxial bending in the vertical members. In such cases, distributed plasticity and fiber models undoubtedly provide superior response predictions. After all, it is no coincidence that concentrated plasticity models never win in the blind prediction contests that are carried out worldwide (contrary to the fiber models that are typically employed by the winning teams, very often users of SeismoStruct). However, the predictions of the concentrated plasticity models are not unacceptably bad either. Considering the inaccuracies in the knowledge of the structural configuration that significantly affect the reliability of the entire assessment, personally I believe that these models can be used safely. After all, concentrated plasticity is a level of magnitude more accurate with respect to all the elastic methods; if the latter can be used for assessment, the former is definitely suitable.

- *Lumped plasticity models are very inaccurate and they are not suitable for academic purposes:* Given all the inaccuracies involved in the concentrated plasticity model and, because in research increased accuracy in the analytical predictions is vital, indeed fiber models are much more suitable for academic use.
- *Lumped plasticity is faster and distributed plasticity (although more reliable) is not suitable for practical applications*: Lumped plasticity models are indeed faster than distributed plasticity models in the general case. In the former, in order to get the moment from the curvature one only needs to perform a numerical interpolation in the data set that holds the moment-curvature pairs for the different axial load levels. In fiber models, such calculations are carried for all monitoring points of the section (usually 100 or more) and additional operations are required for the integration of the contributions of all the integration sections of the member. However, in FE applications the formation of the element stiffness matrices is just one of the operations required by the problem, and a large proportion of the analytical time is consumed in other operations, such as the assembly of the global stiffness matrix or the solution of equations, which are common in both cases. What is even more important is that, because of its basic concepts and assumptions, the concentrated plasticity model tends to be less stable numerically. In fiber models, the element stiffness gradually changes, as the monitoring points enter in the inelastic range one after the other. On the contrary, in lumped plasticity, the entire section yields suddenly, considerably changing the stiffness of the member and that of the entire structural model. This abrupt change in the structural stiffness distribution leads to numerical instabilities, ultimately resulting in larger execution times, since with lumped plasticity a larger number of iterations is generally required to achieve convergence. What is even more important is that very often a premature end of the analysis occurs, due to convergence difficulties, and the entire analysis fails to run. That said, the argument above applies mostly to pushover analysis rather than nonlinear dynamic analysis. In pushover analysis maybe in 60% or 70% of the steps, some parts of the structure experience inelastic deformations. On the contrary in dynamic analysis this percentage is generally smaller, since the time-steps of large acceleration are a small proportion of the total duration of the records (could be 10% or 20% in most cases). Therefore, concentrated plasticity is generally faster in nonlinear dynamic analysis, but

not in pushover analysis. Since in most practical applications pushover analysis is employed, the statement that distributed plasticity models are not suitable for practical applications, because they are slow, is not valid.

- *With the gradual increase of computer capacity, the lumped plasticity models will eventually become obsolete.* This might happen but in a very slow and gradual process. Concentrated plasticity has been around for several decades, it has proven very useful, it is the approach employed by most of the packages with nonlinear capabilities worldwide, and academics and practicing engineers alike are more accustomed to it than fiber models (for the time being, at least). My prediction is that in some (probably many) years' time, it will eventually become obsolete. However, at that time engineers will not be using today's fiber models either, which will probably have been replaced by an evolution of them, or even by entirely new models.
- *Equations for the capacity of members in the different standards have been developed with lumped plasticity in mind, hence engineers cannot use distributed plasticity models, when using these standards.* This is a rather absurd statement, and to be honest it took me quite some time to understand its meaning. Usually, this is the last resort of the advocates of lumped plasticity, and surprisingly it is an argument that convinces quite a few. In all assessment methodologies the process is standard: the engineer (i) calculates the demand on the elements for the specified seismic action from linear or nonlinear structural analysis; (ii) calculates the capacity of these elements for the selected performance level from the expressions that are given directly by the codes; and (iii) compares the demands against the capacities. Structural analysis is about the calculation of the demands and just that. It has nothing to do with the equations for the capacity and how these have been derived. After all, the capacities of existing members (against which the calculated demands are compared) depend entirely on their actual characteristics, the materials, the section dimensions and the reinforcement. The actual strength could be measured if the member was tested in a laboratory. The expressions available in the codes and in the literature are simply attempts to estimate this measured strength.

Coming to the final verdict regarding which of the two approaches is better, it can be said that distributed plasticity models are certainly more suitable for research. However, for practical applications this is not always clear. My personal experience shows that fiber models seem to be more practical, mainly because they are numerically more stable. All the same, this applies mainly to pushover analysis, which of course is the standard in the assessment methodologies nowadays. If one employs nonlinear dynamic analysis with 7, 9, or 11 records, there would possibly be some significant speed gains if one uses concentrated plasticity models. Ultimately, it is up to the engineer to use his/her judgment and choose the method that seems more suitable for each project.

## 8.4 Modeling of Shear Walls

Because of their large size and stiffness, structural walls assume a dominant role in the dynamic behavior of any building. Ideally, shear walls in structural analysis would be modeled with a mesh of three or four nodes bidirectional shell or plate elements.

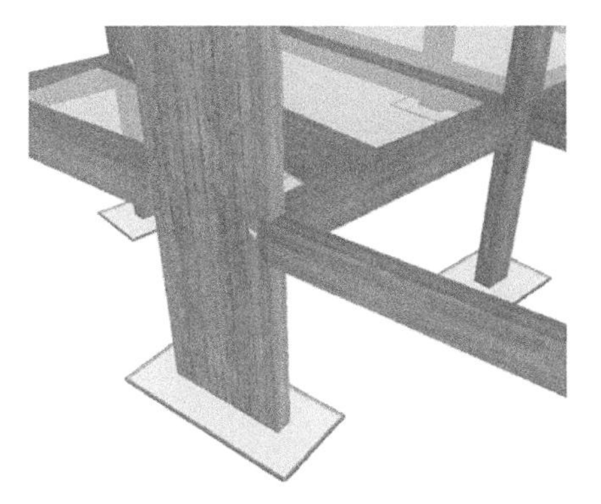

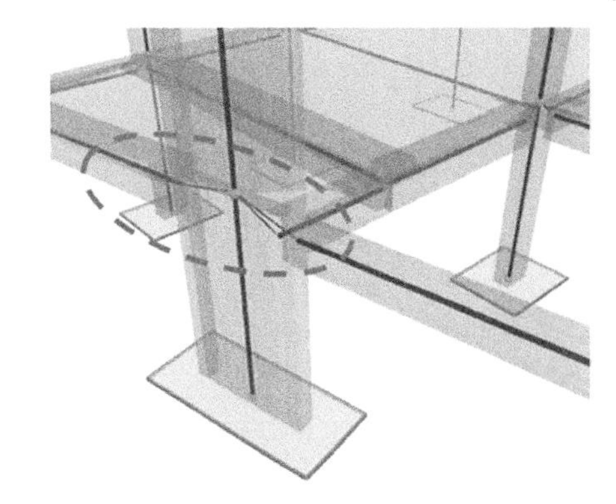

**Figure 8.4** Shear wall modeling with rigid offsets (nonlinear procedures) *Source:* Stelios Antoniou.

Unfortunately, in professional software packages plane elements usually support only linear elastic modeling. However, under significant seismic loading damage is typically expected at these shear walls, thus these plane elastic elements do not provide an accurate option for modeling. In such cases, the modeling with an equivalent nonlinear one-dimensional element is necessary. In order to model the volume of the wall, which is not negligible (contrary to columns of typical sizes), horizontal rigid links should be employed to connect the wall nodes to the adjacent beams (Figure 8.4).

Rigid offsets are also employed for the modeling of RC core walls, typically found around stairs or elevator shafts. Because employing a single element for the entire core along the center of gravity of the section would not accurately represent its torsional rigidity, each side of the core is modeled with one inelastic element, and then these elements are horizontally connected with each other with rigid links at their top and bottom nodes.

In this way, the inelastic response of the core wall, as well as its torsional stiffness are modeled adequately, as discussed in Beyer et al. (2008a, 2008b, and 2008c) (Figure 8.5). The advantages of this approach are clearer in cases of flexible floors in the perimeter of the core that do not prevent the out of plane deformations of the side walls. This type of modeling is also more practical, when carrying out performance checks after the structural analysis.

## 8.5 Modeling of Slabs

Gravity loading usually assumes a secondary role in the seismic assessment of buildings, since the seismic load combination is typically the most demanding. Furthermore, slabs are marginally affected by the earthquake loads, and their role in the seismic assessment is limited to their diaphragmatic action, and the transfer of the gravity loads (usually the dead and live loads, and sometimes the snow loads that are included in the seismic combinations) to the adjacent load carrying members.

The action of floors as diaphragms plays a very important role in the overall structural behavior, since it provides the horizontal links that collect and transmit the inertia forces from the location of masses to the vertical structural components, ensuring that these components act together in resisting the horizontal seismic action. This is particularly relevant in cases of complex and nonuniform layouts of the vertical members, or vertical members with different horizontal deformability.

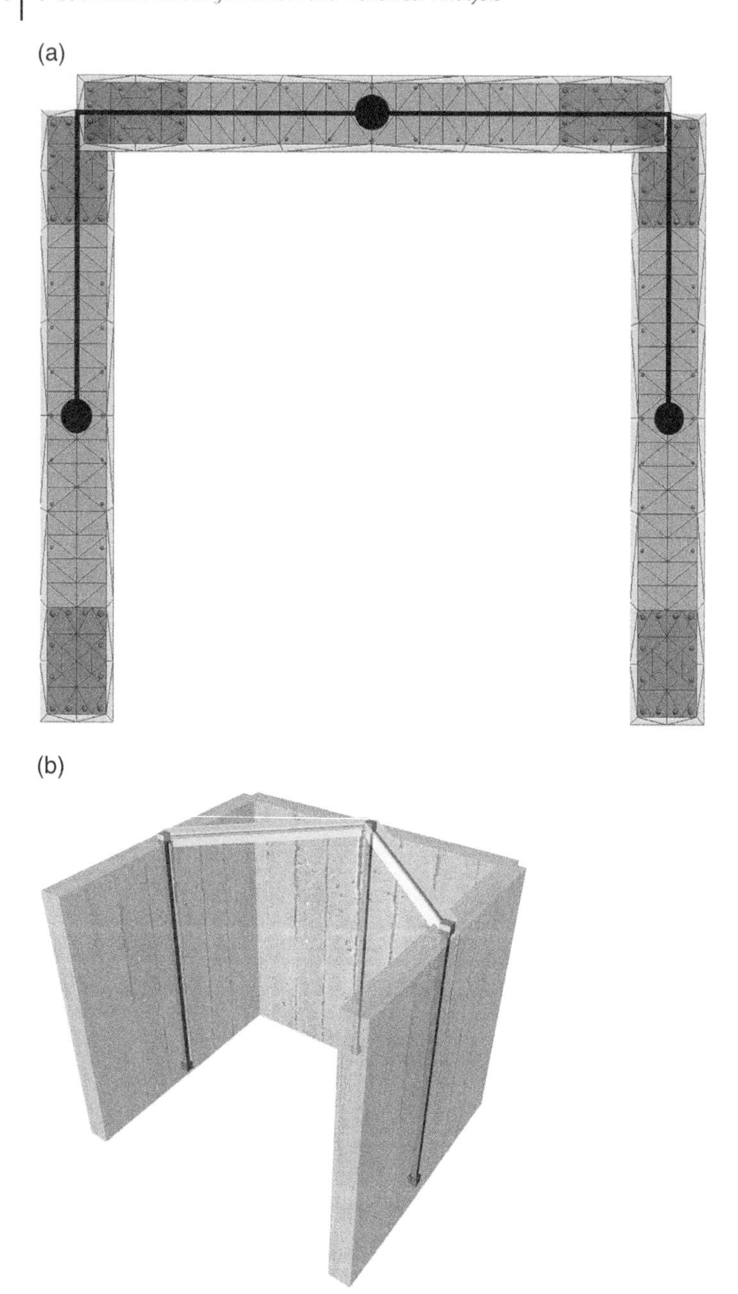


Figure 8.5 (a–b) Core walls modeling with rigid links (nonlinear procedures). *Source:* Stelios Antoniou.

According to ASCE 41, the diaphragms shall be classified as:

- *Flexible*, when the maximum horizontal deformation of the diaphragm along its length is more than twice the average story drift of the vertical members immediately below the diaphragm.

- *Rigid*, when the maximum lateral deformation of the diaphragm is less than half the average story drift of the vertical members immediately below the diaphragm.
- *Stiff*, if they are neither flexible nor rigid.

Whereas in newly constructed buildings it is common for slabs to have adequate thickness so as to be fully rigid, in older construction this is not always the case and thinner slabs may also be classified as stiff diaphragms (flexible concrete slabs are very rare).

The mathematical modeling of rigid diaphragms must account for the effects of torsion. This is usually done with rigid links provided by the software packages, which connect all the nodes of the slab with each other. FE packages have different methods for defining rigid links. The ones that are mostly used are the penalty functions or penalty augmentation method, which create fictitious elastic structural elements of very large stiffness (called the penalty elements) that connect the appropriate degree-of-freedom, and the Lagrange multipliers method, which introduces additional unknowns in the system of equations that represent the constraint forces.[2]

It is important for engineers to keep in mind that binding together the end nodes of the beams with the rigid diaphragm to simulate the concrete slab can cause unintended spurious axial forces in the beams. This may lead to an overestimation of their resisting bending moment in fiber elements, due to the interaction of the bending moments with the axial forces (in lumped plasticity models this is not a significant problem because the moment curvature curves do not depend on the axial force at the actual loading step).

Unlike rigid diaphragms, the mathematical models of the stiff or flexible diaphragms account for their actual flexibility. This is done either by discretizing the slab to smaller elastic plate elements (the slabs are expected to remain elastic during large seismic events), or with the Penalty Functions method with small values of the penalty coefficients. In both options, sensitivity analysis with a trial and error approach might be needed in order to consider the correct flexibility of the floor.

## 8.6 Modeling of Stairs

One of the main puzzles in structural modeling, linear and nonlinear alike, is the modeling of stairs – that is whether to include them in the structural model and how. The effect of the inclusion of stairs in the structural model can be easily shown with a simple example, where two similar building configurations are presented, with the only difference being that the first variant has structural walls in both horizontal directions and core walls around the stairs. Both structural models were created with and without the stairs and a simple eigenvalue analysis was carried out in the four models (Figure 8.6).

Clearly, in the model with walls, the effect of the stairs is negligible. However, modeling the stairs in an otherwise flexible model does indeed make a difference, in this particular example it actually decreased the fundamental period from 0.37 to 0.30 seconds.

Without delving into more detail, the following rule of the thumb usually applies:

- When there are large shear walls in the structure, the rigidity of the stairs does not have a significant effect on the overall structural stiffness, and there is no need to model them.

---

2 The direct Master–Slave method, which eliminates the slave DOFs before solving the equations is rarely used, because it leads to numerical instabilities.

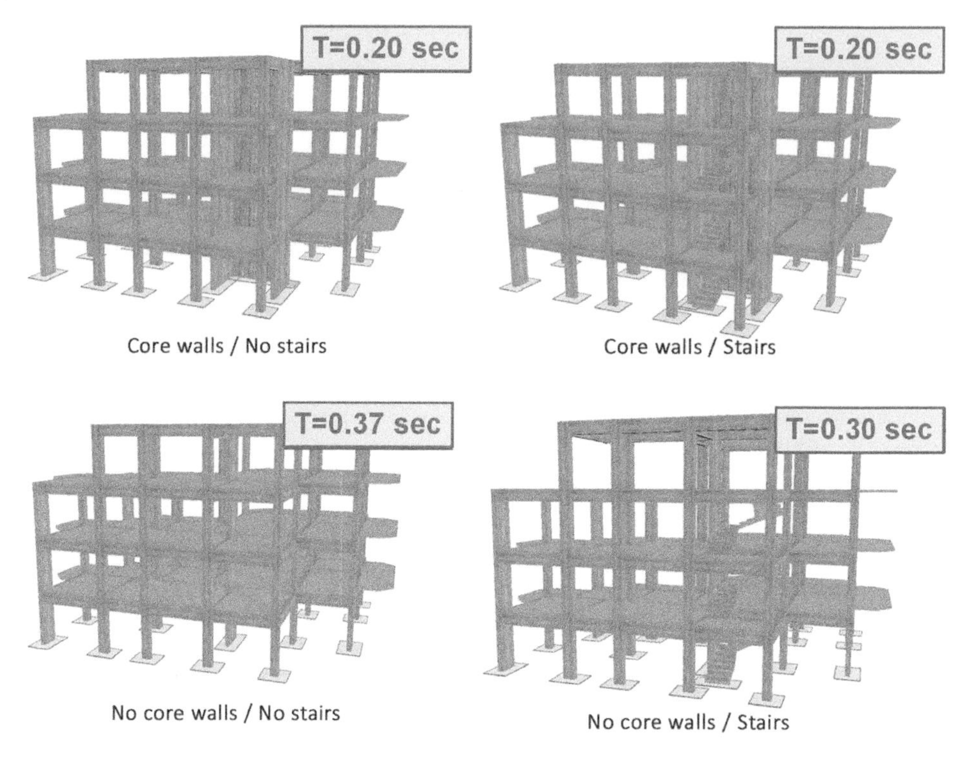


Figure 8.6 Fundamental period of a building with and without core walls and stairs. *Source:* Stelios Antoniou.

- When there are no large shear walls, stairs can play an important role in the structural response, and it is advisable to include them in the structural model.

In the latter case, it is important to note that many times the stairs are directly supported by the columns in their perimeter at mid-height between the floor levels, creating short columns that are very vulnerable in seismic loading, and usually fail in a brittle manner. In such cases, the modeling of the stairs and the explicit modeling of the short columns that are formed is essential (Figure 8.7).

Similarly to the slabs, stairs are not expected to sustain serious damage during an earthquake; hence, if they are modeled, this is typically done with elastic elements.

## 8.7 Modeling of Infills

New reinforced concrete framed structures are generally designed without considering the effect of infill panels. This is not unreasonable. Due to the presence of the large shear walls that are required by modern code provisions, their effect is usually limited. What is more, this is typically on the safe side, since engineers nowadays are aware of the problems caused by structural irregularities and they place infills in a symmetrical fashion (and if not, they take additional measures to protect the framing system).

However, existing buildings are weaker and more flexible; consequently, their stiffness distribution and their capacity are significantly affected by the presence of infills. When the

Figure 8.7 Brittle failure of a short column that was created by the constraints imposed by the stairs. *Source:* Stelios Antoniou.

masonry layout is symmetric in plan and in elevation, it acts positively and leads to higher levels of safety against seismic actions, as the infills undertake a large proportion of the seismic loading. In such cases, they may be omitted from the structural model.[3]

Unfortunately, this is not always the case. The asymmetric layout of infills can have a very adverse effect on the structural response. The most common case of such problems is the structural configuration of pilotis, where large openings are left at the ground level for architectural reasons and/or to create free space for circulation or car parking. This configuration creates a very weak, soft story at the ground level and significantly increases the vulnerability of the building. When the vertical members are small, lightly reinforced, and with small ratios of transverse reinforcement, pilotis lead to a large concentration of the deformations at the ground floor with early brittle failures (Figure 8.8).

Another typical unfavorable effect is the formation of short columns, due to the constraints imposed in the vertical structural members from the presence of infills that do not cover the entire height of the floor. The columns are free to deform above the level of the infills, but the limited free length leads to large deformations and early failures, usually by crushing in a brittle manner (Figure 8.9).

Finally, the adverse effect of irregularities in plan should also be taken into account. This can occur due to the absence of infills along some of the sides of the building – for instance, because of shops' display windows. In all these cases of asymmetric infill layout, their effect in the structural response should always be considered.

### 8.7.1 A Simple Example: Infilled Frame vs. Bare Frame

To illustrate the importance of infills in structural response, the simple SeismoStruct (2023) example of Section 4.18 is extended here, in order to compare the cases of the bare frame and the fully infilled frame. The role of infills is obvious even with eigenvalue analysis; the

3 Even in these cases, modeling infills is desirable, since this can lead to lighter retrofit measures.

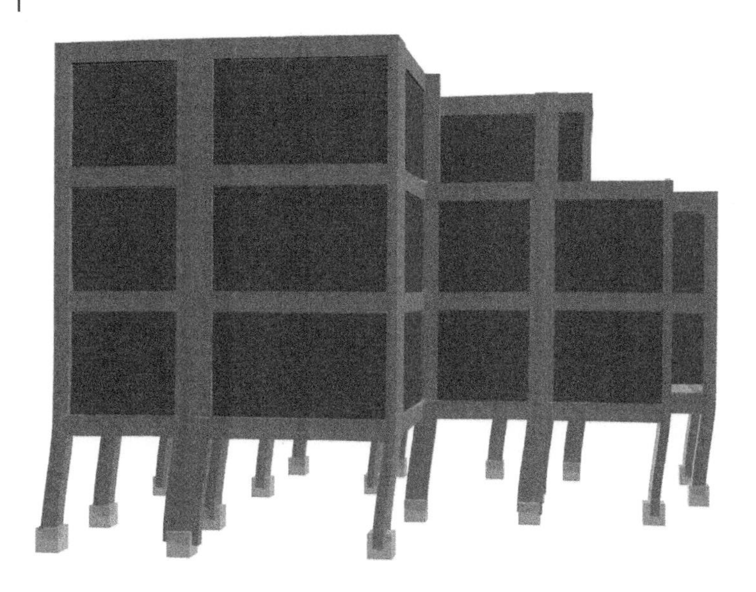

Figure 8.8 Large deformation concentrations at the ground level, caused by the irregular layout of infills in elevation.

fundamental period of the bare frame is 0.31 seconds and that of the infilled frame 0.23 seconds.

If the entire nonlinear static procedure NSP is carried out (Figure 8.10), with the calculation of the target displacement and the execution of the capacity checks (Eurocode 8 has been employed for the assessment and the checks were performed only in shear, which is usually the most critical check), one gets a completely different picture for the two models:

- The infills contribute significantly to the lateral structural capacity, doubling it from 2.800 kN to almost 5.600 kN. Furthermore, the building's initial stiffness, which is represented by the different slope of the curves at zero displacement, is also considerably larger.

Figure 8.9 Short columns formed due to the partial height of infills. *Source:* Stelios Antoniou.

- The infilled frame, after it has reached its maximum capacity, exhibits a steep descending branch, which reflects the gradual deterioration of the infills. After a significant number of infills have failed, the response gets very close to that of the bare frame.
- The values of the target displacement for the different limit states of the infilled frame is roughly 25% that of the bare frame.
- The demand-to-capacity ratios are considerably larger in the bare frame; for instance, there are more than 10 failed members in shear in the Significant Damage limit state, versus none for the infilled frame.

Similar conclusions are drawn with nonlinear dynamic analysis (Figure 8.11). The maximum displacement at the top in the infilled frame is 30% that of the bare frame; consequently, more member failures occur in the bare model. An interesting observation is that gradually small torsional effects develop in the infilled frame during the time-history, which are not observed in the case of the bare frame. These are triggered by the asymmetric placement of the infills, but more importantly by the early failure of the infills at the weaker side of the building, which result in a gradually increasing irregularity in plan. All the same, despite this small torsion, the infilled frame is still less vulnerable than the bare one due to the general positive effect on the infills.

### 8.7.2 Another Example: Partially Infilled Frame (Soft Story) vs. Bare Frame

As discussed before, the asymmetric layout of the infills in plan or in elevation has a very adverse effect on the structural response. In particular, if in the previous example the brick walls are removed from the ground level of the infilled frame, a much weaker structure is formed.

In pushover analysis, the partly infilled model starts with a larger stiffness, with respect to the bare frame, and reaches a slightly larger lateral capacity (2.900 kN with respect to the 2.750 kN of the bare frame). However, this capacity is reached at a much lower deformation level, approximately 50% that of the bare frame (0.04 m versus 0.08 m). What is more, the

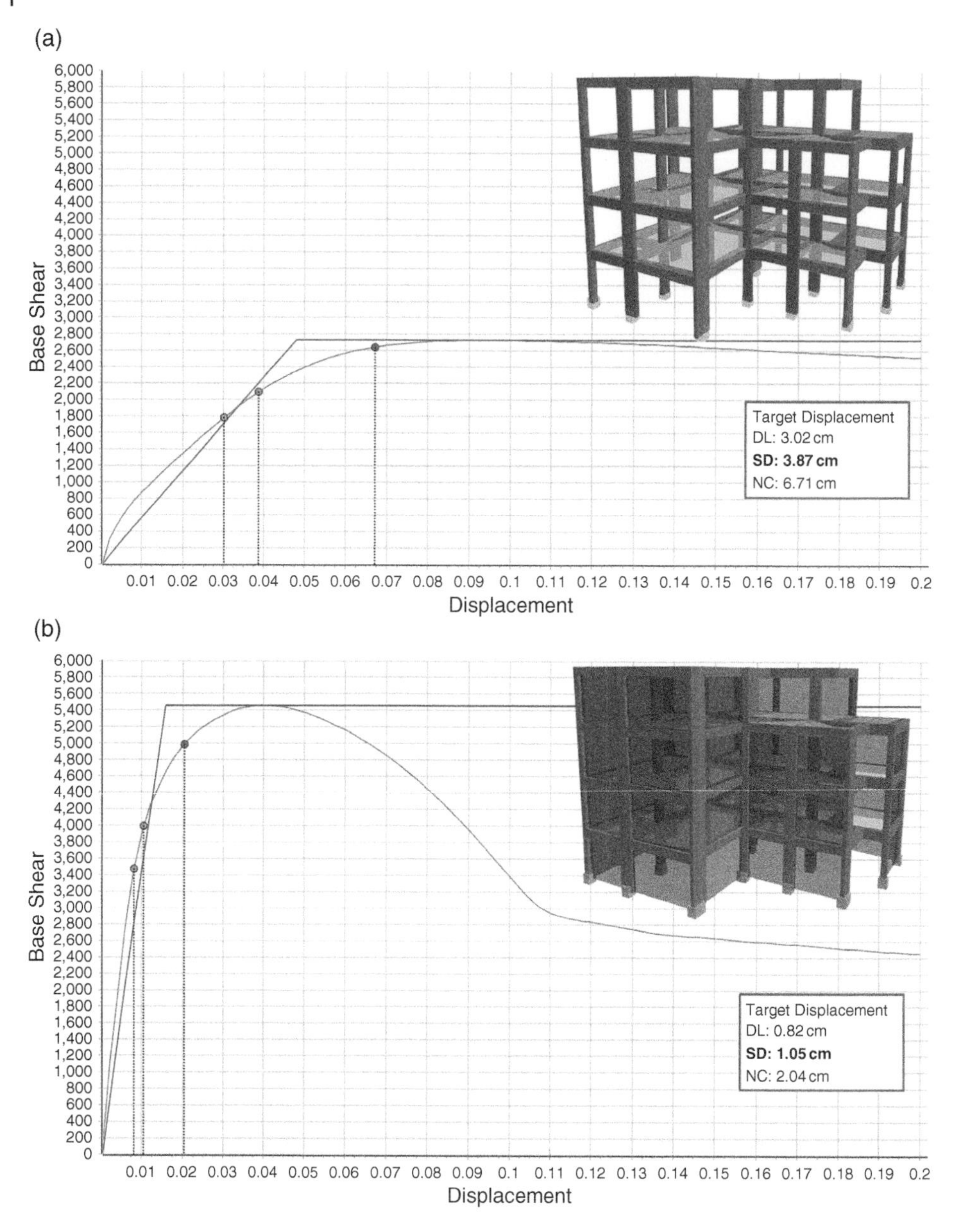


Figure 8.10 (a–b) Nonlinear Static Procedure of the building with and without infills. *Source:* Stelios Antoniou.

infilled model exhibits smaller ductility, and its strength degrades quickly after the peak is reached and gradually becomes lower than that of the bare frame. This is attributed to the early failure of most of the columns of the ground story, as deformations concentrate there.

With dynamic analysis the displacements at the top of the building are larger at the bare frame (max. 5.48 cm vs. 3.67 cm), which points toward the opposite direction of what one would have expected, i.e., that buildings with soft ground stories are more vulnerable. With a closer look at the plot of the deformations at the critical ground floor, however, the

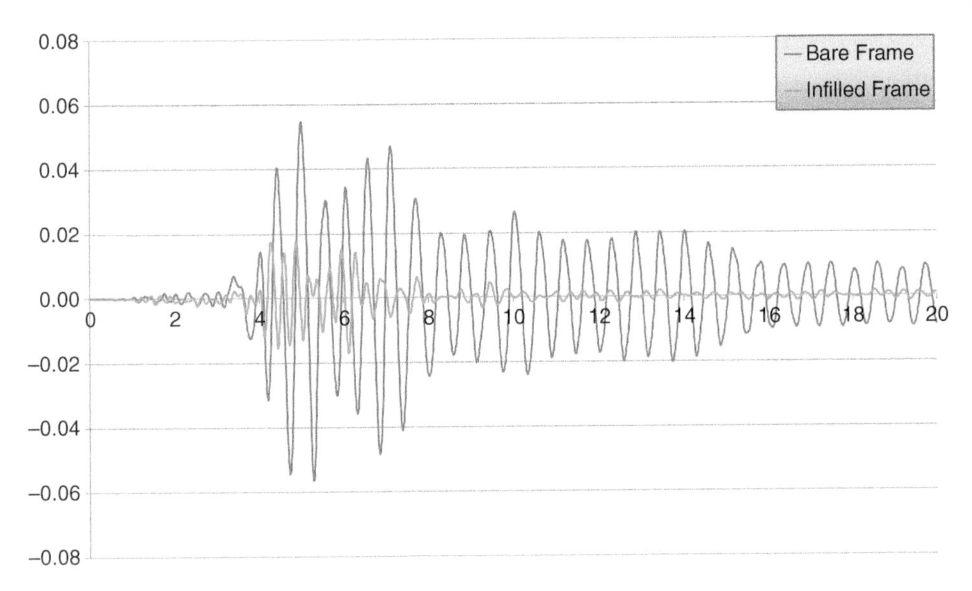


Figure 8.11 Displacement of the control node vs. time, nonlinear dynamic analysis of the building with and without infills. *Source:* Stelios Antoniou.

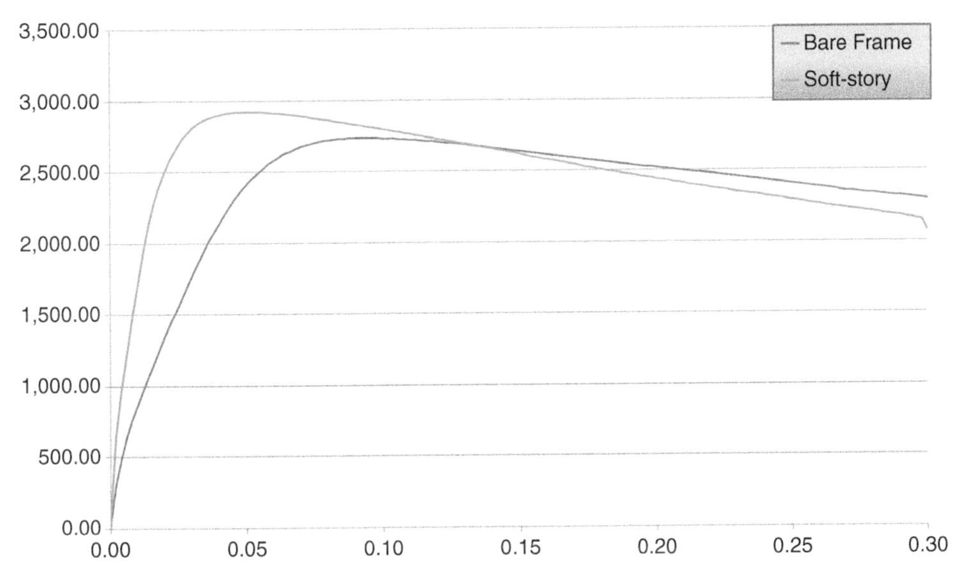


Figure 8.12 Pushover analysis of the bare and the partially infilled building. *Source:* Stelios Antoniou.

adverse effect of the infills in the upper levels becomes clear. At the ground floor the maximum deformation of the bare frame is 1.66 cm (30% of the total), whereas that of the partially infilled frame is 3.15 cm (86% of the total), i.e., almost all the deformations along the building height are concentrated at the ground level, while the upper stories behave largely as a rigid block (see Figures 8.13 and 8.14). Obviously, infills considerably increase the vulnerability of the building, so it is unsafe to ignore them in the structural analysis!

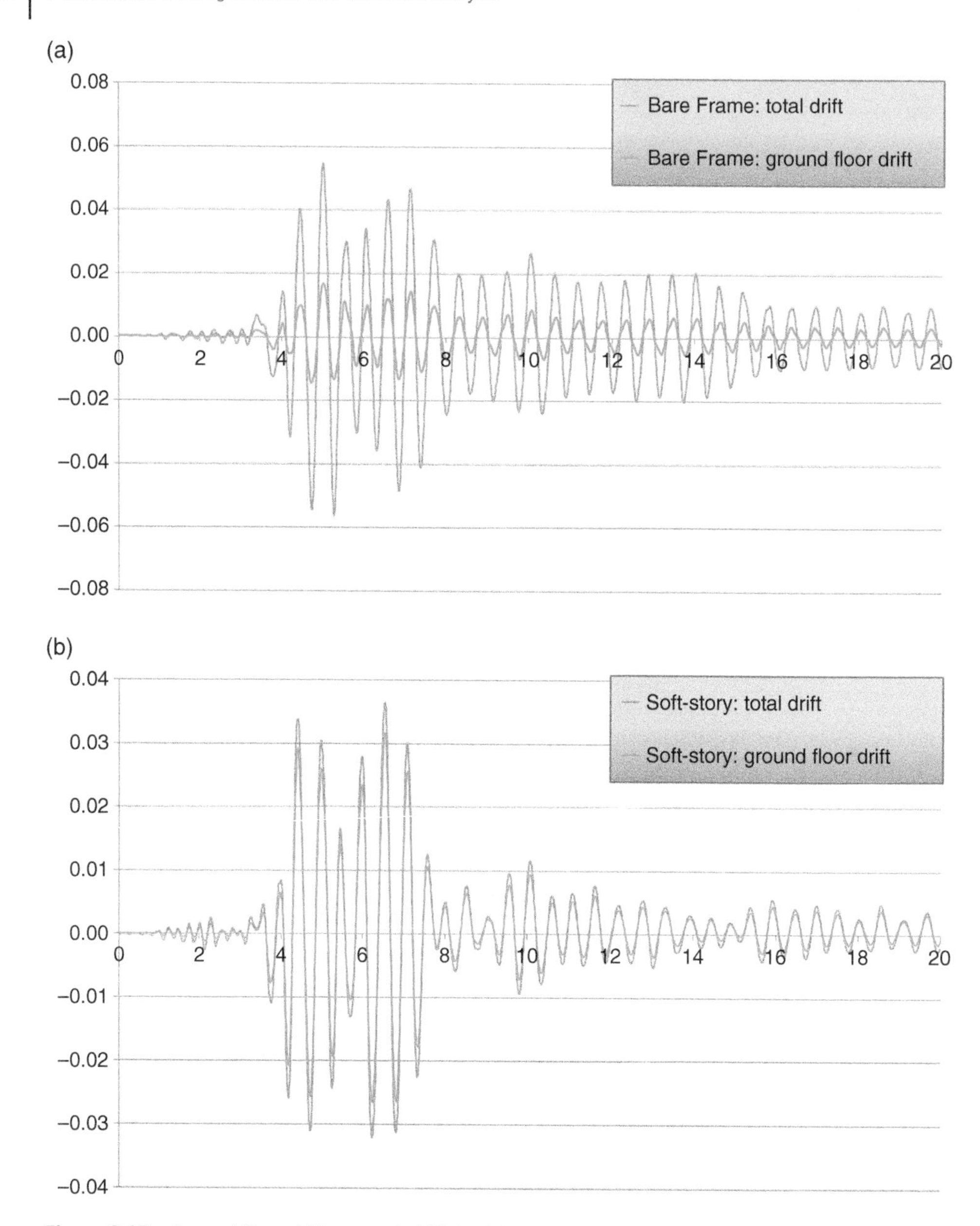


**Figure 8.13** Ground floor drift vs. total drift for (a) the bare and (b) the partially infilled frame. *Source:* Stelios Antoniou.

### 8.7.3 Problems in the Modeling of Infills

Because we can't always know whether infills will have a positive or negative effect, it is generally advisable to include them in the structural model. All the same, infills are not treated in a unified manner in the different standards worldwide. For instance, ASCE 41 generally requires the modeling of infills, EC8 leaves this issue vague and up to the judgment of the engineer, the Italian Code NTC-18 recommends that their contribution to the

**Figure 8.14** Deformed shape at the time-step of the maximum displacement for the bare and the partially infilled frame. *Source:* Stelios Antoniou.

response of the structural system should be considered only if it has a negative effect whilst other codes, like the Greek Interventions Code, propose the analysis of both the infilled and the bare frame in order to be on the safe side.

Yet, including the effect of infills in the structural model is not always easy. There are often significant uncertainties in the estimation of the mechanical properties on the infills, and large variabilities are expected even within the same building. Furthermore, even if the strength of the bricks and the mortar are known, it is usually difficult to calibrate the infill panel models in the analysis, especially their hysteretic response in nonlinear dynamic analysis. The presence of openings, doors, or windows, in the infills further adds to the complexity of the process.

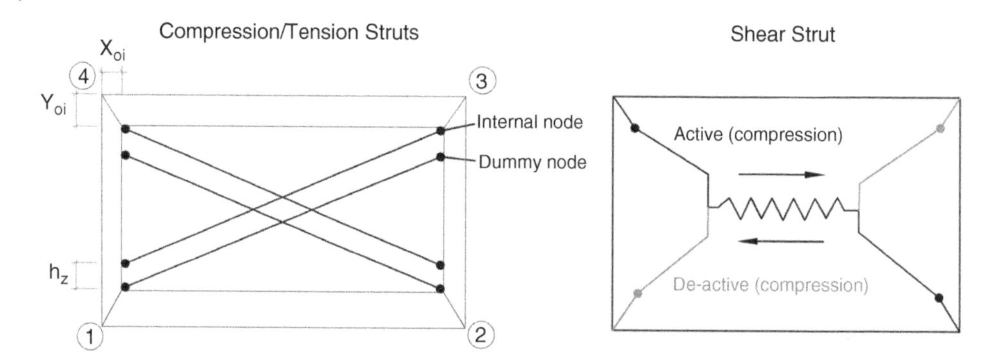


**Figure 8.15** Infill modeling by means of two diagonal struts and one shear spring per direction (Crisafulli 1997).

Several models have been proposed to describe the response of infill panels. These are generally separated into two groups, namely the micro-models and the macro-models. The first group includes models that divide the infill in a number of shell or brick sub-elements that consider the infill characteristics in more detail. The response of the interface with the surrounding frame, where separation or sliding may occur, is modeled using tie-links or interface elements. Infill micro-models, although more accurate, are not generally employed in practical applications, as they are numerically very demanding, and their use is only justified in certain research activities.

The second group includes more simplified models that are mostly based on the physical understanding of the response of the infill, and constitute a good compromise between efficiency and accuracy. Although there is a general consensus that the response of masonry panels in infilled frames subjected to lateral loads can be described with reasonable accuracy with struts along the two diagonal panel diagonals, there is a wide range of models with different levels of sophistication from the single diagonal model to multiple strut models. In some models, additional shear springs exist that describe the response of the infilled frame system, when horizontal shear sliding occurs in the masonry.

One of the most sophisticated models was proposed by Crisafulli (Crisafulli 1997; Crisafulli et al. 2000). The model is implemented as a four-node panel element. It considers the compressive and shear behavior of the masonry separately, using a double truss mechanism and a shear spring in each direction, and it is capable of accounting for the volume of the surrounding frame elements, by means of a set of four internal nodes that are connected with rigid offset to the external element nodes. The model is able to adequately represent the different modes of failure observed for masonry infills, and it is currently implemented in SeismoStruct (2023) and other nonlinear analysis packages worldwide (Figure 8.15).

## 8.8 Modeling of Beam-Column Joints

In new buildings, the beam-column joints have adequate reinforcement (longitudinal and transverse) and confinement, and they are expected to remain elastic, and in most of the cases effectively rigid, because of their large stiffness. In such cases, modeling is done with rigid offset of the members' ends to the joint (Figure 8.16).

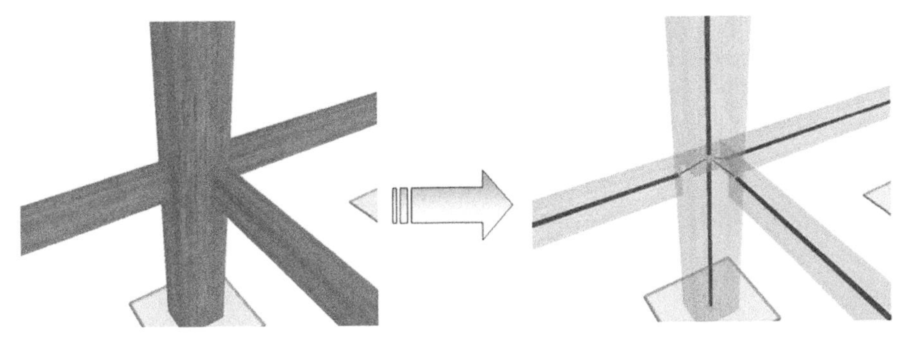

Figure 8.16 Modeling of beam-column joints with rigid offsets, SeismoBuild screenshot (2023). *Source:* Stelios Antoniou.

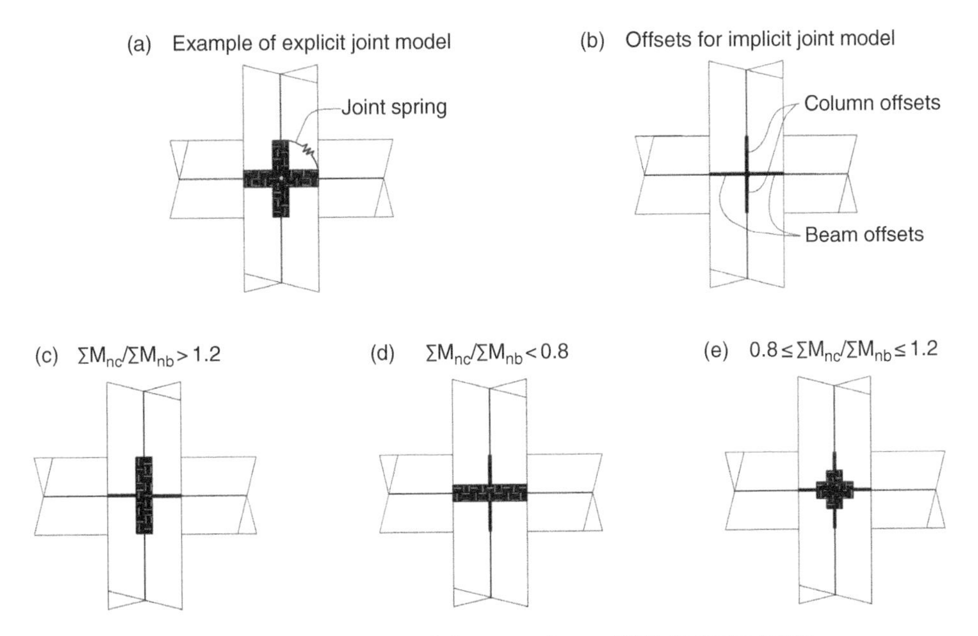


Figure 8.17 Advanced beam-column modeling according to ASCE 41 (ASCE 2017).

In existing buildings, however, beam-column joints suffer from inadequate reinforcement (mainly stirrups) and inadequate lap splices, leading to a response that is neither rigid nor elastic. In such cases, a more elaborated kind of modeling may be more accurate. For instance, special inelastic elements have been derived for the modeling of the joints or alternatively, link elements (elastic or even inelastic) can be attached at the end of the rigid link to model the additional flexibility (Figure 8.17).

The problem with such advanced ways of modeling is that, even though they lead to a more accurate representation of the structural response, they pose difficulties in their application. For example, there are well-tested and reliable joint elements only for 2D analysis, and it is not easy to calibrate the hysteretic curves of the inelastic link connected to the rigid offsets. Furthermore, such representations significantly increase the analytical effort.

Therefore, such advanced modeling techniques are mostly used in research. In practical applications, they are adopted only when the hysteretic behavior of the beam-column joint considerably affects the structural response on the global level. In all other cases, a simple modeling with rigid offsets (although inaccurate) is generally acceptable.

## 8.9 Modeling of Bar Slippage

The loss of bond between the rebar and the concrete is a significant parameter in the seismic structural response and generally it should be considered in the analytical calculations. For new buildings bar slippage is not an important issue, since in the typical cases ribbed rebar with adequate lapping lengths are used; thus, the loss of bond is negligible. This is not the case, however, with existing buildings, where smooth rebar with very small lapping lengths are very common, and bar slippage is indeed a problem (Figure 8.18).

There are two ways to model bar slippage in a structural model:

- The bar slippage is accounted for in the joint deformation of the beam-column joints. This is the most accurate structural representation; however, the solution is computationally very demanding, and the link curves are difficult to calibrate, hence this type of modeling is only justified in special cases, and/or in research activities.
- Simply reduce the stiffness of the reinforcing steel in a crude, uneducated manner (e.g., by 20%–30%).

Note that these two methods are proposed for the analytical calculations, in order to accurately model the structural response and calculate the demand on the different structural members. Unless the bar slippage is indeed a big problem and significantly affects the load distribution throughout the building, it is not always justified that it be included in the analytical model. On the contrary, the effects of bar slippage should *always* be accounted for in the checks and in the calculation of the chord rotation capacity through the relevant expressions that are provided by the codes (see next chapter).

**Figure 8.18** Smooth rebar with very short lap splices. *Source:* Stelios Antoniou.

## 8.10 Shear Deformations

As is the case with the modeling of bar slippage, in practical applications it is difficult to model the effect of the shear deformations on the global structural response and in most cases this is omitted, since in typical structural configurations it is small or negligible. Moreover, it is either difficult to calibrate the curve models to account for the shear deformations (in plastic-hinge elements) or altogether impossible to include them (in fiber-based models).

Yet, the shear capacity of all members should *always* be calculated and accounted for in the checks. In fact, the shear capacity check is the most critical check for existing buildings in the vast majority of cases, especially in older buildings that have low transverse reinforcement ratios.

## 8.11 Foundation Modeling

Two things need to be considered in the nonlinear modeling of the foundation system:

- The foundation flexibility. The foundation system should be modeled considering the degree of fixity provided at the base of the structure. The possible options are to either model the footings flexibility with elastic (or inelastic) link elements or – in the case of relatively rigid supports – to fully fix the support nodes.
- Foundation checks and acceptance criteria (i.e., whether the foundation system performs elastically -as is expected- or fails).

The dynamic response of footings is a very complex problem requiring skill in soil mechanics, foundation engineering, structural dynamics and soil structure interaction. The accurate modeling of the stiffness and strength of a footing can be a daunting task, which becomes even more challenging for jacketed footings, and foundation systems with connecting beams or strip footing. Moreover, it is noted that in recent earthquakes there have been very few cases of failures at the foundation level, and these were mostly attributed to reasons irrelevant to the vibration and the structural response (e.g., soil liquefaction or slope stability). For this reason, foundation modeling and checking usually assume a secondary role in the assessment and strengthening process.

In general, the foundation is expected to perform elastically during a seismic event, and the action effects acting on the foundation components should be derived on the basis of capacity design considerations. In the presence of a rigid basement, foundation modeling can be ignored, and it can be safely assumed that the building is fixed at the level of the slab above the basement. For the case of strengthened footings with jackets, these can be treated in the analysis as plain nonjacketed members of the same size, ignoring the reinforcement of the existing footing and distributing the calculated (for the entire footing) reinforcement on the jacket only.

## 8.12 How Significant Are Our Modeling Decisions?

The short answer to this question is "very significant." Obviously, wrong modeling decisions can result in significantly flawed results, very far from the actual seismic demand on the members during the design earthquake. However, all the modeling parameters and

**Table 8.1** Modeling of different structural members.

| Structural element | Include in the analysis | Easy/difficult to model | Remarks |
|---|---|---|---|
| Columns/beams | Always | Easy | Model reinforcement in NL |
| Walls/core walls | Always | Easy | Model reinforcement in NL |
| Slabs (new structures) | Always | Easy | As rigid diaphragm |
| Slabs (existing structures) | Always if it has rigidity | Relatively Easy | As rigid diaphragm or as 2D linear elements |
| Stairs | Sometimes | Not-so-easy | Only when it affects the lateral stiffness of the building (lack of large vertical members) |
| Infills | Yes and No | Difficult | • Symmetric distribution of infills: safe to ignore them<br>• Asymmetric distribution of infills: unsafe to ignore them, should be considered |
| Beam – column joints (new structures) | Yes | Easy | |
| Beam – column joints (existing structures) | Yes | Difficult | Capacity checks for the joints should be performed |
| Bar slippage (new structures) | No | – | |
| Bar slippage (existing structures) | Yes | Very difficult | Should be considered in the capacity checks (chord rotation) |
| Shear deformations | Yes | Very difficult | Shear capacity checks should always be performed |
| Geometric nonlinearities | Yes | Easy | |
| Foundation modeling | Yes | Extremely difficult | In certain cases, it is ok to ignore foundations |

*Source:* Stelios Antoniou.

details are not equally important. It is not the same to incorrectly model the hysteretic behavior of a shear wall, as opposed to a small slab or a weak insignificant infill.

Table 8.1 provides a concise summary of all the issues that have been raised in this chapter, regarding the significance of each modeling decision that engineers must make and the relative ease or difficulty with which this can be considered.

## References

[ASCE] American Society of Civil Engineers (2017). *Seismic Evaluation and Retrofit of Existing Buildings (ASCE/SEI 41–17), 2017*. Virginia: Reston.

[NIST] National Institute of Standards and Technology (2010). *Nonlinear Structural Analysis for Seismic Design, A Guide for Practicing Engineers, GCR 10–917-5*, prepared by the. Gaithersburg, Maryland: NEHRP Consultants Joint Venture, a partnership of the Applied

Technology Council and the Consortium of Universities for Research in Earthquake Engineering, for the National Institute of Standards and Technology.

[NIST] National Institute of Standards and Technology (2013). *Nonlinear Structural Analysis for Seismic Design, A Guide for Practicing Engineers, GCR 14-917-27*, prepared by the. Gaithersburg, Maryland: NEHRP Consultants Joint Venture, a partnership of the Applied Technology Council and the Consortium of Universities for Research in Earthquake Engineering, for the National Institute of Standards and Technology.

Beyer, K., Dazio, A., and Priestley, M.J.N. (2008a). Inelastic wide-column models for U-shaped reinforced concrete wall. *Journal of Earthquake Engineering* 12 (1): 1–33, Imperial College Press.

Beyer, K., Dazio, A., and Priestley, M.J.N. (2008b). *Seismic Design of Torsionally Eccentric Buildings with U-Shaped RC Walls*. Pavia, Italy: ROSE School.

Beyer, K., Dazio, A., and Priestley, M.J.N. (2008c). Elastic and inelastic wide-column models for RC non rectangular walls. In: *Proceedings of the Fourteenth World Conference on Earthquake Engineering*. Beijing, China: WCEE.

Calabrese, A., Almeida, J.P., and Pinho, R. (2010). Numerical issues in distributed inelasticity modeling of RC frame elements for seismic analysis. *Journal of Earthquake Engineering* 14 (1): 38–68.

Clough, R.W. and Johnston, S.B. (1966). Effect of stiffness degradation on earthquake ductility requirements proceedings. *Second Japan National Conference on Earthquake Engineering* 1966: 227–232.

Correia, A.A. and Virtuoso, F.B.E. (2006). Nonlinear Analysis of Space Frames. In: *Proceedings of the Third European Conference on Computational Mechanics: Solids, Structures and Coupled Problems in Engineering* (ed. M. Soares et al.). Lisbon, Portugal: Spinger.

Crisafulli F.J. (1997). Seismic behaviour of reinforced concrete structures with masonry infills. PhD thesis. University of Canterbury, New Zealand.

Crisafulli, F.J., Carr, A.J., and Park, R. (2000). Analytical modeling of infilled frame structures – a general overview. *Bulletin of the New Zealand Society for Earthquake Engineering* 33 (1): 30–47.

Giberson, M.F. (1967). The response of nonlinear multi-story structures subjected to earthquake excitation, PhD dissertation. California Institute of Technology, Pasadena, CA, 232.

Hellesland, J. and Scordelis, A. (1981). *Analysis of RC Bridge Columns under Imposed Deformations*, 545–559. Delft: IABSE Colloquium.

Mari, A. and Scordelis, A. (1984). *SESM Report 82-12: Nonlinear Geometric Material and Time Dependent Analysis of Three-Dimensional Reinforced and Prestressed Concrete Frames*. Berkeley: Department of Civil Engineering, University of California.

Neuenhofer, A. and Filippou, F.C. (1997). Evaluation of nonlinear frame finite-element models. *Journal of Structural Engineering* 123 (7): 958–966.

Scott, M.H. and Fenves, G.L. (2006). Plastic hinge integration methods for force-based beam–column elements. *ASCE Journal of Structural Engineering* 132 (2): 244–252.

SeismoBuild (2023). SeismoBuild – A computer program for the linear and nonlinear analysis of Reinforced Concrete Buildings. www.seismosoft.com.

SeismoStruct (2023). SeismoStruct – A computer program for static and dynamic nonlinear analysis of framed structures. www.seismosoft.com.

Spacone, E., Ciampi, V., and Filippou, F.C. (1996). Mixed formulation of nonlinear beam finite element. *Computers & Structures* 58 (1): 71–83.

# 9

# Checks and Acceptance Criteria

## 9.1 Introduction

Once structural analysis has been completed for the selected seismic hazard level(s), the specific demands for the imposed actions of relevance are identified for all the structural components of the building. The demand parameters of interest are specified by the employed standard, and typically include peak forces and deformations in structural and nonstructural components, peak or residual story drifts, and floor accelerations. Other demand parameters, such as cumulative deformations or dissipated energy, may also be checked to help confirm the accuracy of the analysis and/or to assess cumulative damage effects. There may also be other performance limits (e.g., the onset of structural damage), which might have major implications on lifecycle cost and functionality of the building, and should be checked.

The next step is to calculate the appropriate acceptance criteria i.e., the members' capacities, for the selected performance level. In performance-based engineering, the recommendations for modeling the structure and carrying out the analyses must always be accompanied by an additional set of guidelines for employing the results of the analyses in determining whether a structure meets the specified acceptance criteria.

The performance is quantitatively checked by comparing the calculated values of demand parameters (in short the 'demand') with the acceptance criteria ("capacity") for the desired performance level. The acceptance criteria and the corresponding safety factors for seismic performance may vary depending on the type of the employed structural analysis (nonlinear methods, which are more accurate, require checks that are easier to fulfill), on the knowledge of the structure, and how uncertainties associated with the demands and acceptance criteria are handled. The building is considered to satisfy the requirements of the selected performance level, if the demands do not exceed the capacities for all actions of relevance, for all structural elements.

In this chapter, the main checks that should be carried out during an assessment or strengthening methodology will be described for the linear and the non-linear methods alike. Within this context, the definition of several new concepts that are not present in the procedures for the design of new buildings will be given. The actions applied on the structural components can generally be classified as "deformation-controlled" actions (for members that can tolerate inelastic deformations), and "force-controlled" actions (for non-ductile members whose capacity is governed by strength). Typical force-controlled actions are the shear acting on any member, or the axial force acting on columns. A typical deformation-controlled action is the bending moment or the end chord-rotations developed on beams. Before selecting component acceptance criteria, each component should be classified as primary or secondary, depending on the level on which the component contributes to the resistance against seismic forces. Finally, the choice of the appropriate material strengths (the expected or the lower-bound strengths) that should be employed in the calculation of the component capacities for the force and the deformation-controlled actions will also be explained.

## 9.2 Primary and Secondary Members

According to ASCE 41-17, Section 1.2, a *component* is a part of an architectural, mechanical, electrical, or structural system of a building. Components are the different parts of the building or the structural model, such as beams, columns, and diaphragms or infills of any kind (ASCE 2017). What is investigated and modeled during the analytical procedures is the mechanical properties (strength, deformability, and toughness) of the components and their interconnection.

The components of a structure may be classified as *Primary* or *Secondary*.

According to ASCE 41-17, Section 7.5.1.1:

- A structural component that is required to resist seismic forces and accommodate deformations for the structure to achieve the selected performance level shall be classified as *primary*.
- A structural component that accommodates seismic deformations and is not required to resist seismic forces for the structure to achieve the selected performance level shall be classified as *secondary*.

Roughly the same definitions are also employed within the Eurocodes' framework (CEN 2004, Section 4.2.2). The components, on which the engineer *relies* to resist the specified earthquake effects, are designated as primary. The components, on which the engineer *does not rely* to resist the specified earthquake effects, are designated as secondary. Typically, the secondary designation is used, when a component does not add considerably or reliably to the earthquake resistance and the lateral structural stiffness.

In general, stricter acceptance criteria are specified for primary components with respect to the secondary components. The designation of primary and secondary components has been introduced to allow some additional flexibility in the evaluation and retrofit process. Some examples of secondary components are:

- Where a nonstructural component does not contribute significantly or reliably to resist earthquake effects in any direction (e.g., a gypsum partition).
- Where a structural component does not contribute significantly or reliably to resist earthquake effects in any direction (e.g., a slab–column interior frame with large walls in the perimeter).

- When a component, intended in the original design of the building to be primary, is deformed beyond the point where it can be relied on to resist earthquake effects (e.g., coupling beams adjacent to large walls). In such cases, the engineer may designate these beams as secondary, allowing them to be deformed beyond their useful limits, provided that the sustained damage does not result in the loss of gravity load capacity.

All the other components are considered as primary.

## 9.3 Deformation-Controlled & Force-Controlled Actions

According to ASCE 41-17, Section 1.2, an *action* is an internal moment, shear, torque, axial force, deformation, displacement, or rotation corresponding to a displacement caused by a structural degree of freedom (ASCE 2017). In other words, actions are all the loads and deformations acting on a particular component.

The response of any component that is subjected to a specific action can be idealized by the Generalized Force-Deformation Curve, shown in Figure 9.1. The response starts with the linear, elastic part of the curve between point A (unloaded element) and an effective yield point B. The slope from point B to point C is typically a small percentage (0% to 10%) of the elastic slope and is included to represent phenomena such as strain hardening. Point C is the maximum component strength, and corresponds to the deformation after which significant strength degradation occurs (line CD). Beyond point D, the element responds with a substantially reduced strength (residual strength), until point E, after which no strength remains in the component. At deformations greater than point E, the element's seismic strength is essentially zero.

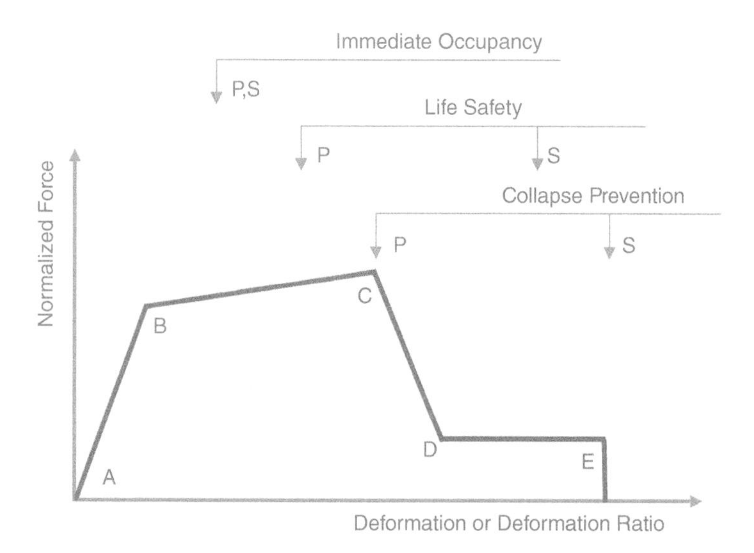


**Figure 9.1** Generalized Force-Deformation relations for depicting modeling and acceptance criteria (adapted from ASCE 2017, Fig. C7–3). *Source:* Stelios Antoniou.

All actions shall be classified as either:

- *Deformation-controlled actions* (US nomenclature) or *ductile actions* (European nomenclature), which have ductile failure mechanisms and perform well in the inelastic range.
- *Force-controlled actions* (US nomenclature) or *brittle actions* (European nomenclature), which have nonductile, brittle type of failure mechanisms.

Deformation-controlled actions have force vs. deformation curves similar to Type 1 in Figure 9.2. Force-controlled actions have force vs. deformation curves similar to Type 3 in Figure 9.2. There is also a third type of curve, Type 2, which describes structural behavior that is somewhere between the full deformation and the full force control actions.

The classification as a deformation or force-controlled action is not up to the discretion of the engineer. The deformation-controlled actions are explicitly specified by the standards, whereas all other actions are considered force-controlled. An example of a typical deformation-controlled action is the flexural behavior of members without significant axial load (Type 1 curve). An example of a force-controlled action is the shear behavior of any member (Type 3 curve). Any given component may have a combination of both deformation and force-controlled actions – for instance the shear acting on a member is force-controlled action, whereas the bending can be a deformation-controlled action.

The different standards specify which actions are to be checked for each component type, and engineers are not free to carry out checks simply based on their engineering judgment. Table C7-1 in ASCE 41-17 provides some examples of possible deformation and force-controlled actions in common framing systems (Table 9.1). The acceptability of force and deformation actions must be evaluated for every component of the structure. Engineers are not free to check only specific members i.e., those that they feel are the most important.

## 9.4 Expected Vs. Lower-Bound Material Strengths

In the force vs. deformation curves of Figure 9.2, $Q_y$ represents the yield strength of the component. Because the consequences of the exceedance of the capacity are generally different in force and deformation-controlled actions (with brittle and ductile behavior, respectively), it is permitted to treat them differently when calculating the members' strengths. With deformation-controlled actions $Q_y$ is expressed as the expected (mean)

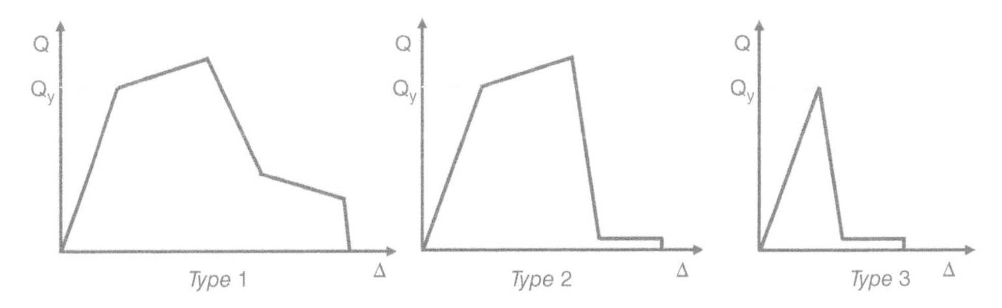


**Figure 9.2** Component force vs. deformation curves (adapted from ASCE 2017, Fig. 7-4). *Source:* Stelios Antoniou.

Table 9.1 Deformation-controlled and force controlled actions (ASCE 2017, Table C7-1).

| Component | Deformation-controlled action | Force-controlled action |
|---|---|---|
| Moment frames | | |
| • Beams | Moment ($M$) | Shear ($V$) |
| • Columns | – | Axial load ($P$), $V$ |
| • Joints | – | $V^a$ |
| Shear walls | $M$, $V$ | $P$ |
| Braced frames | | |
| • Braces | $P$ | |
| • Beams | – | $P$ |
| • Columns | – | $P$ |
| • Shear link | $V$ | $P$, $M$ |
| Connections | $P$, $V$, $M^b$ | $P$, $V$, $M$ |
| Diaphragms | $M$, $V^c$ | $P$, $V$, $M$ |

[a] Shear may be a deformation-controlled action in steel moment frame construction.
[b] Axial, shear, and moment may be deformation-controlled actions for certain steel and wood connections.
[c] If the diaphragm carries lateral loads from vertical-force-resisting elements above the diaphragm level, then $M$ and $V$ shall be considered force-controlled actions.

strength of the component $Q_{CE}$. On the contrary, for force-controlled actions, a lower-bound estimate of the component strength is used $Q_{CL}$, in order to be on the safe side. Typically, $Q_{CL}$ is the mean minus one standard deviation of the resistance of a component.

At this point, it is important to make a crucial distinction between existing and new materials. In the existing materials of the structure (e.g., concrete and steel rebar in existing beams and columns) the material strengths can be directly estimated from experimental tests that provide the mean value $\mu$ and the standard deviation $\sigma$. By contrast, new materials (e.g., concrete and steel rebar of new shear walls or of jackets for the strengthening of existing members) do not exist during the analytical process and will only be constructed at a later stage. Therefore, what is known is their characteristic or nominal value, from which the mean value may be estimated by multiplying it with a factor $\gamma$ that is prescribed in the codes (Table 9.2). The mean strength value (for existing materials) and the characteristic value times $\gamma$ (for new materials) are considered as the expected strength of the materials. Likewise, the mean minus one standard deviation (for existing materials) and the characteristic value (for new materials) are considered as the lower-bound strength of the materials (Table 9.3, see also ASCE 41: Table 7.6).

As a simple example, one may consider the strengthening of a reinforced concrete building with jackets. In the process, the designer has to accommodate both existing and new materials (concrete and steel reinforcement). For the concrete of the columns, a set of tests is carried out and the measured values (in MPa) of the compressive strengths are for instance (16.48, 19.65, 22.17, 19.87, 20.54, 21.63), which results in a mean strength $\mu$ equal to 20.06 MPa and a standard deviation $\sigma$ equal to 2.01 MPa. The lower-bound compressive strength is equal to $\mu - \sigma = 18.05$ MPa, whereas the expected strength is equal to the mean

**Table 9.2** Multiplying factors between lower-bound and expected strengths (ASCE 2017, Table 10-1).

| Material property | Factor |
|---|---|
| Concrete compressive strength | 1.50 |
| Reinforcing steel tensile and yield strength | 1.25 |
| Connector steel yield strength | 1.50 |

**Table 9.3** Lower-bound vs. expected strengths for existing and new materials.

| | Lower-bound strength | Expected strength |
|---|---|---|
| Existing materials | mean - $\sigma$ | mean |
| New materials | characteristic value | (characteristic value) • $\gamma$ |

*Source:* Stelios Antoniou.

value, i.e., 20.06 MPa. On the contrary the concrete of the jackets is a new material, and the lower-bound strength is equal to its nominal value, e.g., for C25/30 it is 25 MPa, whereas its mean value is 1.50*25 = 37.5 MPa.[1]

What should be remembered regarding the lower-bound and the expected strengths of the materials is that the latter are used for the calculation of the strengths in deformation-controlled actions, such as bending. By contrast, for the brittle, more dangerous force-controlled actions the lower-bound values of the strengths are employed, in order to be on the safe side.

## 9.5 Knowledge Level and Knowledge Factor

Another important concept in the assessment of existing structures that is not present in the design process is the *knowledge level* with which the properties of the building components are known when calculating component capacities, and the corresponding *knowledge factors* (US nomenclature) or *confidence factors* (European nomenclature).

The concepts were described in detail in Chapter 2. Here it will only be reiterated that there are three knowledge levels for good, average, or poor knowledge of the structural configuration. The knowledge factors in ASCE 41 and the Turkish code are equal or smaller to unity and operate on the capacity side, i.e., they are employed to decrease the capacity, if the knowledge level is not satisfactory. The confidence factors instead are equal or larger to unity and operate on the demand side, i.e., they are used to increase the demand in the checks that are carried out.

1 In the assessment and strengthening procedures the distinction between existing and new materials is something that often confuses engineers, who are accustomed to the design of new structures, where inevitably there are only new materials and they are treated in a universal manner.

## 9.6 Capacity Checks

As in the standard design methodologies, the requirement of the capacity checks is that the component strength is larger than the demand on the component, i.e.:

$$Q_C \geq Q_U \text{ for ASCE 41, or } R_d \geq E_d \text{ for Eurocode 8}$$

$Q_U$ or $E_d$ is the design value of the action effect for the seismic design situation for the selected hazard level. $Q_c$ or $R_d$ is the corresponding resistance of the element, considering specific material parameters (e.g., lower-bound, nominal, or mean value of the strength), based on the type of analysis (linear or nonlinear), the type of the action (ductile or brittle) and the selected performance level.

Note two important differences, with respect to the typical design procedures, in which all the checks are carried out in terms of forces using the characteristic material strengths:

- When calculating the components' strengths for the deformation-controlled actions (e.g., bending) the mean or expected strength of the materials is employed, rather than the characteristic or lower-bound strength. However, for the force-controlled actions (e.g., shear), the characteristic or lower-bound strengths of the components are always used.[2]
- When employing the nonlinear analytical procedures, the capacity checks of the deformation-controlled actions are performed in terms of deformations, rather than forces. For instance, the bending checks are done using the chord rotation or plastic hinge rotation of the member, rather than the bending moments, as one would do in a typical design procedure. On the contrary, in the linear procedures, the checks are still performed in terms of forces.

### 9.6.1 Capacity Checks for Linear Methods – ASCE 41

When using the linear methods of analysis, the checks are always performed in terms of forces for both the deformation and the force-controlled actions.

#### 9.6.1.1 Component Demands

For the deformation-controlled actions the component demand $Q_{UD}$ is calculated from the set of linear analyses.

For the force-controlled actions the components demand $Q_{UF}$ is calculated based on *capacity design considerations* (i.e., estimate of the maximum action that can be developed in a component, based on a limit-state analysis), taking into account the expected strength of the components that deliver forces to the component under consideration, in order to make sure that a failure in the force-controlled action is avoided. Alternatively, in the members that are not expected to experience large inelasticity more simplified expressions may be used that adapt the results extracted from the linear analyses (see ASCE 41, Section 7.5.2.1.2 and specifically Eq. [7.35]).

2 Such an explicit distinction between the expected and lower-bound material strengths applies only to ASCE 41. In Eurocode 8, in both cases the mean material values are employed, however in the brittle actions these values are further divided by the material partial safety factors $\gamma_c$ and $\gamma_s$.

#### 9.6.1.2 Component Capacities

For the deformation-controlled actions the capacity of the components shall be based on the expected strengths $Q_{CE}$. It is equal to $\kappa \cdot Q_{CE}$ for the existing materials and $Q_{CE}$ for the new materials ($\kappa$ is the knowledge factor). On the contrary, for force-controlled actions, the capacity of the components shall be based on the lower-bound strengths $Q_{CL}$, and are equal to $\kappa \cdot Q_{CL}$ for the existing materials and $Q_{CL}$ for the new materials (Table 9.4).

For the deformation-controlled actions, some kind of nonlinear response is anticipated; hence it is permitted for the actions to exceed the actual strength of the component. This is achieved through the so-called ***m-factors***. The m-factors are capacity modification factors that account for the inelasticity expected in the ductile members at the selected structural performance level. They are generally larger than 1.0, and they effectively increase the capacity of the member. They are an indirect measure of the nonlinear deformation capacity of the component, and they are somewhat similar to the well-known q-factors, with the important distinction that m-factors are component-based, rather than structure-based.

The values of the m-factors depend on the selected performance level, the type of member, the member's reinforcement, the lap splices, and the axial and shear forces that act on the member. They are provided directly from tables in ASCE 41 (Table 9.5). The values between the limits listed in the tables are determined using linear interpolation functions.

Consequently, the capacities for the deformation-controlled actions shall be defined as the product of the m-factors, the κ-factors, and the expected strengths, $Q_{CE}$.

$$Q_C > Q_U \xrightarrow{\text{Deformation-controlled actions}} m\kappa Q_{CE} > Q_{UD} \tag{9.1}$$

On the contrary, for force-controlled actions, a nonlinear response is not allowed, and the actions cannot exceed the actual strength of the components. Therefore, the capacities for force-controlled actions shall be defined as the product of the κ-factors, and lower-bound strengths, $Q_{CL}$.

$$Q_C > Q_U \xrightarrow{\text{Force-controlled actions}} \kappa Q_{CL} > Q_{UF} \tag{9.2}$$

**Table 9.4** Component capacities for existing and new materials and for deformation and force-controlled actions. Linear Procedures (ASCE 2017, Table 7-6).

| Parameter | Deformation controlled | Force controlled |
|---|---|---|
| Existing material strength | Expected mean value with allowance for strain hardening | Lower bound value (approximately mean value minus 1σ level) |
| Existing action capacity | $\kappa Q_{CE}$ | $\kappa Q_{CL}$ |
| New material strength | Expected material strength | Specified material strength |
| New action capacity | $Q_{CE}$ | $Q_{CL}$ |

**Table 9.5** Example of a table in ASCE 41 that provides values for the m-factors for the linear procedures (ASCE 2017, Table 10-10a).

| | | | $m$-Factors[a] | | | | |
|---|---|---|---|---|---|---|---|
| | | | | Performance level | | | |
| | | | | Component type | | | |
| | | | | Primary | | Secondary | |
| $\left(\frac{N_{UD}}{A_g f'_{cE}}\right)$ | $\rho_t$ | $V_{yE}/V_{ColOE}$ | IO | LS | CP | LS | CP |
| Columns not controlled by inadequate development or splicing along the clear height[b] | | | | | | | |
| ≤0.1 | ≥0.0175 | ≥0.2 <0.6 | 1.7 | 3.4 | 4.2 | 6.8 | 8.9 |
| ≥0.7 | ≥0.0175 | ≥0.2 <0.6 | 1.2 | 1.4 | 1.7 | 1.4 | 1.7 |
| ≤0.1 | ≤0.0005 | ≥0.2 <0.6 | 1.5 | 2.6 | 3.2 | 2.6 | 3.2 |
| ≥0.7 | ≤0.0005 | ≥0.2 <0.6 | 1.0 | 1.0 | 1.0 | 1.0 | 1.0 |
| ≤0.1 | ≥0.0175 | ≥0.6 <1.0 | 1.5 | 2.7 | 3.3 | 6.8 | 8.9 |
| ≥0.7 | ≥0.0175 | ≥0.6 <1.0 | 1.0 | 1.0 | 1.0 | 1.0 | 1.0 |
| ≤0.1 | ≤0.0005 | ≥0.6 <1.0 | 1.3 | 1.9 | 2.3 | 1.9 | 2.3 |
| ≥0.7 | ≤0.0005 | ≥0.6 <1.0 | 1.0 | 1.0 | 1.0 | 1.0 | 1.0 |
| ≤0.1 | ≥0.0175 | ≥1.0 | 1.3 | 1.8 | 2.2 | 6.8 | 8.9 |
| ≥0.7 | ≥0.0175 | ≥1.0 | 1.0 | 1.0 | 1.0 | 1.0 | 1.0 |
| ≤0.1 | ≤0.0005 | ≥1.0 | 1.1 | 1.0 | 1.1 | 1.7 | 2.1 |
| ≥0.7 | ≤0.0005 | ≥1–0 | 1.0 | 1.0 | 1.0 | 1.0 | 1.0 |
| Columns controlled by inadequate development or splicing along the clear height[b] | | | | | | | |
| ≤0.1 | ≥ 0.0075 | | 1.0 | 1.7 | 2.0 | 5.3 | 6.8 |
| ≥0.7 | ≥ 0.0075 | | 1.0 | 1.0 | 1.0 | 2.8 | 3.5 |
| ≤0.1 | ≤0.0005 | | 1.0 | 1.0 | 1.0 | 1.4 | 1.6 |
| ≥0.7 | ≤0.0005 | | 1.0 | 1.0 | 1.0 | 1.0 | 1.0 |

[a] Values between those listed in the table shall be determined by linear interpolation.
[b] Columns are considered to be controlled by inadequate development or splicing where the calculated steel stress at the splice exceeds the steel stress specified by Eq. (10–1a) or (10–1b). Acceptance criteria for columns controlled by inadequate development or splicing shall never exceed those of columns not controlled by inadequate development or splicing.

## 9.6.2 Capacity Checks for Nonlinear Methods – ASCE 41

### 9.6.2.1 Component Demands

For the deformation-controlled actions the quantities checked are deformations (rather than forces), as these are directly calculated from the nonlinear analysis.

For the force-controlled actions, the component demands are the forces that are again calculated from the nonlinear analysis. Note that the demands are *not* to be determined from capacity design considerations as in the linear methods, since inelasticity is explicitly accounted for by the nonlinear analysis method, and the capacity design concept is no longer needed.

### 9.6.2.2 Component Capacities

For the deformation-controlled actions, the component capacities are taken as *permissible inelastic deformation limits* that are determined considering all coexisting forces and deformations at the target displacement. The values of the deformation limits are taken from tables that are provided for the different types of members, and depend on the selected performance level, the type of member, the member reinforcement, the lap splices, and the axial and shear forces that act on the member (Table 9.6).

For force-controlled actions, the component capacities are taken as lower-bound strengths that are determined considering all coexisting forces and deformations at the target displacement. Contrary to the deformation-controlled actions, the checks for the force-controlled actions are performed in terms of forces (Table 9.7).

## 9.6.3 Capacity Checks for Linear Methods – Eurocode 8, Part 3

Contrary to ASCE 41, where the inelasticity in the ductile mechanisms is taken into account by m-factors, in Eurocode 8 inelasticity is considered by the selected behavior factor q. The philosophy of both factors is the same, i.e., to account for the capability of ductile members to deform beyond their yield point, and the values of both are equal or larger to 1.00. However, there are two main differences between the q and the m-factors. The most important difference is that, whereas the m-factors are member specific (i.e., different m-factors may be assigned to the different structural components), the q-factor is based on the entire capability of the building to absorb energy. Secondly, the m-factors operate on the capacity side of the inequality, effectively increasing the strengths, whereas the q-factor is employed to decrease the demand on the components.

### 9.6.3.1 Component Demands

For the ductile actions, the component demand $E_{d,\,D}$ is calculated from the linear analysis using a behavior factor q that is consistent with the ductility of the building. For existing reinforced concrete buildings, a value of q equal to1.50 may be adopted. Higher values of q can only be employed if they are suitably justified with reference to the local and global available ductility of the structure.

For brittle actions, the component demand $E_{d,B}$ is calculated in a similar fashion to ASCE 41, i.e., based on capacity design considerations, taking into account the expected strength of the components that deliver force to the component under consideration.

### 9.6.3.2 Component Capacities

For the ductile actions, the capacity of the components shall be based on the expected material strengths divided by the confidence factor $R_{d,m}$/CF for existing materials, and on the characteristic strengths $R_{d,k}$ for new materials (see CEN 2005, Section 2.2.1[5]P).

Table 9.6 Numerical acceptance criteria for nonlinear procedures, RC beams (ASCE 2017, Table 10-7).

| Conditions | | | Modeling parameters[a] | | | Acceptance criteria[a] Plastic rotation angle (radians) | | |
|---|---|---|---|---|---|---|---|---|
| | | | Plastic rotation angle (radians) | | Residual strength ratio | Performance level | | |
| | | | a | b | c | IO | LS | CP |
| Condition i. Beams controlled by flexure[b] | | | | | | | | |
| $\frac{\rho-\rho'}{\rho_{bal}}$ | Transverse reinforcement[c] | $\frac{V^d}{b_w d\sqrt{f'_{cE}}}$ | | | | | | |
| ≤0.0 | C | ≤3 (0.25) | 0.025 | 0.05 | 0.2 | 0.010 | 0.025 | 0.05 |
| ≤0.0 | C | ≥6 (0.5) | 0.02 | 0.04 | 0.2 | 0.005 | 0.02 | 0.04 |
| ≥0.5 | C | ≤3 (0.25) | 0.02 | 0.03 | 0.2 | 0.005 | 0.02 | 0.03 |
| ≥0.5 | C | ≥6 (0.5) | 0.015 | 0.02 | 0.2 | 0.005 | 0.015 | 0.02 |
| ≤0.0 | NC | ≤3 (0.25) | 0.02 | 0.03 | 0.2 | 0.005 | 0.02 | 0.03 |
| ≤0.0 | NC | ≥6 (0.5) | 0.01 | 0.015 | 0.2 | 0.0015 | 0.01 | 0.015 |
| ≥0.5 | NC | ≤3 (0.25) | 0.01 | 0.015 | 0.2 | 0.005 | 0.01 | 0.015 |
| ≥0.5 | NC | ≥6 (0.5) | 0.005 | 0.01 | 0.2 | 0.0015 | 0.005 | 0.01 |
| Condition ii. Beams controlled by shear[b] | | | | | | | | |
| Stirrup spacing ≤ $d/2$ | | | 0.0030 | 0.02 | 0.2 | 0.0015 | 0.01 | 0.02 |
| Stirrup spacing > $d/2$ | | | 0.0030 | 0.01 | 0.2 | 0.0015 | 0.005 | 0.01 |
| Condition iii. Beams controlled by inadequate development or splicing along the span[b] | | | | | | | | |
| Stirrup spacing ≤ $d/2$ | | | 0.0030 | 0.02 | 0.0 | 0.0015 | 0.01 | 0.02 |
| Stirrup spacing > $d/2$ | | | 0.0030 | 0.01 | 0.0 | 0.0015 | 0.005 | 0.01 |
| Condition iv. Beams controlled by inadequate embedment into beam-column joint[b] | | | | | | | | |
| | | | 0.015 | 0.03 | 0.2 | 0.01 | 0.02 | 0.03 |

*Note:* $f'_{cE}$ in lb./in.$^2$ (MPa) units.

[a] Values between those listed in the table should be determined by linear interpolation.

[b] Where more than one of conditions i, ii, iii, and iv occur for a given component, use the minimum appropriate numerical value from the table.

[c] "C" and "NC" are abbreviations for conforming and nonconforming transverse reinforcement, respectively. Transverse reinforcement is conforming if, within the flexural plastic hinge region, hoops are spaced at ≤ $d/3$, and if, for components of moderate and high ductility demand, the strength provided by the hoops ($V_s$) is at least 3/4 of the design shear. Otherwise, the transverse reinforcement is considered nonconforming.

[d] $V$ is the design shear force from NSP or NDP.

**Table 9.7** Component capacities for existing and new materials and for deformation and force-controlled actions. Nonlinear procedures (ASCE 41 2017, Table 7-7).

| Parameter | Deformation controlled | Force controlled |
|---|---|---|
| Deformation capacity (existing component) | $\kappa\times$Deformation limit | N/A |
| Deformation capacity (new component) | Deformation limit | N/A |
| Strength capacity (existing component) | N/A | $\kappa\times$QCL |
| Strength capacity (new component) | N/A | QCL |

The brittle actions consider the same values $R_{d,m}$/CF and $R_{d,k}$. However, for the primary elements these are further divided by the material partial factors $\gamma_c$ and $\gamma_s$ for concrete and steel (the recommended values are 1.50 and 1.15, respectively), providing values close or equal to the lower-bound strengths.

## 9.6.4 Capacity Checks for Nonlinear Methods – Eurocode 8, Part 3

### 9.6.4.1 Component Demands

For the ductile actions the component demands are the deformations (rather than forces), as these are calculated from the nonlinear analysis.

For the brittle actions the component demands are forces again calculated from the nonlinear analysis. Note that the demands are *not* to be determined from capacity design considerations as in the linear methods, since inelasticity is explicitly accounted for by the nonlinear analysis method, and the capacity design concept is not needed anymore.

### 9.6.4.2 Component Capacities

For the ductile actions the component capacities are taken as *permissible inelastic deformation limits* that are determined considering all coexisting forces and deformations on the member at the target displacement. The values of the deformation limits are calculated from expressions provided in the code, using expected material strengths divided by the confidence factor $R_{d,m}$/CF for existing materials and the characteristic strengths $R_{d,k}$ for new materials (see CEN 2005, Section 2.2.1[5]P).

For the brittle actions, the checks are performed in terms of forces, and the component capacities are taken as strengths that are determined considering all coexisting forces and deformations on the member at the target displacement. For brittle actions the same material strength values $R_{d,m}$/CF and $R_{d,k}$ are considered, but, as in the linear methods, for the primary elements these are further divided by the material partial factors $\gamma_c$ and $\gamma_s$ for concrete and steel (the recommended values are 1.50 and 1.15, respectively).

# 9.7 Main Checks to Be Carried Out in an Assessment Procedure

In the current section, basic information is provided regarding the main checks that should be carried out in the assessment of a RC building. The Eurocodes (and specifically Eurocode 8, Part 3), as well as ASCE 41 and the framework of the US standards will be covered briefly. A more detailed description is provided for both standards in Appendices A.1 and A.2, respectively.

### 9.7.1 Bending Checks

Bending, with the exception of vertical members loaded with very large axial forces, is a deformation-controlled action with a ductile failure mechanism. Hence, in linear analysis the checks are performed in terms of bending moments, and in the nonlinear procedures they are carried out in terms of deformations. In nonlinear methods, the chord rotation (in Eurocode 8) or the plastic hinge rotation (in ASCE 41), are selected as the deformation quantity to be checked.[3] Most of the other standards adopt the chord rotation as the quantity to be checked, with the exception of the Turkish code TBDY, which specifies material strains.

#### 9.7.1.1 Eurocode Framework (EC8: Part 1 and EC8: Part 3) – Nonlinear Methods

According to CEN (2005), in the nonlinear procedures the bending capacity for the limit state of Near Collapse (NC) is checked in terms of the value of the ultimate chord rotation capacity (elastic plus inelastic part), which is calculated from the following equation:

$$\theta_{um} = \frac{1}{\gamma_{el}} \cdot 0.016 \cdot \left(0.3^{\nu}\right)\left[\frac{\max(0.01;\omega')}{\max(0.01;\omega)} f_c\right]^{0.225} \cdot \left(\min\left(9, \frac{L_V}{h}\right)\right)^{0.35} 25^{\left(\alpha\rho_{sx}\frac{f_{yw}}{f_c}\right)} \left(1.25^{100\rho_d}\right) \quad \text{EN1998-3 Eq. (A.1)}$$

where $\gamma_{el}$ is equal to 1.5 for primary and 1.0 for secondary elements, and $L_V$ is the ratio between the bending moment M and the shear force V at the end of the member. The other parameters of the equation are defined in Appendix A.1. For walls, the value given by the (A.1) is multiplied by 0.58.

The total chord rotation capacity at ultimate may also be calculated as the sum of the chord rotation at yielding and the plastic part of the chord rotation capacity calculated by EN 1998-3 Eq. (A.3), the expression is given in Appendix A.1 of the current document.

The chord rotation capacity that corresponds to the limit state of Significant Damage (SD) is assumed to be three-fourths of the ultimate chord rotation, as calculated from the equation above.

The chord rotation capacity that corresponds to the limit state of Damage Limitation (DL) is given by the chord rotation at yielding:

For rectangular beams and columns:

$$\theta_y = \phi_y \frac{L_V + \alpha_V z}{3} + 0.0014\left(1 + 1.5\frac{h}{L_V}\right) + \frac{\varepsilon_y}{d - d'} \frac{d_{bL} f_y}{6\sqrt{f_c}} \quad \text{EN1998-3 Eq. (A.10a)}$$

For walls or rectangular T- or barbelled section:

$$\theta_y = \phi_y \frac{L_V + \alpha_V z}{3} + 0.0013 + \frac{\varepsilon_y}{d - d'} \frac{d_{bL} f_y}{6\sqrt{f_c}} \quad \text{EN1998-3 Eq. (A.11a)}$$

3 Large RC walls that are controlled by shear are an exception in ASCE 41; their bending checks are carried out in terms of the lateral drift ratio, rather than the hinge rotation.

The relevant input parameters are explained in Appendix A.1.

Alternative expressions also exist (Eqs. (A.10 b) and (A.11b) of EN 1998–3; these expressions are also presented in Appendix A.1).

#### 9.7.1.2 US Framework (ASCE 41 and ACI 318) – Nonlinear Methods

Contrary to the European standards, in ASCE the plastic part of the member hinge rotation is employed for the bending checks in the nonlinear procedures. This is given in tables:

- For beams according to Table 10.7 of ASCE 41.
- For columns according to Table 10.8 of ASCE 41 (see Table 9.8 of the current document).
- For walls controlled by flexure according to Table 10.19 of ASCE 41.

The deformation capacity of walls controlled by shear is defined in terms of the inter-story drift ratio as indicated in Table 10.20 of ASCE 41-17.

### 9.7.2 Shear Checks

Shear is the most common, most important, and typically the most critical check in the assessment methodologies. As explained in detail in Chapter 2, the main reason for this is the significant lack of transverse reinforcement in older construction.

#### 9.7.2.1 Eurocodes Framework (EC8, Part 1, and EC8, Part 3)

The shear capacity of beams, columns and walls is calculated through the following expression according to Appendix A of EN1998–3 (CEN 2005), as controlled by the stirrups, and accounting for the reduction due to the plastic part of ductility demand.

$$V_R = \frac{1}{\gamma_{el}}\left[\frac{h-x}{2L_V}\min\left(N;0.55A_c f_c\right)+\left(1-0.05\min\left(5;\mu_\Delta^{pl}\right)\right)\cdot \left[0.16\max\left(0.5;100\rho_{tot}\right)\left(1-0.16\min\left(5;\frac{L_V}{h}\right)\right)\sqrt{f_c}A_c+V_w\right]\right] \quad \text{EN1998-3 Eq. (A.12)}$$

where $\gamma_{el}$ is equal to 1.15 for primary seismic elements and to 1.0 for secondary ones. The other parameters are defined in Appendix A.1.

The shear strength of a concrete wall should also not be taken greater than the value corresponding to failure by web crushing, $V_{R,max}$, which under cyclic loading is calculated according to the following expression:

$$V_{R,max} = \frac{0.85\left(1-0.06\min\left(5;\mu_\Delta^{pl}\right)\right)}{\gamma_{el}}\left(1+1.8\min\left(0.15;\frac{N}{A_c f_c}\right)\right)\left(1+0.25\max\left(1.75;100\rho_{tot}\right)\right)\cdot \left(1-0.2\min\left(2;\frac{L_V}{h}\right)\right)\sqrt{f_c}b_w z \quad \text{EN1998-3 Eq. (A.15)}$$

**Table 9.8** Numerical acceptance criteria for nonlinear procedures – Reinforced concrete columns (ASCE 41 2017, Table 10-8).

| **Modeling parameters** | **Acceptance criteria** | | |
|---|---|---|---|
| | **Plastic rotation angle (radians)** | | |
| | **Performance level** | | |
| Plastic rotation angles, a and $b$ (radians)<br>Residual strength ratio, $c$ | IO | LS | CP |
| Columns not controlled by inadequate development or splicing along the clear height[a] | | | |
| $a = \left(0.042 - 0.043\frac{N_{UD}}{A_g f'_{cE}} + 0.63\rho_t - 0.023\frac{V_{yE}}{V_{ColOE}}\right) \geq 0.0$ | 0.15 a<br>$\leq 0.005$ | 0.5 $b^b$ | 0.7 $b^b$ |
| $For\ \frac{N_{UD}}{A_g f'_{cE}} \leq 0.5 \left\{ b = \frac{0.5}{5 + \frac{N_{UD}}{0.8 A_g f'_{cE}} \frac{1}{\rho t} \frac{f'_{cE}}{f_{ytE}}} - 0.01 \geq a^a \right.$ | | | |
| $c = 0.24 - 0.4\frac{N_{UD}}{A_g f'_{cE}} \geq 0.0$ | | | |
| Columns controlled by inadequate development or splicing along the clear height[c] | | | |
| $a = \left(\frac{1 \rho_t f_{ytE}}{8 \rho_l f_{ylE}}\right) \begin{matrix} \geq 0.0 \\ \leq 0.025^d \end{matrix}$ | 0.0 | 0.5 $b$ | 0.7 $b$ |
| $b = \left(0.012 - 0.085\frac{N_{UD}}{A_g f'_{cE}} + 12\rho_t^{\ e}\right) \begin{matrix} \geq 0.0 \\ \geq a \\ \leq 0.06 \end{matrix}$ | | | |
| $c = 0.15 + 36\rho_t \leq 0.4$ | | | |

*Notes:* $\rho_t$ shall not be taken as greater than 0.0175 in any case nor greater than 0.0075 when ties are not adequately anchored in the core. Equations in the table are not valid for columns with $\rho_t$ smaller than 0.0005. $V_{yE}/V_{ColOE}$ shall not be taken as less than 0.2.

NUD shall be the maximum compressive axial load accounting for the effects of lateral forces as described in Eq. (7-34). Alternatively, it shall be permitted to evaluate NUD based on a limit-state analysis.

[a] $b$ shall be reduced linearly for $N_{UD} / (A_g f'_{cE}) > 0.5$ from its value at $N_{UD} / (A_g f'_{cE}) = 0.5$ to zero at $N_{UD} / (A_g f'_{cE}) = 0.7$ but shall not be smaller than a.

[b] $N_{UD} / (A_g f'_{cE})$ shall not be taken as smaller than 0.1.

[c] Columns are considered to be controlled by inadequate development or splices where the calculated steel stress at the splice exceeds the steel stress specified by Eq. (10-1a) or (10-1b). Modeling parameter for columns controlled by inadequate development or splicing shall never exceed those of columns not controlled by inadequate development or splicing.

[d] $a$ for columns controlled by inadequate development or splicing shall be taken as zero if the splice region is not crossed by at least two tie groups over its length.

[e] $\rho_t$ shall not be taken as greater than 0.0075.

#### 9.7.2.2 US Framework (ASCE 41 and ACI 318)

For the components with low ductility demands, the shear strengths are calculated following the procedures for effective elastic response for the typical shear design of ACI 318: Chapter 22 (ACI 2019). For the components with moderate to high ductility demands, the shear strengths are calculated according to the procedures for ductile components of ACI 318: Chapter 18, employing capacity design considerations.

In particular for beam–column moment frames, the shear capacity of columns can also be calculated through the following expression:

$$V_{Col} = k_{nl} V_{Col0} = k_{nl}\left[\alpha_{Col}\left(\frac{A_v f_{ytL/E} d}{s}\right) + \lambda\left(\frac{6\sqrt{f'_{cL/E}}}{M_{UD}/V_{UD}d}\sqrt{1+\frac{N_{UG}}{6A_g\sqrt{f'_{cL/E}}}}\right)0.8A_g\right]\left(\text{lb/in.}^2\ \text{units}\right)$$

$$V_{Col} = k_{nl} V_{Col0} = k_{nl}\left[\alpha_{Col}\left(\frac{A_v f_{ytL/E} d}{s}\right) + \lambda\left(\frac{0.5\sqrt{f'_{cL/E}}}{M_{UD}/V_{UD}d}\sqrt{1+\frac{N_{UG}}{0.5A_g\sqrt{f'_{cL/E}}}}\right)0.8A_g\right]\left(\text{MPa units}\right)$$

$$\text{ASCE 41-17 Eq. } (10\text{-}3)$$

It is noteworthy that in ASCE 41, specific rules are provided for cases of large stirrup spacing and inadequate anchoring of the hoops, which are not so uncommon in existing buildings and can be very unfavorable when resisting seismic hysteretic shear (see ASCE 2017, section 10.3.4). In members where the longitudinal spacing of the transverse reinforcement exceeds half the member effective depth, the transverse reinforcement should be assumed to have reduced effectiveness in resisting shear by a factor of $2{\cdot}(1 - s/d)$. Furthermore, in members where the longitudinal spacing of transverse reinforcement exceeds the component effective depth, transverse reinforcement should be assumed ineffective in resisting shear.

### 9.7.3 Beam-Column Joints

This is a force-controlled action. and the checks are carried out in terms of forces, typically in terms of the joint shear strength. Among the different codes, only ASCE 41 provides some explicit rules on how to model the flexibility and nonlinear behavior of the beam-column joints, although the information provided is somehow limited (e.g., it gives details only for the 2D case).

In ASCE 41 the checks are carried out in terms of the joint shear strength (Eq. [10.4]). Eurocode 8, Part 3, instead refers directly to the corresponding clauses for new buildings in EN 1998-1, which is quite conservative since it requires capacity design considerations. CEN (2004) specifies checks for the shear joint strength, the existence of horizontal hoops, and the existence of vertical reinforcement.

## References

[ACI] American Concrete Institute ACI 318 (2019). ACI CODE-318-19: Building Code Requirements for Structural Concrete and Commentary. ACI Committee 318.

[ASCE] American Society of Civil Engineers (2017). *Seismic Evaluation and Retrofit of Existing Buildings (ASCE/SEI 41–17), 2017*. Virginia: Reston.

CEN (2004). European Standard EN 1998-1: 2004. In: *Eurocode 8: Design of Structures for Earthquake Resistance, Part 1: General Rules*. Comité Européen de Normalisation, Brussels: Seismic Actions and Rules for Buildings.

CEN (2005). European Standard EN 1998-3: 2005. In: *Eurocode 8: Design of Structures for Earthquake Resistance, Part 3: Assessment and Retrofitting of Buildings*. Brussels: Comité Européen de Normalisation.

# 10

# Practical Example: Assessment and Strengthening of a Six-Story RC Building

## 10.1 Introduction

In the current chapter, the assessment and strengthening of an existing multi-story reinforced concrete building will be presented. The full assessment procedure will be described in detail, from the assembly of the as-built information and the structural modeling, through to the selection of the limit states and the seismic hazard levels, until the execution of the code-based checks and the output of the deliverables. Subsequently, the building will be strengthened and the procedure will be carried out again, in order to confirm that the proposed retrofit measures comply with the acceptance criteria that correspond to the selected performance level.

The building under consideration has a typical design for residential use with limited anti-seismic specifications of the 1980s in Southern Europe. The EC8, Part 3 (CEN 2005) methodology will be followed with the Nonlinear Static (pushover) Procedure NSP, which nowadays is considered the standard method for assessment and retrofit. The SeismoBuild package (2023) is used, since it employs SeismoStruct's (2023) well-tested analytical solver, but also permits the implementation of the entire procedure in a series of easy steps.

## 10.2 Building Description

The building under consideration has five typical floors, a penthouse, and no basement. The building is irregular in plan, and its layout can be inscribed in a rectangle of 24.00 m by 17.50 m approximately (Figure 10.1). The height of all stories is 2.90 m with the exception of the ground floor, which is 3.20 m, and the total building height is almost 18.00 m. The 3D model of the building is depicted in Figure 10.2.

The load-bearing system consists of frames in both horizontal directions, and a lightly reinforced U-shaped wall core in the perimeter of the elevator shaft. The vertical members are small and relatively weak, with low longitudinal and even lower transverse reinforcement ratios. The beams are also small and lightly reinforced with relatively small spans.

*Seismic Retrofit of Existing Reinforced Concrete Buildings*, First Edition. Stelios Antoniou.

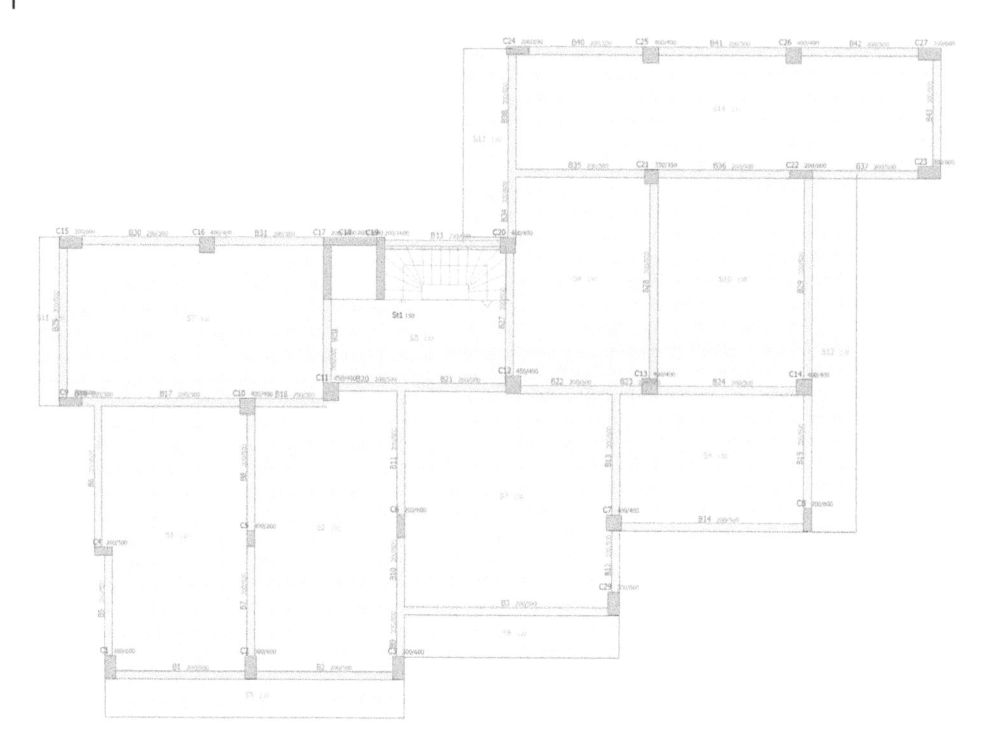


Figure 10.1 Building layout of the typical floor. *Source:* Stelios Antoniou.

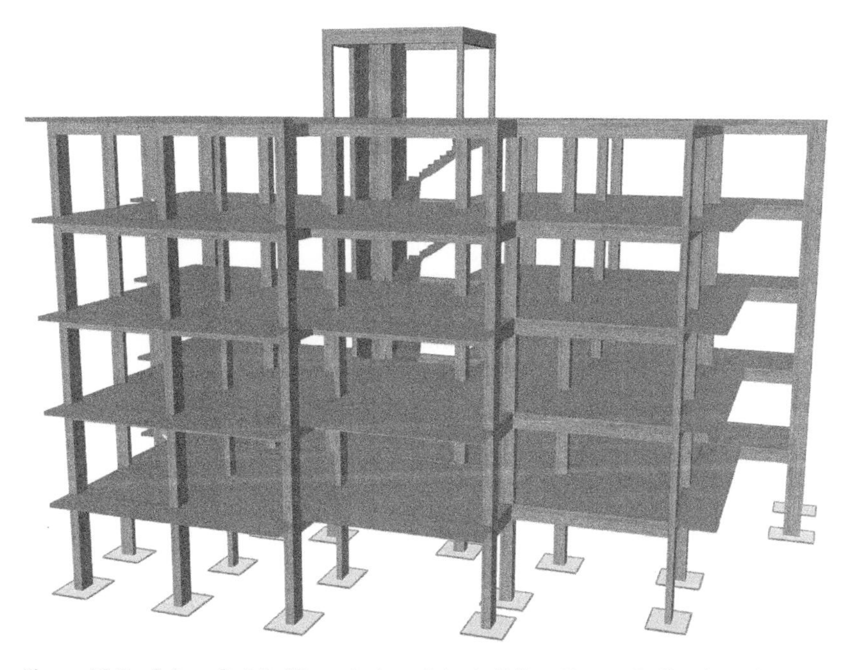

Figure 10.2 SeismoBuild's 3D rendering of the building. *Source:* Stelios Antoniou.

The slabs are thin (13 cm thick), and the infills are made of ceramic perforated bricks and are evenly distributed throughout the building, both in plan and in elevation.

## 10.3 Knowledge of the Building and Confidence Factor

Prior to any analysis, a site visit and a comprehensive survey are conducted, in order to obtain all the necessary as-built information, as described in Chapter 3. This includes the geometry, the identification of the structural system, the member dimensions, the material properties, the reinforcement, the nonstructural components that affect the seismic building response, and the foundation and the soil conditions.

A good knowledge of the structure should be based on the construction drawings for both the original construction and (if any) subsequent modifications, as well as a sufficient number of checks for the geometry, the member sizes and the reinforcement, which confirm that the available information corresponds to the actual reality. Information on the mechanical properties of the construction materials should be acquired based on testing of material specimens, as well as from the original test reports.

The thorough or insufficient knowledge of the structural configuration, and the reliability of the extracted information, determines the knowledge level and the confidence factors employed in the assessment, as well as the suitable procedures (linear or nonlinear) for the structural analysis.

### 10.3.1 Geometry

The dimensions and the exact location of the columns and beams of the structure should be determined. The possession of the original design or as-built drawings are extremely helpful, however generally they are not enough; the validity and the accuracy of their information should be confirmed with a detailed visual inspection. With the identification of the building's geometry, the load transfer paths and the configuration of the load-bearing system (for both gravity and lateral loads) may be established. With the exception of the penthouse, all columns have dimensions of 35/35, 40/40, or 60/20, and all the beams are 50/20.

### 10.3.2 Reinforcement

During the site visit, a thorough scan of the reinforcement of the primary structural members should be conducted, especially for the vertical ones. Even in the cases when detailed information of the members' reinforcement can be retrieved from drawings or reinforcement tables, this information should be systematically checked and confirmed on site. Recall that the transverse reinforcement (especially in columns), whose role in seismic resistance is universally acknowledged nowadays, was considered of secondary importance before the gradual development of earthquake engineering in the 1970s and later. Furthermore, the transverse reinforcement in old construction was placed manually, stirrup by stirrup, and there are high chances of missing stirrups or very large spacing between them, due to poor workmanship and lack of supervision. Hence, engineers should be particularly cautious when checking the transverse reinforcement of existing buildings.

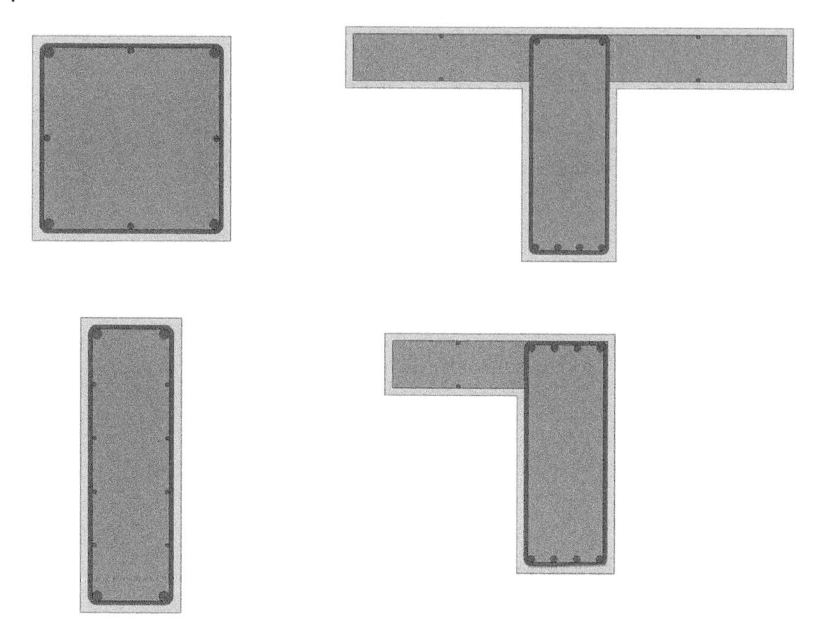

Figure 10.3 Typical cross sections of the building members. *Source:* Stelios Antoniou.

In the building under consideration, the transverse reinforcement consists of ⌀8 mm hoops in both columns and beams. In the columns, the spacing is 25 cm, and in the beams it is taken equal to 20 cm.[1] The longitudinal reinforcement of the square columns is considered to be 4⌀20 + 4⌀12, and that of the rectangular columns is equal to 4⌀20 + 8⌀8. The beam reinforcement is 4⌀14 (lower reinforcement at midspan and upper reinforcement at the edges). Typical member cross sections are shown in Figure 10.3.

The walls in the vicinity of the elevator shaft are relatively thin (20 cm thick), and lightly reinforced (a grid of #⌀8/30 cm horizontal and vertical rebars at every wall side, with additional ⌀20 rebars in every wall corner). The combination of their large size, with respect to the other vertical members, and their low reinforcement ratios makes these walls the most vulnerable structural components, as will be explained in subsequent sections.

Smooth S220 rebars have been used for columns, but ribbed S400 reinforcement was employed in the beams. There was no detailing for earthquake resistance (for instance, the hoops were not bent inside the concrete core at 135° angles). In all members, the cover was taken as 15 mm, whereas an absolute lap length of 0.50 m was considered for all the columns (but in the beams adequate lap splices were considered).

### 10.3.3 Material Strengths

Although it is possible to find the grades of the materials used in the construction from the original drawings (if these are available), carrying out additional testing, at least for concrete, can be very beneficial. The concrete compressive strength is usually considerably

1 Recall that in existing buildings typically the beams have higher transverse reinforcement, due to the larger shear forces that develop from the gravity forces.

different from the strength indicated in the original technical documents, and sometimes it varies within the same construction (e.g., for different floors).[2] Moreover, a combination of destructive and nondestructive methods is cheap and relatively easy to execute without disturbing the residents much. Therefore, it is strongly advisable to carry out on site testing and measure the actual values of the concrete strength.

By contrast, measuring the actual steel strength on site is not usually required. The yield strength is generally known with high reliability from the grade of the material, since with respect to concrete the steel production is a more standardized process. Furthermore, it is difficult and disruptive to take rebar specimens from the building and then repair the damage. The testing itself is also much more expensive.

The amount of required testing varies and depends on the target knowledge level (refer to Chapter 3 for more details on the information needed in each case). Note that some standards (e.g., ASCE-41 or the Greek code) also provide conservative default values, in accordance with the standards at the time of construction of the building. Even though these values can be used by the engineer without the need for further investigation, it is preferable to carry out testing, which in the majority of cases will provide higher material strengths, and lead to the need for fewer interventions.

In the current project, uniform concrete strength is employed throughout the building. With destructive and nondestructive tests the mean compressive concrete strength for both columns and beams was found equal to $\mu = 20.20\,\text{MPa}$ with a standard deviation equal to $\sigma = 1.70\,\text{MPa}$. Thus, the lower-bound compressive strength is equal to $\mu - \sigma = 18.50\,\text{MPa}$.

**Cube vs. Cylinder Concrete Strength**

It should be pointed out that in the case of concrete the engineer should not confuse the cylindrical strength, which is typically used as input in structural analysis programs, with the cube strength, which is typically used to report the concrete strength in the experimental tests and, with which engineers are more accustomed.

The Stahl I mild steel grade was employed for the columns, as indicated in the original drawings (mean strength 244.44 MPa, lower-bound strength 220.00 MPa), and the Stahl III grade was employed for the beams (mean strength 444.44 MPa, lower-bound strength 400.00 MPa)

The information obtained about the building configuration (geometry, dimensions, reinforcement, details, material properties) permits the use of knowledge level KL3, which implies a very good knowledge of the structure. Both linear and nonlinear analytical methods may be used with KL3 and a value of 1.00 is assumed for the confidence factor.

2 Note that usually the measured concrete strengths are larger than those indicated in the technical drawings and documents of the original design.

## 10.4 Seismic Action and Load Combinations

The standard Eurocode 8 elastic response spectrum is employed in the analysis (CEN 2004). A design ground acceleration equal to 0.14g was selected. A type 1 spectrum (surface-wave magnitude $M_S$ greater than 5.50) was employed and soil type D was chosen. The shape of the spectrum is described in Section 3.2.2.2 of EN 1998-1, the only point worth mentioning here is the fact that the spectrum is scaled up and down for different levels of seismic action. The spectrum is the 5% damped spectrum with a return period of 475 years (probability of exceedance of 10% in 50 years), which corresponds to the limit state of the Significant Damage.

The spectrum for the seismic action with return period equal to 225 years (probability of exceedance of 20% in 50 years), which is to be used with the Damage Limitation limit state, is derived from the 475 year spectrum by its multiplication with a scaling factor equal to 0.779, in accordance to Section 2.1 of EN 1998-1 (CEN 2004). Similarly, the 2475-year spectrum, which is employed with the limit state of Near Collapse, can be calculated from the 475-year spectrum by multiplying it with a factor of 1.733 (Figure 10.4).

The dead loads $G$ of the structural system are automatically calculated by the program for all structural components (e.g., for a typical 13 cm slab with an average value of the specific weight of the reinforced concrete equal to $24\,kN/m^3$, the dead load is $3.12\,kN/m^2$). The additional dead loads $G'$ (e.g., from floors and tiles) is taken as 1.50 kN (calculated for an average floor thickness of around 6–7 cm and an average specific weight between 20 and $22\,kN/m^3$). The value for the live load Q was taken from the EN 1991-1-1 recommendations (CEN 2002), $Q = 2.00\,kN/m^2$ for the internal slabs and $Q = 2.50\,kN/m^2$ for the cantilevers (Figure 10.5).

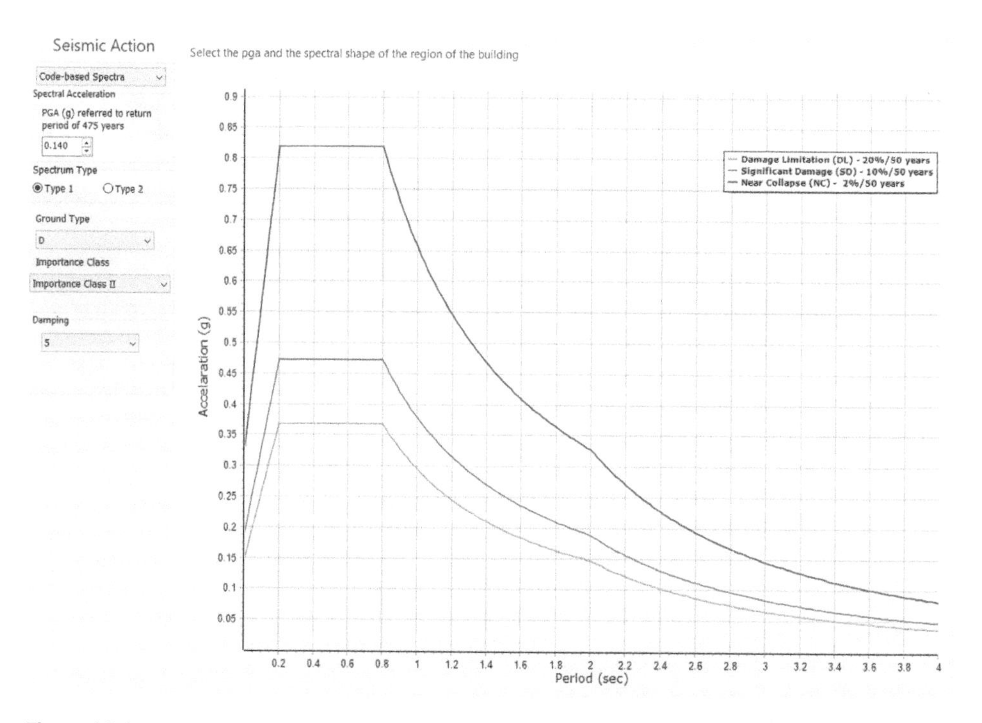


**Figure 10.4** Horizontal elastic response spectra for the limit states of Damage Limitation, Significant Damage and Near Collapse (SeismoBuild screenshot). *Source:* Stelios Antoniou.

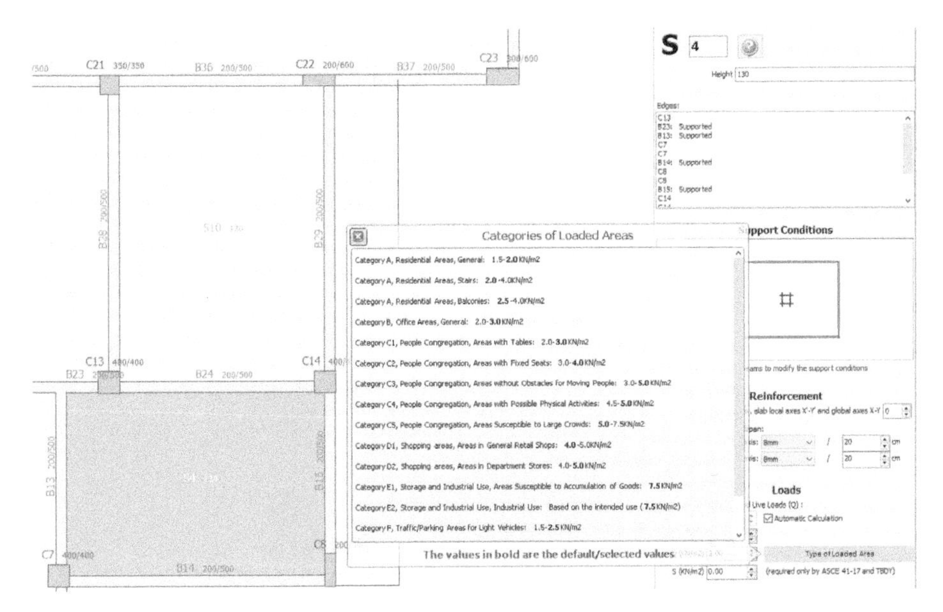

Figure 10.5 Screenshots from SeismoBuild with the live loads of slabs (default Eurocode values). *Source:* Stelios Antoniou.

The loads from the partition walls were considered as additional linear loads on the beams. A self-weight of 3.60 kN/m$^2$ was considered for the external walls and the thick internal walls (25–30 cm of thickness, including the mortar), leading to an approximate value of the load applied on the beams right below them equal to 3.60 kN/m$^2$ × 2.50 m = 9.00 kN/m. Instead, a self-weight of 2.40 kN/m$^2$ was considered for the thinner partition walls (15–20 cm of thickness, including the mortar), leading to an approximate value of the load applied on the beam equal to 2.40 kN/m$^2$ × 2.50 m = 6.00 kN/m. In both cases, an average value of 2.50 m was considered for the height of the walls.

The seismic loading combination was taken from Section 6.4.3.4 of EN 1990:2002 (CEN 2002) as follows:

$$G + G' + A_{Ed} + \sum_j \psi_{2j} Q_{kj}$$

The value of the partial factor $\psi_2$ is equal to 0.30, according to *EN 1990:2002: Table A1.1 – Recommended values of factors for buildings.*

## 10.5 Structural Modeling

In SeismoBuild, only the parameters regarding the geometry, the reinforcement and the material properties of the building are input by the user without any consideration about the formation of the finite element meshing, which is automatically created by the program. The entire input is introduced on plan view, and after the completion of the introduction of the structural components, a space nonlinear model of the structure is created, following the recommendations of Chapter 8 (Figure 10.6).

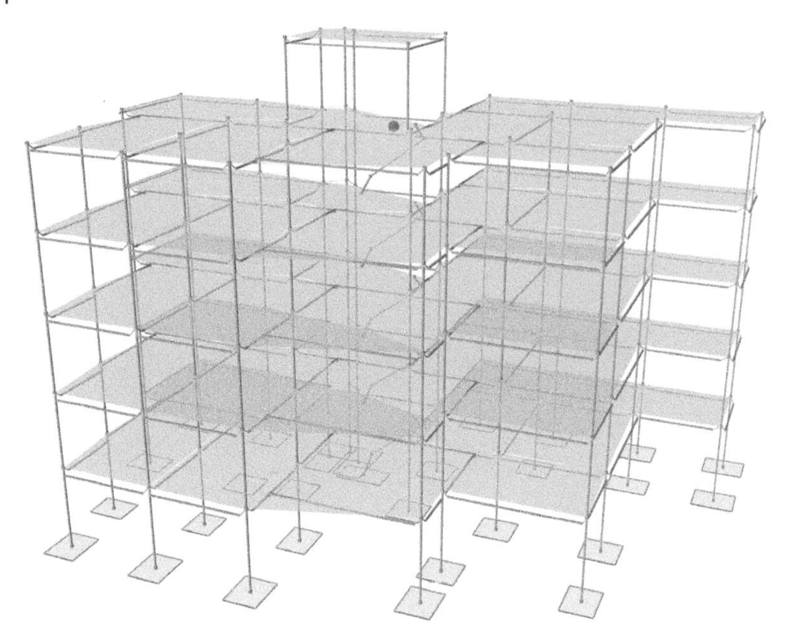

**Figure 10.6** FE model of the building. *Source:* Stelios Antoniou.

The beams and columns are modeled as one-dimensional nonlinear finite elements, whereas the in-plane rigidity of the slabs is considered as horizontal diaphragms. The U-shaped core wall close to the elevator shaft is modeled as three one-dimensional nonlinear elements, connected with rigid links at every floor level, according to the relevant recommendations in Chapter 8, and Beyer et al. (2008a, 2008b, 2008c).

The model also accounts for the reduced deformability of the joint regions (end zones in beams or columns), through the use of the rigid offsets in SeismoBuild's frame elements. It was decided to model the stairs, although it was found that their omission does not significantly affect the global dynamic characteristics of the building. The modeling of the stairs was done with linear elastic frame elements (Figure 10.7).

A fiber type of model was employed for all the frame elements. In particular, SeismoBuild's force-based plastic hinge infrmFBPH element was employed, which has two integration sections at the element's two edges. This approach was preferred to lumped plasticity modeling, because it is able to model the axial load vs. bending moment interaction and to allow for the straightforward representation of biaxial loading, and the interaction between the flexural strength in orthogonal directions. Moreover, distributed plasticity models tend to be more stable numerically in nonlinear, and in particular pushover analysis. Constant longitudinal and transverse reinforcement was assumed for all the vertical members, whereas for the beams it was possible to define different reinforcement profiles along the length of the member. In Figure 10.8 the discretization to fibers of typical beam and column cross-sections is depicted.

For the modeling of the nonlinear behavior of concrete and steel, the default materials from SeismoBuild's internal library were employed, i.e. the model by Mander et al. (1988) for concrete, and that by Menegotto and Pinto (1973) for the reinforcement (Figure 10.9). These material stress–strain laws are applied to every reinforcing steel and concrete fiber

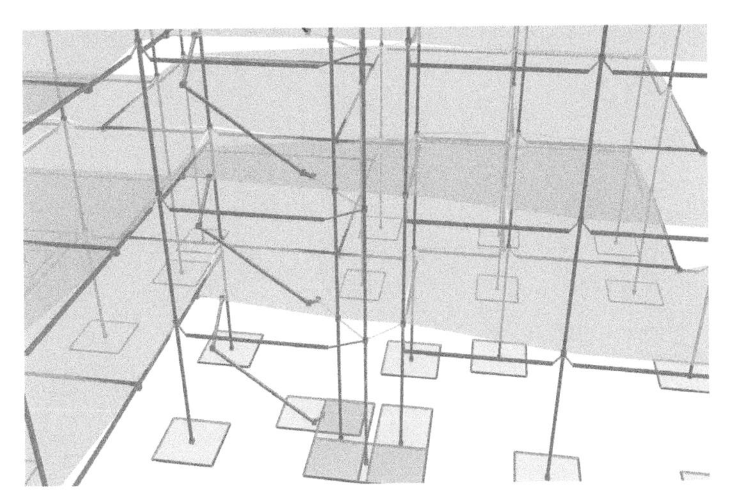

**Figure 10.7** Close view of the FE discretization in the vicinity of the stairs and the walls around the elevator shaft. One can see (i) the offsets at the edges of the structural members, (ii) the rigid links that connect the three sides of the U-shaped core, and (iii) the modeling of the stairs with linear elastic frame elements. *Source:* Stelios Antoniou.

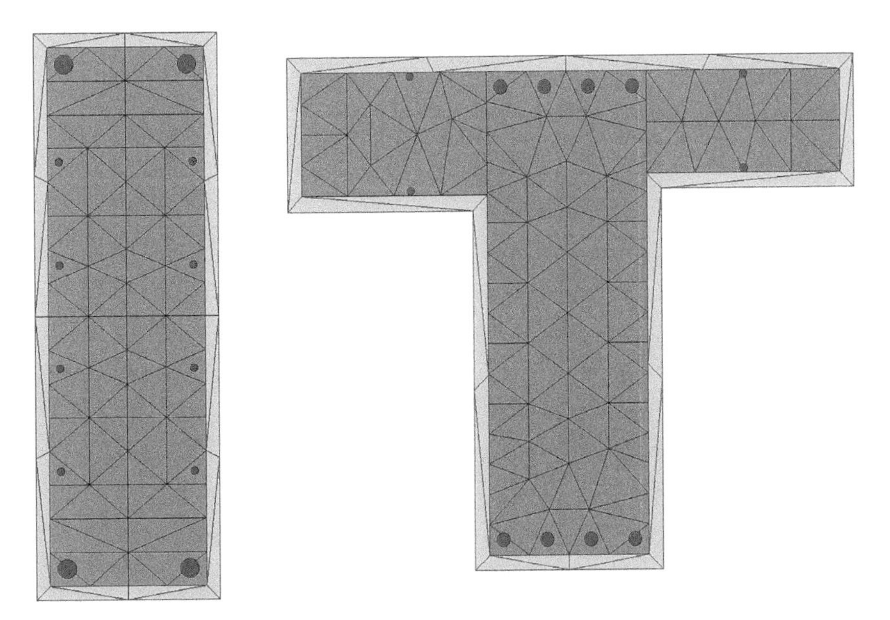

**Figure 10.8** Discretization to fibers of one typical beam and one typical column cross-section. *Source:* Stelios Antoniou.

separately, and their contributions are integrated at every step and iteration of the analysis to derive the axial load and the bending moments around the two local section axes.

The modeling of the infill walls was omitted for reasons of simplicity, but also because their even distribution in plan and elevation guarantees that their effect in the dynamic structural response is not adverse. Regarding the modeling of the building's mass, an accurate representation of the structural masses is automatically considered by the program, by

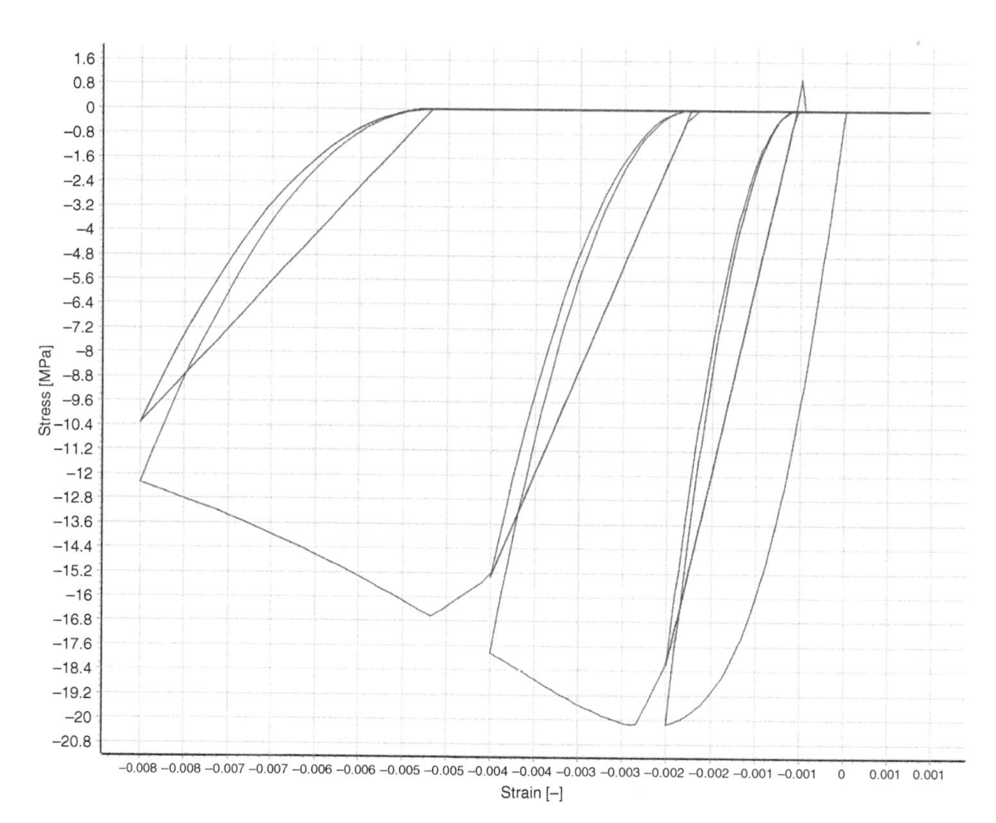


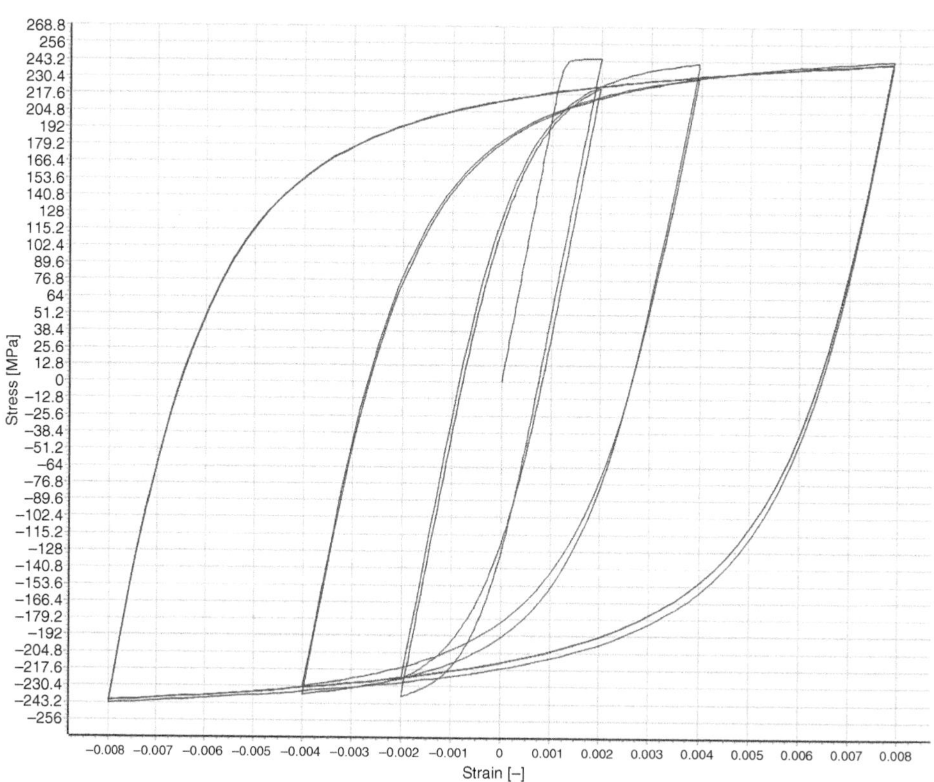


**Figure 10.9** Nonlinear material models for concrete and steel. *Source:* Stelios Antoniou.

discretizing the slabs in smaller areas and distributing their masses to the adjacent beams and columns. Finally, the geometric nonlinearities and the P-delta effects are considered automatically by the program.

## 10.6 Eigenvalue Analysis

The elastic dynamic characteristics of the building can be estimated with eigenvalue analysis. Eigenvalue analysis is needed to determine the lateral force profiles in the linear procedures, and for the calculation of the target displacement in the nonlinear static procedure.

Because of the irregularities in plan, and in particular the asymmetric location of the core wall at the elevator shaft, the first translational mode of vibration along the *X* horizontal direction is not clearly translational, but rather it is mixed with the rotational RZ components (Figure 10.10). By contrast, the core wall is located approximately at the center of the building perpendicular to the *Y*-axis, and the mode along *Y* is almost purely translational, as shown in Figure 10.11.

Similar observation can also be made from Table 10.1, where the period and percentages of the effective modal masses for the first 10 modes of vibration are given.

From the same table it is clear that the 10 eigenvalues that are sought for from the analysis are enough, since they account from more than 90% of the total mass in both horizontal directions (in fact, even the first five modes would have been enough).

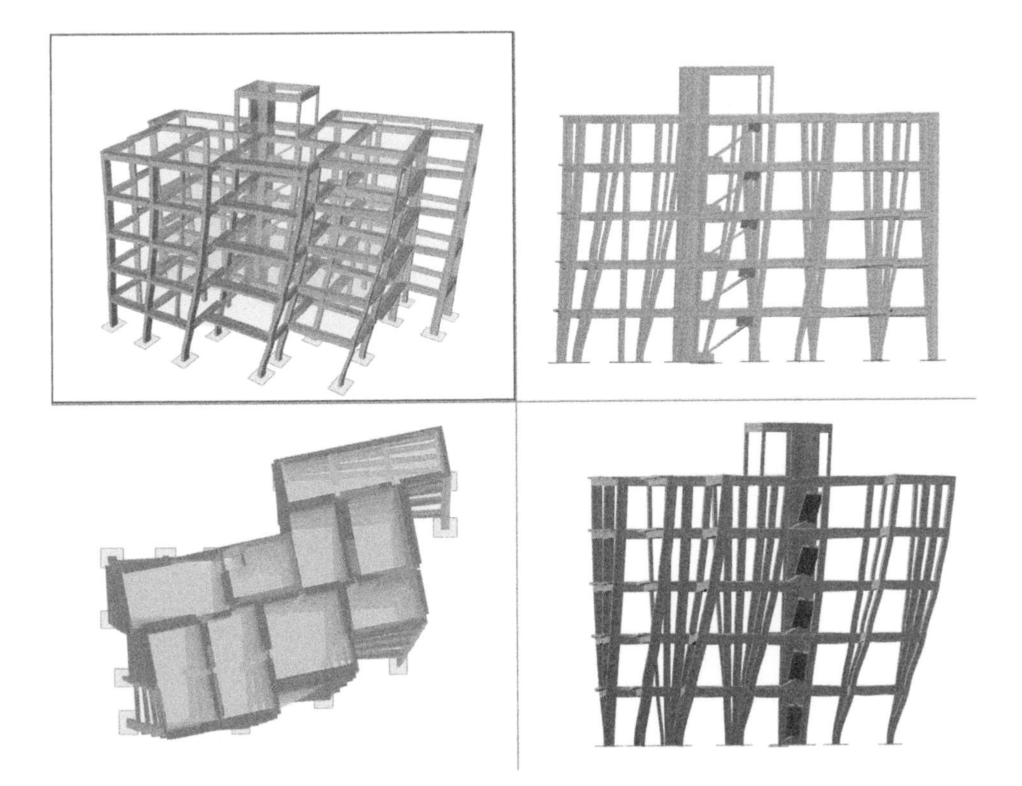

**Figure 10.10** Fundamental mode in the *X* + RZ global directions (T = 0.662 sec). *Source:* Stelios Antoniou.

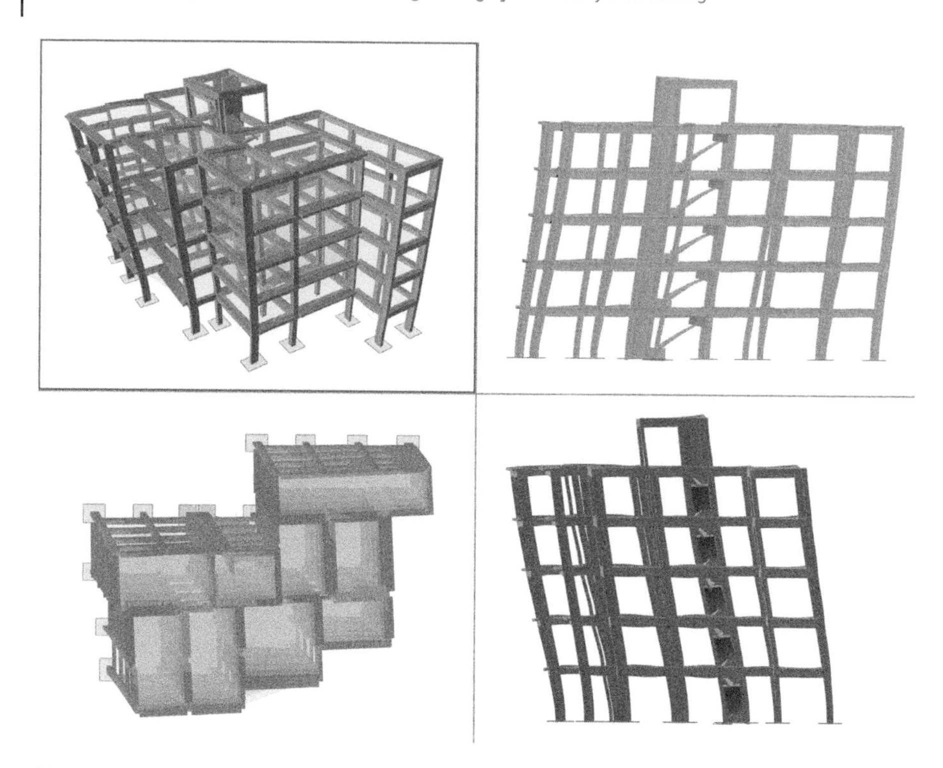

**Figure 10.11** Fundamental mode in the *Y* global direction (T = 0.557 sec). *Source:* Stelios Antoniou.

**Table 10.1** Periods and effective modal masses for the first 10 modes of vibration.

| Mode | Period (sec) | [Ux] (%) | [Uy] (%) | [Rz] (%) |
|---|---|---|---|---|
| 1 | 0.6622 | 20.64 | 24.46 | 41.34 |
| 2 | 0.5574 | 32.85 | 46.21 | 1.96 |
| 3 | 0.4440 | 29.45 | 9.76 | 43.97 |
| 4 | 0.2168 | 3.09 | 2.88 | 3.63 |
| 5 | 0.1716 | 4.53 | 7.78 | 0.11 |
| 6 | 0.1491 | 2.18 | 0.51 | 4.52 |
| 7 | 0.1278 | 2.82 | 2.10 | 0.34 |
| 8 | 0.1223 | 0.05 | 0.02 | 1.70 |
| 9 | 0.0912 | 1.75 | 2.27 | 0.00 |
| 10 | 0.0901 | 0.06 | 0.22 | 0.55 |
| TOTAL | | 97.43 | 96.22 | 98.13 |

*Source:* Stelios Antoniou.

## 10.7 Nonlinear Static Procedure

Since the inelasticity in the structure is expected to be significant, pushover analysis is a more appropriate method for assessment as compared to the linear methods, which require low levels of damage, evenly distributed in the building. According to EN 1998-3, Section 4.4.2 (CEN 2005), the use of the linear methods is permitted only when the ratio $\rho_{max}/\rho_{min}$ between the maximum and the minimum demand to capacity ratio (the abbreviation DCR is employed in the rest of this chapter) does not exceed a value in the range of 2–3, which is not the case in this particular example.

### 10.7.1 Lateral Load Patterns

The lateral load patterns in pushover analysis are expected to approximate the inertia forces in the building during an earthquake. Within the Eurocode context, 16 loading combinations are considered. The loads are applied in a single horizontal (+X, −X, +Y and −Y) direction for each analysis, but a 5% accidental eccentricity is also considered (±X±eccY and±Y±eccX). Both the uniform and the modal patterns are employed, according to Section 4.3.3.4.2.2 of EN 1998-1 (CEN 2004; Figure 10.12). In the former, the lateral forces are proportional to the mass distribution regardless of elevation. In the latter,

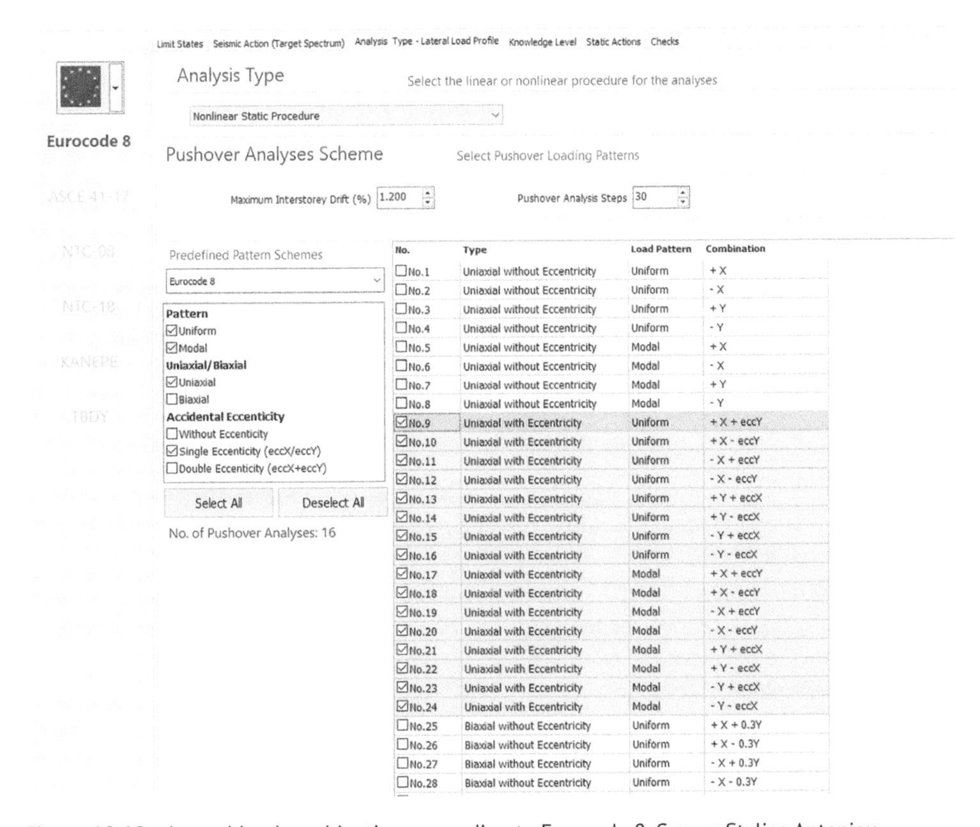


**Figure 10.12** Lateral load combinations according to Eurocode 8. *Source:* Stelios Antoniou.

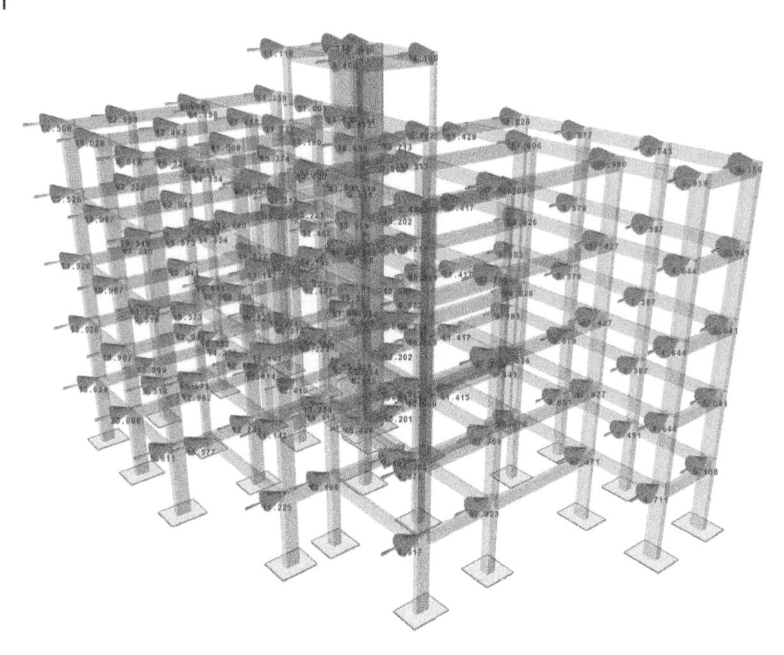

Figure 10.13 Lateral force distribution for the Uniform +Y-eccX load combination. *Source:* Stelios Antoniou.

the lateral forces are consistent with a force distribution determined by elastic analysis. In both cases, the pushover loads are automatically calculated by SeismoBuild (2023) (Figure 10.13).

### 10.7.2 Selection of the Control Node

The Control Node is typically located at the center of mass of the top story or roof of the building. For buildings with a penthouse, as is the case with the building under consideration, the control node is the center of mass of the slab at the floor (rather than the roof) of the penthouse as in Figure 10.14.

### 10.7.3 Capacity Curve and Target Displacement Calculation

Using SeismoBuild's nonlinear capabilities, the 16 pushover analyses are carried out with the calculated lateral load distributions, employing a response control scheme. Because of the limited strength and ductility of the building, the yield of the structure was reached at relatively low deformation levels. Therefore, it was decided to push the building in both horizontal directions at a displacement equal to 1.20% of the total height of the building (or more correctly the height of the control node), at approximately 18 cm.

The target displacement calculation is carried out with the methodology described in Appendix B of EN 1998-1. In Figure 10.15 the capacity curve of one of the 16 analyses (Uniform, +X+eccY) is shown, together with the bi-linearization of the curve and the calculation of the target displacements for the three limit states.

Although the estimation of the target displacement is carried out automatically by SeismoBuild for all pushover analyses, the calculation is explained below for display purposes.

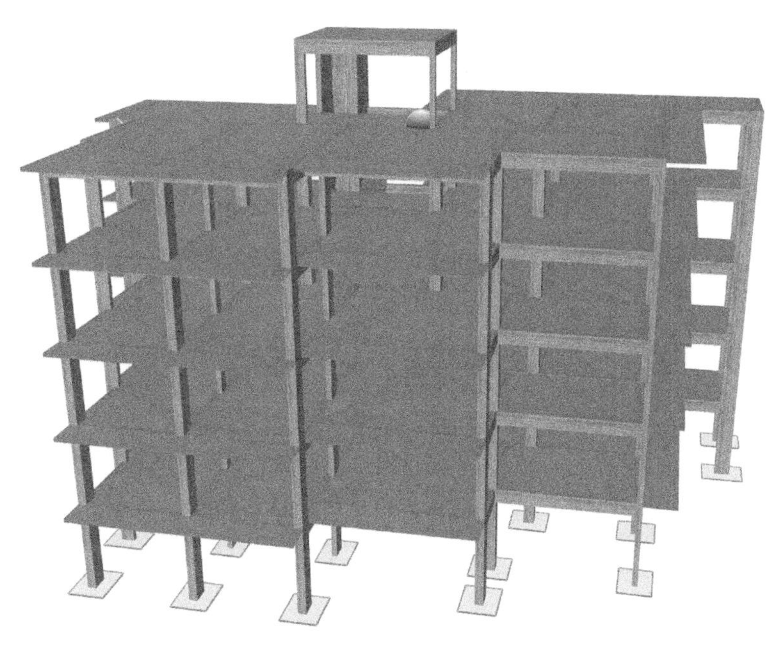

**Figure 10.14** Location of the control node of the building at the penthouse floor level. *Source:* Stelios Antoniou.

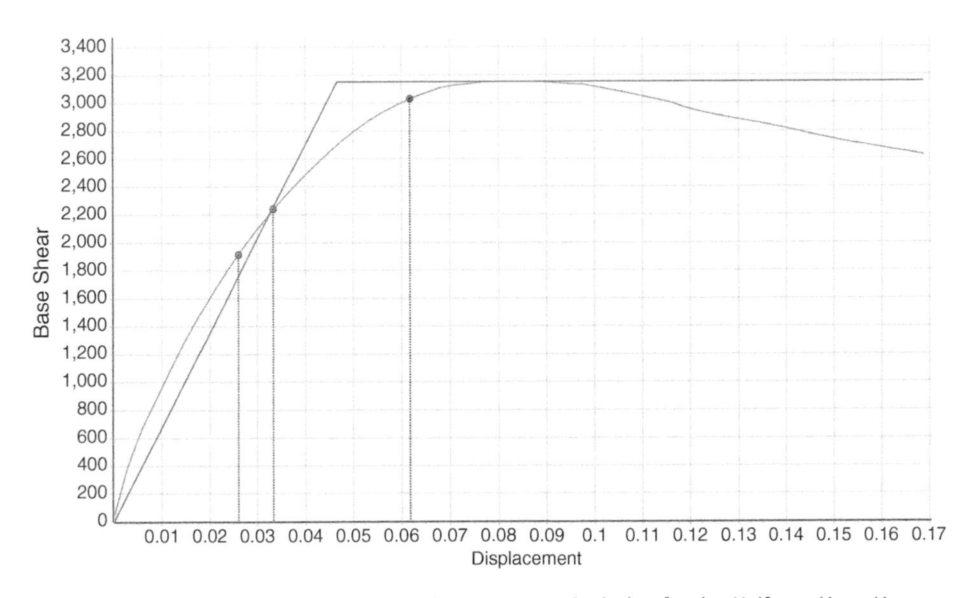


**Figure 10.15** Capacity curve and target displacement calculation for the *Uniform+X+eccY* combination. *Source:* Stelios Antoniou.

From the results of the eigenvalue analysis the mass of an equivalent single degree of freedom (SDOF) system m* is determined from:

$$m^* = \Sigma m_i \ \Phi_i = 611.54 \ \text{tonnes} \qquad \text{EN 1998 – 1 (B.2)}$$

The factor for the transformation to an equivalent SDOF system is:

$$\Gamma = \frac{m^*}{\Sigma m_i \ \Phi_i^2} = 0.7968 \qquad \text{EN 1998 – 1 (B.3)}$$

The elastic stiffness (i.e. the tangent stiffness at zero deformation) of the capacity curve is:

$$K_{el} = 122{,}281.49 \ \text{kN / m}$$

The idealized elasto-perfectly plastic force–displacement relationship of Figure 10.15 is calculated according to Section B.3 of Appendix B with an iterative procedure, in order to fulfill EN 1998-1, Eq. (B.6), and the areas under the actual and the idealized force–deformation curves are equal.

The effective stiffness of the elastic part is $K_{ef} = 67{,}563.27 \frac{\text{kN}}{\text{m}} < K_{el}$ and the yield strength is $F_b$ = 3153.29 kN. The period $T^*$ of the idealized equivalent SDOF system is determined from

$$T^* = 2\pi\sqrt{\frac{m^* \cdot d_y^*}{F_y^*}} = 2\pi\sqrt{\frac{m^*}{K_{ef}}} = 0.5978 \text{sec} \qquad \text{EN 1998 – 1 (B.7)}$$

$$F_y^* = \frac{F_b}{\Gamma} = 3957.44 \, \text{kN} \qquad \text{EN 1998 – 1 (B.4)}$$

$$d_y^* = \frac{d_n}{\Gamma} = \frac{0.0467}{0.7968} = 0.0585 \text{m} \qquad \text{EN 1998 – 1 (B.5)}$$

$$K_{ef} = \frac{F_y^*}{d_y^*} = 67{,}563.27 \, \text{kN / m}$$

The target displacement dt* of the idealized equivalent SDOF system is estimated with the procedure described in Section B.5. For the Significant Damage limit state, the spectral acceleration is extracted from the elastic spectrum of Figure 10.4 for $T^*$ = 0.5978 sec and it is equal to $S_e(T^*) = 0.4725\text{g} = 4.635 \frac{\text{m}}{\text{sec}^2}$.

$$d_{et}^* = S_e\left(T^*\right)\left[\frac{T^*}{2\pi}\right]^2 = 0.0419 \text{m} \qquad \text{EN 1998 – 1 (B.8)}$$

Because $T^* > T_C = 0.40 \, \text{sec}$, $d_t^* = d_{et}^* = 0.0419 \text{m}$

Finally, the target displacement $d_t$ of the MDOF structure is calculated from Eq. (B.5) as:

$$d_t = d_t^* \cdot \Gamma = 0.0334 \text{m} \qquad \text{EN 1998 – 1 (B.5)}$$

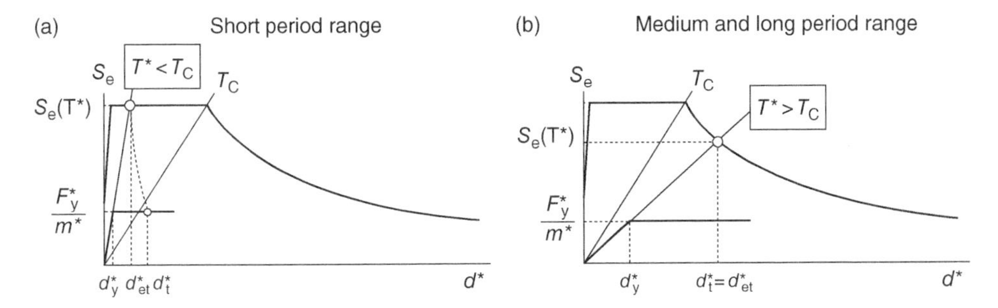


Figure 10.16 Implementation of the capacity spectrum method in Eurocode 8 for (a) the short period range and (b) the medium and long period range (EN 1998-1, Figure B.2).

The target displacements for the other two limit states are calculated in a similar fashion. For the Damage Limitation limit state, the target displacement is 0.0261 m and for Near Collapse it is 0.0618 m. Note that what is different between the limit states is the spectral acceleration as read from the spectrum (see Figure 10.4).

Eurocode 8 employs a variation of the Capacity Spectrum Method (Freeman et al. 1975; Freeman 1998) for the calculation of the target displacement. The method compares the capacity of the lateral force-resisting system of the building (representing the strength) with response spectra values (representing the demand). The values are plotted in an Acceleration-Displacement Response Spectrum (ADRS) format, in which spectral accelerations are plotted against spectral displacements and periods T are represented by radial lines. The relation between the different quantities can be visualized in Figure 10.16.

A similar approach (a variation of the CSM) is employed by the Italian standards NTC-18. On the contrary, in ASCE-41, as well as the Greek and the Turkish codes, the so-called equation method is applied. In this particular example and for the Life Safety performance level (which is the one closer to the SD limit state of the Eurocode) the target displacement would be:

$$\delta_t = C_0\, C_1\, C_2\, S_a\, \frac{T_e^2}{4\pi^2}\, g = 0.0607\,\text{m} \qquad \text{ASCE 41 - 17 (7.28)}$$

$\mathbf{S_a}$ is the spectral acceleration, which is equal to 0.4725 g (assuming the same spectral shape).

$\mathbf{T_e}$ is the effective fundamental period, which is equal to 0. 5978 sec (assuming the same procedure for the bi-linearization of the capacity curve and the same effective fundamental period).

$\mathbf{C_0}$ is a modification factor that relates spectral displacements with the likely roof displacement, and it is equal to 1.42 in this particular case.

$\mathbf{C_1}$ relates maximum inelastic displacements to displacements calculated from the linear response. It corresponds to the ductility demand μ of the SDOF model. It is equal to 1.02.

$\mathbf{C_2}$ is a modification factor that represents the effect of the hysteresis shape on the maximum displacement response. It depends on the framing system and the selected performance level, and is equal to 1.00.

Note that the difference of approximately 40% between the target displacements in ASCE-41 and the Eurocodes should not be surprising or unexpected. There are several small differences between the two methodologies, including the bi-linearization of the curves, the expression for the calculation of the target displacement, the damage related to the different limit states, the expressions for the calculation of the capacities, the material strengths and the safety factors employed in the calculations. The methodologies (as well as the corresponding methodologies in the other standards) usually provide similar results in terms of the damage distribution and the demand-to-capacity ratios.

### 10.7.4 Safety Verifications

At the pushover step that corresponds to the target displacement of each limit state, a set of member action effects and displacements are extracted. This is the seismic demand expected on the structural elements during the design earthquake action (for the specific hazard level). This set of action effects is then compared with the member capacities for the selected limit state. If the capacity exceeds the demand for any member, the member is safe; otherwise, it requires strengthening.

According to Appendix A of EN1998-3:2005, the safety verification checks that should be performed are the following:

- Flexural strength with and without axial force, according to Section A.3.2. In nonlinear analysis the bending checks are carried out in terms of the deformation capacity, employing the chord rotation capacity of the members.
- Shear strength, according to Section A.3.3.
- Checks for the beam-column joints, according to Section A.3.4.

All the checks should be carried out for both primary and secondary seismic elements. However, less stringent rules apply to the latter.

### 10.7.5 Chord Rotation Checks

Since the problematic lap splices were explicitly considered in the chord rotation capacity of the columns, and the deformation demand was generally higher for them, it is not surprising that several failures are observed in the vertical members, contrary to the beams that are generally safe.

For the limit state of Damage Limitation, the maximum DCR for the columns/walls is 1.91, and for the beams it is 0.84.

For the limit state of Significant Damage, the maximum DCR for the columns/walls is 1.19, and for the beams it is 0.78.

For the limit state of Near Collapse, the maximum DCR for the columns/walls is 1.95, and for the beams it is 1.13 (Figure 10.17).

One interesting observation is that, whereas the Near Collapse limit state gives the largest DCRs, the Damage Limitation limit state has higher DCRs than the Significant Damage limit state (max. DCR of DL = 1.91, max. DCR of SD = 1.19). This is not unexpected, and is attributed to the more stringent requirements of the DL limit state, i.e. requirement for lower damage, indicated by the yield chord rotation, with respect to three-quarters of the ultimate chord rotation capacity that is required in the SD limit state.

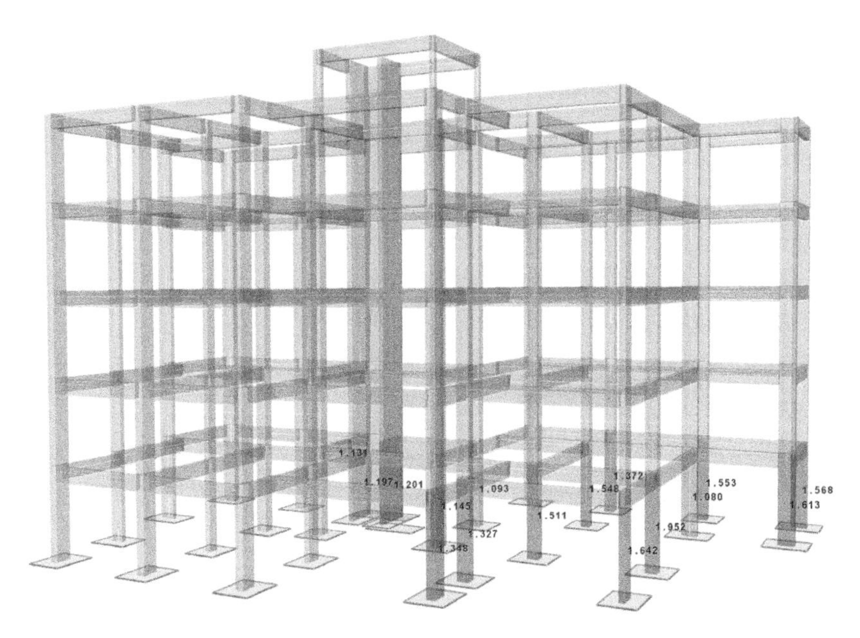


Figure 10.17 Chord rotation checks for the Near Collapse limit state. *Source:* Stelios Antoniou.

### 10.7.6 Example of the Calculation of Chord Rotation Capacity

The calculation of the capacity and the corresponding checks in chord rotation will be shown for column C20 of the ground floor for the Significant Damage limit state. The location of C20 is depicted in Figure 10.18. The first of the 16 analyses will be checked (Uniform $+X + eccY$), and the calculations will be made for local axis (2) in the "Start" (lower) end section of the member.

The chord rotation capacity of C20 on the ground floor in the Significant Damage limit state is three-fourths of the ultimate chord rotation capacity $\theta_{um}$:

$$\theta_{um} = \frac{1}{_{el}} \cdot 0.016 \cdot \left(0.3^{\nu}\right)\left[\frac{\max\left(0.01;\omega'\right)}{\max\left(0.01;\omega\right)} f_c\right]^{0.225} \cdot \left(\min\left(9, \frac{L_V}{h}\right)\right)^{0.35} 25^{\left(\alpha\rho_{sx}\frac{f_{yw}}{f_c}\right)}\left(1.25^{100\rho_d}\right)$$

EN 1998 – 3 (A.1)

- $\gamma_{el}$ is equal to 1.5 for primary seismic elements.
- Confidence factor = CF = 1.00.
- $f_c = f_{cm}/CF = 20.20/1.00 = 20.20$ MPa is the concrete compressive strength.
- $f_{yw} = f_{ywm}/CF = 244.44/1.00 = 244.44$ MPa is the stirrup yield strength.
- $\nu = N/(b \cdot h \cdot f_c) = 0.3302$, where (*i*) $b = 0.40$ m is the width of the compression zone, (ii) $h = 0.40$ m is the depth of the cross-section, (iii) $N = 1067.2$ kN is the axial force (positive for compression).

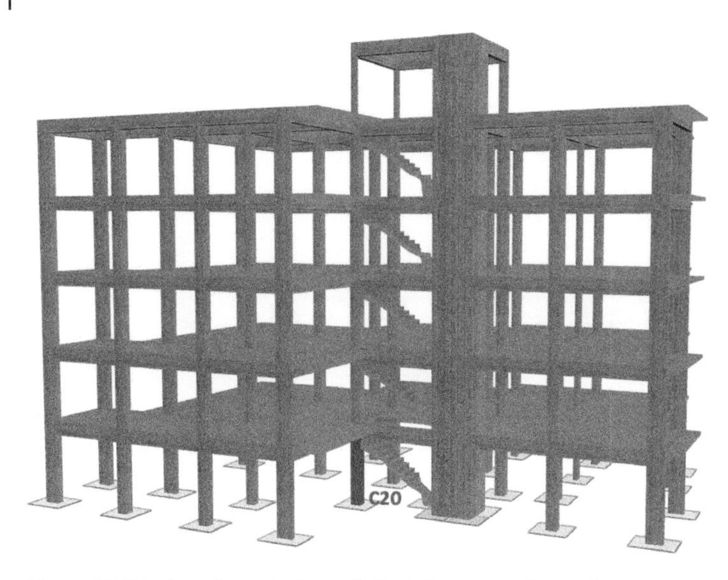


Figure 10.18 Location of column C20 of the ground level. *Source:* Stelios Antoniou.

- $L_V = M/V - l_o = 1.381$ m is the ratio between bending moment and shear, $M = 150.23$ kNm, $V = 79.88$ kN, and $l_o = 0.50$ m. Note that the shear span is reduced by the lap length $l_o$ as the ultimate condition is controlled by the region right after the end of the lap (see Section A.3.2.2(5)).
- $\omega = 6.045\%$ is the mechanical reinforcement ratio of the longitudinal reinforcement in tension.
- $\omega' = 7.890\%$ is the mechanical reinforcement ratio of the longitudinal reinforcement in compression.
- $\rho_{sx} = 0.1005\%$ is the ratio of transverse steel parallel to the direction of loading.
- $\rho_d = 0$ is the steel ratio of diagonal reinforcement.
- $\alpha = 0$ is the confinement effectiveness factor (stirrups not closed at 135° provide no confinement).

In members without detailing for earthquake resistance, the value given by expression (A.1) is divided by 1.20, according to Section A.3.2.2(3).

The result is also multiplied by $0.019{*}(10 + \min[40, l_o/d_{bL}]) = 0.78375$, according to Section A.3.2.2(5). $d_{bL}$ is the (mean) diameter of the lapped bars equal to 16 mm and $l_o = 500$ mm.

Based on the input above, the ultimate chord rotation capacity is $\theta_{um} = 0.0151$.

The chord rotation capacity of C20 on the ground floor in the SD limit state is $¾{\cdot}\theta_{um} = 0.0113$.

### 10.7.7 Shear Checks

In existing buildings, usually the most critical check is for shear, due to the lack of adequate transverse reinforcement. This is also the case in the building of the example. As explained in Chapter 6, when shear is the critical check, the most critical limit state will always be the one with the larger ground motion, i.e. the higher spectrum. Since in Eurocode 8, Part 3,

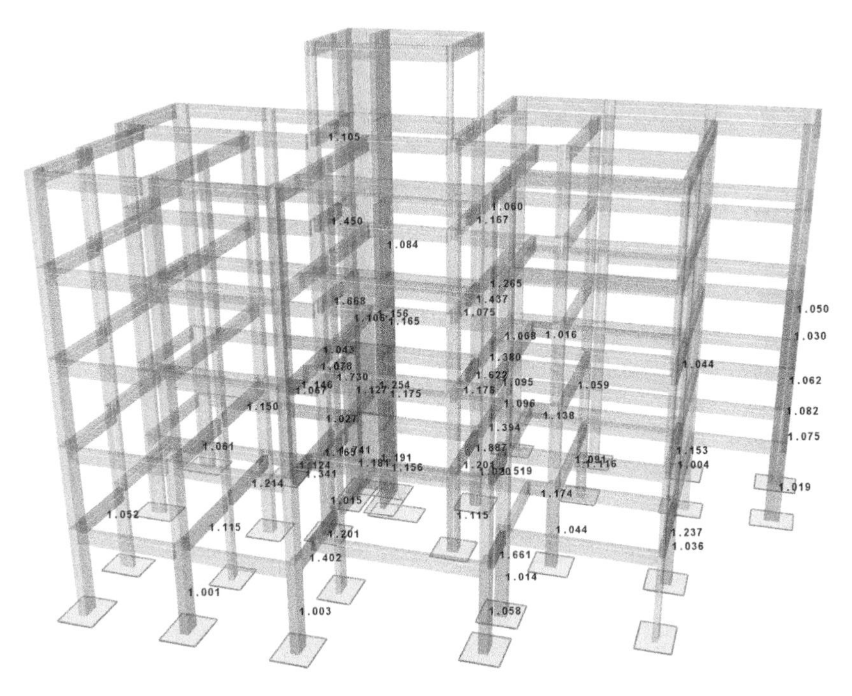


Figure 10.19 Shear checks for the Damage Limitation limit state. *Source:* Stelios Antoniou.

specific hazard levels are ascribed to each limit state (225-year earthquake for the DL limit state, 475-year earthquake for the SD limit state and 2475-year earthquake for the NC limit state), expectedly the demand-to-capacity ratios DCR increase with increasing limit state.

For the limit state of Damage Limitation the maximum DCR for the columns/walls is 1.25, and for the beams it is 1.89. In total, 73 members have failed (Figure 10.19).

For the limit state of Significant Damage the maximum DCR for the columns/walls is 1.280, and for the beams it is 2.06. In total, 99 members have failed (Figure 10.20).

For the limit state of Near Collapse the maximum DCR for the columns/walls is 1.283, and for the beams it is 2.21. In total, 154 members have failed (Figure 10.21).

### 10.7.8 Example of the Calculation of Shear Capacity

The shear capacity is always given by the Eq. (A.12), which provides similar capacities for all limit states (small differences are expected, due to differences regarding the ductility, the depth of the compression zone, the member action effects and the $L_V$ ratio). For the current example, column C20 at the ground floor level will be checked (Figure 10.18) in local axis (2) and at the start of the member (lower edge). The calculations are for the Significant Damage limit state and the uniform $+X + eccY$ analysis.

$$V_R = \frac{1}{\gamma_{el}}\left[\frac{h-x}{2L_V}\min\left(N;0.55A_c f_c\right)+\left(1-0.05\min\left(5;\mu_\Delta^{pl}\right)\right)\cdot\left[\begin{matrix}0.16\max\left(0.5;100\rho_{tot}\right)\\ \left(1-0.16\min\left(5;\dfrac{L_V}{h}\right)\right)\sqrt{f_c}A_c+V_w\end{matrix}\right]\right]$$

EN1998 – 3(A.12), with units in MN and meters

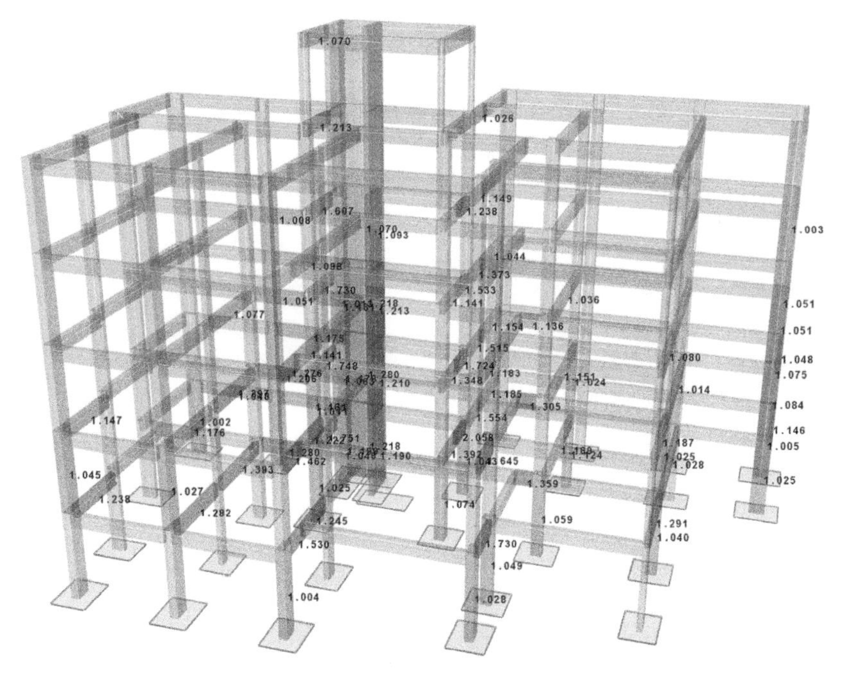

Figure 10.20 Shear checks for the Significant Damage limit state. *Source:* Stelios Antoniou.

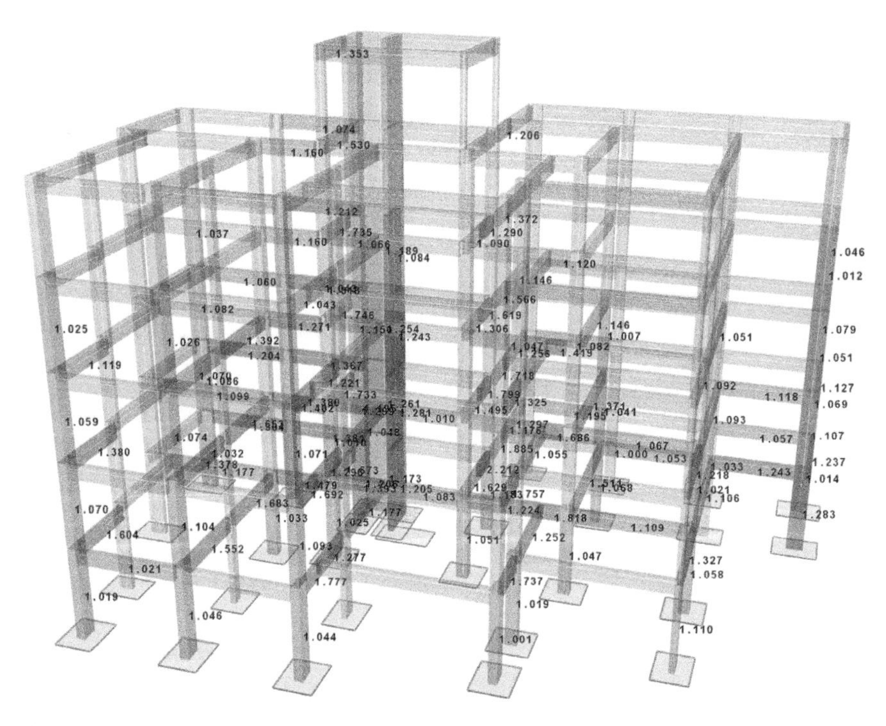

Figure 10.21 Shear checks for the Near Collapse limit state. *Source:* Stelios Antoniou.

- Confidence factor = $CF = 1.00$.
- $f_c = f_{cm}/(CF \cdot \gamma_c) = 20.20/(1.00 \cdot 1.50) = 13.47$ MPa is the concrete compressive strength; for primary elements $f_c$ should be divided by the partial factor for concrete $\gamma_c$, in accordance with EN 1998-1: 2004, 5.2.4.
- $\gamma_{el}$ is equal to 1.15 for primary seismic elements.
- $H = 0.40$ m is the depth of the cross-section.
- $\frac{h-x}{2L_V} = \frac{dx}{L}$ is the depth of the inclined strut = 0.0034.
- $N = 1067.20$ kN = 1.0672 MN is the axial force (positive for compression).
- $A_c = b_w \cdot d = 0.1484\,\text{m}^2$ is the cross-section area.
- $\mu_\Delta^{pl} = 0.00$.
- $\rho_{tot} = 1.068\%$ is the ratio of the total longitudinal reinforcement.
- $L_V = M/V = 1.394$ m is the ratio between bending moment and shear, M = 8.60 kNm, $V = 6.17$ kN.
- $V_w$ is the contribution of transverse reinforcement to the shear resistance; for rectangular sections $V_w$ is equal to:

$$V_w = \rho_w b_w z f_{yd} = 29.23\,\text{kN} = 0.02923\,\text{MN} \qquad \text{EN 1998 - 3 (A.13)}$$

- $\rho_w = A_{sw}/(s \cdot b_w) = 0.101\%$ is the ratio of the transverse reinforcement.
- $z = d\text{-}d'' = 0.342$ m is the length of the internal lever arm, as specified in Section A.3.2.4(2).
- $f_{yd} = f_{yw}/(CF \cdot \gamma_s) = 244.44/(1.00 \cdot 1.15) = 212.56$ MPa is the stirrup yield strength; for primary elements $f_{yd}$ should be divided by the partial factor for steel $\gamma_s$, in accordance with EN 1998-1: 2004, 5.2.4.
- $b_w = 0.40$ m is the section width.

Based on the input above, the shear for the Significant Damage limit state is $V_R = 0.06439$ $MN = 64.39$ kN.

### 10.7.9 Beam-Column Joint Checks

The diagonal compression induced in the joint by the diagonal strut mechanism should not exceed the compressive strength of concrete in the presence of transverse tensile strains. The expressions for the calculation of the shear demand and the shear capacity of the joints are provided in EN 1998-1:2004.

For the limit state of Damage Limitation the maximum DCR is 1.33.

For the limit state of Significant Damage the maximum DCR is 1.48.

For the limit state of Near Collapse the maximum DCR is 1.51 (Figure 10.22).

### 10.7.10 Example of the Checks for Beam-Column Joints

The calculation of the shear capacity and the corresponding demand of one joint for the Near Collapse limit state will be presented in this section. The results of the Modal $+Y-eccX$ analysis are employed for one of the most critical joints, the joint of column C4 of the ground floor and the adjacent beams B5 and B6. The exact location of C4, B5, and B6 is shown in Figure 10.23.

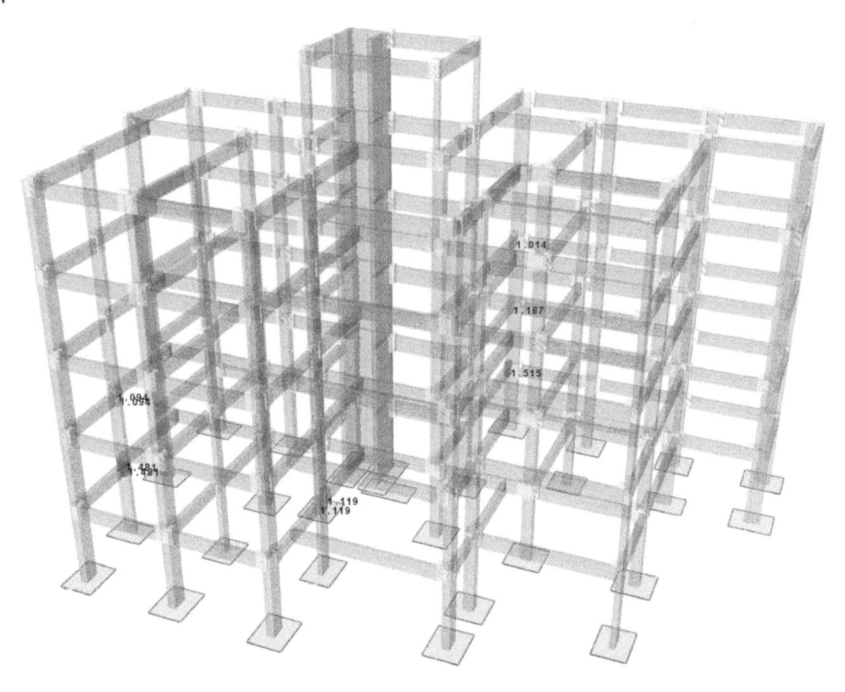


**Figure 10.22** Beam-column checks for the Near Collapse limit state. *Source:* Stelios Antoniou.

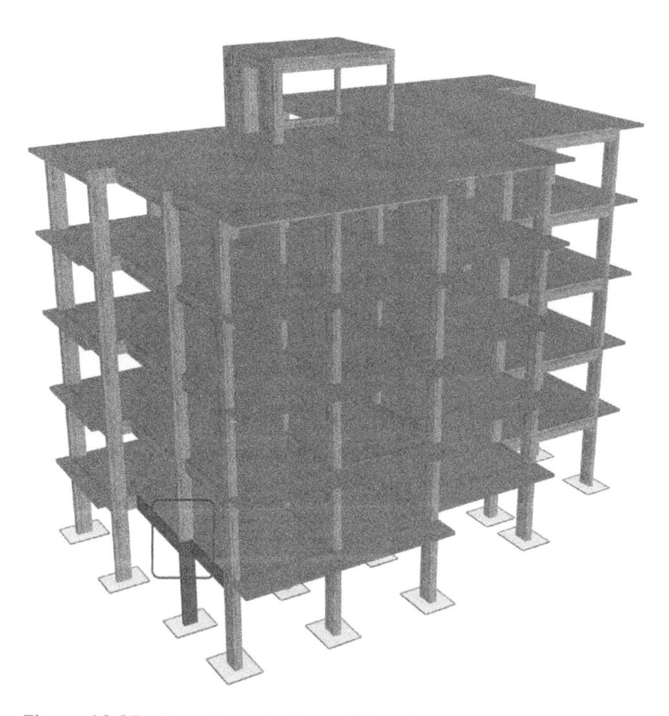

**Figure 10.23** Location of column C4 and beams B5 and B6 on the ground level. *Source:* Stelios Antoniou.

The diagonal compression induced in the joint must not exceed the compressive strength of concrete in the presence of transverse tensile strains. For interior beam-column joints[3] the horizontal shear force acting on the concrete core of the joints is given by the following expression:

$$V_{jhd} = \gamma_{\mathrm{Rd}} \cdot (A_{s1} + A_{s2}) \cdot f_{\mathrm{yd}} - V_{\mathrm{C}}$$

However, running nonlinear analysis allows for the calculation of the demand directly from the analysis, rather than with capacity design considerations, even for brittle types of failure. The option to consider rebar stresses from analyses rather than the yielding stresses for the calculation of joints horizontal shear force demand in nonlinear analysis is available in SeismoBuild and it is the selected one in the current example. Hence:

$$V_{jhd} = (\Sigma A_{1i} \cdot \sigma_{1i} + \Sigma A_{2i} \cdot \sigma_{2i}) - V_{\mathrm{C}} = 388.52\ \mathrm{kN}$$

- $A_{s1} = 615.75\,\mathrm{mm}^2$ is the area of the beam top reinforcement.
- $A_{s2} = 615.75\,\mathrm{mm}^2$ is the area of the beam bottom reinforcement.
- $A_{1i}$ and $\sigma_{1i}$ are the area and the stress of the *i* rebar of the beam top reinforcement, and $A_{2i}$ and $\sigma_{2i}$ are the area and the stress of the *i* rebar of the beam bottom reinforcement. $\sigma_{1i}$ and $\sigma_{2i}$ are calculated from the analysis.
- $\Sigma(A_{s1,i}{*}\sigma_{1,i}) = 261.30\,\mathrm{kN}$ and $\Sigma(A_{s2,i}{*}\sigma_{2,i}) = 133.85\,\mathrm{kN}$
- $f_{yd} = f_{yw}/(CF \cdot \gamma_s) = 444.44/(1.00 \cdot 1.15) = 386.47\,\mathrm{MPa}$ is the design yield strength of the beam's longitudinal reinforcement
- $V_C = 6.63$ kN is the shear force in the column above the joint (lowest compatible values under the most adverse conditions under seismic actions).
- This shear force should not exceed the compressive strength of the joint concrete:

$$V_{jhd} \leq \eta \cdot f_{\mathrm{cd}} \cdot \sqrt{1 - \frac{\nu_{\mathrm{d}}}{\eta}} \cdot b_{\mathrm{j}} \cdot h_{\mathrm{jc}} \qquad \text{EN 1998} - 1\ (5.33)$$

- $\eta = 0.6 \cdot (1 - f_{ck}/250) = 0.55$.
- $f_{ck} = f_{ck}/CF = 20.20/1.00 = 20.20\,\mathrm{MPa}$
- $f_{cd} = f_{cm}/(CF \cdot \gamma_c) = 20.20/(1.00 \cdot 1.50) = 13.47\,\mathrm{MPa}$ is the concrete compressive strength.
- $b_j = 0.30\,\mathrm{m}$, see EN 1998-1, Eq. (5.34).
- $\nu_d = N_{Ed}/(A_c{*}f_{cd}) = 0.2868$ is the normalized axial force in the column above the joint.
- $N_{Ed} = 386.23\,\mathrm{kN}$
- $A_c = 0.10\,\mathrm{m}^2$
- The shear joint capacity is $262.42\,\mathrm{kN} < 388.52\,\mathrm{kN}$, and the joint fails in diagonal compression.

3 Despite the fact that the joint is on the exterior side of the building, it is considered interior, because there are two beams at the opposite sides of the column. In the perpendicular direction, the joint is indeed considered exterior in the calculations.

## 10.8 Strengthening of the Building

As explained in detail in Chapter 5, based on whether there will be interventions in all or some vertical members of the building, the solutions for strengthening an RC building can be categorized into two main groups:

- Interventions in all vertical members employing methods such as jackets or FRP wrapping.
- Interventions only in certain locations, preferably in the perimeter, with methods such as new shear walls, steel braces or base isolation.

### 10.8.1 Strengthening with Jackets

In the current example, the first approach will be employed and all the columns of the building will be upgraded with RC jackets. Simultaneously, the beams that fail to fulfill the verification checks will also be strengthened with jackets; approximately, 20–25% of the total beams of the building will be upgraded, as shown on the plan view of the typical strengthened floor in Figure 10.24.

All the column jackets are 10 cm wide, and the beams jackets are 7.5 cm wide. Obviously, the jackets have larger longitudinal and more importantly transverse reinforcement ratios,

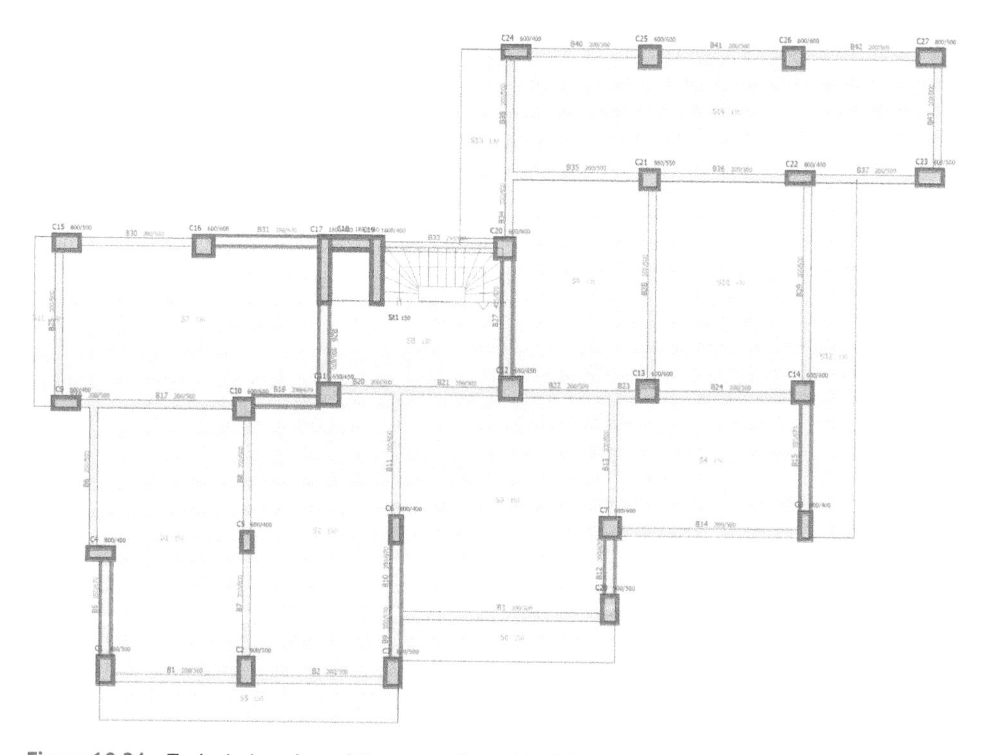


**Figure 10.24** Typical plan view of the strengthened building. *Source:* Stelios Antoniou.

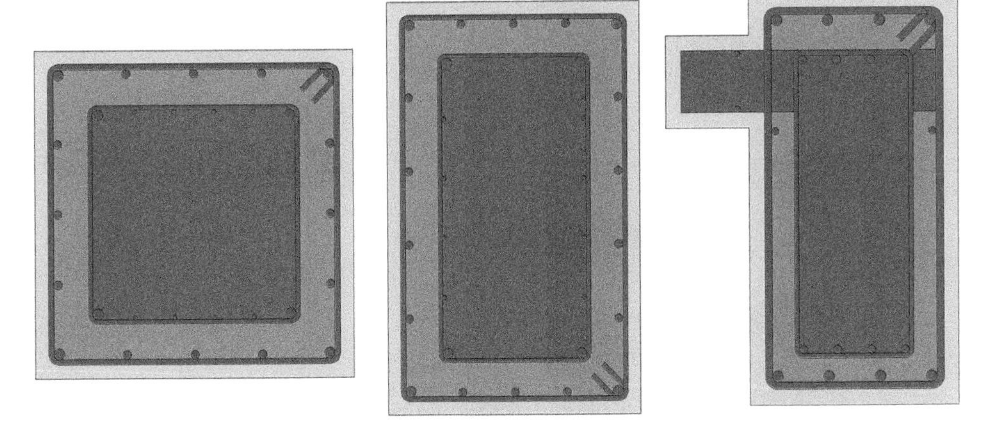

Figure 10.25 Typical cross sections of the jacketed columns and beams. *Source:* Stelios Antoniou.

with respect to the existing members, and they provide increased strength and increased ductility. Typical column cross-sections are shown in Figure 10.25.

All the checks of the new building will be performed for the Significant Damage and the Damage Limitation limit states. The Near Collapse limit state with the corresponding seismic hazard with a 2475-year return period, which is the most critical check in shear, was deemed extremely conservative, as it would lead to either extensive and unnecessarily expensive interventions or alternatively to the decision to do nothing.

All members of the strengthened building were adequate in the chord rotation checks for all the limit states. For the limit state of Damage Limitation the maximum DCR is 0.53, and for the limit state of Significant Damage it is just 0.21. Again the DCR for the DL limit state was larger, as it was for the unstrengthened building (Figure 10.26).

In shear the DCRs were generally larger. For the limit state of Damage Limitation, the maximum DCR for the columns/walls is 0.87, and for the beams it is 0.94. For the limit state of Significant Damage, the maximum DCR for the columns/walls is 0.96, and for the beams it is $1.01 \cong 1.00$ (Figure 10.27).

In the checks of the beam-column joints, all members fulfill the acceptance criteria. For the limit state of Damage Limitation the maximum DCR is 0.56, and for the limit state of Significant Damage the maximum DCR is 0.69 (Figure 10.28).

### 10.8.2 Designing the Interventions

As explained in Chapter 6, the seismic retrofit of a building is not a direct procedure, but rather it is an incremental process, whereby different schemes are proposed and tested and corrections are gradually made, until the building is fully compliant. Usually, the final scheme of the strengthening interventions is the result of a series of trials of different schemes.

The process starts with an initial design of the retrofit measures. This is based on the obtained as-built information and the seismic evaluation of the existing building, with

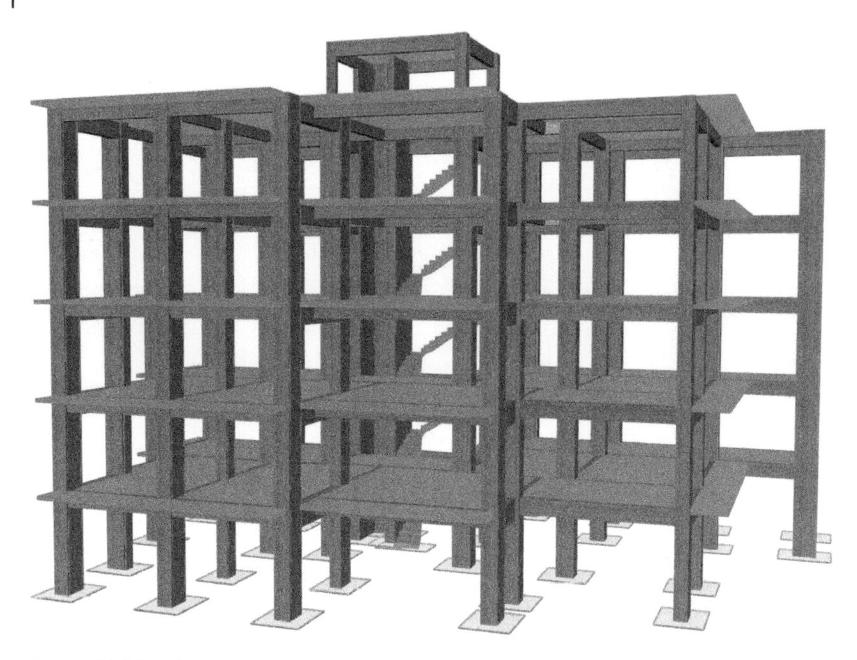

**Figure 10.26** Chord rotation checks for the strengthened building for the damage limitation limit state. The members with DCRs lower than 0.40 are identified with dark green; the members with DCRs between 0.40 and 1.00 are identified with light green. *Source:* Stelios Antoniou.

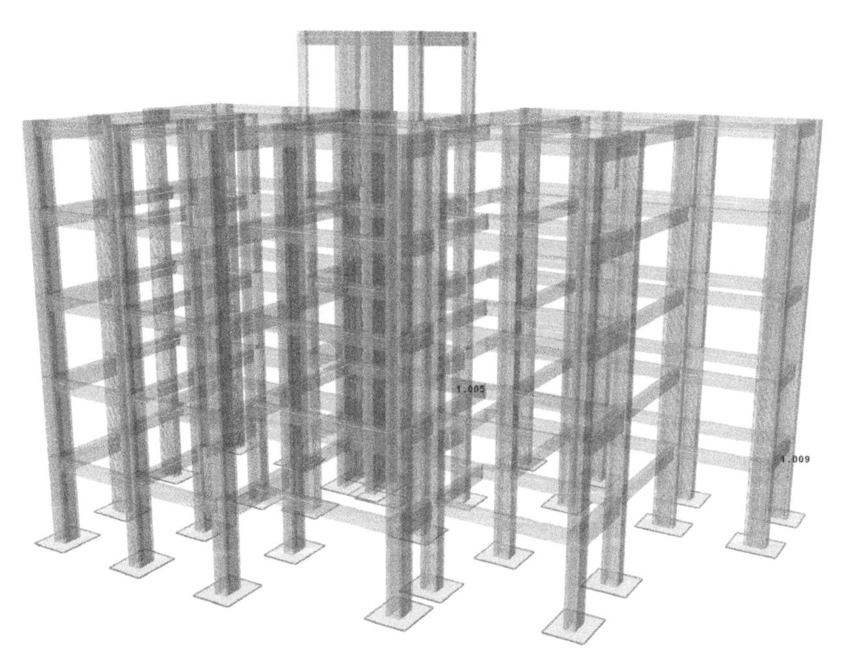


**Figure 10.27** Shear checks for the strengthened building for the significant damage limit state. *Source:* Stelios Antoniou.

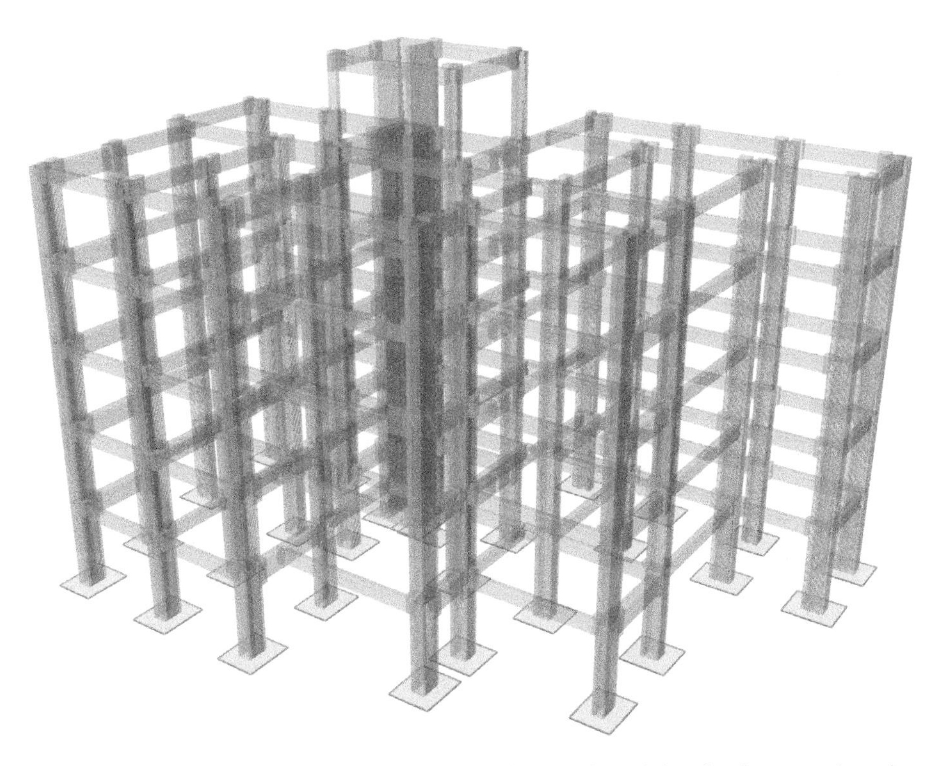

Figure 10.28 Checks for diagonal compression in beam-column joints for the strengthened building for the significant damage limit state. The joints with DCRs lower than 0.40 are identified with dark green; the joints with DCRs between 0.40 and 1.00 are identified with light green. *Source:* Stelios Antoniou.

which the engineer identifies its main seismic deficiencies. Each successive strengthening scheme is designed based on the observations and the results of the previous iteration, and its performance is assessed through linear or nonlinear analysis. If the proposed retrofit measures fail to comply with the acceptance criteria for the selected performance level(s), they are redesigned and an alternative retrofit strategy (possibly with different performance levels and/or seismic hazard levels) is adopted.

This process is repeated until the design is in compliance with the selected performance objective(s). In this process, engineering judgment is required, and some level of experience is definitely helpful, especially when the nonlinear methods are employed.

In the building under consideration, the initial scheme included the strengthening of all the vertical members with jackets. It was quickly recognized that the most critical check was the shear check of beams and columns in the Significant Damage limit state, which is why subsequently I will make reference only to this check. The plan view of the initial interventions scheme, and the results of the shear checks are shown in Figure 10.29. The subsequent three iterations of the design of the interventions are shown in Figure 10.30, together with the corresponding checks. The final design is depicted in Figure 10.31.

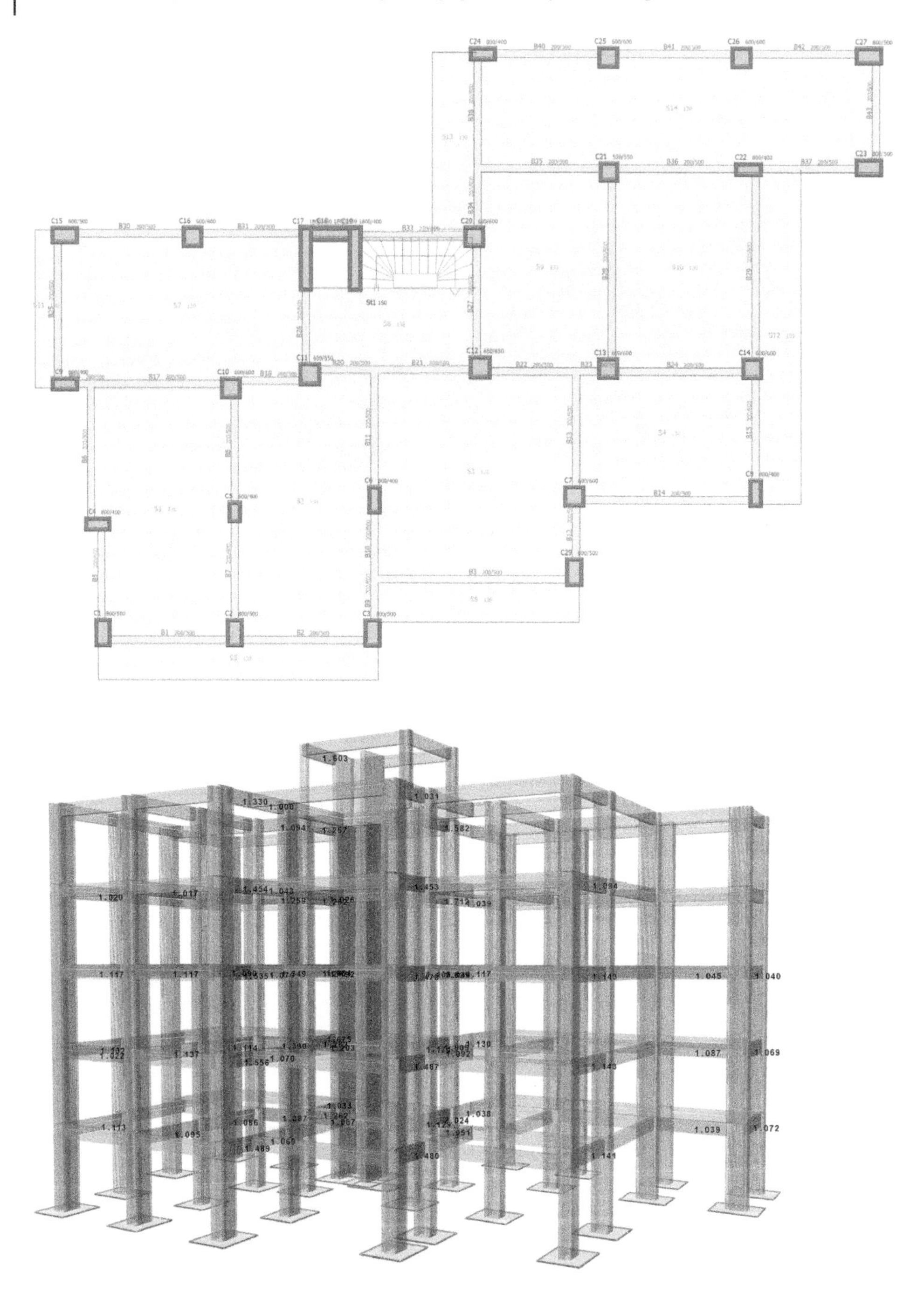

Figure 10.29 The initial interventions scheme for the strengthening of the building, and the corresponding checks in shear (SD limit state). The maximum DCR is equal to 2.03. *Source:* Stelios Antoniou.

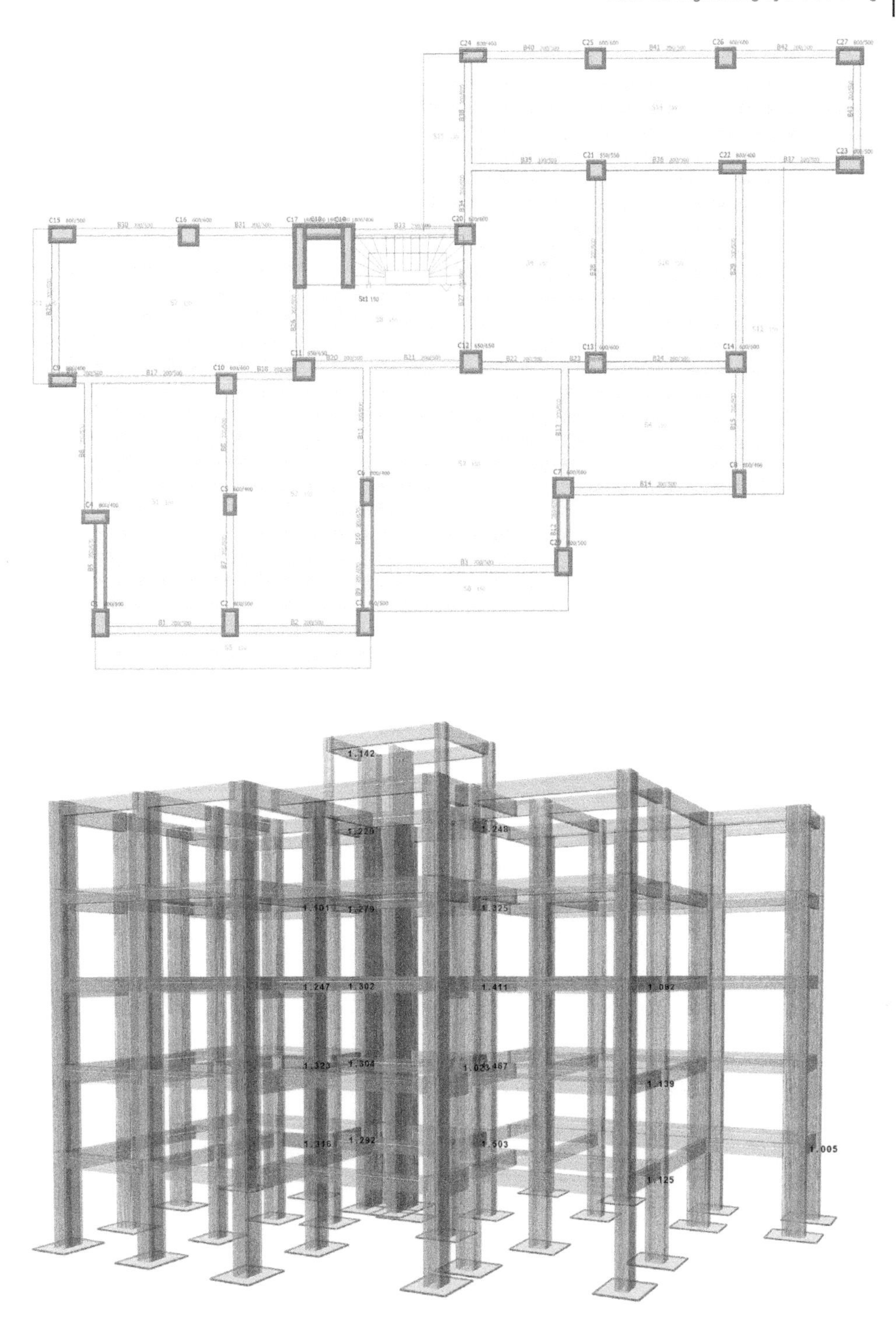


**Figure 10.30** The first, second, and third iteration for the strengthening of the building, and the corresponding checks in shear (SD limit state). The maximum DCRs were 1.50, 1.11, and 1.15, respectively. *Source:* Stelios Antoniou.

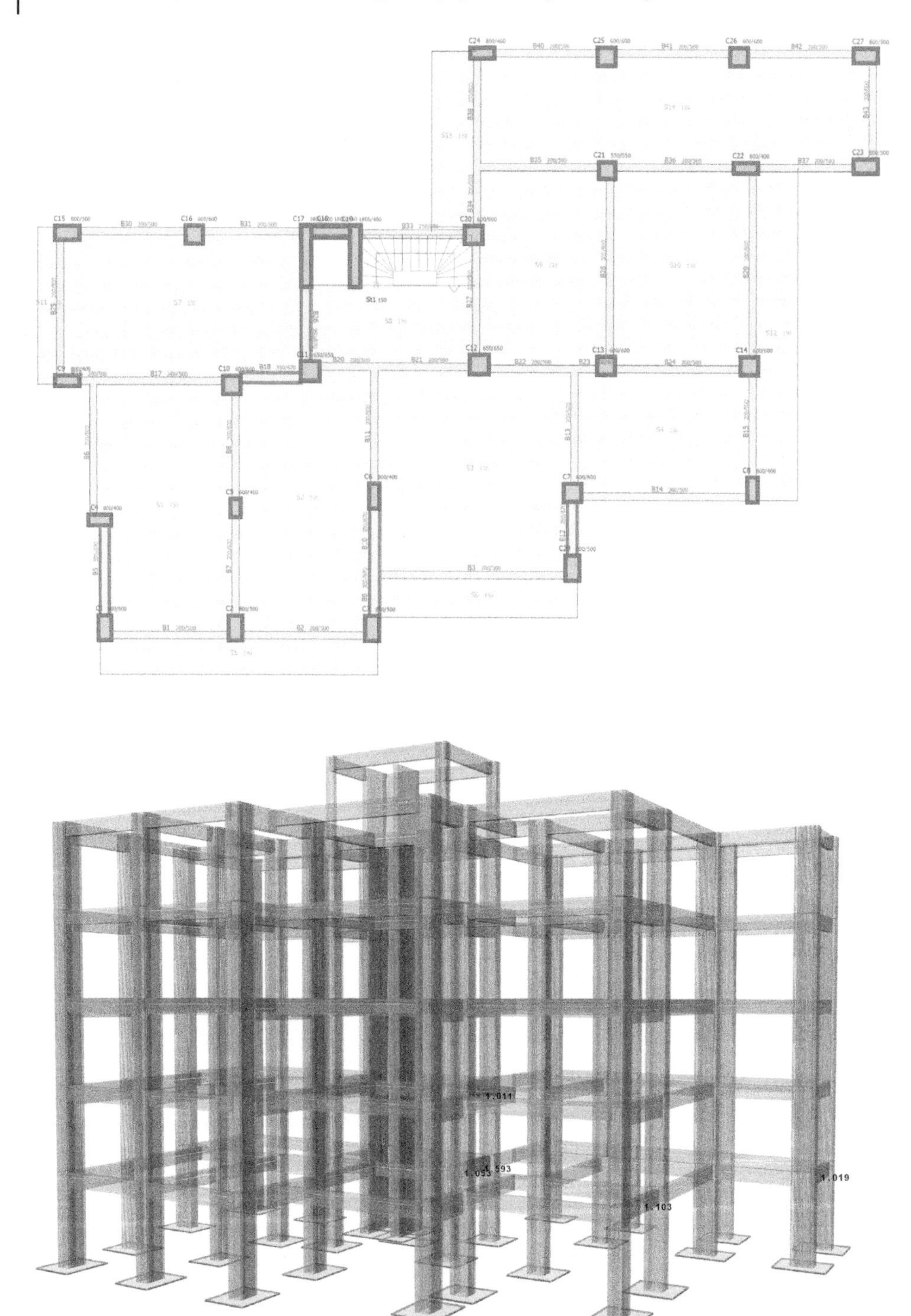

Figure 10.30 (Continued)

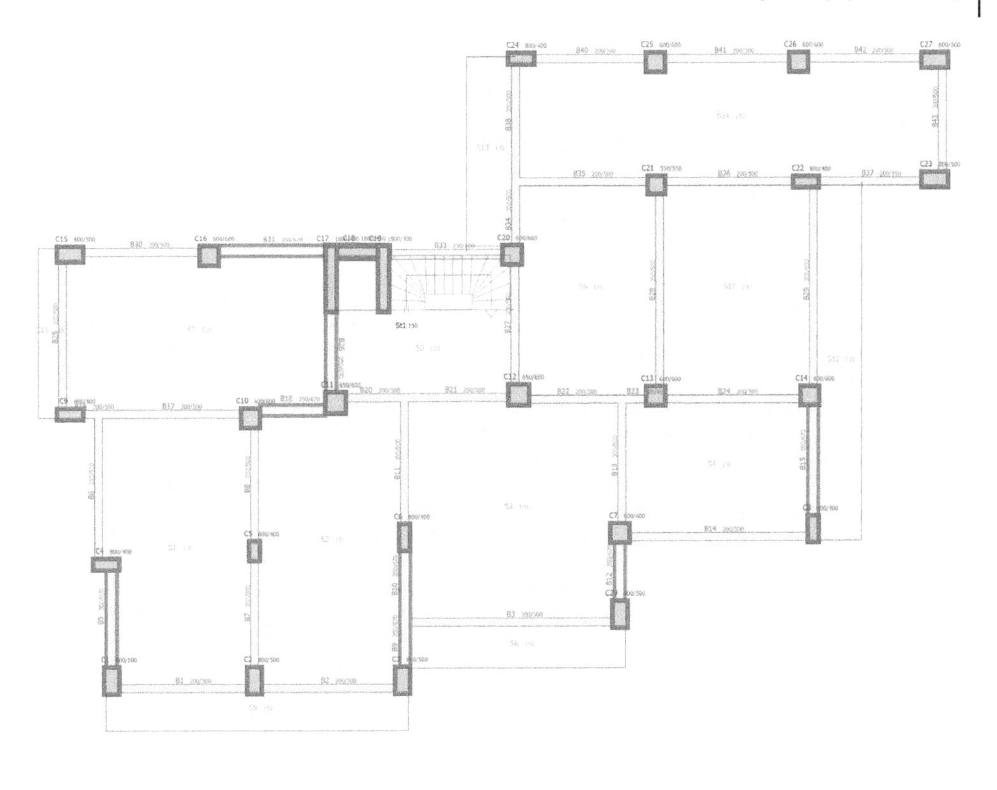


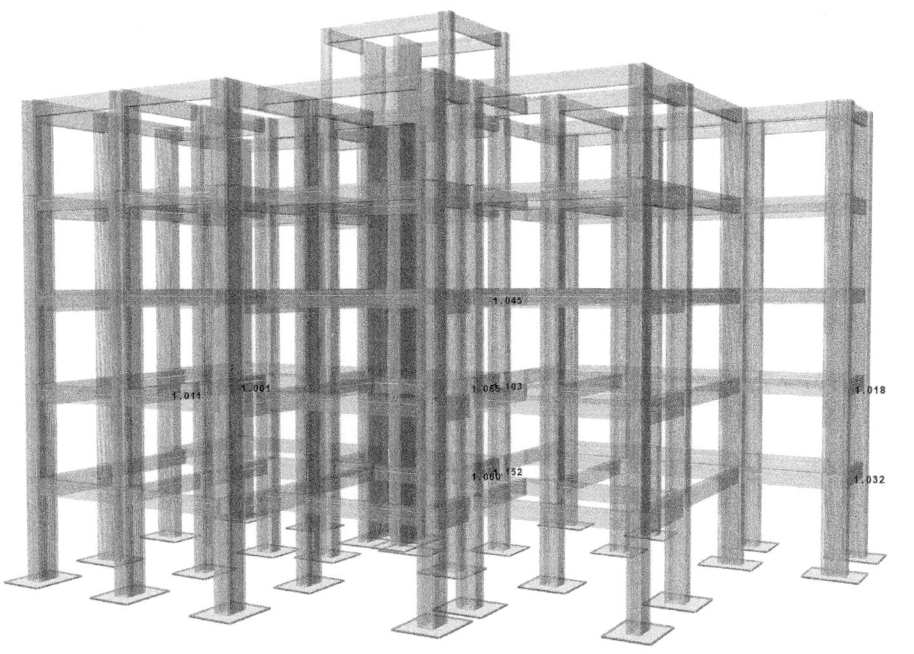


Figure 10.30 (Continued)

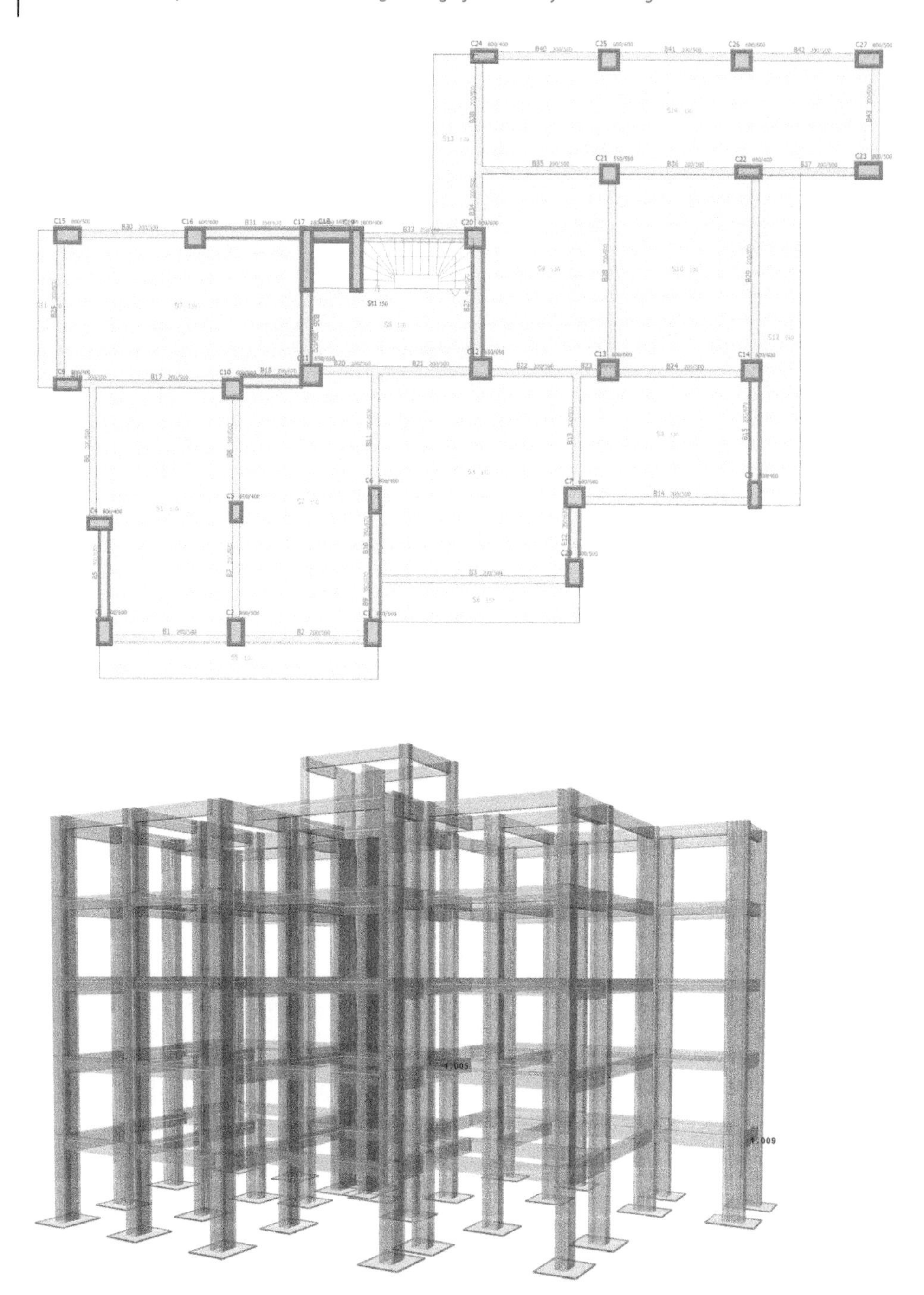

Figure 10.31 The final design of the strengthening interventions, and the corresponding checks in shear (SD limit state). The maximum DCR is equal to 1.01. *Source:* Stelios Antoniou.

### 10.8.3 Deliverables

Once the design of the retrofit measures meets the acceptance criteria for the selected performance objectives, construction documents are prepared. These include:

- The technical report. SeismoBuild automatically creates the technical report with the description of the model and the program assumptions, the target displacement calculations, the results from the analyses, the estimation of the member capacities, and the member checks (Figure 10.32).
- The drawings that should be sent to the construction site for the implementation of the retrofit works. Again, SeismoBuild automatically creates detailed CAD drawings, plan views, cross-section details and tables of reinforcement, including the strengthening interventions (Figure 10.33).

### 10.8.4 Strengthening with Shear Walls

An alternative strengthening strategy could be with the use of external shear walls in the perimeter of the building. This scheme is shown in Figure 10.34 and, as explained in Chapter 5, it can be very beneficial in the cases when the structural retrofit is not accompanied by a radical architectural renovation. This is because the strengthening is carried out mainly with interventions in the external side of the building (areas highlighted in green in Figure 10.35), which cause very limited disruption to the operation of the building (e.g., for the demolition of the plaster, in order to place the dowels that connect the new walls with the existing building). These are accompanied by some interventions that take place in the building perimeter again, but require the demolition of the external infilled walls and cause disturbance, and interventions close to the elevator and the stairs, but not inside the apartments (areas highlighted in red in Figure 10.35). The comparison of these works with the damage needed for the construction of jackets in all the building columns is extremely advantageous, if the building is occupied during the duration of the retrofit. Hence, it constitutes a major reason for the selection of this type of scheme with concentrated and localized damage, and decreased nonstructural damage.

All members comply with the chord rotation checks for both limit states. For the Damage Limitation limit state, the maximum DCR is 0.49 for the columns/walls and 0.45 for the beams. For the Significant Damage limit state, the maximum DCR is 0.29 for the columns/walls and 0.26 for the beams.

In shear, in the Damage Limitation limit state there are no exceedances for the walls and the columns (maximum DCR equal to 0.87), but there is exceedance in one beam (DCR = 1.06). Similarly, in the Significant Damage limit state the maximum DCR for walls/columns is equal to 0.95, and there is exceedance of the shear capacity in a small number of beams (maximum DCR = 1.22) (Figure 10.36). It is up to the engineer to decide if these exceedances are important, and whether the corresponding members should be upgraded and how (e.g., with FRP fabrics, anchored with FRP strings, which is a cleaner method, or with jackets).

In the checks for the beam-column joints, all joints fulfill the relevant acceptance criteria. For the limit state of damage limitation, the maximum DCR is 0.66, and for the limit state of significant damage the maximum DCR is 0.78 (Figure 10.37).

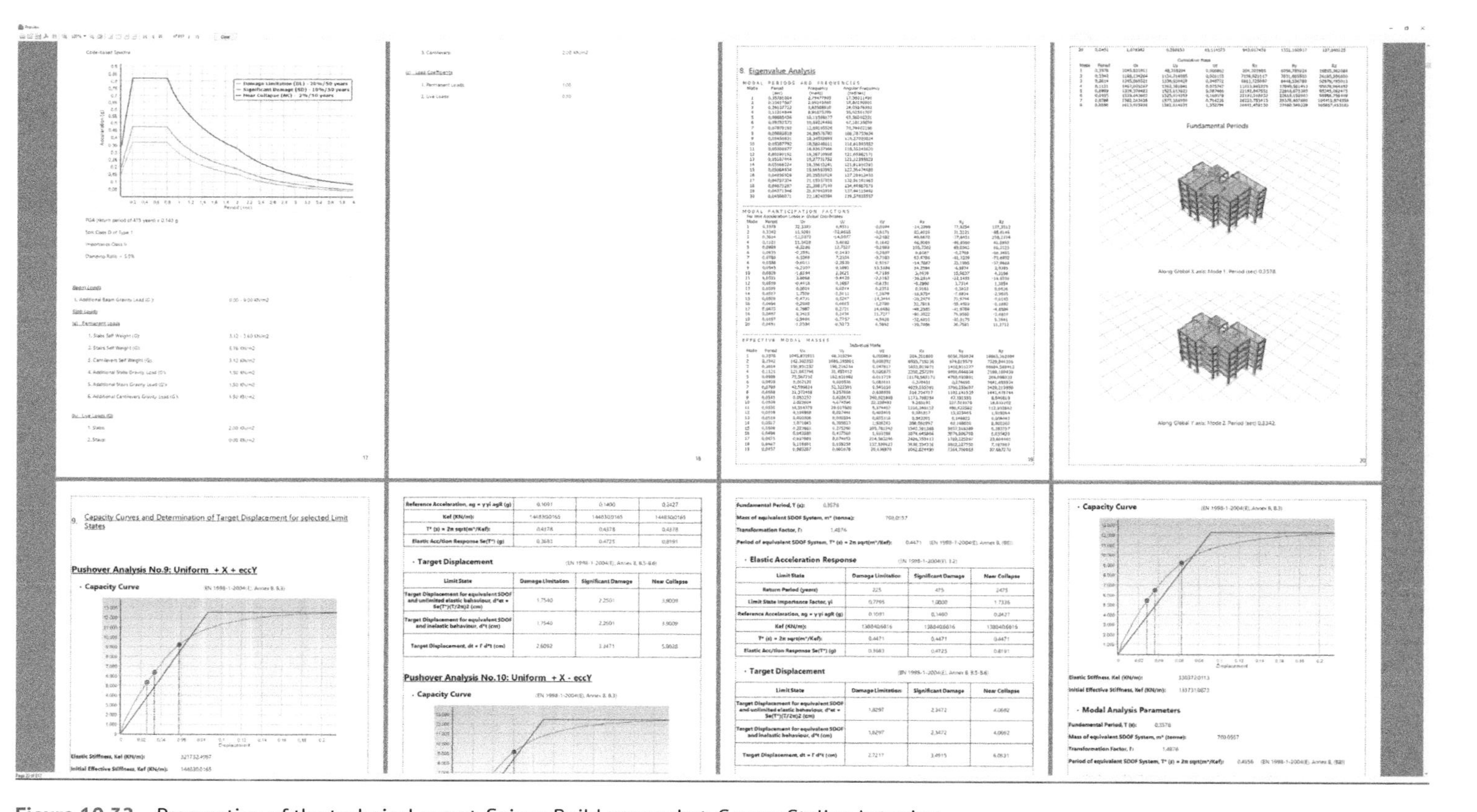


Figure 10.32 Preparation of the technical report, SeismoBuild screenshot. *Source:* Stelios Antoniou.

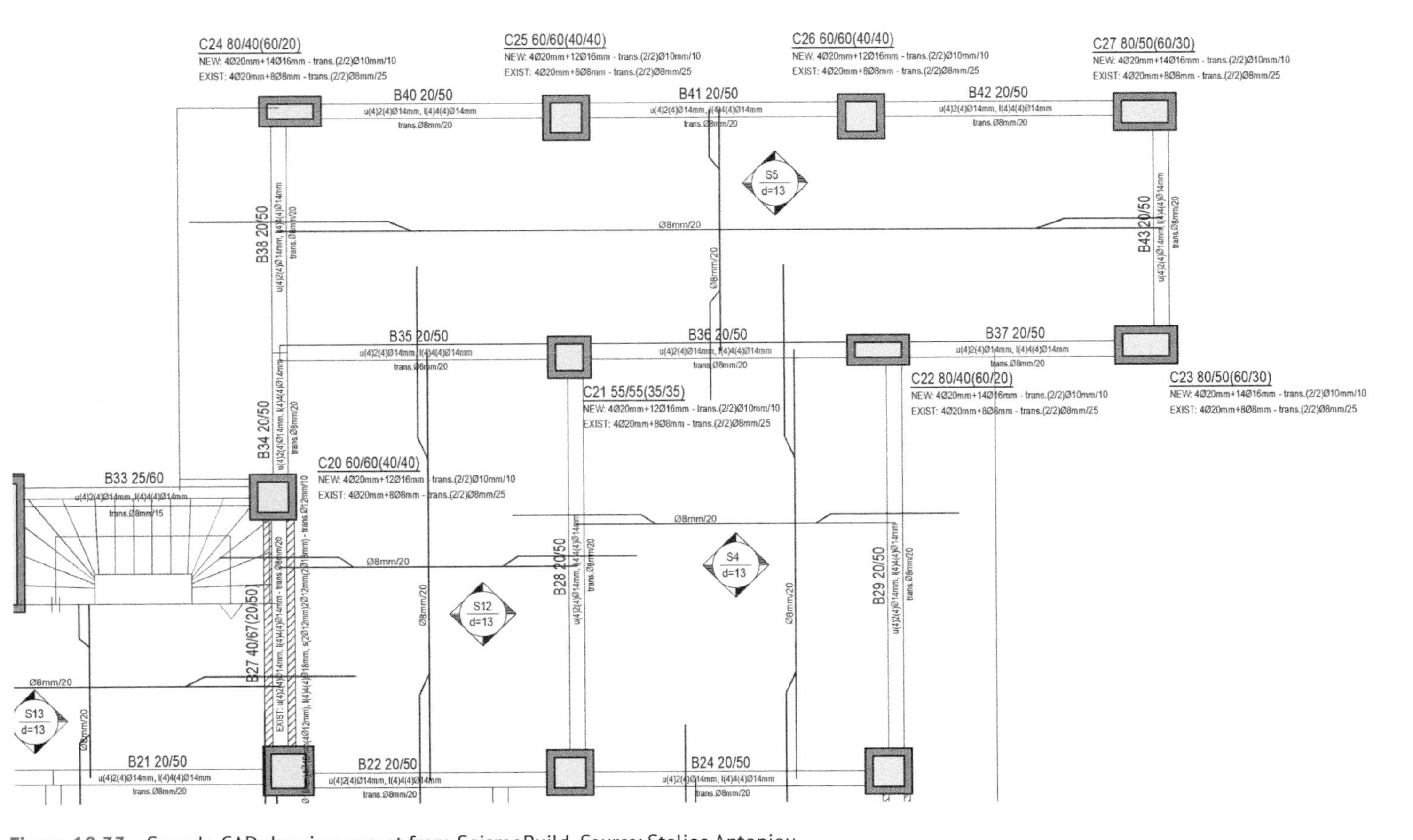


Figure 10.33 Sample CAD drawing export from SeismoBuild. *Source:* Stelios Antoniou.

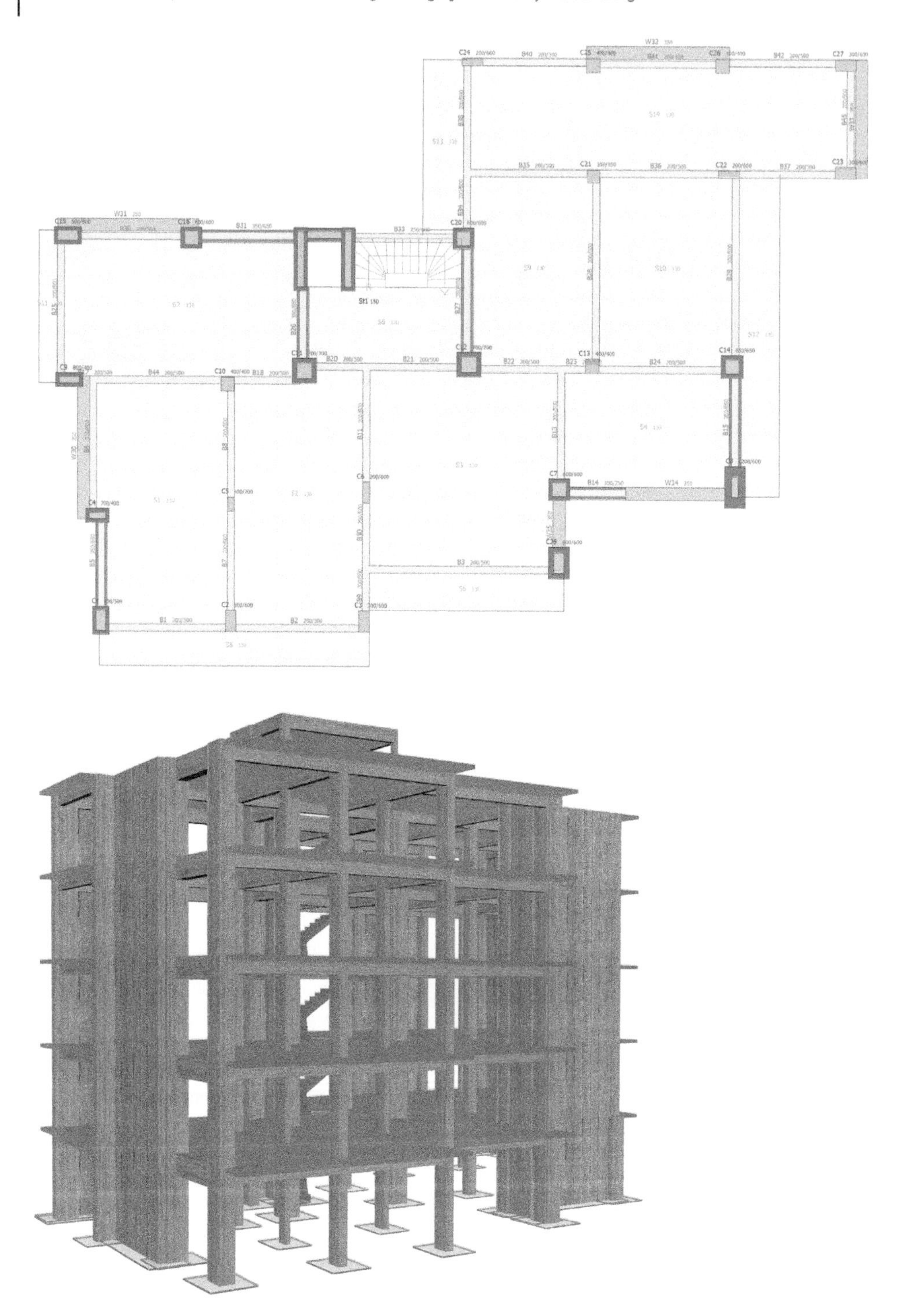

Figure 10.34 Plan view and 3D model of the strengthening scheme with RC walls. *Source:* Stelios Antoniou.

**Figure 10.34** (Continued)

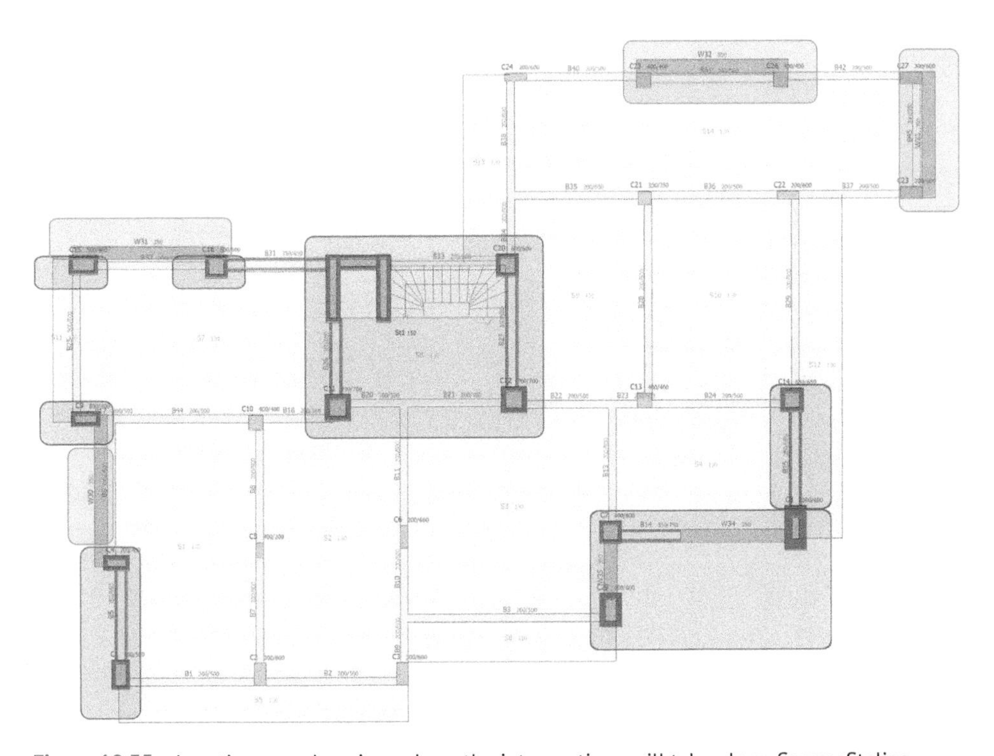

**Figure 10.35** Locations on plan view, where the interventions will take place. *Source:* Stelios Antoniou.

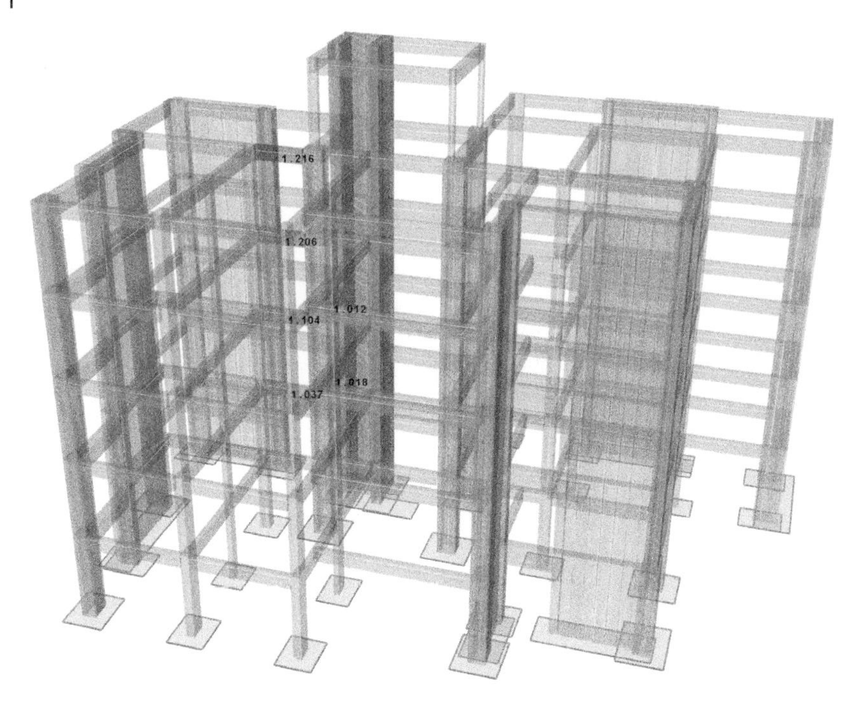


**Figure 10.36** Shear checks in the significant damage limit state for the building strengthened with RC walls. *Source:* Stelios Antoniou.

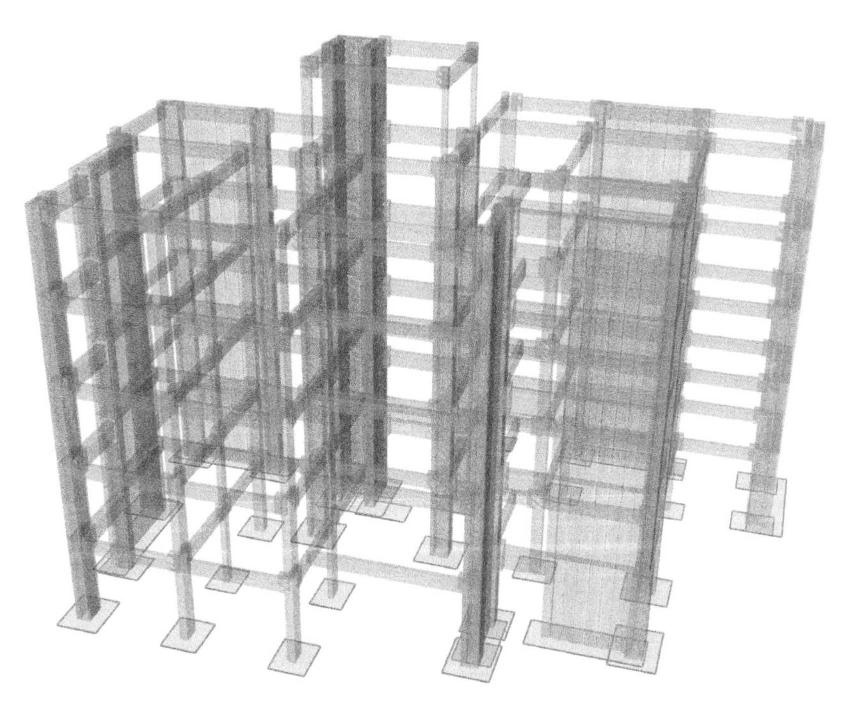

**Figure 10.37** Beam-column joint checks in the significant damage limit state for the building strengthened with RC walls. *Source:* Stelios Antoniou.

## References

Beyer, K., Dazio, A., and Priestley, M.J.N. (2008a). Inelastic wide-column models for U-shaped reinforced concrete wall. *Journal of Earthquake Engineering* 12 (1): 1–33. Imperial College Press.

Beyer, K., Dazio, A., and Priestley, M.J.N. (2008b). *Seismic Design of Torsionally Eccentric Buildings with U-Shaped RC Walls*. Pavia, Italy: ROSE School.

Beyer K, Dazio A. and Priestley M.J.N. (2008c). Elastic and inelastic wide-column models for RC non rectangular walls. *Proceedings of the Fortieth World Conference on Earthquake Engineering*. Beijing, China.

CEN (2002). *Eurocode 1: Actions on Structures – Part 1-1: General Actions – Densities, Self-Weight, Imposed Loads for Buildings*. Brussels: Comité Européen de Normalisation.

CEN (2004). *European Standard EN 1998-1: 2004. Eurocode 8: Design of Structures for Earthquake Resistance, Part 1: General Rules*. Comité Européen de Normalisation, Brussels: Seismic Actions and Rules for Buildings.

CEN (2005). *European Standard EN 1998-3: 2005. Eurocode 8: Design of Structures for Earthquake Resistance, Part 3: Assessment and Retrofitting of Buildings*. Brussels: Comité Européen de Normalisation.

Freeman, S.A. (1998). Development and use of Capacity Spectrum Method. *Proceedings, Sixth U.S. National Conference on Earthquake Engineering* [computer file], Earthquake Engineering Research Inst., Oakland, California, 12 pages.

Freeman, S.A., Nicoletti, J.P., and Tyrell, J.V. (1975). Evaluation of existing buildings for seismic risk – A case study of Puget Sound Naval Shipyard Bremerton, Washington. In: *Proceedings of the United States National Conference on Earthquake Engineering*, 113–122. Berkeley.

Mander, J.B., Priestley, M.J.N., and Park, R. (1988). Theoretical stress-strain model for confined concrete. *Journal of Structural Engineering* 114 (8): 1804–1826.

Menegotto, M. and Pinto, P.E. (1973). Method of analysis for cyclically loaded R.C. plane frames including changes in geometry and non-elastic behaviour of elements under combined normal force and bending. In: *Symposium on the Resistance and Ultimate Deformability of Structures Acted on by Well Defined Repeated Loads*, 15–22. Zurich, Switzerland: International Association for Bridge and Structural Engineering.

SeismoBuild (2023). SeismoBuild - A computer program for the linear and nonlinear analysis of Reinforced Concrete Buildings. www.seismosoft.com.

SeismoStruct (2023). SeismoStruct - A computer program for static and dynamic nonlinear analysis of framed structures. www.seismosoft.com.

# Appendix A

# Standards and Guidelines

## A.1 Eurocodes

In Appendix A.1 the main parameters used for the structural assessment according to the Eurocodes framework, and in particular Eurocode 8, Part-1 and Eurocode 8, Part-3, are presented.

### A.1.1 Performance Requirements

According to EN1998-3 (2005) Section 2.1, the performance requirements refer to the state of damage in the structure that is defined through three discrete limit states, namely Near Collapse (NC), Significant Damage (SD) and Damage Limitation (DL).

#### A.1.1.1 Limit State of near Collapse (NC)

The structure is heavily damaged, with low residual lateral strength and stiffness, although the vertical elements are still capable of sustaining the vertical loads. Most nonstructural components have collapsed. Large permanent drifts are present. The structure would probably not survive another earthquake, even of moderate intensity. The level of protection is checked by choosing a seismic action with a return period of 2.475 years corresponding to a probability of exceedance of 2% in 50 years.

#### A.1.1.2 Limit State of Significant Damage (SD)

The structure is significantly damaged, but some residual lateral strength and stiffness remain, and the vertical elements are capable of carrying the vertical loads. Nonstructural components are damaged, although partitions and infills have not failed out-of-plane. Moderate permanent drifts are present, and the structure can sustain aftershocks of only moderate intensity. The structure can be repaired, but it is likely that this turns out to be not worth it economically. The level of protection is checked by choosing a seismic action with a return period of 475 years corresponding to a probability of exceedance of 10% in 50 years.

#### A.1.1.3 Limit State of Damage Limitation (DL)

The structure is only lightly damaged, with structural elements not succumbing to significant yielding and retaining their strength and stiffness properties. Nonstructural

*Seismic Retrofit of Existing Reinforced Concrete Buildings*, First Edition. Stelios Antoniou.

components, such as partitions and infills, may show distributed cracking, but the damage could be repaired economically. Permanent drifts are negligible. The structure does not need any immediate repair measures. The level of protection is checked by choosing a seismic action with a return period of 225 years corresponding to a probability of exceedance of 20% in 50 years.

The Eurocodes National Annexes specify whether to employ all three limit states, two of them, or just one.

### A.1.2 Information for Structural Assessment

In order to choose the admissible type of analysis and the appropriate confidence factor values, the following three knowledge levels are defined:

- KL1: Limited knowledge
- KL2: Normal knowledge
- KL3: Full knowledge

The following factors are used to determine the knowledge level:

- The geometry, i.e. the geometrical properties of the structural system and the non-structural elements, e.g., masonry infill panels, that may affect the structural response.
- The details, which include the amount and detailing of reinforcement in reinforced concrete sections, the connection of floor diaphragms to lateral resisting structure, the bond and mortar of masonry infill walls.
- The materials, that is, the mechanical properties of the constituent materials.

#### A.1.2.1 KL1: Limited Knowledge

The limited knowledge level corresponds to a state of knowledge, where the overall structural geometry and member sizes are known from survey or from original outline construction drawings used for both the original construction and any subsequent modifications, as well as a sufficient sample of dimensions of both overall geometry and member sizes checked on site. If significant discrepancies from the outline construction drawings are found, a fuller dimensional survey is performed. The structural details are not known from detailed construction drawings and are assumed based on simulated design, in accordance with the usual practice at the time of construction. Limited inspections, performed in the most critical elements, should prove that the assumptions correspond to the actual situation. Information on the mechanical properties of the construction materials is not available, and default values are assumed in accordance with the standards at the time of construction, accompanied by limited in-situ testing in the most critical elements.

Structural evaluation based on this state of knowledge is performed through linear analysis methods only, either static or dynamic.

#### A.1.2.2 KL2: Normal Knowledge

The normal knowledge level corresponds to a state of knowledge, where the overall structural geometry and member sizes are known from extended survey or from outline construction drawings used for both the original construction and any subsequent modifications, as well as a sufficient sample of dimensions of both overall geometry and member sizes. The structural details are known from an extended in-situ inspection or from incomplete detailed construction drawings, in combination with limited in-situ inspections in the most critical elements, which confirms that the available information corresponds to the actual situation. Information on the mechanical properties of the construction materials is available from extended in-situ testing or from original design specifications and limited in-situ testing.

Structural evaluation based on this state of knowledge is performed through linear or nonlinear analysis methods, either static or dynamic.

#### A.1.2.3 KL3: Full Knowledge

The full knowledge level corresponds to a state of knowledge, where the overall structural geometry and member sizes are known from a comprehensive survey or from the complete set of outline construction drawings used for both the original construction and subsequent modifications, as well as a sufficient sample of both overall geometry and member sizes checked on site. The structural details are known from comprehensive in-situ inspection or from a complete set of detailed construction drawings in combination with limited in-situ inspections in the most critical elements, which prove that the available information corresponds to the actual situation. Information on the mechanical properties of the construction materials is available from comprehensive in-situ testing or from original test reports and limited in-situ testing.

Structural evaluation based on this state of knowledge is performed through linear or nonlinear analysis methods, either static or dynamic.

#### A.1.2.4 Confidence Factors

Table A.1 presents a summary of the recommendations of EN 1998-3 (2005) regarding the knowledge levels and the corresponding methods of analysis and confidence factors.

### A.1.3 Safety Factors

The values of the safety factors employed in the safety verifications are defined at different parts of Eurocode 8, usually after each specific expression. The material partial factors $\gamma_c$ and $\gamma_s$ for concrete and steel are defined in Eurocode 2 (the recommended values are 1.50 and 1.15, respectively).

### A.1.4 Capacity Models for Assessment and Checks

All the member checks (chord rotation capacity and shear capacity) should be carried out for all the elements of every floor, according to Annex A of EN1998-3(2005).

#### A.1.4.1 Deformation Capacity

The deformation capacity of beams, columns and walls is defined in terms of the chord rotation $\theta$. The deformation capacity of beams and columns is highly influenced by the lack of appropriate seismic resistant detailing in longitudinal reinforcement, as well as by the bar type, that is whether the bars are smooth or/and made of cold-worked brittle steel. Inadequate development of splicing along the span (beams) and height (columns), and

**Table A.1** Knowledge levels and the corresponding methods of analysis and confidence factors CF (Table 3.1 of EN 1998-3: 2005).

| Knowledge level | Geometry | Details | Materials | Analysis | CF |
|---|---|---|---|---|---|
| KL1 | From original outline construction drawings with sample **visual** survey **or** from **full** survey | Simulated design in accordance with relevant practice **and** from **limited** in-situ inspection | Default values in accordance with standards of the time of construction **and** from **limited** in-situ testing | Lateral Force procedure or Modal response spectrum analysis | $CF_{KL1}$ |
| KL2 | | From incomplete original detailed construction drawings with **limited** in-situ inspection **or** from **extended** in-situ inspection | From original design specifications with **limited** in-situ testing **or** from **extended** in-situ testing | All | $CF_{KL2}$ |
| KL3 | | From original detailed construction drawings with **limited** in-situ inspection **or** from **comprehensive** in-situ inspection | From original test reports with **limited** in-situ testing **or** from **comprehensive** in-situ testing | All | $CF_{KL3}$ |

*Note:* The values ascribed to the confidence factors to be used in a country may be found in its National Annex. The recommended values are $CF_{KL1} = 1.35$, $CF_{KL2} = 1.20$, and $CF_{KL3} = 1.00$.

inadequate embedment into beam-column joints can control the member's response to seismic action, drastically limiting its capacity in respect to the situation, in which the reinforcement is considered fully effective. These problems that affect the deformation capacity are taken into consideration.

The value for the chord rotation capacity for the limit state of Near Collapse is the value of the total chord rotation capacity (elastic plus inelastic part) at ultimate, which is given by Eqs. (A.1) and (A.3):

$$\theta_{um} = \frac{1}{\gamma_{el}} \cdot 0.016 \cdot \left(0.3^{\nu}\right) \left[ \frac{\max\left(0.01; \omega'\right)}{\max\left(0.01; \omega\right)} f_c \right]^{0.225} \cdot \left( \min\left(9, \frac{L_V}{h}\right)\right)^{0.35} 25^{\left(\alpha \rho_{sx} \frac{f_{yw}}{f_c}\right)} \left(1.25^{100\rho_d}\right)$$

EN 1998 – 3 (A.1)

- $\gamma_{el}$ is equal to 1.5 for primary seismic elements and to 1.0 for secondary seismic ones.
- $L_V$ is the ratio between bending moment, $M$, and shear force, $V$.
- $h$ is the depth of the cross-section,
- $\nu = N/(b \cdot h \cdot f_c)$, where $b$ = width of compression zone and $N$ = axial force, positive for compression.

- $\omega$ and $\omega$' are the mechanical reinforcement ratio of the longitudinal reinforcement in tension (including the web reinforcement) and compression, respectively.
- $f_c$ and $f_{yw}$ are the concrete compressive strength (MPa) and the stirrup yield strength (MPa).
- $\rho_{sx}$ is the ratio of transverse steel parallel to the direction of loading.
- $\rho_d$ is the steel ratio of diagonal reinforcement.
- $\alpha$ is the confinement effectiveness factor.
- In walls, the value given by the Eq. (A.1) is multiplied by 0.58.

The total chord rotation capacity at ultimate of concrete members under cyclic loading may be also calculated as the sum of the chord rotation at yielding and the plastic part of the chord rotation capacity calculated from the following expression:

$$\theta_{um}^{pl} = \theta_{um} - \theta_y = \frac{1}{\gamma_{el}} \cdot 0.0145 \cdot \left(0.25^{\nu}\right)\left[\frac{max\left(0.01;\omega'\right)}{max\left(0.01;\omega\right)}\right]^{0.3} \cdot f_c^{0.2} \cdot \left(min\left(9,\frac{L_V}{h}\right)\right)^{0.35} 25^{\left(\alpha\rho_{sx}\frac{f_{yw}}{f_c}\right)}\left(1.275^{100\rho_d}\right) \quad \text{EN 1998 - 3 (A.3)}$$

where $\gamma_{el}$ is equal to 1.8 for primary seismic elements and to 1.0 for secondary seismic ones.

In walls the value of $\theta_{um}^{pl}$ given by the Eq. (A.3) is multiplied by 0.6.

The chord rotation capacity corresponding to the limit state of Significant Damage is assumed to be ¾ of the ultimate chord rotation, calculated from the equations above.

The chord rotation capacity that corresponds to the limit state of Damage Limitation is given by the chord rotation at yielding, evaluated as:

For rectangular beams and columns:

$$\theta_y = \varphi_y \frac{L_V + \alpha_V z}{3} + 0.0014\left(1 + 1.5\frac{h}{L_V}\right) + \frac{\varepsilon_y}{d - d'} \frac{d_{bL} f_y}{6\sqrt{f_c}} \quad \text{EN 1998 - 3 (A.10a)}$$

For walls or rectangular T- or barbelled sections:

$$\theta_y = \varphi_y \frac{L_V + \alpha_V z}{3} + 0.0013 + \frac{\varepsilon_y}{d - d'} \frac{d_{bL} f_y}{6\sqrt{f_c}} \quad \text{EN 1998 - 3 (A.11a)}$$

or from alternative and equivalent expressions for rectangular beams and columns

$$\theta_y = \varphi_y \frac{L_V + \alpha_V z}{3} + 0.0014\left(1 + 1.5\frac{h}{L_V}\right) + \varphi_y \frac{d_{bL} f_y}{8\sqrt{f_c}} \quad \text{EN 1998 - 3 (A.10b)}$$

For walls or rectangular T- or barbelled sections:

$$\theta_y = \varphi_y \frac{L_V + \alpha_V z}{3} + 0.0013 + \varphi_y \frac{d_{bL} f_y}{8\sqrt{f_c}} \quad \text{EN 1998 - 3 (A.11b)}$$

- $\varphi$ is the yield curvature of the end section.
- $z$ is the length of the internal lever arm.
- $\alpha_V$ is equal to unity, if shear cracking is expected to precede flexural yielding at the end section, or zero otherwise.
- $f_y$ and $f_c$ are the steel yield stress and the concrete compressive strength, respectively.
- $\varepsilon_y$ is the steel strain at yield, equal to $f_y/E_s$.
- $d$ and $d'$ are the depths to the tension and compression reinforcement, respectively.
- $d_{bL}$ is the mean diameter of the tension reinforcement.

According to Annex A of EN1998-3, the chord rotation capacity is highly influenced by a number of different factors, such as the type of the longitudinal bars. If cold-worked brittle steel is used, the plastic part of chord rotation is divided by 2, whereas if smooth (plain) longitudinal bars are applied, Section A.3.2.2(5) of Annex A is employed, which also takes into consideration whether the longitudinal bars are well lapped or not. In the case of members with a lack of appropriate seismic resistant detailing, the values given by expressions (A.1) and (A.3) are divided by 1.20. Furthermore, if the deformed longitudinal bars have straight ends lapped starting at the end section of the member, the plastic part of the chord rotation is calculated with the value of the compression reinforcement ratio, $\omega'$, doubled over the value applying outside the lap splice. In addition, in sections where the reinforcement lap length $l_o$ is less than the minimum lap length for ultimate deformation $l_{ou,min}$, the plastic part of the chord rotation capacity, as given in (A.3), is multiplied by the ratio $l_o/l_{ou,min}$. For more information about the calculation of $l_{ou,min}$ you may refer to Section A.3.2.2(4) of Annex A, while the value for chord rotation at yielding, $\theta_y$ accounts for the effect of the lapping in accordance with Section A.3.2.4(3) of Annex A.

#### A.1.4.2 Shear Capacity

The shear capacity of beams, columns, and walls is calculated through the following expression according to Annex A of EN1998-3:2005, as controlled by the stirrups, accounting for the reduction due to the plastic part of ductility demand.

$$V_R = \frac{1}{\gamma_{el}}\left[\frac{h-x}{2L_V} min\left(N; 0.55A_c f_c\right) + \left(1-0.05min\left(5; \mu_\Delta^{pl}\right)\right)\cdot \left[0.16max\left(0.5; 100\rho_{tot}\right)\left(1-0.16min\left(5; \frac{L_V}{h}\right)\right)\sqrt{f_c}A_c + V_w\right]\right] \quad \text{EN 1998} - 3\ (\text{A.12})$$

- $\gamma_{el}$ is equal to 1.15 for primary seismic elements and to 1.00 for secondary ones
- $h$ is the depth of the cross-section (equal to the diameter D for circular sections).
- $x$ is the compression zone depth.
- $A_c$ is the cross-section area.
- $\rho_{tot}$ is the ratio of the total longitudinal reinforcement.
- $V_w$ is the contribution of transverse reinforcement to the shear resistance; for rectangular sections $V_w$ is equal to:

$$V_w = \rho_w b_w z f_{yd} \quad \text{EN 1998} - 3\ (\text{A.13})$$

- $\rho_w$ is the ratio of the transverse reinforcement.
- $z$ is the length of the internal lever arm, as specified in Section A.3.2.4(2).

- $f_{yw}$ is the stirrup yield strength (MPa); for primary elements, $f_{yw}$ should further be divided by the partial factor for steel $\gamma_s$, in accordance with EN 1998-1 (2004, Section 5.2.4).

The shear strength of a concrete wall should not be greater than the value corresponding to failure by web crushing, $V_{R,max}$, which under cyclic loading is calculated according to Section A3.3.1(2) of Annex A of EN1998-3:2005:

$$V_{R,max} = \frac{0.85\left(1 - 0.06min\left(5;\mu_{\Delta}^{pl}\right)\right)}{\gamma_{el}}\left(1 + 1.8min\left(0.15;\frac{N}{A_c f_c}\right)\right) \left(1 + 0.25max\left(1.75;100\rho_{tot}\right)\right)\cdot\left(1 - 0.2min\left(2;\frac{L_V}{h}\right)\right)\sqrt{f_c}b_w z \qquad \text{EN 1998 - 3 (A.15)}$$

If in a concrete column the shear span ratio ($L_V/h$) at the end section with the maximum of the two end moments is less or equal to 2, the shear strength is not taken greater than the value corresponding to the failure by web crushing along the diagonal of the column after flexural yielding, $V_{R,max}$, which under cyclic loading is calculated according to Section A3.3.1(3) of Annex A of EN1998-3 (2005) from the following expression:

$$V_{R,max} = \frac{4/7\left(1 - 0.02min\left(5;\mu_{\Delta}^{pl}\right)\right)}{\gamma_{el}}\left(1 + 1.35\frac{N}{A_c f_c}\right) \left(1 + 0.45\left(100\rho_{tot}\right)\right)\sqrt{min\left(40;f_c\right)}b_w z\ sin2\delta \qquad \text{EN 1998 - 3 (A.16)}$$

where $\delta$ is the angle between the diagonal and the axis of the column ($tan\delta = h/2L_V$).

#### A.1.4.3 FRP Wrapping

According to Section A.4.4.2(9) of Annex A of EN1998-3 (CEN 2005), in members with their plastic hinge region fully wrapped in an FRP jacket over a length at least equal to the member depth, the cyclic resistance $V_R$, may be calculated from expression (A.12) of Eurocode 8: Part 3, also considering in $V_w$ the contribution of the FRP jacket to shear resistance. The contribution of the FRP jacket to $V_w$ is computed through the following expression:

$$V_{w,f} = 0.5\rho_f b_w z f_{u,fd} \qquad \text{EN 1998 - 3 (A.33)}$$

where $\rho_f$ is the geometric ratio of the FRP, z the length of the internal lever arm and $f_{u,fd}$ the design value of the FRP ultimate strength.

### A.1.5 Target Displacement Calculation in Pushover Analysis

The target displacement is defined with a variation of the Capacity Spectrum Method CSM, according to Annex B of EN1998-1 (CEN 2004), as the seismic demand derived from the elastic response spectrum, in terms of the displacement of an equivalent single-degree-of-freedom (SDOF) system. The SDOF displacement is then used to calculate the target displacement of the multi-degree of freedom (MDOF) system.

The following relation between normalized lateral forces $F_i$ and normalized displacements $\Phi_i$ is assumed:

$$F_i = m_i \Phi_i.$$

where $m_i$ is the mass in the $i^{th}$ story.

Displacements are normalized in such a way that $\Phi_n = 1$, where $n$ is the control node, consequently $F_n = m_n$.

#### A.1.5.1 Transformation to an Equivalent Single Degree of Freedom (SDOF) System

The mass of an equivalent SDOF system $m^*$ is determined as:

$$m^* = \sum m_i \Phi_i = \sum F_i$$

And the transformation factor is given by:

$$\Gamma = \frac{m^*}{\sum m_i \Phi_i^2} = \frac{\sum F_i}{\sum \left( \frac{F_i^2}{m_i} \right)}$$

The force $F^*$ and displacement $d^*$ of the equivalent SDOF system are computed as:

$$F^* = \frac{F_b}{\Gamma}$$

$$d^* = \frac{d_n}{\Gamma}$$

where $F_b$ and $d_n$ are, respectively, the base shear force and the control node displacement of the MDOF system.

#### A.1.5.2 Determination of the Idealized Elasto-Perfectly Plastic Force-Displacement Relationship

The yield force $F_y^*$, which also represents the ultimate strength of the idealized SDOF system, is equal to the base shear force at the formation of the plastic mechanism. The initial stiffness of the elastic perfectly plastic force–displacement relationship in the idealized system is determined in such a way that the areas under the actual and the idealized force–deformation curves are equal, as shown in Figure A.1:

**Figure A.1** Determination of the idealized elasto-perfectly plastic force–displacement relationship (Figure B.1 of EN 1998-1).

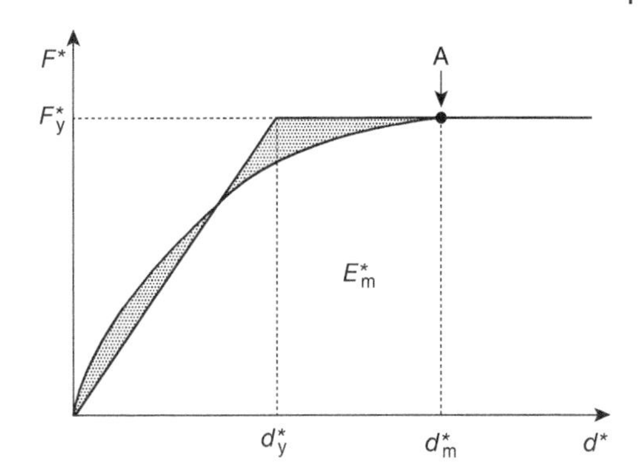


Based on this assumption, the yield displacement on the idealized SDOF system $d_y^*$ is given by:

$$d_y^* = 2\left(d_m^* - \frac{E_m^*}{F_y^*}\right)$$

where $E_m^*$ is the actual deformation energy up to the formation of the plastic mechanism.

#### A.1.5.3 Determination of the Period of the Idealized Equivalent SDOF System

The period $T^*$ of the idealized equivalent SDOF system is determined by:

$$T^* = 2\pi\sqrt{\frac{m^* d_Y^*}{F_Y^*}}$$

#### A.1.5.4 Determination of the Target Displacement for the Equivalent SDOF System

The target displacement of the structure with period $T^*$ and unlimited elastic behavior is given by:

$$d_{et}^* = S_e\left(T^*\right)\left[\frac{T^*}{2\pi}\right]^2$$

where $S_e(T^*)$ is the elastic acceleration response spectrum at the period $T^*$.

For the determination of the target displacement $d_t^*$ for structures in the short-period range and for structures in the medium and long-period ranges different expressions should be used as indicated below. The corner period between the short- and medium-period range is $T_C$.

- For $T^* < T_C$ (short period range)
  If $F_y^* / m^* \geq S_e(T^*)$, the response is elastic and thus: $d_t^* = d_{et}^*$
  If $F_y^* / m^* < S_e(T^*)$, the response is nonlinear and: $d_t^* = \frac{d_e t^*}{q_u}(1 + (q_u - 1)\frac{T_C}{T^*}) \geq d_{et}^*$

$q_u$ is the ratio between the acceleration in the structure with unlimited elastic behavior $S_e(T^*)$ and the structure with limited strength $F_y^*/m^*$:

$$q_u = \frac{S_e(T^*)m^*}{F_Y^*}$$

- For $T^* \geq T_C$ (medium and long period range)

$$d_t^* = d_{et}^*$$

$d_t^*$ must not exceed $3d_{et}^*$.

#### A.1.5.5 Determination of Target Displacement for the MDOF System

The target displacement of the MDOF system is given by:

$$d_t = \Gamma d_t^*$$

Note that the target displacement corresponds to the displacement of the control node.

## A.2. ASCE 41-17

In Appendix A.2 the parameters used for the structural assessment, according to the *American Seismic Evaluation and Retrofit of Existing Buildings Code, ASCE/SEI 41-17* (ASCE 2017), are presented. Apart from ASCE 41, several references will also be made to "*ACI 318, Building Code Requirements for Structural Concrete and Commentary*," (ACI 2019) the US document with the requirements for design and construction of new RC buildings.

### A.2.1 Performance Requirements

According to ASCE 41-17 Section 2.2, the objectives of the assessment or redesign (Table C2.2) consist of combinations of both a performance level and a seismic action, given an "acceptable probability of exceedance within the life cycle of the building" (design earthquake), as shown in Table A.2.

Table A.2 Building performance levels in ASCE 41 (ASCE 2017, Table C2-2).

| | Target Building Performance Levels | | | |
|---|---|---|---|---|
| **Seismic hazard level** | **Operational performance level (1-A)** | **Immediate Occupancy performance level (1-B)** | **Life Safety performance level (3-C)** | **Collapse Prevention performance level (5-D)** |
| 50%/50 years | a | b | c | d |
| BSE-1E (20%/50 years) | e | f | g | h |
| BSE-2E (5%/50 years) | i | j | k | l |
| BSE-2N (ASCE 7 $MCE_R$) | m | n | o | p |

The target building performance levels refer to the state of damage in the structure defined through four limit states, namely Operational Level (1-A), Immediate Occupancy (1-B), Life Safety (3-C), and Collapse Prevention (5-D).

#### A.2.1.1 Performance Level of Operational Level (1-A)

The building is expected to sustain minimal or no damage both to the structural and non-structural components. The building is suitable for its normal occupancy and use, although possibly in a slightly impaired mode. All the essential utilities, such as the power and water supply, are functioning, and only some nonessential systems might not be functioning correctly. The building poses an extremely low risk to human life.

#### A.2.1.2 Performance Level of Immediate Occupancy (1-B)

The building is expected to sustain minimal or no damage to the structural elements and only minor damage to the nonstructural components. Although it would be safe to reoccupy a building meeting this performance level immediately after a major earthquake, nonstructural systems might not function, either because of the lack of electrical power or because of internal damage to equipment. Therefore, although immediate re-occupancy is possible, it might be necessary to perform some cleanup and repair and await the restoration of utility service before the building can function normally again. The risk to life safety is very low.

#### A.2.1.3 Performance Level of Life Safety (3-C)

The performance level of Life Safety does not necessarily mean that there will be no injuries to occupants or people in the immediate vicinity of the building. However, it means that only a few, if any, are expected to be serious enough to require skilled medical attention. Buildings meeting this level may experience extensive damage to structural and nonstructural components. Repairs may be required before re-occupancy occurs, and repair, although technically feasible, may be deemed economically impractical. The risk to life safety is low.

#### A.2.1.4 Performance Level of Collapse Prevention (5-D)

The building may experience extensive damage to structural and nonstructural components, and nonstructural falling hazards that cause serious injury or death may occur. However, the total or partial building collapse is prevented and major nonstructural falling hazards that can cause serious injury or death to large numbers of people are not likely either. Extensive repairs may be required before re-occupancy of the building occurs, and repair may be deemed economically unjustified. In the performance level of Collapse Prevention, there is a risk to life safety.

### A.2.2 Information for Structural Assessment

In order to choose the admissible type of analysis and the appropriate knowledge factor values, the following three knowledge levels are defined:

- Minimum knowledge
- Usual knowledge
- Comprehensive knowledge

The factors determining the obtained data reliability level are:

i) The geometry, i.e. the geometrical properties of the structural system and the non-structural elements, e.g. masonry infill panels, that may affect the structural response;
ii) The details, which include the amount and detailing of reinforcement in reinforced concrete sections, the connection of floor diaphragms to lateral resisting structure, the bond and mortar of masonry infill walls.
iii) The materials, that is the mechanical properties of the constituent materials.

#### A.2.2.1 Minimum Knowledge

The minimum data collection requirements correspond to a state of knowledge, where information is obtained from design drawings with sufficient information to analyze component demands and calculate component capacities. The design drawings show the configuration of the gravity load system and seismic-force-resisting system in sufficient detail. Information is verified by a visual condition assessment.

In the absence of sufficient information from design drawings, incomplete or nonexistent information is supplemented by a comprehensive condition assessment, including destructive and nondestructive investigation. In the absence of material test records and quality assurance reports, default material properties are used according to Section 10.2.2.5 of ASCE 41-17.

#### A.2.2.2 Usual Knowledge

The usual data reliability level corresponds to a state of knowledge, where information is obtained from design drawings with sufficient information to analyze component demands and calculate component capacities. The design drawings show the configuration of the gravity load system and seismic-force-resisting system in sufficient detail. Information is verified by a visual condition assessment.

In the absence of sufficient information from design drawings, incomplete or nonexistent information is supplemented by a comprehensive condition assessment, including destructive and nondestructive investigation. In the absence of material test records and quality assurance reports, default material properties are used according to Section 10.2.2.5 of ASCE 41-17.

#### A.2.2.3 Comprehensive Knowledge

The comprehensive data reliability level corresponds to a state of knowledge, where information is obtained from construction documents including design drawings, specifications, material test records, and quality assurance reports covering original construction and subsequent modifications to the structure. Information is verified by a visual condition assessment.

In the cases where construction documents are incomplete, missing information is supplemented by comprehensive condition assessment, including destructive and nondestructive investigation. In the absence of material test records and quality assurance reports, material properties are determined by comprehensive material testing in accordance to Section 10.2.2.4.2 of ASCE 41-17.

### A.2.3 Safety Factors

In ASCE 41-17 the safety factors are directly incorporated in the member's strengths and deformation limits, as provided in the different tables of the document.

### A.2.4 Capacity Models for Assessment and Checks

All the member checks (plastic hinge rotation capacity and shear capacity) should be carried out for all the elements of every floor, according to Chapter 10 of ASCE 41-17 (ASCE

41 2017), Chapters 18 and 22 of ACI 318-19 (ACI 318 2019), and Chapter 11 of ACI 440 (2017), taking into account Table 7.7 of ASCE 41-17.

#### A.2.4.1 Deformation Capacity

For nonlinear analysis, the deformation capacity of beams, columns, and walls that are controlled by flexure is defined in terms of the plastic hinge rotation, according to Section 10.4.2.2. The deformation capacity is highly influenced by the lack of appropriate seismic resistant detailing in longitudinal reinforcement, as well as whether there are smooth bars. Inadequate development of splicing along the span (beams) and height (columns) and inadequate embedment into beam-column joints can control the members' response to seismic action, drastically limiting its capacity, as compared to the situation in which the reinforcement is considered fully effective. These limitations are all taken into consideration.

The ultimate total hinge rotation capacity of concrete members under cyclic loading is calculated as the sum of the hinge rotation at yielding and the plastic part of the hinge rotation capacity:

$$\theta = \theta_y + \theta_p$$

The rotation capacity at yield $\theta_y$ is calculated as described below:

- For beams and columns from the equation:

$$\theta_y = \frac{M_y L_s}{3EI_{\text{eff}}}$$

The effective stiffness value $EI_{\text{eff}}$, is calculated according to Table 10.5 of ASCE 41-17.

- For walls from Eq. ((10.5) of ASCE 41-17:

$$\theta_{yE} = \left(\frac{M_{yE}}{(EI)_{\text{eff}}}\right) l_p \qquad \text{ASCE 41 – 17 (10-5)}$$

The plastic part of the hinge rotation capacity is calculated as indicated below:

- For beams according to Table 10-7 of ASCE 41-17 (Table A.3 in the current document).
- For columns according to Table 10-8 of ASCE 41-17 (Table A.4 in the current document).
- For walls controlled by flexure according to Table 10-19 of ASCE 41-17 (Table A.5 in the current document).

The deformation capacity of walls controlled by shear is defined in terms of the interstory drift ratio as indicated in Table 10.20 of ASCE 41-17 (Table A.6 in the current document).

#### A.2.4.2 Shear Capacity

The Shear capacity of columns is calculated through the following expression according to Section 10.4.2.3 of ASCE 41-17.

$$V_{Col} = k_{nl} V_{Col0} = k_{nl} \left[ \alpha_{Col} \left( \frac{A_v f_{ytL/E} d}{s} \right) + \lambda \left( \frac{6\sqrt{f'_{cL/E}}}{M_{UD} / V_{UD} d} \sqrt{1 + \frac{N_{UG}}{6A_g \sqrt{f'_{cL/E}}}} \right) 0.8A_g \right]$$

$$\left(\text{lb / in.2 units}\right)$$

**Table A.3** Modeling Parameters and numerical acceptance criteria for nonlinear procedures – Reinforced concrete beams in ASCE 41 (ASCE 2017, Table 10-7).

| Conditions | | | Modeling parameters[a] | | | Acceptance criteria[a] | | |
|---|---|---|---|---|---|---|---|---|
| | | | | | | Plastic Rotation Angle (radians) | | |
| | | | Plastic rotation angle (radians) | | Residual strength ratio | Performance level | | |
| | | | *a* | *b* | *c* | IO | LS | CP |
| Condition i. Beams controlled by flexure[b] | | | | | | | | |
| $\frac{\rho-\rho'}{\rho_{bal}}$ | Transverse reinforcement[c] | $\frac{V^d}{b_w d\sqrt{f'_{cE}}}$ | | | | | | |
| ≤0.0 | C | ≤3 (0.25) | 0.025 | 0.05 | 0.2 | 0.010 | 0.025 | 0.05 |
| ≤0.0 | C | ≥6 (0.5) | 0.02 | 0.04 | 0.2 | 0.005 | 0.02 | 0.04 |
| ≥0.5 | C | ≤3 (0.25) | 0.02 | 0.03 | 0.2 | 0.005 | 0.02 | 0.03 |
| ≥0.5 | C | ≥6 (0.5) | 0.015 | 0.02 | 0.2 | 0.005 | 0.015 | 0.02 |
| ≤0.0 | NC | ≤3 (0.25) | 0.02 | 0.03 | 0.2 | 0.005 | 0.02 | 0.03 |
| ≤0.0 | NC | ≥6 (0.5) | 0.01 | 0.015 | 0.2 | 0.0015 | 0.01 | 0.015 |
| ≥0.5 | NC | ≤3 (0.25) | 0.01 | 0.015 | 0.2 | 0.005 | 0.01 | 0.015 |
| ≥0.5 | NC | ≥6 (0.5) | 0.005 | 0.01 | 0.2 | 0.0015 | 0.005 | 0.01 |

| | | | | | | |
|---|---|---|---|---|---|---|
| Condition ii. Beams controlled by shear[b] | | | | | | |
| Stirrup spacing $\leq d/2$ | 0.0030 | 0.02 | 0.2 | 0.0015 | 0.01 | 0.02 |
| Stirrup spacing $> d/2$ | 0.0030 | 0.01 | 0.2 | 0.0015 | 0.005 | 0.01 |
| Condition iii. Beams controlled by inadequate development or splicing along the span[b] | | | | | | |
| Stirrup spacing $\leq d/2$ | 0.0030 | 0.02 | 0.0 | 0.0015 | 0.01 | 0.02 |
| Stirrup spacing $> d/2$ | 0.0030 | 0.01 | 0.0 | 0.0015 | 0.005 | 0.01 |
| Condition iv. Beams controlled by inadequate embedment into beam-column joint[b] | | | | | | |
| | 0.015 | 0.03 | 0.2 | 0.01 | 0.02 | 0.03 |

Note: $f'_{cE}$ in lb/in.$^2$ (MPa) units.

[a] Values between those listed in the table should be determined by linear interpolation.

[b] Where more than one of conditions i, ii, iii, and iv occur for a given component, use the minimum appropriate numerical value from the table.

[c] "C" and "NC" are abbreviations for conforming and nonconforming transverse reinforcement, respectively. Transverse reinforcement is conforming if, within the flexural plastic hinge region, hoops are spaced at $\leq d/3$, and if, for components of moderate and high ductility demand, the strength provided by the hoops ($V_s$) is at least 3/4 of the design shear. Otherwise, the transverse reinforcement is considered nonconforming.

[d] $V$ is the design shear force from NSP or NDP.

**Table A.4** Modeling parameters and numerical acceptance criteria for nonlinear procedures – Reinforced concrete columns other than circular with spiral reinforcement or seismic hoops as defined in ACI 318 (ASCE 2017, Table 10-8).

| Modeling parameters | Acceptance criteria | | |
|---|---|---|---|
| | Plastic rotation angle (radians) | | |
| | Performance level | | |
| **Plastic rotation angles, a and *b* (radians) Residual strength ratio, *c*** | **IO** | **LS** | **CP** |
| Columns not controlled by inadequate development or splicing along the clear height[a] | | | |
| $a=\left(0.042-0.43\frac{N_{UD}}{A_f f'_{cE}}+0.63\rho_t-0.023\frac{V_{yE}}{V_{ColOE}}\right)\ge 0.0$ | 0.15 a $\le 0.005$ | $0.5\,b^b$ | $0.7\,b^b$ |
| For $\frac{N_{UD}}{A_g f'_{cE}}\le 0.5\left\{b=\frac{0.5}{5+\frac{N_{UD}}{0.8A_g f'_{cE}}\frac{1}{\rho t}\frac{f'_{cE}}{f_{ytE}}}-0.01\ge a^a\right.$ | | | |
| $c=0.24-0.4\frac{N_{UD}}{A_g f'_{cE}}\ge 0.0$ | | | |
| Columns controlled by inadequate development or splicing along the clear height[c] | | | |
| $a=\left(\frac{1\rho_t f_{ytE}}{8\rho_l f_{ylE}}\right)\begin{matrix}\ge 0.0\\ \le 0.025^d\end{matrix}$ | 0.0 | 0.5 *b* | 0.7 *b* |
| $b=\left(0.012-0.085\frac{N_{UD}}{A_g f'_{cE}}+12\rho_t^{\,e}\right)\begin{matrix}\ge 0.0\\ \ge a\\ \le 0.06\end{matrix}$ | | | |
| $c=0.15+36\rho_t\le 0.4$ | | | |

Notes: $\rho_t$ shall not be taken as greater than 0.0175 in any case nor greater than 0.0075 when ties are not adequately anchored in the core. Equations in the table are not valid for columns with $\rho_t$ smaller than 0.0005.

$V_{yE}/V_{ColOE}$ shall not be taken as less than 0.2.

$N_{ud}$ shall be the maximum compressive axial load accounting for the effects of lateral forces as described in Eq. (7-34). Alternatively, it shall be permitted to evaluate $N_{ud}$ based on a limit-state analysis.

[a] *b* shall be reduced linearly for $N_{UD}/(A_g f'_{cE})>0.5$ from its value at $N_{UD}/(A_g f'_{cE})=0.5$ to zero at $N_{UD}/\left(A_g f'_{cE,}\right)=0.7$ but shall not be smaller than a.

[b] $N_{UD}/(A_g f'_{cE})$ shall not be taken as smaller than 0.1.

[c] Columns are considered to be controlled by inadequate development or splices where the calculated steel stress at the splice exceeds the steel stress specified by Eq. (10-1a) or (10-1b). Modeling parameter for columns controlled by inadequate development or splicing shall never exceed those of columns not controlled by inadequate development or splicing.

[d] *a* for columns controlled by inadequate development or splicing shall be taken as zero if the splice region is not crossed by at least two tie groups over its length.

[e] $\rho_t$ shall not be taken as greater than 0.0075.

**Table A.5** Modeling parameters and numerical acceptance criteria for nonlinear procedures – Reinforced concrete structural walls controlled by flexure (ASCE 2017, Table 10-19).

| Conditions | | | Plastic hinge rotation (radians) | | Residual strength ratio | Acceptable plastic hinge rotation[d] (radians) Performance level | | |
|---|---|---|---|---|---|---|---|---|
| | | | *a* | *b* | *c* | IO | LS | CP |
| i. Structural walls and wall segments | | | | | | | | |
| $\frac{(A_s - A'_s) f_{yE} + P}{t_w l_w f'_{cE}}$ | $\frac{V}{t_w l_w \sqrt{f'_{cE}}}$ | Confined boundary[a] | | | | | | |
| ≤0.1 | ≤4 | Yes | 0.015 | 0.020 | 0.75 | 0.005 | 0.015 | 0.020 |
| ≤0.1 | ≥6 | Yes | 0.010 | 0.015 | 0.40 | 0.004 | 0.010 | 0.015 |
| ≥0.25 | ≤4 | Yes | 0.009 | 0.012 | 0.60 | 0.003 | 0.009 | 0.012 |
| ≥0.25 | ≥6 | Yes | 0.005 | 0.010 | 0.30 | 0.0015 | 0.005 | 0.010 |
| ≤0.1 | ≤4 | No | 0.008 | 0.015 | 0.60 | 0.002 | 0.008 | 0.015 |
| ≤0.1 | ≥6 | No | 0.006 | 0.010 | 0.30 | 0.002 | 0.006 | 0.010 |
| ≥0.25 | ≤4 | No | 0.003 | 0.005 | 0.25 | 0.001 | 0.003 | 0.005 |
| ≥0.25 | ≥6 | No | 0.002 | 0.004 | 0.20 | 0.001 | 0.002 | 0.004 |
| ii. Structural wall coupling beams[b] | | | | | | | | |
| Longitudinal reinforcement and transverse reinforcement[c] | | $\frac{V}{t_w l_w \sqrt{f'_{cE}}}$ | | | | | | |
| Nonprestressed longitudinal reinforcement with conforming transverse reinforcement | | ≤3 | 0.025 | 0.050 | 0.75 | 0.010 | 0.025 | 0.050 |
| | | ≥6 | 0.020 | 0.040 | 0.50 | 0.005 | 0.020 | 0.040 |
| Nonprestressed longitudinal reinforcement with nonconforming transverse reinforcement | | ≤3 | 0.020 | 0.035 | 0.50 | 0.006 | 0.020 | 0.035 |
| | | ≥6 | 0.010 | 0.025 | 0.25 | 0.005 | 0.010 | 0.025 |
| Diagonal reinforcement | | NA | 0.030 | 0.050 | 0.80 | 0.006 | 0.030 | 0.050 |

[a] A boundary element shall be considered confined where transverse reinforcement exceeds 75% of the requirements given in ACI 318 and spacing of transverse reinforcement does not exceed 8 $d_b$. It shall be permitted to take modeling parameters and acceptance criteria as 80% of confined values where boundary elements have at least 50% of the requirements given in ACI 318 and spacing of transverse reinforcement does not exceed $8d_b$. Otherwise, boundary elements shall be considered not confined.
[b] For coupling beams spanning 8 ft 0 in., with bottom reinforcement continuous into the supporting walls, acceptance criteria values shall be permitted to be doubled for LS and CP performance.
[c] Nonprestressed longitudinal reinforcement consists of top and bottom steel parallel to the longitudinal axis of the coupling beam. Conforming transverse reinforcement consists of (a) closed stirrups over the entire length of the coupling beam at a spacing ≤ $d/3$, and (b) strength of closed stirrups $V_s \geq 3/4$ of required shear strength of the coupling beam.
[d] Linear interpolation between values listed in the table shall be permitted.

**Table A.6** Modeling parameters and numerical acceptance criteria for nonlinear procedures – Reinforced concrete structural walls controlled by shear (ASCE 2017, Table 10-20).

| Conditions | | Total drift ratio (%), or chord rotation (radians)[a] | | | Strength ratio | | Acceptable total drift (%) or chord rotation (radians)[a] | | |
|---|---|---|---|---|---|---|---|---|---|
| | | | | | | | Performance level | | |
| | | $d$ | $e$ | $g$ | $c$ | $f$ | IO | LS | CP |
| i. Structural walls and wall segments[b] | | | | | | | | | |
| $\frac{(A_s - A_s')f_{yE} + P}{t_w l_w f_{cE}'} \leq 0.05$ | | 1.0 | 2.0 | 0.4 | 0.20 | 0.6 | 0.40 | 1.5 | 2.0 |
| $\frac{(A_s - A_s')f_{yE} + P}{t_w l_w f_{cE}'} > 0.05$ | | 0.75 | 1.0 | 0.4 | 0.0 | 0.6 | 0.40 | 0.75 | 1.0 |
| ii. Structural wall coupling beams[c] | | | | | | | | | |
| Longitudinal reinforcement and transverse reinforcement[d] | $\frac{V}{t_w l_w \sqrt{f_{cE}'}}$ | | | | | | | | |
| Nonprestressed longitudinal reinforcement with conforming transverse reinforcement | $\leq 3$ | 0.02 | 0.030 | | 0.60 | | 0.006 | 0.020 | 0.030 |
| | $\geq 6$ | 0.016 | 0.024 | | 0.30 | | 0.005 | 0.016 | 0.024 |
| Nonprestressed longitudinal reinforcement with nonconforming transverse reinforcement | $\leq 3$ | 0.012 | 0.025 | | 0.40 | | 0.006 | 0.010 | 0.020 |
| | $\geq 6$ | 0.008 | 0.014 | | 0.20 | | 0.004 | 0.007 | 0.012 |

[a] For structural walls and wall segments, use drift; for coupling beams, use chord rotation; refer to Figs. 10-5 and 10-6.
[b] For structural walls and wall segments where inelastic behavior is governed by shear, the axial load on the member must be $\leq 0.15\, Ag f_{cE}'$, otherwise, the member must be treated as a force-controlled component.
[c] For coupling beams spanning ≤8 ft 0 in., with bottom reinforcement continuous into the supporting walls, acceptance criteria values shall be permitted to be doubled for LS and CP performance.
[d] Nonprestressed longitudinal reinforcement consists of top and bottom steel parallel to the longitudinal axis of the coupling beam. Conforming transverse reinforcement consists of (a) closed stirrups over the entire length of the coupling beam at a spacing $\leq d/3$ and (b) strength of closed stirrups $V_s \geq 3/4$ of required shear strength of the coupling beam.

$$V_{Col} = k_{nl} V_{Col0} = k_{nl} \left[ \alpha_{Col} \left( \frac{A_v f_{ytL/E} d}{s} \right) + \lambda \left( \frac{0.5\sqrt{f'_{cL/E}}}{M_{UD} / V_{UD} d} \sqrt{1 + \frac{N_{UG}}{0.5 A_g \sqrt{f'_{cL/E}}}} \right) 0.8 A_g \right]$$

$(\text{Mpa units})$

ASCE 41 – 47 (10-3)

According to ACI 318-19, the shear strength is calculated from the following expression:

$$V_n = V_c + V_s$$ ACI 18 – 19 (22.5.1.1)

For non-prestressed members, $V_c$ shall be calculated in accordance with Table 22.5.5.1. The shear strength provided by the transverse reinforcement is computed from the following expression:

$$V_s = \frac{A_v f_{yt} d}{s}$$ ACI 18 – 19 (22.5.8.5.3)

Readers are advised to refer to the relevant publications for the definition of the other parameters and further details on the expressions.

#### A.2.4.3 FRP Wrapping

The contribution of the FRP jacket to the shear resistance is computed through the following expression multiplied by a reduction factor $\psi_f$, as described in Section 11.4 of ACI 440 (2017):

$$V_f = \frac{A_{fv} f_{fe} \left( sin a + cos a \right) d_{fv}}{s_f}$$ ACI 440 (11.4a)

where

$$A_{fv} = 2 n t_f w_f$$ ACI 440 (11.4b)

and

$$f_{fe} = \varepsilon_{fe} E_f$$ ACI 440 (11.4d)

The total shear strength provided by the sum of the FRP shear reinforcement and the steel shear reinforcement should be limited as indicated in the equation below:

$$V_s + V_f \leq 8\sqrt{f'_c} b_w d, \text{in in} - \text{lb units}$$

$$V_s + V_f \leq 0.66\sqrt{f'_c} b_w d, \text{in SI units}$$ ACI 440 (11.4.3)

Readers are advised to refer to the relevant publications for the definition of the other parameters and further details on the expressions.

### A.2.5 Target Displacement Calculation in the Nonlinear Static Procedure

The target displacement $\delta_t$ (Section 7.4.3.3 of ASCE 41-17) is calculated by taking into account all the relevant factors affecting the displacement of a building that responds inelastically. It is permitted to consider the displacement of an elastic SDOF system with a fundamental period equal to the fundamental period of the building that is subjected to the seismic actions, for which the verification is made. An appropriate correction is needed, in order to derive the corresponding displacement of the building, assuming that it is responding as an elastic-perfectly plastic system.

For buildings with rigid diaphragms at each floor level, the target displacement shall be calculated in accordance with Eq. (7.28) of ASCE 41.

$$\delta_t = C_0 C_1 C_2 S_\alpha \left( \frac{T_e^2}{4\pi^2} \right) g \qquad \text{ASCE 41 – 17 (7-28)}$$

where $S_\alpha$ is the response spectrum acceleration at the effective fundamental period and damping ratio of the building in the direction under consideration, and $C_0$, $C_1$, and $C_2$ are modification factors that are defined as follows:

$C_0$: Modification factor that relates the spectral displacement of the equivalent SDOF system with the roof displacement of the building MDOF system calculated using the appropriate value from Table A.7.

$C_1$: Modification factor to relate expected maximum inelastic displacements to displacements calculated for linear elastic response. For periods less than 0.2 sec, $C_1$ should be taken greater than the value at $T = 0.2$ sec.

$C_1 = 1.0$ *for* $T \geq 1$ *sec*, and

$$C_1 = 1 + \frac{\mu_{strength} - 1}{\alpha T_e^2} \text{ for } 0.2 \sec \leq T < 1 \sec, \qquad \text{ASCE 41 – 17 (7-29)}$$

$\alpha$ is the site class factor and it is equal to 130 for site class A or B, 90 for site class C and 60 for site class D, E, or F. $T_e$ is the fundamental period of the building in the direction under consideration, and $\mu_{strength}$ is the ratio of the elastic strength demand to the yield strength, calculated in accordance with Eq. (7.31) of ASCE 41-17.

$C_2$: Modification factor to represent the effect of pinched hysteresis shape, cyclic stiffness degradation, and strength deterioration on the maximum displacement response. For periods greater than 0.7 sec, $C_2 = 1.0$.

**Table A.7** Values for the modification factor $C_0$ (ASCE 2017, Table 7-5).

| | **Shear buildings** | | **Other buildings** |
|---|---|---|---|
| **Number of stories** | **Triangular load pattern (1.1, 1.2, 1.3)** | **Uniform load pattern (2.1)** | **Any load pattern** |
| 1 | 1.0 | 1.0 | 1.0 |
| 2 | 1.2 | 1.15 | 1.2 |
| 3 | 1.2 | 1.2 | 1.3 |
| 5 | 1.3 | 1.2 | 1.4 |
| 10+ | 1.3 | 1.2 | 1.5 |

$$C_2 = 1 + \frac{1}{800}\left(\frac{\mu_{strength} - 1}{T_e}\right)^2 \qquad \text{ASCE 41 - 17 (7-30)}$$

The strength ratio $\mu_{strength}$ is calculated according to the following equation:

$$\mu_{strength} = \frac{S_\alpha}{V_y / W} C_m \qquad \text{ASCE 41 - 17 (7-31)}$$

$C_m$ is the effective mass factor with values between 0.80 and 1.00, according to Table 7.4 of ASCE 41-17.

Readers are advised to refer to the Code for the definition of the other parameters and further details on the expressions.

#### A.2.5.1 Determination of the Idealized Elasto-Perfectly Plastic Force-Displacement Relationship

The nonlinear force–displacement relationship that relates the base shear with the displacement of the control node is replaced by an idealized curve for the determination of the equivalent lateral stiffness $K_e$ and the corresponding yield strength $V_y$ of the building.

The idealized capacity curve (force–displacement relationship) is bilinear, with the slope of the first branch equal to $K_e$ and the slope of the second branch equal to $\alpha_1.K_e$. The two lines that compose the bilinear curve can be defined graphically, on the criterion of approximately equal areas of the sections defined above and below the intersection of the actual and the idealized curves (Figure A.2).

The equivalent lateral stiffness $K_e$ is determined as the secant stiffness that corresponds to a base shear force equal to 60% of the effective yield strength $V_y$, the latter defined by the intersection of the lines above. The normalized inclination ($\alpha_1$) of the second branch is determined by a straight line passing through the point ($V_d$, $\Delta_d$) and a point at the intersection with the first line segment such that the areas above and below the actual curve are approximately balanced. ($V_d$, $\Delta_d$) is a point on the actual force–displacement curve at the

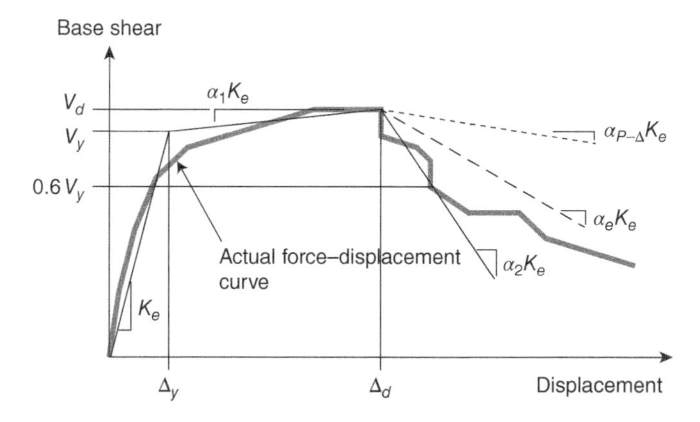


Figure A.2 Idealized force–displacement curve (Figure 7-3 of ASCE 41-17).

calculated target displacement, or at the displacement corresponding to the maximum shear, whichever is least.

### A.2.5.2 Determination of the Fundamental Period

The effective fundamental period in the direction under consideration is estimated based on the idealized force–displacement curve.

The value $T_e$ of the effective fundamental period is derived by the following expression:

$$T_e = T_i \sqrt{\frac{K_i}{K_e}} \qquad \text{ASCE 41 - 17 (7-27)}$$

where $T_i$ is the elastic fundamental period in the direction under consideration, and is derived by eigenvalue analysis, $K_i$ is the corresponding elastic lateral stiffness, and $K_e$ is the effective lateral stiffness in Figure A.2.

## References

[ACI] American Concrete Institute ACI 318 (2019). *ACI CODE-318-19: Building Code Requirements for Structural Concrete and Commentary. ACI Committee 318*. American Concrete Institute.

[ACI] American Concrete Institute ACI 440 (2017). ACI PRC-440.2-17: Guide for the Design and Construction of Externally Bonded FRP Systems for Strengthening Concrete Structures. ACI Committee 440.

[ASCE] American Society of Civil Engineers (2017). *Seismic Evaluation and Retrofit of Existing Buildings (ASCE/SEI 41–17)*. Reston, Virginia: ASCE.

CEN (2004). *European Standard EN 1998-1: 2004. Eurocode 8: Design of Structures for Earthquake Resistance, Part 1: General Rules, Seismic Actions and Rules for Buildings*. Brussels: Comité Européen de Normalisation.

CEN (2005). *European Standard EN 1998-3: 2005. Eurocode 8: Design of Structures for Earthquake Resistance, Part 3: Assessment and Retrofitting of Buildings*. Brussels: Comité Européen de Normalisation.

# Appendix B

# Poor Construction and Design Practices in Older Buildings

In Appendix B, some of the most characteristics cases of poor design and construction practices, which I have encountered in the past twenty years in this field, will be presented. The main objective is to make the reader aware of the difficulties in the accurate and reliable inspection of an existing building, and to understand what to expect in older construction. Ultimately, he/she should not take any information for granted without further checking all the relevant parameters more thoroughly.

## B.1 Stirrup Spacing

The photographs depict columns with very large stirrup spacing. In all photographs the spacing is larger than 60 cm, and in some cases it is even larger than 1.00 m (Figures B.1–B.6).

## B.2 Lap Splices

The photographs depict cases of columns with lap splices less than 20–30 cm (Figures B.7–B.10).

## B.3. Member Alignment

The photographs depict columns with offset equal to 15–20 cm or more between two adjacent floors (Figures B.11–B.14).

## B.4 Pipes inside RC Members

The photographs depict columns, in which pipelines have been passed. In most cases, the pipelines are corroded, resulting in the gradual degradation of the concrete core (Figures B.15–B.20).

*Seismic Retrofit of Existing Reinforced Concrete Buildings*, First Edition. Stelios Antoniou.

**Figure B.1** Column with very large stirrup spacing. *Source:* Stelios Antoniou.

**Figure B.2** Column with very large stirrup spacing. *Source:* Stelios Antoniou.

**Figure B.3** Column with very large stirrup spacing. *Source:* Stelios Antoniou.

**Figure B.4** Column with very large stirrup spacing. *Source:* Stelios Antoniou.

**Figure B.5** Column with very large stirrup spacing. *Source:* Stelios Antoniou.

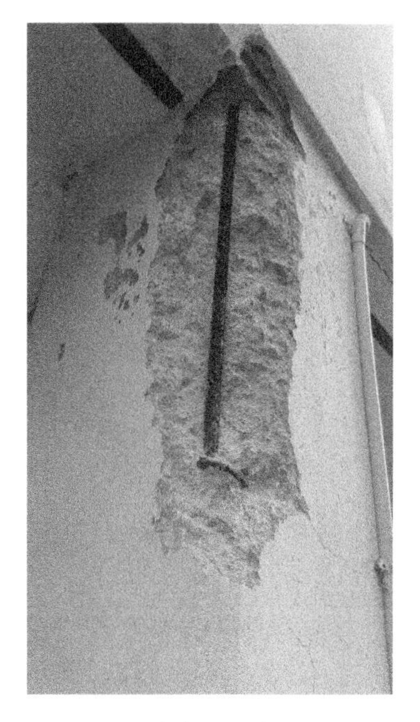

**Figure B.6** Column with very large stirrup spacing. *Source:* Stelios Antoniou.

**Figure B.7** Column with very short lap splices. *Source:* Stelios Antoniou.

**Figure B.8** Column with very short lap splices. *Source:* Stelios Antoniou.

**Figure B.9** Column with very short lap splices. *Source:* Stelios Antoniou.

**Figure B.10** Column with very short lap splices. *Source:* Stelios Antoniou.

**Figure B.11** Column with bad alignment in the floor level. The upper column has been displaced by 20cm. *Source:* Stelios Antoniou.

**Figure B.12** Column with bad alignment in the floor level. *Source:* Stelios Antoniou.

Figure B.13 Column with bad alignment in the floor level. *Source:* Stelios Antoniou.

Figure B.14 Column with bad alignment in the floor level. The upper column has been displaced by 20cm in one horizontal direction and by 10cm in the other. *Source:* Stelios Antoniou.

## B.5 Bad Casting of Concrete

The photographs depict cases of extremely bad casting of concrete. The first photograph shows the casting of a thin RC jacket (5 cm of thickness), which should have been constructed with shotcrete (Figures B.21 and B.22).

## B.6 Footings

The photographs depict extremely deficient footings. In all cases, the footings are extremely small and there is no reinforcement. In Figures B.23 and B.27 there is no footing altogether, in (a) only a thin 5 cm layer of concrete was poured on the rock, which was excavated in the shape of a cone! (Figures B.23–B.32).

**Figure B.15** Column with a metallic downspout inside its core. The corrosion of the downspout has caused serious damage to the concrete of the column. *Source:* Stelios Antoniou.

**Figure B.16** Beam-column joint crossed by a drainage pipe. *Source:* Stelios Antoniou.

**Figure B.17** Column with a downspout inside its core. *Source:* Stelios Antoniou.

**Figure B.18** Column with a downspout inside its core. *Source:* Stelios Antoniou.

Figure B.19 Column with a downspout inside its core. *Source:* Stelios Antoniou.

Figure B.20 Beam with several pipes passing from its core. *Source:* Stelios Antoniou.

Figure B.21 Bad casting of concrete. The concrete could not pass from a jacket that is just 5 cm wide. *Source:* Stelios Antoniou.

Figure B.22 Bad casting of concrete in a cylindrical column. *Source:* Stelios Antoniou.

**Figure B.23** Non-existent footing. Just 5 cm of concrete has been cast on top of the rock below that had been dug in the shape of a cone. *Source:* Stelios Antoniou.

**Figure B.24** A footing without reinforcement. *Source:* Stelios Antoniou.

**Figure B.25** Very small footings without reinforcement. *Source:* Stelios Antoniou.

**Figure B.26** Non-existent footing. *Source:* Stelios Antoniou.

**Figure B.27** Non-existent footing in a building built on soft soil. *Source:* Stelios Antoniou.

**Figure B.28** Very small footing without reinforcement. *Source:* Stelios Antoniou.

**Figure B.29** Very small footing without reinforcement. *Source:* Stelios Antoniou.

**Figure B.30** Very small footing without reinforcement. *Source:* Stelios Antoniou.

**Figure B.31** Very small footing without reinforcement. *Source:* Stelios Antoniou.

**Figure B.32** Very small footing without reinforcement. *Source:* Stelios Antoniou.

# Appendix C

# Methods of Strengthening

In Appendix C, characteristic photographs from the most common strengthening methods will be shown, so that readers can see and better understand the construction details of the retrofit. All photographs are from projects that I was involved in directly or indirectly. Consequently, the sample is not 100% representative of all the methods that have been presented in Chapter 4, since it has more photographs from the most common methods, which we have employed most.

## C.1 Reinforced Concrete Jackets

Figures C.1–C.30

**Figure C.1** RC jackets in beams and slabs. *Source:* Stelios Antoniou.

**Figure C.2** RC jackets in beams and columns. *Source:* Stelios Antoniou.

**Figure C.3** RC jackets in beams and slabs. *Source:* Stelios Antoniou.

**Figure C.4** Strengthening of a soft ground story with RC jackets. *Source:* Stelios Antoniou.

**Figure C.5** Connection of the new and the existing reinforcement with steel plates. *Source:* Stelios Antoniou.

**Figure C.6** Connection of the existing column and the jacket with dowels. *Source:* Stelios Antoniou.

**Figure C.7** RC jacket in a column. *Source:* Stelios Antoniou.

**Figure C.8** RC jacket in a column. *Source:* Stelios Antoniou.

**Figure C.9** RC jacket in a column. *Source:* Stelios Antoniou.

**Figure C.10** RC jacket in a column. *Source:* Stelios Antoniou.

Figure C.11 RC jacket in a column. *Source:* Stelios Antoniou.

Figure C.12 3-sided jackets in an expansion joint. *Source:* Stelios Antoniou.

Figure C.13 RC jackets in beams and columns. *Source:* Stelios Antoniou.

Figure C.14 RC jacket in a column. *Source:* Stelios Antoniou.

Figure C.15 Extension of an RC jacket in the beam-column joint. *Source:* Stelios Antoniou.

Figure C.16 Extension of an RC jacket in the beam-column joint. *Source:* Stelios Antoniou.

Figure C.17 RC jackets in beams and columns. *Source:* Stelios Antoniou.

**Figure C.18** RC jackets in the beam-column region. *Source:* Stelios Antoniou.

**Figure C.19** Strengthening of a cantilever with RC jackets. *Source:* Stelios Antoniou.

**Figure C.20** Welding of the stirrups. *Source:* Stelios Antoniou.

**Figure C.21** Connection of the new and the existing reinforcement with steel plates. *Source:* Stelios Antoniou.

**Figure C.22** RC jacket in a column of a soft ground story. *Source:* Stelios Antoniou.

**Figure C.23** Strengthening of a wall with a one-sided RC jacket. *Source:* Stelios Antoniou.

**Figure C.24** Closing of the stirrups with welding at the upper side of the beams. *Source:* Stelios Antoniou.

**Figure C.25** RC jackets in a ribbed slab. *Source:* Stelios Antoniou.

**Figure C.26** RC jackets in a masonry building. *Source:* Stelios Antoniou.

**Figure C.27** RC jackets in a masonry building. *Source:* Stelios Antoniou.

**Figure C.28** Application of shotcrete. *Source:* Stelios Antoniou.

Figure C.29 Application of shotcrete. *Source:* Stelios Antoniou.

Figure C.30 Application of shotcrete. *Source:* Stelios Antoniou.

## C.2 New Shear Walls

Figures C.31–C.36

Figure C.31 New shear wall in an existing building. *Source:* Stelios Antoniou.

**Figure C.32** New shear wall with the simultaneous strengthening of the foundation with strip footings. *Source:* Stelios Antoniou.

**Figure C.33** New shear wall at the basement of an existing building. *Source:* Stelios Antoniou.

Figure C.34 Closer view of the reinforcement of a new shear wall. *Source:* Stelios Antoniou.

Figure C.35 New shear wall in an existing building. *Source:* Stelios Antoniou.

**Figure C.36** Passing the reinforcement of a new shear wall through the existing slab and beam. *Source:* Stelios Antoniou.

## C.3 Fiber-Reinforced Polymers

### C.3.1 FRP Wrapping of Columns

Figures C.37–C.48

**Figure C.37** FRP wrapping of columns in a soft ground story. *Source:* Stelios Antoniou.

Figure C.38 FRP wrapping of a column after the placement of the quartz sand to improve adhesion with the overlay. *Source:* Stelios Antoniou.

Figure C.39 FRP wrapping of a column. *Source:* Stelios Antoniou.

Figure C.40 FRP wrapping of a column. *Source:* Stelios Antoniou.

Figure C.41 Preparation of the concrete for the application of FRP wrapping. *Source:* Stelios Antoniou.

**Figure C.42** FRP wrapping of columns in a soft ground story. *Source:* Stelios Antoniou.

**Figure C.43** FRP wrapping of a column. *Source:* Stelios Antoniou.

**Figure C.44** FRP wrapping of a column after the placement of the quartz sand to improve adhesion with the overlay. *Source:* Stelios Antoniou.

**Figure C.45** Placement of the fabric in FRP wrapping. *Source:* Stelios Antoniou.

**Figure C.46** Placement of the epoxy resin in FRP wrapping. *Source:* Stelios Antoniou.

**Figure C.47** Placement of the FRP wrap. *Source:* Stelios Antoniou.

**Figure C.48** Placement of the FRP wrap. *Source:* Stelios Antoniou.

### C.3.2 FRP Fabrics in Slabs

Figure C.49

Figure C.49 Strengthening of a slab in bending with FRP fabrics. *Source:* Stelios Antoniou.

### C.3.3 FRP Wraps for Shear Strengthening

Figures C.50–C.54

Figure C.50 FRP strings opened as a tuft for the anchorage of the FRP fabrics. *Source:* Stelios Antoniou.

**Figure C.51** FRP strings and FRP fabrics for the strengthening of a beam in shear. *Source:* Stelios Antoniou.

**Figure C.52** FRP strings for the anchorage of FRP fabrics in the strengthening of a beam in shear. *Source:* Stelios Antoniou.

**Figure C.53** Strengthening of beams in shear with FRP fabrics. *Source:* Stelios Antoniou.

**Figure C.54** Strengthening of a beam in shear with FRP fabrics. *Source:* Stelios Antoniou.

### C.3.4 FRP Laminates

Figures C.55–C.68

Figure C.55 Strengthening of a slab in bending with FRP laminates (lower slab side). *Source:* Stelios Antoniou.

Figure C.56 Strengthening of a slab in bending with FRP laminates (lower slab side). *Source:* Stelios Antoniou.

**Figure C.57** Strengthening of beams and slabs in bending with FRP laminates (lower slab side).
*Source:* Stelios Antoniou.

**Figure C.58** Strengthening of a slab in bending with FRP laminates (upper slab side).
*Source:* Stelios Antoniou.

**Figure C.59** Strengthening of a slab in bending with FRP laminates (upper slab side).
*Source:* Stelios Antoniou.

Figure C.60 Strengthening of a slab in bending with FRP laminates (upper slab side). *Source:* Stelios Antoniou.

Figure C.61 Strengthening of a slab in bending with FRP laminates (upper slab side). *Source:* Stelios Antoniou.

Figure C.62 Strengthening of a slab in bending with FRP laminates (upper slab side). *Source:* Stelios Antoniou.

Figure C.63 Mechanical anchorage at the end of the FRP laminates. *Source:* Stelios Antoniou.

**Figure C.64** Strengthening of a slab in bending with FRP laminates (lower slab side). *Source:* Stelios Antoniou.

**Figure C.65** Placement of the FRP laminates (upper slab side). *Source:* Stelios Antoniou.

**Figure C.66** Placement of the FRP laminates (lower beam side). *Source:* Stelios Antoniou.

**Figure C.67** Placement of the FRP laminates (upper slab side). *Source:* Stelios Antoniou.

**Figure C.68** Mechanical anchorage at the end of the FRP laminates. *Source:* Stelios Antoniou.

### C.3.5 FRP Strings

Figures C.69–C.74

**Figure C.69** Cleaning of the slit for the placement of FRP strings. *Source:* Stelios Antoniou.

Figure C.70 Preparation before the placement of the FRP strings. *Source:* Stelios Antoniou.

Figure C.71 Placement of the epoxy resin. *Source:* Stelios Antoniou.

Figure C.72 Placement of the epoxy resin. *Source:* Stelios Antoniou.

Figure C.73 Final layout of the FRP strings for the strengthening in bending at the slab support region. *Source:* Stelios Antoniou.

Figure C.74 An FRP string opened as a tuft for the anchorage of FRP fabrics. *Source:* Stelios Antoniou.

## C.4 Steel Braces

Figures C.75–C.81

**Figure C.75** Steel truss inside a building. *Source:* Stelios Antoniou.

**Figure C.76** External steel truss in the perimeter of a building. The adjacent existing beams are strengthening with steel plates. *Source:* Stelios Antoniou.

**Figure C.77** External steel truss in the perimeter of a building. *Source:* Stelios Antoniou.

**Figure C.78** Steel truss inside a building. The adjacent existing beams are strengthening with steel plates. *Source:* Stelios Antoniou.

**Figure C.79** Steel truss inside a building

**Figure C.80** Steel truss inside a building. *Source:* Stelios Antoniou.

**Figure C.81** Steel truss inside a building

## C.5 Steel Jackets

Figures C.82 and C.83

**Figure C.82** Strengthening of a column with steel jackets. *Source:* Stelios Antoniou.

**Figure C.83** Strengthening of a column with steel jackets. *Source:* Stelios Antoniou.

## C.6 Steel Plates

Figures C.84–C.89

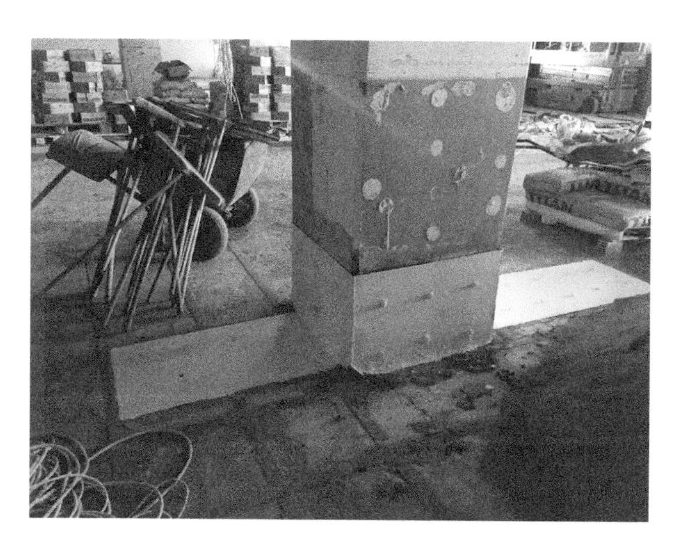

**Figure C.84** Strengthening of a beam in bending with steel plates anchored at the base of the column. *Source:* Stelios Antoniou.

Figure C.85 Strengthening of a beam in bending with steel plates anchored at the base of the column. *Source:* Stelios Antoniou.

Figure C.86 Strengthening of beams in bending with steel plates (upper side). *Source:* Stelios Antoniou.

**Figure C.87** Strengthening of a beam in bending with steel plates (upper side). *Source:* Stelios Antoniou.

**Figure C.88** Strengthening of a beam in bending with one steel plate (lower side). *Source:* Stelios Antoniou.

Figure C.89 Strengthening of a beam in bending with one steel plate (lower side). *Source:* Stelios Antoniou.

## C.7 Infills

Figures C.90–C.93

Figure C.90 Application of textile-reinforced mortars (TRM) in an infilled wall. *Source:* Stelios Antoniou.

**Figure C.91** Application of TRMs in an infilled wall. Note the FRP strings that are used for the anchorage of the system. *Source:* Stelios Antoniou.

**Figure C.92** TRMs in an infilled wall. *Source:* Stelios Antoniou.

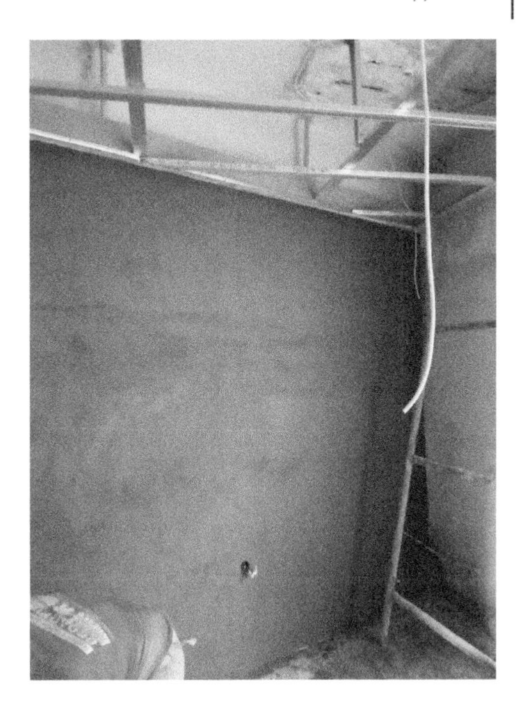

**Figure C.93** TRMs in an infilled wall. *Source:* Stelios Antoniou.

## C.8 Foundations

Figures C.94–C.107

**Figure C.94** Strengthening of the foundation with RC jackets and connecting beams. *Source:* Stelios Antoniou.

**Figure C.95** Strengthening of the foundation with RC jackets and connecting beams. *Source:* Stelios Antoniou.

**Figure C.96** RC jacket and connecting beams in an individual footing. *Source:* Stelios Antoniou.

**Figure C.97** Strengthening of the foundation with strip footings. *Source:* Stelios Antoniou.

**Figure C.98** Strengthening of the foundation of a building constructed at different levels with strip footings. *Source:* Stelios Antoniou.

**Figure C.99** Strengthening of the foundation with strip footings. The short columns that remain above the footings are strengthened with jackets. *Source:* Stelios Antoniou.

**Figure C.100** Strengthening of the foundation with strip footings and new walls at the basement. *Source:* Stelios Antoniou.

**Figure C.101** Strengthening of the foundation with RC jackets and connecting beams. *Source:* Stelios Antoniou.

**Figure C.102** Strengthening of the foundation with strip footings. *Source:* Stelios Antoniou.

**Figure C.103** Strengthening of the foundation with strip footings. *Source:* Stelios Antoniou.

**Figure C.104** Strengthening of the foundation with strip footings. *Source:* Stelios Antoniou.

**Figure C.105** Strengthening of the foundation with strip footings. *Source:* Stelios Antoniou.

**Figure C.106** Foundation strengthening after the casting of concrete. *Source:* Stelios Antoniou.

**Figure C.107** Strengthening of the foundation with a new footing and connecting beams. *Source:* Stelios Antoniou.

## C.9 Dowels and Anchorages

Figures C.108–C.111

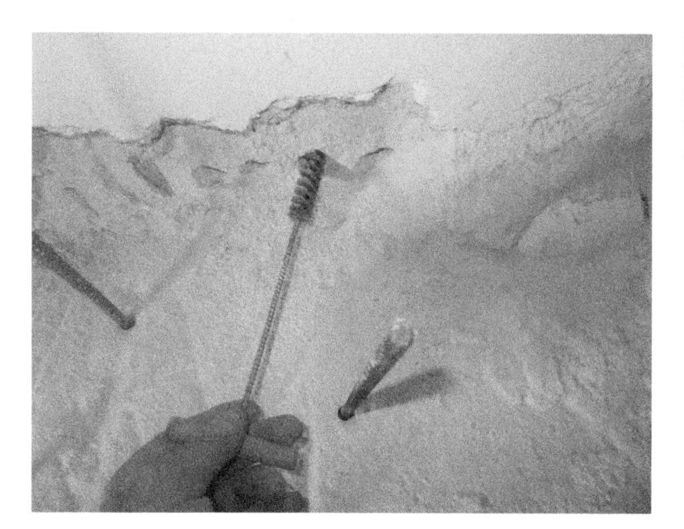

**Figure C.108** Cleaning of the hole in the concrete before the placement of the dowel. *Source:* Stelios Antoniou.

**Figure C.109** Grid of holes in an existing concrete base for the anchorage of a new wall. *Source:* Stelios Antoniou.

Figure C.110 Placement of the anchorages. *Source:* Stelios Antoniou.

Figure C.111 Placement of the anchorages. *Source:* Stelios Antoniou.

## C.10 Demolition with Concrete Cutting

Figures C.112–C.119

**Figure C.112** Concrete cutting in a parapet. *Source:* Stelios Antoniou.

**Figure C.113** Concrete cutting in walls. *Source:* Stelios Antoniou.

**Figure C.114** Cutting of a concrete beam. *Source:* Stelios Antoniou.

**Figure C.115** Part of a concrete strip footing that has been cut off. *Source:* Stelios Antoniou.

**Figure C.116** Cutting of a concrete column. *Source:* Stelios Antoniou.

**Figure C.117** Removal of a 40tn concrete slab. *Source:* Stelios Antoniou.

Figure C.118 Removal of parts of a concrete wall that weight 2.5-3tn each. *Source:* Stelios Antoniou.

Figure C.119 Removal of parts of a concrete wall. *Source:* Stelios Antoniou.

## C.11 Reinforcement Couplers

Figures C.120 and C.121

**Figure C.120** Extension of the reinforcement of a column with couplers. *Source:* Stelios Antoniou.

**Figure C.121** Reinforcement couplers. *Source:* Stelios Antoniou.

## C.12 Epoxy Injections

Figures C.120–C.125

**Figure C.122** Epoxy injections in a concrete slab. *Source:* Stelios Antoniou.

**Figure C.123** Ports for the epoxy injections and epoxy paste on the surface of a concrete slab. *Source:* Stelios Antoniou.

**Figure C.124** Epoxy injections in a concrete slab. *Source:* Stelios Antoniou.

**Figure C.125** Epoxy injections in a concrete slab. *Source:* Stelios Antoniou.

# Index

## a



*Seismic Retrofit of Existing Reinforced Concrete Buildings*, First Edition. Stelios Antoniou.

## o



## p



## r